THEORY OF
TURBULENT
PLASMA

STUDIES IN SOVIET SCIENCE

PHYSICAL SCIENCES

1973

DENSIFICATION OF METAL POWDERS DURING SINTERING
V. A. Ivensen
THE TRANSURANIUM ELEMENTS
V. I. Goldanskii and S. M. Polikanov
GAS-CHROMATOGRAPHIC ANALYSIS OF TRACE IMPURITIES
V. G. Berezkin and V. S. Tatarinskii
A CONFIGURATIONAL MODEL OF MATTER
G. V. Samsonov, I. F. Pryadko, and L. F. Pryadko
COMPLEX THERMODYNAMIC SYSTEMS
V. V. Sychev
CRYSTALLIZATION PROCESSES UNDER HYDROTHERMAL CONDITIONS
A. N. Lobachev
MIGRATION OF MACROSCOPIC INCLUSIONS IN SOLIDS
Ya. E. Geguzin and M. A. Krivoglaz

1974

THEORY OF PLASMA INSTABILITIES
Volume 1: Instabilities of a Homogeneous Plasma
A. B. Mikhailovskii
THEORY OF PLASMA INSTABILITIES
Volume 2: Instabilities of an Inhomogeneous Plasma
A. B. Mikhailovskii
NONEQUILIBRIUM STATISTICAL THERMODYNAMICS
D. N. Zubarev
REFRACTORY CARBIDES
G. V. Samsonov
WAVES AND SATELLITES IN THE NEAR-EARTH PLASMA
Ya. L. Al'pert

1975

ENVIRONMENTAL HAZARDS OF METALS:
Toxicity of Powdered Metals and Metal Compounds
I. T. Brakhnova
DOMAIN ELECTRICAL INSTABILITIES IN SEMICONDUCTORS
V. L. Bonch-Bruevich, I. P. Zvyagin, and A. G. Mironov

1976

THE ROTATING DISC ELECTRODE
Yu. V. Pleskov and V. Yu. Filinovskii

1977

THEORY OF TURBULENT PLASMA
V. N. Tsytovich

A Continuation Order Plan is available for this series. A continuation order will bring delivery of each new volume immediately upon publication. Volumes are billed only upon actual shipment. For further information please contact the publisher.

STUDIES IN SOVIET SCIENCE

THEORY OF TURBULENT PLASMA

V. N. Tsytovich

P. N. Lebedev Physics Institute
Academy of Sciences of the USSR
Moscow, USSR

Translated from Russian by
David L. Burdick

CONSULTANTS BUREAU • NEW YORK AND LONDON

Library of Congress Cataloging in Publication Data

TSytovich, Vadim Nikolaevich.
  Theory of turbulent plasma.

  (Studies in Soviet science)
  Translation of Teoriĭa turbulentnoĭ plazmy.
  Includes bibliographical references.
  1. Plasma turbulence. I. Title. II. Series.
QC718.5.T8T7913   530.4'4   75-31720
ISBN 978-1-4684-7925-6   ISBN 978-1-4684-7923-2 (eBook)
DOI 10.1007/978-1-4684-7923-2

Born in 1929, *Vadim Nikolaevich Tsytovich* is a senior scientific member of the
P. N. Lebedev Physics Institute, Academy of Sciences of the USSR, and a Doctor
of Physico-Mathematical Sciences. Until 1956 his scientific interests were in the
field of quantum electrodynamics. From 1957 to 1960 he worked with Veksler on
problems related to collective acceleration methods. He has been engaged in plasma theory studies and astrophysics since 1959.

The original Russian Text, published by Atomizdat in Moscow in 1971, has been
corrected by the author for the present edition. This translation is published under
an agreement with the Copyright Agency of the USSR (VAAP).

Теория турбулентной плазмы.
В. Н. Цытович
TEORIYA TURBULENTNOI PLAZMY
V. N. Tsytovich

© 1977 Consultants Bureau, New York
Softcover reprint of the hardcover 1st edition 1977

A Division of Plenum Publishing Corporation
227 West 17th Street, New York, N.Y. 10011

Preface to the English Translation

Even though the English edition of this book appears some time after the Russian version was published, and although many new studies will have been made in a field which is as rapidly expanding as the theory of turbulent plasma, these new results will not require any fundamental changes or additions to the accepted views, except perhaps the problem of strong Langmuir turbulence. One may now be able to formulate with greater clarity some of the factors which were emphasized in the Preface to the Russian edition with respect to turbulent broadening of resonances, correlation effects (§2.5 and §2.8), and the theory of turbulence spectra (Chapter 3). The problem is a more strict and exact representation of the nature of weak plasma turbulence. Unfortunately, it has been hitherto current to suppose that the weak-turbulence approximation corresponds to the use of expansions of the nonlinear interactions in terms of the turbulence energy. This is not valid, in general. The effective turbulent collisions, which are related to the nonlinear interactions and which lead to randomization of the collective plasma motions, make it essential to include the terms which depend on the turbulence energy in all the energy denominators. This in fact indicates the breakdown of expansions in terms of the turbulence energy.

The real problem lies in the effects which are described by the ladder diagrams in the graphical representation of these processes; they are analogous to the processes which are taken into account in the natural widths of the spectral lines in quantum electrodynamics. In this case, however, the effective collisions which broaden the resonances now depend on the turbulence energy. It should be emphasized that the correct treatment of these reso-

nances in turbulence theory does not imply an actual departure
from the limits on the applicability of weak turbulence theory,
nor does it imply the availability of a theory of strong turbulence
(or the so-called weak-coupling approximation). A strict and
systematic theory for weak turbulence cannot be constructed without a correct description of the turbulent widths of the resonances,
and for the time being this is provided in a closed and final form
only within the theory of weak turbulence. This turbulent broadening of the resonances allows us to give a very different physical
interpretation to the Landau damping of the plasma's collective
motions and to provide a description of the stochastic heating of
plasma particles and the irreversiblity of processes in a turbulent
plasma. But in addition to these aspects of producing a theory of
turbulence, the resonance broadening effects have great practical
value; they change the effectiveness of various nonlinear interactions which determine the turbulence spectra, the plasma heating,
the plasma radiation, the particle acceleration, and so on. For
instance, there are significant changes in the magnitude of the
anomalous plasma resistance to an external electric field [see
Eq. (4.258)], in the skin effect, and other features, all of which
are directly measurable in experiments. Thus, when comparing
weak-turbulence theory with experimental results, one must carefully analyze all the indicated effects of turbulent broadening;
that is, one must compare them with the results of a systematic theory
of weak turbulence. In this regard one must use a certain amount
of care when relating to all those theoretical constructions which
take into account only the so-called quasilinear processes while
neglecting the nonlinear effects; this is because the basis of the
quasilinear equations themselves is rooted in the concept of effective turbulent collisions, and these are often nonlinear in character.

One of the great simplifications which helps one to treat turbulent processes in plasma in a more clear manner is the notion
of elementary excitations, that is, plasmons. This representation
can be used to describe the behavior of a number of integral
plasma characteristics, which contain integrals of resonance denominators which are smeared because of the turbulent collisions.

In some approximations the resonance curves with finite widths
(due to the turbulent collisions) can be approximated as δ-func-
tions. Then one can speak of an approximate relation between

the frequency and the wave number of the excitation, or between the plasmon's energy and momentum. This allows us to use the plasmon concept in conditions of weak turbulence. It is clear that this can be done in principle only when the effective turbulent frequency is smaller than the plasmon frequency. One must always keep the approximate character of these ideas in mind when using them. In particular, all the energy and momentum conservation rules for plasmons in nonlinear interactions must be satisfied only to within the accuracy of the effective turbulent collisions indicated here.

This preface runs ahead a bit, and in some measure just emphasizes the points which the reader will find in this book or in the author's review article "The Development of Plasma Turbulence Concepts," Uspekhi Fizicheskikh Nauk (1972), FIAN Preprint No. 131 (1971). However, our primary aim in this preface is to provide a view of the general ideology now available for describing turbulent plasma processes, and to indicate that more work must be done in both the experimental field and in the theoretical area in order to better compare experiment with theory. Although there is now available a general formulation for the theory of turbulent processes in plasma and for many of these processes (ion-acoustic, Langmuir, and Alfven turbulence) the equations describing their dynamics can be written in rather simple forms, the fact is that in a number of cases there are still no final answers concerning some aspects of the theoretical predictions. This can be traced to the complexity of the integral equations which describe the nonlinear interaction processes. It seems probable that progress in this direction can be achieved in the near future by using numerical methods.

One of the important problems which has received much attention of late is the problem of the anomalous plasma resistance. It affects many aspects of the general theoretical problems, such as the turbulent broadening of resonances, and it touches upon the problem with escaping particles and plasma heating, in whose treatment the three-dimensionality of the processes is one of the most important aspects.

This translation will appear approximately five years after the Russian edition. Since the theory of strong Langmuir turbulence has been the subject of intensive investigation in the interim, it seemed desirable to append a brief review of this topic. This has been done in Chapter 9.

It is hoped that these few comments will aid the reader during his study of this book, and that the English publication will attract more attention to the problems of turbulence theory which still await solution.

V. N. Tsytovich

Preface

In the last few years the physics of turbulent plasma has undergone rapid development, beginning with the first works, in which the term "turbulence" was used in various ways, and ending with the fundamental studies which provide a thorough examination of the turbulent state of plasma. In physics it is usually found that value is not so much contained in specific results for a particular field as it is in the more general outlook and overall view of the problem. Occasionally the older results take on new meaning after the general view of things is perfected. In the case of the physics of turbulent plasma, this general picture is now complete, for the most part.

The first review devoted to the problem of plasma turbulence was written by B. B. Kadomtsev [in "Problems in Plasma Theory," edited by M. A. Leontovich, Volume 4, Moscow, Atomizdat (1964), p. 188; English Translation: "Plasma Turbulence," Academic Press, London (1965)].

This book presents the fundamental physical concepts relative to turbulent processes in plasma, and formulates them within the framework of a logically closed physical theory. Many of the questions which Kadomtsev's book merely touches upon receive a detailed theoretical analysis here. Foremost among them is the theory of turbulence broadening of resonances (Chapter 2), which enables us to establish limitations on the use of expansions in terms of the turbulence energy. It also allows us to relate averages over a statistical ensemble to the method of elementary excitations, and provides at the same time a feeling for the physical meaning behind the elementary excitation method and the limitations on its use [see V. N. Tsytovich, "Nonlinear Effects in

Plasma," Moscow, Nauka (1967) [English Translation: Plenum
Press, New York (1970)] for more on the concept of elementary excitations]. The fundamental problem in turbulence
theory centers on the spectra of stationary and uniform turbulence.
The success achieved in this direction as applied to plasma turbulence has been greater than that achieved in the theory of fluid
turbulence. Our treatment is therefore based on the theory of
stationary or quasistationary turbulence (Chapter 4) and the processes which arise in a stationary, turbulent plasma; among them are the
acceleration and heating of particles (Chapter 5), the radiation
from a turbulent plasma (Chapter 6), the propagation of electromagnetic waves through turbulent plasma (Chapter 7), and the
electromagnetic properties of turbulent plasma (Chapter 8).

These chapters are preceded by an introductory chapter,
which discusses the basics of the physical representations of the
stochastic processes, presented by analogy with fluid turbulence;
it defines both weak and strong turbulence, and discusses the
types of turbulent motions in plasma. Chapter 2 presents the fundamentals of turbulence theory and its primary physical consequences, which enable us to give a physical treatment of turbulence.

Chapter 3 is devoted to the problems of exciting and damping
turbulent fluctuations in plasma; these questions are important
for many applications.

The questions of the dynamics of the turbulent state of plasma
and the establishment of stationary turbulence are independent and
important problems. Therefore we have considered only a few
particular problems of this kind [the stability of stationary turbulence,
second sound, and others have been studied (Chapter 8)].

We should mention that, because of the many different instabilities, a plasma is, as a rule, turbulent under real conditions.
This fact is responsible for the ever-increasing interest in and
widespread application of the results from turbulence theory to
astrophysics. During the course of our treatment we give, where
possible, references to some of the astrophysics consequences,
among which the acceleration of cosmic rays is most important.

It is now becoming even more clear that the collective properties of plasma, about which much has been said in the past, are

related in some manner to the turbulence. At present there are two important directions for the use of collective effects — the turbulent heating of plasma (suggested by E. K. Zavoiskii and Ya. B. Fainberg) and the collective method for particle acceleration (suggested by V. I. Veksler).

Whereas the problem in turbulent heating is concerned with using collective effects to transfer energy to the plasma particles, the problem in collective acceleration is the transferring of momentum. Everything which we now know about turbulence indicates that it should be possible to achieve effective turbulent heating and particle acceleration. Veksler's ideas about collective acceleration have struck a responsive chord, and many laboratories are working toward this end. There is widespread experimentation on the turbulent heating of plasma. The author has been fortunate enough to work under Veksler's direction on one variation of collective acceleration. The author's attempts at a deeper understanding of the nature of collective processes in plasma and plasma turbulence are due in large measure to the constant interest which V. I. Veksler has shown in these problems.

Therefore, with deep appreciation for the memory of one of the most remarkable physicists — V. I. Veksler — the author dedicates this book to him.

Sincere thanks are extended to E. K. Zavoiskii, S. A. Kaplan, and R. Z. Sagdeev for reading the manuscript and making a number of important comments, to L. I. Rudakov and V. G. Makhan'kov for discussions of the problems covered in the book, and to V. A. Liperovskii, M. A. Livshits, A. G. Kulagin, E. N. Krivorutskii, and A. S. Chikhachev for their comments and aid in laying out the manuscript.

Contents

Chapter 1

The Turbulent State of Matter: General Concepts

1.1. A Definition of Turbulence

The term "turbulence" arose in the study of fluid motion. Because of the rapid advances in plasma physics in recent years, our views about the nature of the turbulent state of matter have been greatly extended. On one hand, the turbulent motions possible in a plasma are more varied than those in a fluid, thus making it possible to study turbulence in greater detail in a plasma. On the other hand, in some cases one can follow the processes by which turbulence develops, thereby making it possible to understand the differences between turbulent motions and the processes which give rise to turbulence. It has been learned that turbulence can arise not only in fluids and plasmas, but in solids as well. This permits one to discuss the general concept of the turbulent state of matter, which in turn means that the term "turbulence" now has a much broader meaning.

Using fluids as an example, we shall discuss the important general physical notions related to the nature of turbulence. Let us consider the fluids moving at high velocities in a tube (more precisely, the Reynolds numbers are large [1]). In this situation the motion of an individual fluid element is very complicated. The speed of the element at some specified point in the tube can change in a very complex manner, ranging over a wide spectrum of different values in a sufficiently long period of time.

Such motion cannot be considered turbulent simply because the motion of the individual fluid elements is complicated, since the laminar motion of a fluid moving around some obstacle is also

1

very complex. The basic criterion for turbulence is that measurements of the instantaneous velocities of the fluid be nonreproducible.

Let us assume that at time $t_0 = 0$ the valve is opened and the fluid begins to flow in the tube. Assume that after a time t_1 at some large distance from the valve the average characteristics of the motion have been established. We shall pick some point $\mathbf{r}$ in the region of steady-state motion and follow the fluid flow velocity at this point as a function of time. Starting at some moment $t > t_1$ we obtain a complicated curve (Fig. 1.1). Now we repeat the experiment. The valve is opened and we again measure the fluid velocity at the same point after the same time interval t_1. We repeat this experiment a number of times. We either obtain the same curve (Fig. 1.1) each time, or the curves are different, i.e., the results of the experiment are not reproducible. In the latter case one can say that the fluid is turbulent and the velocity is random and disordered.

The reason for the nonreproducible results of our experiment can be explained in the following manner. If the initial conditions and the boundary conditions for the fluid motion in the tube are known exactly, we can determine the velocities at any subsequent time at any point inside the tube by using the equations which describe fluid motion. However, the exact initial and boundary conditions are unknown. All that is known are the rather average flow characteristics, which are related to the tube dimensions, the way the fluid is admitted to the tube, and so on. It is significant that in the turbulent state small changes in the initial conditions lead to large changes in the motions of the fluid elements. Therefore, the nonreproducibility of the results arises because the system can have different initial conditions for its motion even when the macroscopic parameters are specified.

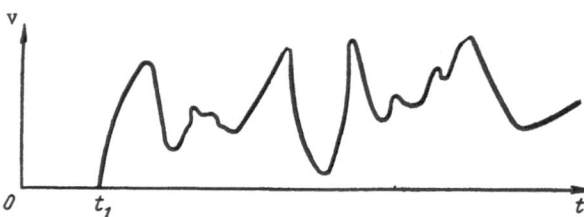

Fig. 1.1. The velocity of an element of turbulent fluid as a function of time.

The situation described above is similar to the problem of the molecular motion of the particles which make up a macroscopic body. If the initial conditions of motion were known for all the particles in the body, one could predict their later motions. But it is meaningless to pose such a problem since it is impossible to have such detailed knowledge about the initial state of the system [1]. Thus, when assigning the macroscopic parameters to a system, its molecular motion must be assumed random and disordered.

The idea which is usually embodied in the notion of turbulence is essentially different from the notion of molecular motion. For when we speak of the motion of some fluid element, we are considering the motion of some macroscopic volume which contains a large number of molecules. All the particles in this volume have the same velocity, and each participates in the overall macroscopic motion; that is, within the confines of this small but macroscopic volume, all are involved in collective motion. In other words, the hydrodynamic motion of the fluid corresponds to specific collective degrees of freedom. The collective motions of macroscopic bodies are extremely varied. A plasma has a large number of collective degrees of freedom. By generalizing the ideas which have been illustrated with fluid turbulence, we can define turbulence in the following manner: turbulent motion of macroscopic bodies is that motion in which the collective degrees of freedom are random and strongly excited.

We should point out that the collective degrees of freedom of large bodies are usually excited even under the conditions of statistical equilibrium, but the energy present in them is very small. For turbulent motion the energy in the collective degrees of freedom must be much greater than the energy which they contain in statistical equilibrium. It is in this specific sense that we shall use the concept of strongly excited collective degrees of freedom.

The concept of turbulence also implies that the number of strongly excited collective degrees of freedom is very large. For if just one collective degree of freedom were excited, the initial conditions prescribed for it would completely determine the subsequent behavior of the system and that behavior would be regular rather than random. Such a situation is possible, at least in principle, when only one degree of freedom is initially excited and the energy is thereafter redistributed between that mode and other modes. The other case, in which many degrees of freedom are excited at the same time, is also possible. In the first case turbulence is not excited immediately, and there will be some finite period of time during which the system makes its transition into a turbulent state. This is what occurs in our example of the fluid flowing in the tube (a time t_1 is needed to establish the turbulent regime).

In order to picture turbulence of a plasma the first requirement is to know what collective motions are possible in a plasma, how they are excited, and finally, at what time the plasma will make its transition into the turbulent state, in which the collective degrees of freedom are random quantities whose values are not reproducible from experiment to experiment. Before turning our attention to these questions, one additional general comment is in order concerning the nature of turbulent pulsations in a plasma.

Plasma is that state of matter in which the atoms or molecules are found in an ionized state. The interactions of electrons and ions are determined by long-range electrical forces. The many forms of collective motion in a plasma are the result of coupling the charged-particle motion to the electromagnetic field. Therefore, the electromagnetic field which accompanies the particle motion is also a random nonreproducible quantity in a turbulent plasma. Measurements have shown that the fields excited in a plasma during the development of turbulence do in fact have a random nature. As an example, Fig. 1.2 shows the electric field for the turbulent pulsations excited by a beam (as a function of time) [2]. The excitation of stochastic fields is a characteristic of plasma turbulence.

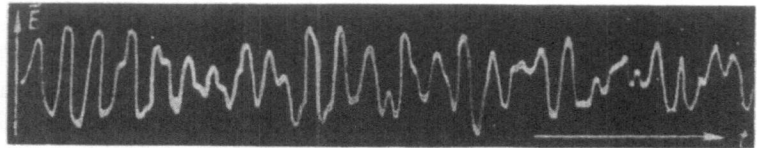

Fig. 1.2. Time dependence of the electric field of turbulent plasma pulsations in the presence of a plasma–beam interaction [2].

1.2. A Statistical Description
of Turbulence

The fact that turbulent processes are random in nature does not mean that the average value of certain quantities cannot be constant, or slowly varying, over a sufficiently long period of time. It is advisable to use a statistical description of the turbulence in order to determine these mean values. Therefore we shall introduce an ensemble of systems which differ only in their initial conditions [3]. To describe the turbulence we shall use the mean values of various quantities obtained by averaging over our statistical ensemble. Because of the way in which the collective degrees of freedom vary with time, it is natural to resort to the ergodic hypothesis, which states that, if one waits long enough, the system will pass through a state which is as close to some previously specified state (corresponding to a given set of initial conditions) as desired. Thus, the initial states of the system exert a weak influence on the average values of the various quantities, and averages taken over the statistical ensemble are identical with the time averages for a given system [3]. These well-known statements of statistical theory, extended to the intense collective motions of a macroscopic body, are the basis for the statistical description of turbulence.

Turbulence is designated as stationary if the average values which characterize the turbulence are not functions of time. The turbulence is uniform if the average values do not depend on position; it is isotropic if the averages at a specific point in space do not depend on the angle relative to any previously designated direction.

The quantities which characterize collective motion, such as the particle velocity or the electric field which accompanies turbulent plasma pulsations, can undergo random turbulent variations. The velocity $\mathbf{v}$ can be divided into two parts, separating the random turbulent component $\mathbf{v}^T$ from the component which experiences no random variations, $\mathbf{v}^R = \langle \mathbf{v} \rangle$:

$$\mathbf{v}(\mathbf{r}, t) = \langle \mathbf{v}(\mathbf{r}, t) \rangle + \mathbf{v}^T(\mathbf{r}, t) = \mathbf{v}^R(\mathbf{r}, t) + \mathbf{v}^T(\mathbf{r}, t). \tag{1.1}$$

The value of $\mathbf{v}^T$ is assumed to be zero when averaged over the statistical ensemble (the average value of $\mathbf{v}$ can always be

assigned to $\mathbf{v}^R$). The intensity of the turbulence can be characterized by the average values of certain quadratic forms such as

$$\langle v_i^T (\mathbf{r}, t) \, v_j^T (\mathbf{r'}, t') \rangle = V_{ij}. \qquad (1.2)$$

Equation (1.2) involves the velocity components along various directions (indexed by i and j) taken at different times and different
points in space. The angular brackets signify an averaging
over the statistical ensemble. The quantity V_{ij} measures the
correlation of velocities of turbulent pulsations at different points
in space and time; here i, j = $\{1, 2, 3\}$ = $\{x, y, z\}$.

The correlation tensor V_{ij} depends on $\mathbf{r} - \mathbf{r'}$ and $t - t'$ only
when the field of turbulent velocities is stationary and uniform, for
then the absolute values of the coordinates and time for one of the
velocities are arbitrary because of the equivalence of all points
and all times. If $\mathbf{v}^T$ is expanded in a Fourier series (see [4],
page 45),

$$\mathbf{v}^T (\mathbf{r}, t) = \int \mathbf{v}_{\mathbf{k}, \omega}^T e^{-i\omega t + i\mathbf{kr}} \, dk d\omega \equiv \int \mathbf{v}_k^T e^{ikx} \, dk, \qquad (1.3)$$

where $k = \{\mathbf{k}, \omega\}$, $x = \{\mathbf{r}, t\}$, $dk = dk \, d\omega$, and $kx = \mathbf{kr} - \omega t$, the correlation tensor can be written as

$$V_{ij} = \int \langle v_{ik}^T v_{jk'}^T \rangle \, dk \, dk' \, e^{ik' (x' - x) + i (k + k') x}. \qquad (1.4)$$

The fact that Eq. (1.4) must be independent of t and r for a
stationary uniform turbulence implies that the mean value of $v_{ik}^T v_{jk'}^T$
is nonzero only when $\omega + \omega' = 0$ and $\mathbf{k} + \mathbf{k'} = 0$, i.e.,

$$\langle v_{i\mathbf{k}, \omega}^T v_{j\mathbf{k'}, \omega'}^T \rangle = V_{ij\mathbf{k}, \omega} \delta (\omega + \omega') \delta (\mathbf{k} + \mathbf{k'}) \equiv V_{ijk} \delta (k + k'), \qquad (1.5)$$

where δ is the well-known Dirac delta function:

$$\delta (x) = \begin{cases} 0, & x \neq 0, \\ \infty, & x = 0, \end{cases} \quad \int \delta (x) \, dx = 1.$$

For isotropic turbulence of an incompressible fluid the tensor
$V_{ijk, \omega}$ reduces to a single quantity, which characterizes the turbulence spectrum. All directions are equivalent in isotropic turbulence. From the definition of a tensor, each index of $V_{ijk, \omega}$ must

transform like a vector when the coordinate system is rotated. But the only vector in this particular case is the vector **k**. From it we can construct the only possible tensor,

$$\frac{k_i k_j}{k^2},\qquad(1.6)$$

except for the trivial unit tensor δ_{ij}. Therefore, for isotropic turbulence

$$V_{ij\, \mathbf{k},\, \omega} = \left(\delta_{ij} - \frac{k_i k_j}{k^2}\right)|V^t_{\mathbf{k},\, \omega}|^2 + \frac{k_i k_j}{k^2}|V^l_{\mathbf{k},\, \omega}|^2.\qquad(1.7)$$

Here the arbitrary coefficients of δ_{ij} and $k_i k_j/k^2$ are taken to be respectively

$$|V^t_{\mathbf{k},\, \omega}|^2 \text{ and } |V^l_{\mathbf{k},\, \omega}|^2 - |V^t_{\mathbf{k},\, \omega}|^2.$$

We shall see that $|V^l_{\mathbf{k},\, \omega}|^2$ and $|V^t_{\mathbf{k},\, \omega}|^2$ describe the longitudinal and transverse turbulent velocity pulsations, respectively. We must keep in mind that, roughly speaking, turbulent pulsations are a superposition of plane waves [see Eq. (1.3)]. For longitudinal pulsations the amplitude of each such plane wave is directed along its direction of propagation, while for transverse pulsations the amplitude is perpendicular to the propagation direction.

It is easy to see that there can be no longitudinal pulsations in an incompressible fluid. This is because from the very condition for incompressibility

$$\operatorname{div} \mathbf{v} = 0\qquad(1.8)$$

we find that

$$i\,(\mathbf{k}\mathbf{v}_{\mathbf{k},\, \omega}) = \frac{1}{(2\pi)^3}\int \operatorname{div} \mathbf{v}\, e^{-i\mathbf{k}\mathbf{r} + i\omega t}\, d\mathbf{r}\, dt = 0,\qquad(1.9)$$

i.e., $\mathbf{v}_{\mathbf{k},\omega}$ is orthogonal to **k**. Thus, for isotropic, uniform, stationary turbulence in an incompressible fluid the tensor of Eq. (1.2) is described by just $|V^t_{\mathbf{k},\omega}|^2$. We can understand the physical significance of this result by computing the mean energy of the turbulent pulsations in a volume of 1 cm^3.

Let

$$W^T = \left\langle \frac{nm\,(v^T)^2}{2} \right\rangle \tag{1.10}$$

Here n is the fluid density and m is the mass of one particle of the fluid. Because n is constant in an incompressible fluid, W^T is determined solely by the mean value of the velocity squared:

$$\langle (v^T)^2 \rangle = \langle v_i^T (\mathbf{r},\,t)\, v_i^T (\mathbf{r},\,t) \rangle = \int \langle v_{i,\,k}\, v_{i,\,k'} \rangle\, e^{i(k+k')x}\, dk\,dk'; \tag{1.11}$$

here we have used Eq. (1.3). After substituting (1.5) and (1.7) into (1.11) we obtain

$$\langle (v^T)^2 \rangle = \int \left(\delta_{ii} - \frac{k_i\,k_i}{\mathbf{k}^2} \right) |V_{\mathbf{k},\,\omega}^t|^2\, dk\,d\omega. \tag{1.12}$$

Therefore for an incompressible fluid

$$W^T = nm \int |V_{\mathbf{k},\,\omega}^t|^2\, dk\,d\omega = \int W_{\mathbf{k},\,\omega}\, dk\,d\omega, \tag{1.13}$$

where

$$W_{\mathbf{k},\,\omega} = nm\,|V_{\mathbf{k},\,\omega}^t|^2 \tag{1.14}$$

is the spectral energy density of the turbulence. Other collective characteristics of the fluid, such as density and temperature, can undergo random variations also.

When discussing plasma turbulence one must keep in mind that most collective plasma motions are accompanied by the appearance of electric and magnetic fields. Usually the turbulent quantities (such as turbulent particle velocities, densities, and so on) can be expressed in terms of these fields. Even very slow, purely hydrodynamic plasma motions† are accompanied by variations in the electric and magnetic fields. In this kind of movement the primary energy in the collective motion is the kinetic

---

†These motions are similar to the turbulent motions of a fluid and are possible when the characteristic times of the pulsations are much longer than the particle's pair-collision times.

energy of the plasma particles; the energy of the electromagnetic field is very small compared to the particle kinetic energy. If the particle motion is coupled to the fields, the total collective motion energy can be determined by the electromagnetic energy. Therefore, in plasma the turbulent motions can be described with the help of the electromagnetic fields connected with them.

Following Eq. (1.1), the electromagnetic field of a plasma can be separated into a regular part E^R and a random part E^T:

$$E(r, t) = E^R(r, t) + E^T(r, t). \tag{1.15}$$

The amplitudes $E_{k, \omega}$ are found by expanding E^T in a Fourier series

$$E_i^T(r, t) = \int E_{ik, \omega} e^{ikr - i\omega t} dk\, d\omega = \int E_{ik} e^{ikx} dk. \tag{1.16}$$

For a uniform, stationary turbulence we find by analogy with (1.4) that

$$\langle E_{ik}^T E_{jk'}^T \rangle = I_{ijk} \delta(k + k') = I_{ijk, \omega} \delta(\omega + \omega') \delta(k + k'). \tag{1.17}$$

Further, for an isotropic turbulence we find that

$$I_{ij\,k, \omega} = |E_{k, \omega}^t|^2 \left(\delta_{ij} - \frac{k_i k_j}{k^2} \right) + |E_{k, \omega}^l|^2 \frac{k_i k_j}{k^2} \tag{1.18}$$

by analogy with Eq. (1.7). Here $|E_{k, \omega}^t|^2$ characterizes the correlation of transverse fields, and $|E_{k, \omega}^l|^2$ is the correlation of longitudinal fields. Following Eq. (1.13), the average energy density of the electric field for turbulent fluctuations takes the form

$$W_E = \frac{\langle E^2 \rangle}{8\pi} = \int \frac{|E_{k, \omega}^t|^2}{4\pi} d k\, d\omega + \int \frac{|E_{k, \omega}^l|^2}{8\pi} dk\, d\omega. \tag{1.19}$$

We see from Eq. (1.19) that $|E_{k,\omega}^t|^2$ and $|E_{k,\omega}^l|^2$ are the spectral densities for the electric field distribution of turbulent fluctuations. According to this equation the intensity of the electromagnetic field for random turbulent fluctuations is in some sense the sum of the intensities of individual plane waves in terms of which the fluctuation field can be expanded. This points up the unimportance of the phase relations which lead to interference terms [Eq. (1.19) contains no interference terms]. One can state that the phases of the waves of turbulent fluctuations are random. The initial values of the wave phases can be considered as quantities which characterize the initial state of the collective degrees of freedom. In this sense, averaging over the statistical ensemble corresponds to averaging over the phases (see [1]).

Let us now discuss the irreversibility of turbulent processes. It is known that the statistical treatment of molecular processes points up their irreversible character, in accord with observation. Since turbulent processes can be described statistically, one naturally arrives at the idea of an irreversible nature for turbulent processes, just as with molecular processes. If molecular friction is related to the transfer of momentum from some molecules to others, then in the turbulent motion of fluids it is possible to transfer momentum from one fluid element to another, or, in other words, from one group of particles to another group. This momentum transfer is accomplished as the result of the rapid movement of separate fluid elements which accompanies the complex motion in a turbulent flow. If one considers the average velocity of the fluid elements in a turbulent flow, which implies that one is not interested in the movement of the individual elements, it is easy to observe that the gradients of this velocity are smaller than the velocity gradients for a laminar flow. Thus, the friction in a turbulent flow is enhanced. Other dissipative processes, such as diffusion and thermal conductivity, can be amplified in the same way.†

The enhanced diffusion, enhanced thermal conduction processes, and enhanced electrical resistance observed in a turbulent plasma are some of the most important stimuli to the study of turbulence. But the many different types of collective plasma motions and their relation to the electromagnetic fields also lead to qualitatively new turbulent dissipative processes in plasma; an example is the irreversible radiation of electromagnetic waves and the acceleration of charged particles.

Let us dwell briefly on the relationship between turbulent motion and molecular motion, and consider the possibility of en-

---

†These facts have served as the foundation for attempts to describe turbulent processes in fluids by analogy with the molecular processes by introducing a mixing length [5], which is similar to the mean free path length in molecular collisions. Kadomtsev [6] has used this approach to turbulence in plasma.

ergy being transferred from the turbulent motions to the thermal motions. We shall assume that at some moment a source responsible for turbulence disappears. The turbulence is damped (by viscosity, for example), and the energy of turbulent motion goes into heat. It is natural to assume that the turbulent processes, which lead to the enhanced dissipative processes of energy and momentum transfer, contain a mechanism for a more intensive transfer of turbulent motion energy into thermal motion energy.

It can therefore be concluded from these considerations (which are treated in greater detail in §1.3) that such a mechanism does in fact exist when stationary turbulence can exist for different powers of the process exciting the turbulence. In fluids stationary turbulence can be set up by various modes of excitation, such as using different fluid velocities in tubes. This effect indicates that the transfer of energy from turbulent motion to thermal motion depends on the intensity of the turbulent motion.

These kinds of effects in a plasma produce heating, which also depends on the intensity of the turbulence; it is called turbulent heating. A number of experiments [2, 7, 8] have been carried out which point out the effectiveness and practical significance of turbulent heating.

In order to understand the turbulent heating process, let us consider the formation of stationary turbulence. We return again to the example of a turbulent fluid.

1.3. The Spectrum of Turbulent Fluctuations in an Incompressible Fluid

In an incompressible fluid the only possible collective motion is vortex motion of the fluid elements. Because of the incompressibility, div $\mathbf{v} = 0$, the fluid motion is determined solely by curl $\mathbf{v}$. Turbulence in an incompressible fluid corresponds to a set of random vortices. One frequently introduces the concept of the scale l of turbulent fluctuations or the characteristic size of the vortex

$$l = \frac{2\pi}{k},$$ (1.20)

where $k = |\mathbf{k}|$ is the wave number appropriate to the turbulent fluctuations.† In the turbulent state (more precisely, in the developed turbulence) one usually finds vortices of different scales. The distribution of intensity over these scales of the turbulent motion is often used to characterize the turbulence spectrum. Assume that the turbulence is isotropic, uniform, and stationary. Returning to Eq. (1.13), we introduce the quantities

$$W_k = \int W_{k\omega}\, d\omega \tag{1.21}$$

and

$$W_k = 4\pi k^2 W_k. \tag{1.22}$$

Then the energy per unit volume of the vortices' turbulent motion‡ can be written as

$$W^T = \int_0^\infty W_k\, dk. \tag{1.23}$$

W_k is usually called the turbulence spectral function. Unlike $W_{k,\omega}$, which characterizes the distribution of turbulent fluctuations per unit $dk\,d\omega$ interval, W_k characterizes the energy of the turbulent fluctuations belonging to dk. That is, it is the distribution of turbulent fluctuations over the turbulence scale.

Let us examine the flow of a fluid in a tube. The average flow velocity will be taken to be much smaller than v_s. At low velocities the movement of a viscous fluid is laminar, but it becomes turbulent at high velocities. Let us trace the transition

---

†The use of the letter k in the future for both the four-vector $\{\mathbf{k},\ \omega\}$ and $|\mathbf{k}|$ should not cause confusion since most of the time the four-vector is encountered only as a subscript on some Fourier components or in their mean squares [see Eq. (1.17)].

‡We note that one is limited to vortex motion only if

$$\frac{nm\langle v^2\rangle}{2} \ll \frac{nmv_s^2}{2} \approx nT, \tag{1.24}$$

$$\frac{\omega}{k} \ll v_s. \tag{1.25}$$

Here v_s is the velocity of sound.

from laminar flow to turbulent flow. Turbulence arises as a result of an instability in the laminar flow [1]. As the average stream velocity $\langle v \rangle$ increases, an instability appears at some critical velocity $\langle v \rangle^*$ (the flow is laminar at lower speeds). Because the fluid velocity is zero at the tube wall, the velocities of the fluid elements must increase as one nears the tube axis. If we are dealing with a cylindrical tube, we can assume that cylindrical layers of fluid move within it, and that each layer experiences some friction from the other layers nearest it because of viscosity. The friction, which transmits momentum from layer to layer, is molecular in origin. As the mean stream velocity increases, the velocity profile is stretched and the velocity gradients increase.

At some value such that $\langle v \rangle = \langle v \rangle^*$ this motion becomes energetically unfavorable; momentum transfer due to layer mixing will be more favorable. The characteristic scale for this motion will be of the order of L, the cross-sectional diameter of the tube. For further increases in $\langle v \rangle$ the large-scale motion also becomes unstable and vortex motion of a somewhat smaller scale appears. This sequence of scale subdivisions means that the turbulence energy will be distributed over a wide range of scales, i.e., over a broad range of k. The mechanism for subdividing the turbulence scale can be nonlinear.

Kolmogorov [9] obtained the asymptotic form of the turbulence spectrum for an incompressible fluid from simple dimensional relationships. He used the following simple physical assumptions:

1. Turbulence is locally isotropic. In other words, the turbulence spectrum for vortex motion can be characterized by a single W_k [see Eq. (1.22)].

2. There is a constant flow of energy from the large-scale fluctuations to the small-scale fluctuations. Such a flow is established for sufficiently small scales of turbulence, considerably smaller than the primary scale

$$L = 2\pi/k_0, \tag{1.26}$$

that is, for k such that

$$k \gg k_0. \tag{1.27}$$

The turbulence spectrum thus obtained is asymptotic, which means that it is valid only for sufficiently large wave numbers k which satisfy Eq. (1.27).

Kolmogorov's second assumption ought to be explained more fully. The constancy of the energy flow means that the energy of turbulent fluctuations is transmitted in cascade fashion only by fluctuations of neighboring scales. The energy of turbulent fluctuations must be transmitted from some wave number to a nearby value of $k + \Delta k$, where Δk is of the same order as k; only then can it be pumped further along. In other words, this is a definite assumption about the coupling between turbulences of different scales, and this nonlinear interaction is most effective when

$$\Delta k \ll k. \tag{1.28}$$

We now obtain the condition for constancy of the energy flow, starting from dimensional considerations. The energy flow has dimensions of energy density divided by time. $W_k k$ has dimensions of energy density. $1/t$ has dimensions of frequency ω; it can be obtained from the wave number k and v_0 only through the product kv_0, where v_0 is some velocity. In other words $W_k k$ has the dimensions of nmv_0^2, or $v_0 \approx (W_k k/nm)^{1/2}$. From this we find the dimensions of the energy current to be $W_k kkv_0 \sim W_k k^2 (W_k k/nm)^{1/2} = Q$.

Since Q is constant, we can find the k dependence of W_k (Fig. 1.3):

$$W_k = \frac{\text{const}}{k^{5/3}}. \tag{1.29}$$

Equation (1.29) gives the well-known Kolmogorov spectrum.

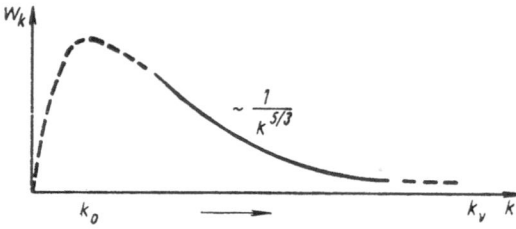

Fig. 1.3. Turbulence spectrum of an incompressible fluid. The arrow shows the direction along which the turbulent energy is transmitted.

In addition to requiring that Eq. (1.27) be satisfied, Eq. (1.29) holds when the fluid viscosity is negligibly small. As a rule viscosity is significant when the gradients are large, i.e., for large k. Let k_ν be that wave number at which viscosity begins to be important. Then the condition under which Eq. (1.29) is valid can be written as

$$k \ll k_\nu. \tag{1.30}$$

When viscosity starts to be important, another quantity with new dimensions enters into the picture, and our previous dimensional derivation is no longer valid. We note that the spectrum described by Eq. (1.29) has been experimentally confirmed with good statistics over a wide range of different k.

It should now be easy to understand the mechanism behind the additional energy absorption which occurs when turbulence develops. Turbulence accomplishes the pumping of fluctuation energy from the generation region, where $k \sim k_0$, into the region of larger wave numbers, in which $k \sim k_\nu$, at least until absorption of the fluctuations due to viscosity begins to be important. Turbulence does not lead to any actual absorption of collective motions in the fluid. This result is reflected in the fact that the energy of turbulent fluctuations is not lost when it is pumped from some scales to others. But the collective-motion energy does move into the region where true molecular absorption becomes very intense.

We can now understand the way in which stationary turbulence is established for different excitation intensities. Stationary turbulence appears because of the energy balance in the system. The turbulence energy goes into large-scale motion because of the external perturbation, and is dissipated in the small-scale motions because of viscosity. When the exciting intensity changes, only the energy flow from the large-scale fluctuations to the small-scale motions is decreased. Viscosity damping of turbulent fluctuations at large k is reflected in the observation that collective hydrodynamic fluid motions at very large k are not possible in general. Fluids sustain collective motions other than hydrodynamic modes.

Hydrodynamic motion is possible in a plasma when collisions are frequent, or, when there is a magnetic field, when collisions are few (see [10-12]). Under these conditions the results discussed above can be carried over to a plasma. There are a large number of collective degrees of freedom in a plasma which have lengths much shorter than the mean free path length. These are characterized by regions of "transparency," where the oscillations are weakly damped, and by regions in which the absorption is strong. The transfer of fluctuation energy into the absorption region can be considered to be the mechanism for both establishing a stationary spectrum and for the turbulent heating of the plasma.

This example contains all the elements from which stationary turbulence in a plasma arises, but the plasma turbulence is far richer in possible modes of interaction and simple dimensional considerations can hardly be used for it.

The first necessary element is to separate the regions for generation of plasma turbulence from those in which its energy is absorbed. If in some wave number region the turbulent fluctuations grow, it must be because the generation growth rates are greater than the fluctuation absorption damping rates; conversely, fluctuation absorption is possible only in wave number neighborhoods wherein this relationship between growth and damping rates is reversed.

The second element required is the presence of a transfer of fluctuation energy from a generation region to an absorption region. Only this type of process can establish the balance of turbulent energy in a plasma and lead to stationary turbulence. This energy transfer process is nonlinear, as a rule.

These effects will be clearer if we become familiar with the possible plasma collective motions. For now we shall be primarily concerned with the linear collective degrees of freedom. We shall begin with an intuitive, qualitative description of them.

1.4. Linear Collective Degrees
of Freedom in Plasma

The collective degrees of freedom in a plasma can be divided into the hydrodynamic (hydromagnetic) and the collisionless degrees of freedom. The latter concept is meant to indicate that, although particle collisions cannot be neglected entirely, the characteristic times for the plasma motions for these degrees of

freedom are much shorter than the average time between collisions. When collisions are frequent (that is, the characteristic
times for the collective degrees of freedom are much longer than
the collision frequencies), the plasma can be treated using the
hydrodynamic equations, or the magnetohydrodynamic equations
when there is a magnetic field present.

In the following we shall adhere to our selected definitions
of the two classes of plasma degrees of freedom, and the separation criterion will be that the collision frequency is either greater
or smaller than the frequency in question, i.e., $\omega \gtrless \nu_{col}$, where the
subscript refers to the collisions. It is necessary to make
such a sharp distinction in the terminology because some low-
frequency collisionless degrees of freedom are similar to the
corresponding hydrodynamic degrees of freedom, and they are
also frequently called hydrodynamic.

1.4.1. An Isotropic Plasma without

External Magnetic Fields

The turbulent motions in a plasma can be described in terms
of their related electromagnetic fields. In an isotropic plasma
these fields can be either longitudinal or transverse.

Let us describe the longitudinal plasma oscillations which
were first examined by Langmuir [13]. We shall consider the case
of linear small-amplitude oscillations. Assume that in a quasi-
neutral plasma a flat layer of electrons Δx thick is shifted by a
distance Δx in the x direction (see Fig. 1.4). This displacement
induces an excess of electrons in the layer and leaves behind an
excess of positive charge. This charge separation produces an
electric field, which creates a force drawing the positive and negative layers of plasma together. As a result the electrons in
the plasma are set into motion. At the moment when the charges
in the positive and negative layers are compensated, the electrons will have their maximum kinetic energy, overshooting the
equilibrium state as a consequence. This leads to oscillations in
the charge density and electric field. We can find the frequency
of these oscillations by noting that the field from a plane charge
layer is

$$E = 4\pi\sigma_*,$$ (1.31)

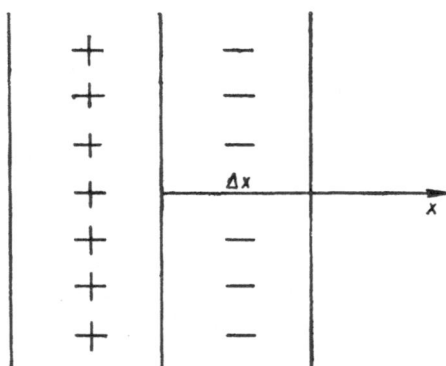

Fig. 1.4. The oscillations of a plasma layer.

where σ_* is the surface charge density, $\sigma_* = en\Delta x$, n is the plasma density. Each electron in the layer experiences a force equal to

$$F = eE = 4\pi e^2\, n\Delta x,\tag{1.32}$$

which is proportional to the deviation from the equilibrium position. Therefore

$$m\frac{d^2\Delta x}{dt^2} = F = m\omega^2\,\Delta x;\qquad \omega^2 = \omega_{pe}^2 = \frac{4\pi ne^2}{m_e},\tag{1.33}$$

where ω_{pe} is the plasma frequency of the electron oscillations. Equation (1.33) is valid for linear oscillations when the electron thermal motion can be neglected. If the thermal motion is included a small correction must be included [14, 15]:

$$\omega^2 = \omega_{pe}^2 + 3k^2 v_{Te}^2,\tag{1.34}$$

where k is the wave number and $v_{Te} = (T_e/m_e)^{1/2}$ is the mean thermal velocity of the plasma electrons.† When $kv_{Te} \ll \omega_{pe}$ the correction to ω_{pe} in Eq. (1.34) is small. The phase velocity of these oscillations is

$$v_p = \frac{\omega}{k} \simeq \frac{\omega_{pe}}{k},\tag{1.35}$$

which varies between v_{Te} and ∞.

---

†Here and in the future, T_e is measured in energy units.

In a collisionless plasma another type of longitudinal oscillation, called ion-acoustic waves, is possible at a much lower frequency. In contrast to Langmuir oscillations, ion-acoustic waves
involve both electrons and ions. Ion-acoustic waves are therefore
rather slow waves, whose phase velocities are much slower than
the mean thermal electron speeds, and the electrons are rapidly
"tuned" to the ion distribution. When the ions are not uniformly
displaced, a space charge develops and creates some distribution
in the potential φ. At each moment of time the electrons, which
will have adjusted to this potential distribution, will have a Boltzmann distribution

$$n_e = n_0 \exp\left(\frac{e\varphi}{T_e}\right). \tag{1.36}$$

In weak ion-acoustic waves the potential $e\varphi$ is small compared with T_e and Eq. (1.36) has the approximate form

$$n_e \approx n_0\left(1 + \frac{e\varphi}{T_e}\right). \tag{1.37}$$

Because of the potential φ the ions are set into motion, which is
governed by the equation (the ions are singly ionized so that $Z = 1$)

$$m_i \frac{dv_i}{dt} = eE = -e\nabla\varphi. \tag{1.38}$$

Because of the continuity equation a change in velocity results in
a change in density:

$$\frac{\partial n_i}{\partial t} + \operatorname{div}(n_i v_i) \simeq \frac{\partial}{\partial t}(n_i - n_0) + n_0 \operatorname{div} v_i = 0. \tag{1.39}$$

For plane waves, where all quantities have their coordinate and
time dependence expressed as $\exp(ikr - i\omega t)$, we find

$$\omega(n_i - n_0) = n_0(kv_i); \quad m_i \omega v_i = ek\varphi;$$

$$n_i = n_0 + n_0\frac{kv_i}{\omega} = n_0\left(1 + \frac{e\varphi k^2}{m_i \omega^2}\right). \tag{1.40}$$

Poisson's equation gives

$$\operatorname{div} E = 4\pi e(n_i - n_e);$$

$$k^2 \varphi = \frac{4\pi n_0 e^2}{m_i}\left(\frac{k^2}{\omega^2} - \frac{m_i}{T_e}\right)\varphi, \tag{1.41}$$

or

$$\omega^2 = \frac{k^2 v_s^2}{1 + k^2 \dfrac{v_{Te}^2}{\omega_{pe}^2}} \; ;$$

(1.42)

$$v_s = \sqrt{\frac{T_e}{m_i}} \; .$$

For $k \ll \omega_{pe}/v_{Te}$,

$$\omega = kv_s,$$

(1.43)

that is, we obtain the usual acoustic oscillations. For $k \gg \omega_{pe}/v_{Te}$ we find that $\omega = \omega_{pi}$, the ion plasma oscillation, where $\omega_{pi}^2 = 4\pi n_0 e^2/m_i$.

The appearance of ion oscillations can be easily understood if we realize that the displacement of the ions from their positions of quasi-neutrality would lead to the same oscillations of the ions as does a displacement of the electrons, but the screening of the space charge by the electrons prevents this from happening. However, there are wavelengths for which screening is impossible, and it is from these that oscillations with $\omega \simeq \omega_{pi}$ come.

The field of a charge placed in a plasma is screened over a distance of the order v_{Te}/ω_{pe}. Let us consider Poisson's equation

$$\Delta\varphi = -4\pi\rho = -4\pi\rho_0 + 4\pi n_0 e \left(e^{\frac{e\varphi}{T_e}} - 1 \right),$$

(1.44)

where ρ_0 is the density of the external charge and the last term in (1.44) describes the electron screening. For small potentials such that $\varphi \ll T_e/e$ we have

$$\Delta\varphi - \frac{4\pi n_0 e^2}{T_e} \varphi = -4\pi\rho_0.$$

(1.45)

The second term on the left of (1.45) is the screening term and will be comparable to the first term [which is of the order of $\varphi/(\Delta x)^2$] at a distance

$$(\Delta x)^2 \sim \frac{T_e}{4\pi n_0 e^2} = \frac{v_{Te}^2}{\omega_{pe}^2} \; .$$

The quantity

$$d_e = \frac{v_{Te}}{\omega_{pe}} \qquad (1.46)$$

is called the Debye radius. It characterizes the distances over which the field of a charge placed into the plasma is screened by the electrons. For oscillations at a frequency $\omega = \omega_{pi}$ the wavelength is shorter than the Debye radius and the electrons cannot screen the ions, and ion oscillations arise.

In general, ion-acoustic vibrations have phase velocities

$$v_p = v_s \left(1 + k^2 d_e^2\right)^{-1/2}, \qquad (1.47)$$

which are always smaller than v_s. In turn, v_s is much smaller than v_{Te}: $v_s = v_{Te}(m_e/m_i)^{1/2}$. For T_e of the same order as T_i, when v_s is then on the order of v_{Ti}, ion-acoustic oscillations are absent because of the strong absorption. These oscillations are only possible when the inequality

$$T_e \gg T_i \qquad (1.48)$$

is satisfied or when

$$v_s \gg v_{Ti}.$$

In addition to longitudinal oscillations, an isotropic plasma can also sustain transverse oscillations. They propagate as waves in a plasma only when

$$\omega > \omega_{pe}. \qquad (1.49)$$

Because the wave frequency is high, the electrons in the wave are displaced according to the following force equation:

$$m_e \frac{dv}{dt} = e\mathbf{E}.$$

For plane waves proportional to $\sim \exp(-i\omega t + i\mathbf{k}\mathbf{r})$ we have

$$\left.\begin{aligned}
-i\omega m_e \, \mathbf{v}_k &= e\mathbf{E}_k; \\
\mathbf{j}_k = en_0 \, \mathbf{v}_k &= -\frac{e^2 n_0}{m_e \, i\omega} \mathbf{E}_k = \sigma \mathbf{E}_k; \\
\varepsilon = 1 + \frac{4\pi i\sigma}{\omega} &= 1 - \frac{\omega_{pe}^2}{\omega^2},
\end{aligned}\right\} \qquad (1.50)$$

where ε is the dielectric constant and σ is the plasma conductivity. For transverse waves the index of refraction is

$$n = \frac{c}{v_p} = \sqrt{\varepsilon}.$$

Equation (1.49) follows from the reality of ε. By substituting $v_p = \omega/k$ into Eq. (1.50) we find

$$\omega^2 = \omega_{pe}^2 + c^2 k^2. \tag{1.51}$$

The phase velocities of transverse waves are greater than the speed of light. Equation (1.51) gives the spectrum for transverse waves in the plasma. Figure 1.5 shows the possible spectra of oscillations for a uniform plasma.

It is advisable to consider what part of the total turbulent energy locked in the collective degrees of freedom of the plasma is due to the energy of the electromagnetic field. We begin with the ion-acoustic oscillations.

The kinetic energy of the ions is

$$W_i = \left\langle n_0 m_i \frac{v_i^2}{2} \right\rangle = n_0 \frac{m_i}{2} \int \langle v_{ik} v_{ik'} \rangle e^{i(k+k')x} \, dk \, dk' =$$

$$= \frac{n_0 e^2}{2m_i} \int \frac{dk\, dk'}{\omega^2} \langle E_{ik} E_{ik'} \rangle \exp\, [i(k+k')x] = \int \frac{\omega_{pi}^2}{\omega^2} \frac{|E_k|^2}{8\pi} \, dk,$$

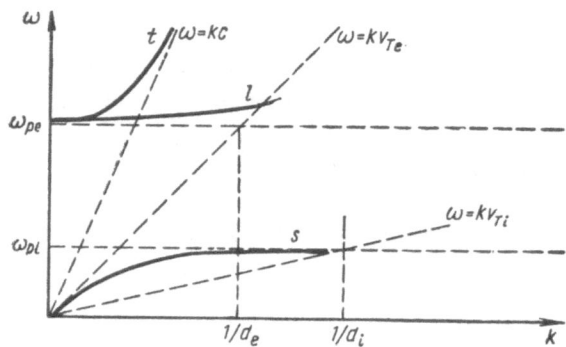

Fig. 1.5. Dispersion relations for turbulent fluctuations in a homogeneous plasma with $H_0 = 0$.

so that as a rough approximation,

$$W_i \approx \frac{\omega_{pi}^2}{\omega^2} W_E; \qquad W_E = \int \frac{|E_k|^2}{8\pi} \, dk. \qquad (1.52)$$

In the region of ion oscillations $\omega \sim \omega_{pi}$; $W_i \sim W_E$, so that the total turbulence energy is divided evenly between the electric field energy and the ion oscillational energy. When

$$\omega = kv_s = kv_{Ti} \sqrt{\frac{T_e}{T_i}}$$

we have

$$W_E \simeq k^2 d_e^2 W_i \ll W_i. \qquad (1.53)$$

It follows from Eq. (1.53) that the greater part of the energy is found in the ion motions. By means of a similar calculation one can easily obtain an expression for the energy of Langmuir and transverse oscillations, to wit

$$W_e \approx \frac{\omega_{pe}^2}{\omega^2} W_E. \qquad (1.54)$$

In Langmuir oscillations the turbulence energy is equal divided between the energies of the field and the electron, but for transverse waves this occurs only when $\omega \rightarrow \omega_{pe}$, that is, because of Eq.(1.51), for very long wavelength waves this term completely dominates the quantity c/ω_{pe}. For very-high-frequency waves, $\omega \gg \omega_{pe}$, and the bulk of the energy is in the electromagnetic field. We must recall that the electromagnetic field in a turbulent plasma participates on equal footing with other parameters, such as the particle velocities, for example. As we shall see presently, even waves for which $\omega \gg \omega_{pe}$ can interact strongly with turbulent plasma fluctuations; they may, in fact, be generated by such fluctuations.

1.4.2. A Homogeneous Plasma in a Strong External

Magnetic Field

The number of possible collective degrees of freedom in a plasma becomes very large in the presence of a homogeneous external magnetic field H_0. We shall examine only the most important of them. Neglecting the Lorentz force from the self-consistent field, the low-amplitude charge motions in the presence of a magnetic field can be described by the equation

$$\frac{d\mathbf{v}}{dt} = \frac{e}{mc}[\mathbf{v}\mathbf{H_0}] + \frac{e}{m}\mathbf{E}. \tag{1.55}$$

If all quantities are written as plane waves $\sim\exp(-i\omega t + i\mathbf{kr})$, then

$$-i\omega\mathbf{v}_k = \omega_H[\mathbf{v}_k\mathbf{h}] + \frac{e}{m}\mathbf{E}_k, \tag{1.56}$$

where $\omega_H = eH_0/mc$ and $\mathbf{h} = \mathbf{H_0}/H_0$. The algebraic equation (1.56) can easily be solved for $\mathbf{v}(\mathbf{H_0}\|z)$:

$$\left.\begin{aligned}
j_x &= en_0 v_x = \frac{e^2 n_0}{m}\cdot\frac{i\omega}{\omega^2-\omega_H^2}\left(E_x + i\frac{\omega_H}{\omega}E_y\right), \\
j_y &= en_0 v_y = \frac{e^2 n_0}{m}\cdot\frac{i\omega}{\omega^2-\omega_H^2}\left(E_y - i\frac{\omega_H}{\omega}E_x\right), \\
j_z &= en_0 v_z = \frac{e^2 n_0}{m\omega}i E_z.
\end{aligned}\right\} \tag{1.57}$$

Unlike the case where $H_0 = 0$, here the field along x causes an electric current to flow in the plasma in both the x and y directions. This phenomenon leads to plasma gyrotropy and anisotropy.

Let us study some new kinds of waves which appear in a plasma when there is a magnetic field. We start with oscillations whose frequencies are much smaller than the ion cyclotron frequency:

$$\omega_{Hi} = \frac{eH_0}{m_i c}.$$

The Alfven waves are of primary interest [16]. Gyrotropic effects are small when $\omega \ll \omega_H$ because the ion contribution to j_x^{gyr} is equal and opposite to the electron contribution:

$$j_{x_i}^{gyr} = \frac{e^2 n_0}{m_i \omega_{Hi}}E_y = -\frac{e^2 n_0}{m_e \omega_{He}}E_y. \tag{1.58}$$

Thus, when $\omega \ll \omega_{Hi}$

$$\mathbf{j}_\perp = -\frac{e^2 n_0\, i\omega}{m_i\, \omega_{Hi}^2}\mathbf{E}_\perp; \qquad j_z = \frac{e^2 n_0\, i}{m_e\, \omega}E_z, \tag{1.59}$$

where ⊥ indicates the projection perpendicular to the external magnetic field.

Maxwell's equations

$$\text{curl } \mathbf{H} = \frac{4\pi}{c}\, \mathbf{j} + \frac{1}{c} \cdot \frac{\partial \mathbf{E}}{\partial t}\, ; \qquad \left.\right\}$$
$$\frac{1}{c} \cdot \frac{\partial \mathbf{H}}{\partial t} = -\text{curl } \mathbf{E} \qquad \tag{1.60}$$

for plane waves give

$$\mathbf{j}\,[\mathbf{kH}_k] = \frac{4\pi}{c}\, \mathbf{j}_k - \mathbf{i}\, \frac{\omega}{c}\, \mathbf{E}_k; \qquad \left.\right\}$$
$$\mathbf{H}_k = \frac{c}{\omega}\, [\mathbf{kE}_k], \qquad \tag{1.61}$$

or, by using Eq. (1.59),

$$c^2(\mathbf{k}_\perp\,(\mathbf{kE}) - \mathbf{k}^2\, \mathbf{E}_\perp) = -\frac{\omega_{pi}^2}{\omega_{Hi}^2}\, \omega^2 \mathbf{E}_\perp - \omega^2\, \mathbf{E}_\perp; \tag{1.62}$$

$$c^2\,(k_z\,(\mathbf{kE}) - k^2 E_z) = \omega_{pe}^2\, E_z - \omega^2\, E_z. \tag{1.63}$$

Since even a very weak field E_z results in a large current in the plasma, E_z must be small for waves with $E_\perp \neq 0$. In fact, it follows from Eq. (1.63) that

$$E_z \sim \frac{k_z\, k_\perp\, c^2 E_\perp}{\omega_{pe}^2} \ll E_\perp.$$

Assume that the x axis is in the kH plane, i.e., $k_y = 0$. The projections of Eq. (1.63) on x and y take the form

$$k_x^2 - k^2 = -k_z^2 = -\omega^2\left(1 + \frac{\omega_{pi}^2}{\omega_{Hi}^2}\right)\frac{1}{c^2}; \tag{1.64}$$

$$-k^2 = -\omega^2\left(1 + \frac{\omega_{pi}^2}{\omega_{Hi}^2}\right)\frac{1}{c^2}. \tag{1.65}$$

These equations can be written as

$$\omega_A = kv_A\,|\cos\theta\,|\,; \tag{1.66}$$

$$\omega_M = kv_A, \tag{1.67}$$

where θ is the angle between k and H_0, and v_A is the Alfven velocity,

$$\frac{v_A^2}{c^2} = \frac{\omega_{Hi}^2}{\omega_{pi}^2} \cdot \frac{1}{1 + \frac{\omega_{Hi}^2}{\omega_{pi}^2}} = \frac{H^2/4\pi n_0 m_i c^2}{1 + \frac{H^2}{4\pi n_0 m_i c^2}} . \tag{1.68}$$

The wave for ω_A is called the Alfven wave, and the wave for ω_M is the magnetoacoustic wave.

In relatively weak magnetic fields, for which

$$\frac{H^2}{4\pi n_0 m_i c^2} \ll 1,$$

there is an approximate expression for v_A, which was first obtained by Alfven [16]:

$$v_A = \frac{H_0}{\sqrt{4\pi n_0 m_i}} . \tag{1.69}$$

The phase velocities of the Alfven and magnetoacoustic waves are either equal to or less than v_A, and in very strong magnetic fields they can be very large. However, because of Eq. (1.68), they cannot exceed the speed of light [17].

For low velocities such that $v_A \approx v_s$ (v_s is the velocity of the ion-acoustic oscillations) Eqs. (1.66) and (1.67) are modified. Let us further assume that the plasma electrons are magnetized, that is, the wavelengths of the waves are much greater than the mean Larmor radius of the electrons rotating in the external magnetic field. In this case it is only the value of k_x which is of significance (since $k_y = 0$). The condition which we have stipulated can be written as

$$k_x r_e \ll 1, \tag{1.70}$$

where $r_e = v_{Te}/\omega_{He}$ is the electrons' Larmor radius when they are moving at a mean thermal velocity of $v_{Te} = \sqrt{T_e/m_e}$. We shall be concerned with a plane wave:

$$E = E_0 \exp(i\mathbf{k r} - i\omega t). \tag{1.71}$$

Because of the magnetization, the field acting on the electrons is approximately given by

$$E \simeq E_0 (1 + i k_x x) \exp (i k_z z - i\omega t). \qquad (1.72)$$

The second term in parentheses in Eq. (1.72) is always small but we shall see that it must be taken into account in certain cases. Because this field depends only on z (in an approximate sense), one can always find a potential φ such that

$$E_z = - \frac{\partial \varphi}{\partial z}$$

or, for a plane wave,

$$E_{z, k} = - i k_z \varphi_k. \qquad (1.73)$$

Because of the electrons' magnetization, the magnetic field prevents a Boltzmann distribution from being set up transverse to the magnetic field, but this is not true along the magnetic field. The potential φ characterizes this Boltzmann distribution of the electrons along the magnetic field. As in the case of the acoustic oscillations, the electron concentration will be given by the expression

$$n_e \approx n_0 \left(1 + \frac{e\varphi}{T_e} \right).$$

The polarization field created by the nonuniform electron distribution will have only the component E_z along the magnetic field. The electron current j_z created by this field can easily be found from the continuity equation

$$\mathrm{div} \, j + \frac{\partial \rho}{\partial t} = 0, \qquad (1.74)$$

which, for plane waves, takes the form

$$k_z j_z = \omega \rho = - \omega e (n_e - n_0) = - \omega n_0 \frac{e^2 \varphi}{T_e}. \qquad (1.75)$$

By using Eq. (1.44) we find that

$$j_z^{(e)} = -\frac{i}{4\pi} \cdot \frac{\omega_{pe}^2}{k_z^2 v_{Te}^2} \omega E_z.$$ (1.76)

Then, by using Eq. (1.57) as written for ions, we find that
Eq. (1.59) is now replaced with the following expression:

$$j_z = \frac{i\omega E_z}{4\pi} \left(\frac{\omega_{pi}^2}{\omega^2} - \frac{\omega_{pe}^2}{k_z^2 v_{Tz}^2} \right).$$ (1.77)

If we consider only those oscillations which do not produce
magnetic fields (i.e., curl E = 0), Maxwell's equations give

$$\frac{1}{c} \cdot \frac{\partial E}{\partial t} + \frac{4\pi}{c} j = 0; \quad i\omega E_k = 4\pi j_k$$ (1.78)

and, by virtue of Eq. (1.77),

$$\omega^2 = \frac{k_z^2 v_s^2}{1 + k_z^2 d_e^2}.$$ (1.79)

This dispersion relation differs from that of (1.42) in that
k^2 is now replaced by k_z^2. This aspect is closely connected with
the magnetization of the plasma electrons. Thus, when $k_z d_e \ll 1$,
we have

$$\omega = \omega_{Ms} = k v_s |\cos \theta|.$$ (1.80)

These oscillations are called magnetic ion-acoustic waves. Magnetic ion-acoustic waves can exist if $v_A \gg v_s$, for it is only in this
approximation that our assumption that the magnetic field oscillations are zero is correct. When $v_A \sim v_s$ the two waves at the
frequencies ω_M and ω_{Ms} are replaced by two other waves, which
are called the fast and slow magnetoacoustic waves. In connection with these new waves one finds that the current j_y has
a component proportional to E_z and j_z has a component proportional to E_y. This last result is easily understood if one notes
that the electron density is modulated by a field

$$\delta n_e = n_e - n_0 = \frac{ei E_z}{k_z T_e} n_0$$ (1.81)

in accord with (1.37). Thus, in addition to the expression (1.59) found earlier, the current j_y,

$$j_y = -\frac{i\,\omega_{pi}^2\,\omega}{4\pi\omega_{Hi}^2}\,E_y = en_0\,(-\,\delta v_y^{(e)} + \delta v_y^{(i)}),\tag{1.82}$$

will contain a correction $\delta j_y = -e\delta n_e\,v_{0y}^{(e)}$ due to the density modulation, where $v_{0y}^{(e)}$ is the rotational velocity of the electrons moving along the Larmor circle, excluding the effect of the wave fields. Using (1.81) we have

$$\delta j_y = -\frac{i\,\omega_{pi}^2}{4\pi k_z}\,m_i\,v_{0y}^{(e)}\,\frac{E_z}{T_e}\,.\tag{1.83}$$

The desired quantity is the current averaged over all the plasma particles. However, since the average value of $v_{0y}^{(e)} = v_0 \sin(\omega_{He}t + \varphi_0)$ is zero, we must take into account the small term in (1.72), which accounts for the finiteness of the particles' Larmor radii. If $v_{0y}^{(e)} = v_0 \sin(\omega_{He}t + \varphi_0)$, then

$$y = -\frac{v_0}{\omega_{He}}\cos\,(\omega_{He}t + \varphi_0); \qquad x = \frac{v_0}{\omega_{He}}\sin\,(\omega_{He}t + \varphi_0) = \frac{v_{0y}^{(e)}}{\omega_{He}}\,.$$

Therefore,

$$\langle\delta j_y\rangle = -\frac{i\,\omega_{pi}^2}{4\pi k_z\,T_e}\,E_z\,ik_x\,m_i\,\langle xv_{0y}^{(e)}\rangle = \frac{\omega_{pi}^2}{4\pi\omega_{Hi}}\tan\theta E_z.\tag{1.84}$$

We note in passing that there are no similar additions to j_x since

$$\langle xv_{0x}^{(e)}\rangle \sim \langle\sin\,(\omega_{He}t + \varphi_0)\cos\,(\omega_{He}t + \varphi_0)\rangle = 0.\tag{1.85}$$

It therefore follows that the Alfven waves have the spectrum $\omega_A = kv_A|\cos\theta|$ not only when $v_A \gg v_s$, but for any relationship between v_A and v_s. One can also show quite readily that the correction to j_z is of the form

$$\delta j_z = -\frac{\omega_{pi}^2}{4\pi\omega_{Hi}}\tan\theta E_y.\tag{1.86}$$

This is simply demonstrated from the nondissipative effects, which lead to the current of Eq. (1.84). We can write down the condition that the average over a period of the work done by the field on the current which this field sets up be zero:

$$\langle Ej \rangle = 0. \tag{1.87}$$

The part $j_\perp$, described by Eq. (1.59), is phase shifted by $\pi/2$ [because of the i in (1.59)] with respect to $E_\perp$ and can be separated from (1.87); then

$$\langle \delta j_y' E_y \rangle + \langle \delta j_z' E_z \rangle = 0. \tag{1.88}$$

It is clear that (1.86) exactly satisfies this equation.

Using the currents we have found, and $k^2 \ll \omega_{pi}^2/c^2$ instead of Eq. (1.65), we find from Eq. (1.61) that

$$-k^2 E_y = -\omega^2 \left(1 + \frac{\omega_{pi}^2}{\omega_{Hi}^2}\right) E_y - \frac{\omega_{pi}^2 \, i\omega}{\omega_{Hi}} E_z \tan\theta; \tag{1.89}$$

$$0 = \omega_{pi}^2 \left(1 - \frac{\omega^2}{k_z^2 v_s^2}\right) E_z + \frac{i\omega\omega_{pi}^2}{\omega_{Hi}} \tan\theta E_y. \tag{1.90}$$

This set of equations leads to the following oscillation spectrum [17, 18]:

$$\omega_\pm^2 = k^2 v_\pm^2; \tag{1.91}$$

$$v_\pm^2 = \frac{1}{2\left(1 + \dfrac{v_{A0}^2}{c^2}\right)} \left\{ v_{A0}^2 + v_s^2 \left(1 + \frac{v_{A0}^2}{c^2}\cos^2\theta\right) \pm \right.$$

$$\left. \pm \sqrt{[v_s^2 (1 + v_{A0}^2 \cos^2\theta/c^2) - v_{A0}^2]^2 + 4v_{A0}^2 v_s^2 \sin^2\theta} \right\}, \tag{1.92}$$

where

$$v_{A0}^2 = c^2 \frac{\omega_{Hi}^2}{\omega_{pi}^2} = \frac{H^2}{4\pi n m_i}.$$

The wave for ω_+ is called the fast magnetoacoustic wave, and that for ω_- is the slow magnetoacoustic wave. In the limit $v_{A0} \gg v_s$, ω_+ goes into ω_M and ω_- transforms into ω_{Ms}.

When $v_{A0}^2 \ll c^2$ we find [19]

$$v_{\pm}^2 = \frac{1}{2} \left[v_{A0}^2 + v_s^2 \pm \sqrt{(v_{A0}^2 - v_s^2)^2 + 4v_{A0}^2 v_s^2 \sin^2 0} \right]. \tag{1.93}$$

Turning to the higher-frequency waves, we have the longitudinal oscillations of a magnetically active plasma. These are the oscillations whose electric field is approximately directed along the wave vector $\mathbf{k}$. Because curl $\mathbf{H} = 0$ for such waves, we have

$$i\omega \mathbf{E}_k = 4\pi \mathbf{j}_k; \quad i\omega k \mathbf{E}_k = 4\pi (\mathbf{j}_k \mathbf{k}). \tag{1.94}$$

We then have from Eq. (1.57) that

$$4\pi (\mathbf{j}_k, \mathbf{k}) = \sum_{\alpha=e,i} \frac{\omega_{p\alpha}^2 \, i\omega}{\omega^2 - \omega_{H\alpha}^2} \frac{k_x^2 + k_y^2}{k} E_k + \sum_{\alpha=e,i} \frac{\omega_{p\alpha}^2 \, k_z^2 E_k}{\omega k}. \tag{1.95}$$

Because $k_x^2 + k_y^2 = k^2 \sin^2 \theta$; $k_z^2 = k^2 \cos^2 \theta$, we find

$$\frac{\omega_{pe}^2 \sin^2 \theta}{\omega^2 - \omega_{He}^2} + \frac{\omega_{pi}^2 \sin^2 \theta}{\omega^2 - \omega_{Hi}^2} + \frac{\omega_{pe}^2 \cos^2 \theta}{\omega^2} = 1. \tag{1.96}$$

For the high-frequency waves the main contribution to ε comes only from the plasma electrons. Equation (1.96) can then be written as

$$\omega^4 - \omega^2 (\omega_{He}^2 + \omega_{pe}^2) + \omega_{He}^2 \omega_{pe}^2 \cos^2 0 = 0. \tag{1.97}$$

We therefore have

$$\omega_{\pm}^2 = \frac{1}{2} \left[\omega_{He}^2 + \omega_{pe}^2 \pm \sqrt{(\omega_{He}^2 - \omega_{pe}^2)^2 + 4\omega_{He}^2 \omega_{pe}^2 \sin^2 \theta} \right]. \tag{1.98}$$

When $\omega_{pe} \gg \omega_{He}$ the two branches of Eq. (1.98) are approximately

$$\omega_+ \simeq \omega_{pe}; \quad \omega_- \approx \omega_{He} |\cos \theta|. \tag{1.99}$$

For angles close to $\pi/2$ we see that for any field

$$\omega_+^2 \simeq \omega_{pe}^2 + \omega_{He}^2. \tag{1.100}$$

As $\theta \to \pi/2$ the branch appropriate to ω_- goes to zero, according to Eq. (1.98), and we must include the contributions from the ions for small ω. We can find the exact solution to (1.96) when $\theta = \pi/2$ by solving this equation

$$\omega^4 - \omega^2 \left(\omega_{He}^2 + \omega_{pe}^2\right) + \omega_{He}\,\omega_{Hi}\left(\omega_{pe}^2 + \omega_{He}\,\omega_{Hi}\right) = 0. \tag{1.101}$$

At high frequencies the last term in (1.101) can be neglected, and we recover Eq. (1.100); however, for small frequencies the first term in (1.101) is negligible and we find that

$$\omega_-^2 \approx \omega_{He}\,\omega_{Hi}\,\frac{\omega_{pe}^2 + \omega_{He}\,\omega_{Hi}}{\omega_{pe}^2 + \omega_{He}^2}. \tag{1.102}$$

In a dense plasma with $\omega_{pe} \gg \omega_{He}$, the frequency ω_- is the geometric mean value of the electron and ion cyclotron frequencies:

$$\omega_- = \sqrt{\omega_{He}\,\omega_{Hi}}. \tag{1.103}$$

This is called the hybrid frequency. For low frequencies at angles much different from $\pi/2$ we need include only the last two terms of the left-hand side of (1.96). Then, if $\omega_{pi} \gg \omega_{Hi}$ we find solutions of the form given in Eq. (1.99), in which the subscript "e" must be replaced with "i":

$$\omega_+^{(i)} = \omega_{pi}; \quad \omega_-^{(i)} = \omega_{Hi}\,|\cos\theta|. \tag{1.104}$$

The branch of the magnetic ion-acoustic oscillations in a dense plasma go into the $\omega_-^{(i)}$ branch.

If the electrons have a Boltzmann distribution (for slow waves), one can use (1.78) for j_z. This alters Eq. (1.96) in the following way for longitudinal waves:

$$\frac{\omega_{pi}^2 \sin^2\theta}{\omega^2 - \omega_{Hi}^2} + \frac{\omega_{pi}^2 \cos^2\theta}{\omega^2} - \frac{\omega_{pe}^2}{k^2 v_{Te}^2} = 1. \tag{1.105}$$

Here we have omitted those terms which give the electrons' contribution to $j_\perp$. The biquadratic equation (1.105) has the following solutions (for $k \ll 1/d_e$):

$$\omega_\pm^2 = \frac{1}{2}\left[\omega_{Hi}^2 + k^2 v_s^2 \pm \sqrt{(\omega_{Hi}^2 + k^2 v_s^2)^2 - 4\cos^2\theta\,\omega_{Hi}^2\,k^2 v_s^2}\right]. \tag{1.106}$$

When $k^2 v_s^2 \ll \omega_{Hi}^2$, we find that

$$\omega_+^2 = \omega_{Hi}^2 + k^2 v_s^2 \sin^2 \theta, \tag{1.107}$$

$$\omega_-^2 = k^2 v_s^2 \cos^2 \theta. \tag{1.108}$$

These equations give us the well-known magnetic ion-acoustic oscillations together with a new branch, Eq. (1.107), whose frequency cannot be smaller than ω_{Hi} (this branch need not be considered for frequencies $\omega \ll \omega_{Hi}$). When $k^2 v_s^2 \gg \omega_{Hi}^2$ we find

$$\omega_+^2 = k^2 v_s^2, \tag{1.109}$$

$$\omega_-^2 = \omega_{Hi}^2 \cos^2 \theta. \tag{1.110}$$

In other words, the magnetic ion-acoustic oscillation branch goes into the longitudinal ion waves of Eq. (1.104), and the branch given by Eq. (1.107) transforms into the unmagnetized acoustic waves. As k increases the unmagnetized acoustic waves change into the ion oscillations with frequency ω_{pi}. This holds for $\omega_{Hi} \ll \omega_{pi}$.

We shall now consider the problem of nonlongitudinal high-frequency waves. This category covers those transverse waves whose frequencies are much higher than all the frequencies characteristic of the plasma. In contrast to an isotropic plasma, $\omega(k)$ is angle dependent for waves whose frequencies are of the order ω_{He} and ω_{pe}. At somewhat lower frequencies, near ω_{Hi}, such waves transform directly into fast magnetoacoustic waves.

Let us examine these waves under assumption that $\omega_{Hi} \ll \omega \ll \omega_{He}$. It was pointed out above that for low frequencies such that $\omega \ll \omega_{He}, \omega_{Hi}$ gyrotropic effects would be small in view of the mutually compensating contributions from the electrons and ions, which rotate in opposite directions. For frequencies such that $\omega_{Hi} \ll \omega \ll \omega_{He}$ the compensation is destroyed, and the gyrotropic terms now make the primary contribution. We readily find from (1.57) that

$$\left. \begin{aligned} i_x &\simeq \frac{\omega_{pe}^2}{4\pi\omega_{He}} E_y, \\ i_y &= -\frac{\omega_{pe}^2}{4\pi\omega_{He}} E_x. \end{aligned} \right\} \tag{1.111}$$

Assuming that E_z is small, which was shown above to be valid when $k \ll \omega_{pe}/c$, we can use Maxwell's equations to replace (1.64) with the following expressions:

$$c^2 k_z^2 E_x = \frac{\omega_{pe}^2 \, i\omega}{\omega_{He}} E_y;$$ (1.112)

$$c^2 k^2 E_y = -\frac{\omega_{pe}^2 \, i\omega}{\omega_{He}} E_x.$$ (1.113)

Multiplying these two equations we obtain

$$\omega^2 = \frac{\omega_{He}^2 \, c^4}{\omega_{pe}^4} k^4 \cos^2 \theta.$$ (1.114)

The kinds of waves into which fast magnetoacoustic waves are transformed carry many names: whistlers, helicons, and howling atmospherics or simply howlers. When $\cos \theta = 1$, that is, for propagation along the field, it is easily seen that

$$E_x = \pm i E_y,$$

i.e., the waves are circularly polarized. If $k \simeq \omega_{pe}/c$ the E_z components begin to be important when $\omega_{pe} \gg \omega_{He}$, and the waves go into the longitudinal electron wave with frequency ω_- [Eq. (1.99)].

Finally, the plasma will sustain the so-called cyclotron waves, for which $\omega_e \simeq \nu\omega_{He}$, $\omega_i \simeq \nu\omega_{Hi}$, where ν is some integer. They propagate primarily perpendicular to the magnetic field and have a wavelength which is of the same order as the Larmor radius. Figure 1.6 shows a schematic representation of the possible modes.

1.4.3. An Inhomogeneous Magnetically Active Plasma

A number of drift waves appear in an inhomogeneous plasma placed in a constant, uniform magnetic field. The frequencies of these drift waves are very small. We shall investigate the changes which occur in three branches treated above (the Alfven wave and two magnetoacoustic waves) when the plasma inhomogeneity is taken into account.

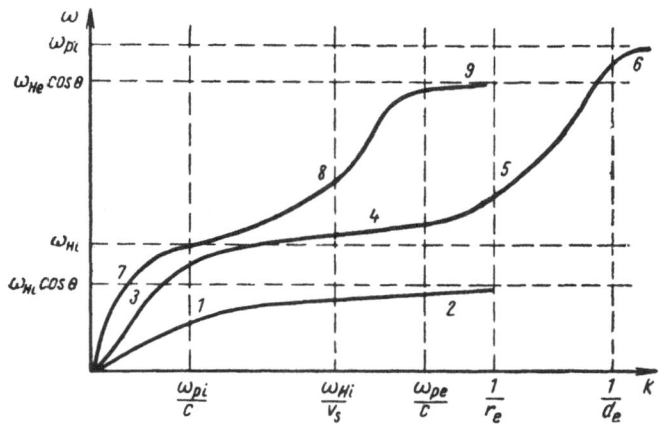

Fig. 1.6. Wave number dependence of frequency for low-frequency fluctuations of a magnetically active plasma at low pressure, such that $\beta = 8\pi p/H_0^2 \ll 1$; $v_A \gg v_s$:

Region of curve	Type of turbulent fluctuation	Dispersion relation
1	Magnetic acoustic oscillations	$\omega = kv_s \mid \cos\theta \mid$
2	Longitudinal ion waves	$\omega = \omega_{Hi} \mid \cos\theta \mid$
3	Alfven waves	$\omega = kv_A \mid \cos\theta \mid$
4	Wave which transforms into ion-cyclotron waves	$\omega^2 = \omega_{Hi}^2 + k^2 v_s^2$
5	Unmagnetized acoustic oscillations	$\omega = kv_s$
6	Ion oscillations	$\omega = \omega_{pi}$
7	Magnetohydrodynamic waves	$\omega = kv_A$
8	Whistlers	$\omega = \dfrac{k^2 c^2 \mid \cos\theta \mid}{\omega_{pe}^2} \omega_{He}$
9	Low-frequency longitudinal electron waves	$\omega = \omega_{He} \mid \cos\theta \mid$

We shall treat the case in which the inhomogeneity is weak; i.e., the wavelengths will be assumed small compared to the characteristic dimension of the inhomogeneity. The unperturbed plasma density n_0 can then be approximated by the expansion

$$n_0 = n_0(0) + y \frac{dn_0}{dy}(0). \tag{1.115}$$

Here and in the following we shall assume that the inhomogeneity exists in just one direction, which is selected to be the y axis. This direction is chosen to be perpendicular to the magnetic field, and we assume that dn_0/dy = const. The characteristic size L of the inhomogeneity can be conveniently described by the quantity

$$L^{-1} = -\frac{1}{n_0(0)} \cdot \frac{dn_0}{dy}(0), \tag{1.116}$$

i.e.,

$$n_0 = n_0(0)\left(1 - \frac{y}{L}\right). \tag{1.117}$$

As pointed out above, the modulation of the density of the electrons due to the Boltzmann distribution set up for them in the field of the waves creates additional currents in a homogeneous plasma [see Eq. (1.84)]. Similarly, currents also arise because of an inhomogeneity in the plasma

$$\delta\mathbf{j} = e\delta n_i \mathbf{v}_{Ti} - e\delta n_e \mathbf{v}_{Te},$$

where $\mathbf{v}_T$ is the velocity connected with the thermal motion of the electrons and ions around the Larmor circle (more precisely, along the helix). In our specific case δn is proportional to y and only this mean value is nonzero:

$$\langle yv_{Tx}\rangle_a = \frac{-1}{\omega_{Ha}} v_{Ta}^2. \tag{1.118}$$

In order to find the change in density produced because of the inhomogeneity we can use the continuity equation

$$\frac{\partial}{\partial t}\delta n_a + \operatorname{div} n_0 \mathbf{v}_a = 0. \tag{1.119}$$

Noting that the derivatives of $\mathbf{v}$ are on the order of $1/\lambda$ (where λ is the wavelength) and that the derivatives of n_0 are much smaller than that (being of the order $1/L$), one can take n_0 to be approximately constant in (1.119). Further, since that part of n_0 which is y-independent will give no contribution on the average, we have

$$\frac{\partial}{\partial t}\delta n_a = n_0(0)\frac{y}{L}\operatorname{div}\mathbf{v}_a. \tag{1.120}$$

Since the dependence on y is slow, one can write the solution to (1.120) as const y exp (ikr −iωt):

$$\omega \delta n_{a,k} = -n_0(0)\frac{y}{L}(\mathbf{k}\mathbf{v}_{ak}). \tag{1.121}$$

Here $\mathbf{v}_\alpha$ are particle velocities (proportional to E) which are related to plasma oscillations [see Eq. (1.57) for example]. Using (1.118) to average the current δj we find that

$$\delta j_x = \frac{en_0(0)}{L}\cdot\frac{(\mathbf{k}\mathbf{v}_i)\,v_{Ti}^2}{\omega\omega_{Hi}} - \frac{en_0(0)\,(\mathbf{k}\mathbf{v}_e)\,v_{Te}^2}{L\omega\omega_{He}}, \tag{1.122}$$

or, expressing this result in terms of the current j when there is no inhomogeneity, we obtain

$$\delta j_x = \frac{1}{\omega L}(\mathbf{k}\mathbf{j}^{(i)})\frac{v_{Ti}^2}{\omega_{Hi}} + \frac{1}{\omega L}(\mathbf{k}\mathbf{j}^{(e)})\frac{v_{Te}^2}{\omega_{He}}. \tag{1.123}$$

The additional current which we have found now suggests that we seek new types of plasma oscillations which are related to the plasma inhomogeneity.

We start our search with the limiting situation in which $v_A \gg v_s$. Let us see how the branches for the frequencies ω_M, ω_A, and ω_{Ms} change when there is inhomogeneity in the plasma. We consider small phase velocities such that $v_p \ll v_A$. In a homogeneous plasma only the purely longitudinal magnetic acoustic vibrations, for which $\omega_{Ms} = kv_s|\cos\theta|$, propagate with these phase velocities. Neglecting the displacement current, the general equation for longitudinal waves is

$$(\mathbf{k}\mathbf{j}_k) = 0. \tag{1.124}$$

If the electrons have a Boltzmann distribution, then

$$4\pi j_{k,z}^{(e)} = -\frac{i\omega_{pe}^2\,k_z\,\omega}{k_z^2 v_{Te}^2\,k}E_k, \tag{1.125}$$

while for ions

$$4\pi j_{k,z}^{(i)} = i\frac{\omega_{pi}^2}{\omega}\cdot\frac{k_z}{k}E_k. \tag{1.126}$$

Finally, only the electron part, Eq. (1.123), makes a contribution to $\delta j_{k,x}$:

$$4\pi\delta j_{k,x} = 4\pi\frac{k_z j_{k,z}^{(e)} v_{Te}^2}{L\omega\omega_{He}}. \tag{1.127}$$

After substituting the various currents in Eqs. (1.125)-(1.127) into Eq. (1.124), we arrive at the dispersion equation [6]

$$\omega^3 - k_z^2 v_s^2 - \omega\omega_D = 0, \tag{1.128}$$

where

$$\omega_D = -\frac{1}{L}\cdot\frac{k_x v_{Te}^2}{\omega_{He}} = k_x v_D; \qquad v_D = -\frac{v_{Te}^2}{L\omega_{He}}. \tag{1.129}$$

ω_D is called the electron drift frequency and v_D is the electron drift velocity. If k_x is not small but k_z is, the spectrum determined by Eq. (1.128) is significantly different from that obtained earlier for the magnetic acoustic oscillations; thus if

$$k_z \ll k_x\frac{v_D}{v_s} \tag{1.130}$$

the spectrum transforms into

$$\omega = \omega_D.$$

This wave is called a drift wave. The wave phase velocity ω_D/k_z increases as k_z decreases. When the phase velocity is comparable to v_A the wave is no longer longitudinal. In order to study the possible plasma motions in this situation, we note that when $v_A \gg v_s$ the current j_x describes Alfven waves and the current j_y gives magnetohydrodynamic waves, for which $\omega_M = kv_A$, and these are by no means related to the acoustic waves. Since the plasma inhomogeneity leads to an additional current along the x axis, the wave for $\omega_M = kv_A$ is not altered by the inhomogeneity. Only the Alfven wave is changed.

In order to study this phenomenon we consider Maxwell's equations without the displacement currents:

$$c^2 k_x(\mathbf{kE}_k) - k^2 c^2 E_{k,x} = -4\pi i\omega j_{k,x}; \tag{1.131}$$

$$c^2 k_z(\mathbf{kE}_k) - k^2 c^2 E_{k,z} = -4\pi i\omega j_{k,z}. \tag{1.132}$$

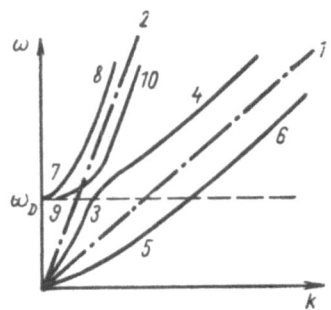

Fig. 1.7. Dispersion relations for drift fluctuations in a magnetically active plasma when $\beta = 8\pi p / H_0^2 \gg m_e/m_i$.

Curve	Type of turbulent fluctuation	Dispersion relation
1	Magnetized acoustic oscillations in a homogeneous plasma	$\omega = k v_s \cos\theta$
2	Alfven waves	$\omega = k v_A \cos\theta$
3	Low-frequency branch of nonelectrostatic drift waves	$\omega = \dfrac{k^2 v_A^2 \cos^2\theta}{\omega_D}$
4	Magnetized acoustic oscillations of an inhomogeneous plasma	$\omega = k v_s \cos\theta + \dfrac{\omega_D}{2}$
5	Low-frequency branch of the electrostatic drift oscillations	$\omega = \dfrac{k^2 v_s^2 \cos^2\theta}{\omega_D}$
6	Magnetized acoustic oscillations in an inhomogeneous plasma	$\omega = k v_s \cos\theta - \dfrac{\omega_D}{2}$
7	Nonelectrostatic drift wave	$\omega = \dfrac{k^2 v_A^2 \cos^2\theta}{\omega_D} + \omega_D$
8	Alfven waves in an inhomogeneous plasma	$\omega = k v_A \cos\theta + \dfrac{\omega_D}{2}$
9	Drift waves	$\omega = \omega_D + \dfrac{k^2 v_s^2 \cos^2\theta}{\omega_D}$
10	Alfven waves in an inhomogeneous plasma	$\omega = k v_A \cos\theta - \dfrac{\omega_D}{2}$

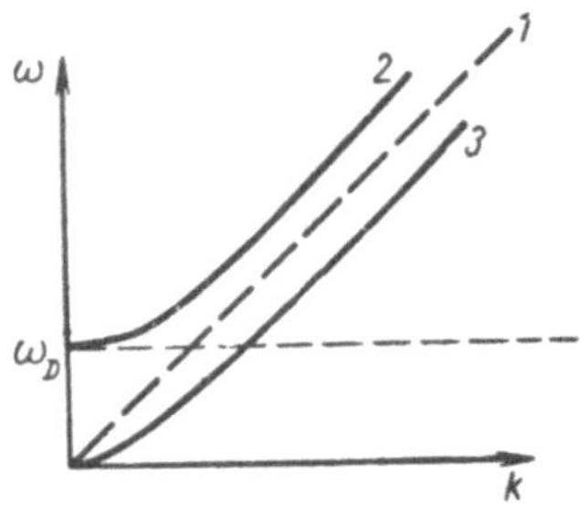

Fig. 1.8. Dispersion relations for drift fluctuations in a magnetically active plasma when $\beta = 8\pi p / H_0^2 \ll m_e/m_i$.

Curve	Type of turbulent fluctuation	Dispersion relation
1	Magnetized acoustic oscillations in a homogeneous plasma	$\omega = k v_s \cos\theta$
2	Magnetized acoustic oscillations in an inhomogeneous plasma	$\omega = k v_s \cos\theta + \dfrac{\omega_D}{2}$
3	Magnetized acoustic oscillations in an inhomogeneous plasma	$\omega = k v_s \cos\theta - \dfrac{\omega_D}{2}$

The current $j_{k,z}$ will remain as in the previous case, being the sum of Eqs. (1.125) and (1.126):

$$- 4\pi i\, \omega j_{k,\,z} = \omega_{pi}^2 \left(1 - \frac{\omega^2}{k_z^2 v_s^2} \right) E_{k,\,z}. \qquad (1.133)$$

When dealing with j_x one must include in addition to (1.127), which accounts for the inhomogeneity, the current of Eq. (1.59), which leads to the Alfven wave in a homogeneous plasma:

$$- 4\pi\, i\omega j_{k,\,x} = \omega_{pi}^2 \left(\frac{k_z v_D\, \omega}{k_z^2 v_s^2} E_{k,\,z} - \frac{\omega^2}{\omega_{Hi}^2} E_{k,\,x} \right). \qquad (1.134)$$

As a result, we obtain the following dispersion equation:

$$(\omega - \omega_D)\, (\omega^2 + \omega\omega_D - k_z^2 v_A^2) = 0. \qquad (1.135)$$

When $k_z v_A \ll \omega_D$ we find that $|\omega| = \omega_D$, i.e., the Alfven branch goes into the drift oscillations.

We should keep in mind that we have considered only the situation where the pressure of the external magnetic field is much greater than the plasma pressure $p = n(T_e + T_i)$. The various spectra possible in an inhomogeneous plasma are shown schematically in Figs. 1.7 and 1.8. This overview has been concerned with just linear collective plasma motions (see §1.5 for treatment of nonlinear motions), and gives a general picture of the nature of the simplest turbulent fluctuations possible in a plasma.

1.5. Nonlinear Collective Plasma
Motions. Strong and Weak Turbulence

The equations of motion for a plasma are nonlinear. It is therefore possible, at least in principle, to have self-consistent nonlinear motions. For instance, let us examine the low-frequency ion-acoustic oscillations without as uming that the amplitude is small. We treat the motion of on dimensional running waves for which all quantities depend only on $x - v_p t$, where v_p is the constant phase velocity of the wave. We then find from Eqs. (1.38) and (1.39) that

$$\left. \begin{aligned} m_i\, \frac{d}{dx} \left(\frac{1}{2} v_i^2 - v_i v_p \right) &= - e\, \frac{\partial \varphi}{\partial x}, \\ \frac{\partial}{\partial x}\, n_i\, (v_p - v_i) &= 0 \end{aligned} \right\} \qquad (1.136)$$

or

$$(v_i - v_p)^2 = v_p^2 - \frac{2e\varphi}{m_i},$$

$$n_i = \frac{\text{const}}{v_p - v} = \frac{n_0 v_p}{\sqrt{v_p^2 - 2e\varphi/m_i}}.$$ (1.137)

Here the constant is found from the condition that $n_i = n_0$ when $\varphi = 0$.

When (1.136) and (1.137) are taken into account, Poisson's equation is

$$\frac{\partial^2 \varphi}{\partial x^2} = -4\pi e (n_i - n_e) = -4\pi n_0 e \left(\frac{v_p}{\sqrt{v_p^2 - \frac{2e\varphi}{m_i}}} - e^{\frac{e\varphi}{T_e}} \right).$$ (1.138)

The first integral of (1.138) is

$$\frac{1}{2} \left(\frac{\partial \varphi}{\partial x} \right)^2 - 4\pi n_0 m_i \left(v_p \sqrt{v_p^2 - \frac{2e\varphi}{m_i}} + \frac{T_e}{m_i} e^{\frac{e\varphi}{T_e}} \right) = \text{const} = \alpha.$$ (1.139)

The solution to Eq. (1.139) gives the velocity profile and the potential φ in the wave. The profile depends on the values of v_p and α, where v_p is the phase velocity and α is related to the wave amplitude. In the $(\partial\varphi/\partial x, \varphi)$ plane Eq. (1.139) gives a series of closed curves describing the periodic waves. This family of curves is bounded by a curve called the separatrix, and the period for it goes to ∞ while the solution itself is transformed into a single pulse. This pulse, called the soliton, arises when

$$\alpha = -4\pi n_0 m_i \left(v_p^2 + \frac{T_e}{m_i} \right),$$

that is, it corresponds to a curve in the plane which passes through the point $\partial\varphi/\partial x = 0$, $\varphi = 0$ [20, 21].

Constant phase velocity nonlinear waves are also possible for high-frequency oscillations such as Langmuir oscillations. For these waves one can neglect the ion drift, and the electron displacements can be found from the continuity equation and the equation of motion:

$$n_e (v_e - v_p) = \text{const}; \quad -\frac{m_e (v_e - v_p)^2}{2} - e\varphi = \text{const}.$$ (1.140)

Poisson's equation then gives the following equation for the potentials:

$$-\frac{d^2}{dx^2} \cdot \frac{e\varphi}{m_e} = \omega_{pe}^2 \left(1 - \frac{1}{\sqrt{1 + 2e\varphi/m_e v_p^2}}\right). \qquad (1.141)$$

We therefore have

$$\left(\frac{d}{dx}\,v\right)^2 + \frac{4\omega_{pe}^2}{v_p^2}\,(v - 2\sqrt{v}) = \text{const} = \alpha, \qquad (1.142)$$

where

$$v = 1 + 2e\varphi/m_e v_p^2.$$

It is easy to see that the conditions $\varphi = 0$, $\partial\varphi/\partial x = 0$ correspond to $\alpha = -4\omega_{pe}^2/v_p^2$ and (1.142) will have only the trivial solution v = 1, dv/dx = 0 for all space. This shows that single pulses (solitons) are not possible and that the only solution is a periodic nonlinear wave.

We will not dwell in detail on the other types of nonlinear waves [21] in a magnetically active plasma. Rather, we note that both soliton and periodic-wave solutions are possible. Two considerations are important for our subsequent discussion. First, since nonlinear motions do not satisfy the superposition principle, a combination of these two solutions cannot be a solution. This means that it is impossible (strictly speaking) to excite many nonlinear degrees of freedom. Since turbulence assumes the transfer of energy from one degree of freedom to others, it also implicitly assumes that these degrees of freedom are independent in a specific sense, that is, the interactions between various degrees of freedom are weak. Thus, either a particular nonlinear motion is present and inhibits other motions (one or more degrees of freedom are excited and turbulence is absent), or the nonlinear motion is unstable and other motions are excited. Then the nonlinear motion represents the initial phase of plasma turbulence, although the final turbulent state can certainly have properties which differ from those of the original nonlinear motion. To be sure, there are examples in which the nonlinearity is not weak, but the interactions between different solutions are weak so that superposition is still approximately valid. Such is the case for solitons, or solutions which are nearly solitons. At the present time it is not completely clear how one would excite a plasma state which involves a large number of weakly-interacting solitons, but the situation cannot be ruled out. Moreover, one-dimensional waves with constant phase velocity represent special nonlinear motions which are unstable relative to three-dimensional excitations.

The second important consideration is that nonlinear collective motions have amplitudes which, as a rule, do not greatly exceed the thermal energies of the plasma particles. Thus, for low-frequency ion-acoustic waves $e\varphi_{max} = 1, 3T_e$ while for the high-frequency waves [see Eq. (1.141)] $e\varphi_{max} = m_e v_p^2/2$. The conditions for the waves to be linear are $e\varphi_{max} \ll T_e$ and $e\varphi_{max} \ll m_e v_p^2/2$ respectively. If we write these conditions in terms of W_B, the energy of the particles in a wave in 1 cm^3, then $W_{B\,max} \simeq n_0 T_e$, and the condition for weak nonlinearity at low frequencies is $W_{B\,max} \ll n_0 T_e$.

Plasma turbulence can also be characterized by the ratio of the turbulence energy W to the plasma particle thermal energy:

$$W/nT_e. \qquad (1.143)$$

This parameter can be either very small, $\eta \ll 1$, or large, $\eta \gtrsim 1$. The presence in the plasma of collective motions for which $\eta \sim 1$ (an example being highly nonlinear waves) does not mean that $\eta \sim 1$ for turbulent motions if the collective motions are regular rather than random in nature.

The problem of distinguishing between weak and strong turbulence must be faced. The parameter η is frequently used to separate weak and strong turbulence ($\eta \ll 1$ for weak and $\eta \gtrsim 1$ for strong turbulence). This separation is not sufficient for an incompressible fluid, whose turbulence is strong while at the same time $\eta \ll 1$. This is because, according to our previous discussion, the condition for incompressibility takes the form $W \ll nm_i v_s^2 = nT$. Further, η can be large while the motion remains nonturbulent. The interaction of the collective motions among themselves is characteristic of turbulence, because without these interactions there is no energy flow.

A state is termed highly turbulent when the dispersion relations and correlations of the turbulent fluctuations are determined by their interactions. A state is then weakly turbulent when the interaction of the fluctuations changes their spectral characteristics only slightly.

If one now examines the possible spectra for linear collective plasma motions it becomes clear that, as their frequencies increase, the limit imposed on the absolute value of the interaction intensity at which the turbulence becomes strong shifts to higher and higher values. This indicates that strong turbulence is more easily observed for low-frequency oscillations than for the high-frequency modes. There is no characteristic frequency for the vortex motions in an incompressible fluid, and the turbulence is strong.

There are also physical arguments which indicate that as a rule high-frequency turbulence is weak. An exception is the Langmuir turbulence, as the linear dispersion of Langmuir waves is small and the nonlinear frequency shift can become comparable with the dispersion effects even for $\eta \ll 1$. Let us assume that the source of the turbulence (i.e., W) is increased. When W increases an energy flow arises along the spectrum. If this energy is absorbed by the plasma, nT_e increases. It is therefore possible that η might even decrease. These energy currents become especially intense for η approximately equal to 1, for then processes for transforming the energy come into play which are of higher order in η. These arguments indicate that weak high-frequency turbulence is most probable when one can speak of the interactions of linear collective motions.

On the other hand, in a turbulent plasma there are interactions between pulsations of the same type and pulsations of different types, such as high-frequency and low-frequency fluctuations. And although such an interaction is weak for the high-frequency fluctuations, it can be strong for the low-frequency modes. Therefore, a weak high-frequency turbulence can significantly alter the dispersion characteristics and the nature of low-frequency collective motions. A good example of such effects is seen when the interaction with high-frequency fluctuations produces a shift in frequencies which is much larger than the characteristic linear frequencies of the low-frequency modes. Thus, at low frequencies a strong turbulence can be connected with both the interactions of fluctuations among themselves and with other fluctuations.†

Finally, it must be emphasized that for a strong turbulence it is immaterial whether it was excited by strong nonlinear waves or by other means, since the information relative to the source of the turbulence would be eliminated by the effects of the interactions which give rise to the fluctuation spectrum. In addition, weak turbulence can be excited along with the strong nonlinear collective motions. Therefore the observation of fluctuations for which $\eta \sim 1$ cannot by itself serve to indicate a strong turbulence.

1.6. Fundamental Problems in the Physics of Turbulent Plasma

The above analysis of some of the problems related to the spectra and type of collective plasma motions enables us to formulate the basic problems concerning plasma turbulence.

1. The study of plasma turbulence is based on the analysis of the possible collective degrees of freedom in the plasma and on their relative roles for any given type of excitation. A fluid is distinguished from a plasma in that it can sustain a relatively small number of collective degrees of freedom. Those collective motions which occur in relatively weak self-consistent electromagnetic fields play an important role in plasma. These motions are termed linear. However, a plasma may also sustain nonlinear collective motions of the type meant by nonlinear self-consistent waves and individual pulses.

2. The investigation of the means for exciting collective plasma motions and turbulence is a very important problem. The general view [1] suggests that the excitation of fluid turbulence is connected with an instability. This connection is not always simple since there can be a quite broad region of transitional regimes in which an instability does not produce turbulence. However, an instability is required in order that turbulence may arise.

It is characteristic that a plasma can have a large number of different instabilities. Therefore, any plasma encountered either

---

† Chapter 8 contains a detailed analysis of the low-frequency electromagnetic properties of a turbulent plasma and Chapter 9 presents an analysis of the problem of strong Langmuir turbulence.

in nature or in the laboratory experiment is frequently turbulent. In many instances the study of physical processes in plasma reduces to the study of plasma turbulence. This side of the problem was not clearly realized in the initial phases of plasma research, just as the first investigations of fluids were concerned with the simplest laminar flows. But gradually the conviction grew that turbulent effects played a fundamental role in plasma. Even if the ideal conditions were realized experimentally, such that the plasma was known to be stable within some approximation, this could represent only an exception, and the study under these conditions was in fact the study of a particular case (but a very important problem in its own right nevertheless). All the many different physical phenomena in a plasma require the concept of turbulence for their explanation. It is this fact which makes the problems of plasma turbulence and the questions of stability fundamental to plasma physics.

The development of the theory initially arose from observations of the various instabilities. But it was not immediately clearly understood that these instabilities led, in the final analysis, to turbulence. The general concept of the role of turbulence developed over the years in connection with the evolution of the experiments on turbulent plasma.

3. We must emphasize that there are no instabilities which from the very outset transform a plasma into the turbulent state. There is, however, a broad class of instabilities whose development involves the simultaneous excitation of many collective degrees of freedom. This takes place if the excitation starts from the initial level, related to the thermal fluctuations, and the random character of the processes is preserved in the initial phase of excitation.

Another class of instabilities includes those which, in the initial stage, lead to the excitation of a single (or a small number) degree of freedom. An example is a monoenergetic beam in a plasma which interacts resonantly with just one plasma wave. Gradually, however, because of the interactions among the waves and with the beam, the wave spectra and the beam velocities spread and the system goes into the turbulent regime (this is the quasilinear stage). Just as in the case presented above, which treats the turbulence of a fluid flowing in a tube, initially only one vortex is excited, and this motion is not turbulent. However, the interactions of the vortices redistribute the energy to vortices of different scales and turbulence then develops. Among the most important problems are the process by which the system goes into the turbulent state, the randomization process, and the interactions among the various oscillations. These problems are closely related to questions involving ergodicity, and although it is clear that any system that is big enough must ultimately devolve into a turbulent regime, the time for randomization of the motion depends critically on the initial conditions.

4. An important problem in the theory of turbulence is the analysis of the possible regions in which turbulent plasma fluctuations are absorbed. It is significant that the collective mechanisms make possible not only the rapid excitation of the oscillations, but also their rapid dissipation. One of the first studies of such mechanisms was done by Landau [15]. The location of absorbing regions along the paths for transforming turbulent energy is important in the theory of turbulence. The possibility of rapid transformation and effective absorption creates the prerequisites needed for effective turbulent heating of a plasma.

5. A fundamental question in the physics of turbulence is related to the nonlinear effects in the interactions of turbulent fluctuations. In the previous discussion of a fluid, the role of such processes in understanding the dynamics of establishing a stationary turbulence was pointed out. Since the number of collective degrees of freedom in a plasma is so much larger than in a fluid, the fundamental role is now played by those problems involving the spectral energy transfer of the turbulent fluctuations and the nonlinear conversion of one type of fluctuation into another. A systematic study of the various nonlinear effects will enable us to isolate the most important of them for a given process and to study the dynamics inherent in the development of turbulence.

Some of the nonlinear transfers cannot determine the dynamics of turbulence because their contributions to the energy relations are small. But they can be important in the interpretation of the observed phenomena. An example of this kind of nonlinear conversion in plasma would be the transformation of turbulent fluctuations into electromagnetic waves, which are easily detected outside the plasma.

6. The study of the stationary spectra for turbulent fluctuations is one of the most important aspects of plasma turbulence. As a

rule the appearance of these spectra is the result of a balance between the effects of building up the transformation and absorption
of turbulence. The essential role in the absorption of turbulent
fluctuations is occupied by the processes of spectral transfer,
which shift the turbulent fluctuations into those ranges of parameter values for which absorption is important. It is this study of
the spectral transfers, and the analysis of the possible regions
for effective absorption of turbulent fluctuations, which is so
important in determining the stationary spectra.

7. Turbulent transport processes have great practical importance. It was shown above in the example of a fluid in turbulent motion that there is a transport of momentum to the walls
which greatly exceeds the momentum transfer encountered in the
molecular viscosity of the fluid. This effect corresponds to turbulent viscosity.

When we are dealing with the transfer of the electron energy from one point to another during the interaction of electrons
with turbulent plasma fluctuations, one can speak of a turbulent
thermal conductivity. If one is concerned with the transfer of
momentum from the electrons (which receive energy from an external electric field) to the ions via the interactions of electrons
with turbulent plasma fluctuations, one can then speak of a turbulent electrical conductivity. Finally, the redistribution of plasma density during interactions with turbulent fluctuations can be
termed turbulent diffusion. The diffusion of a turbulent plasma
is closely related to specific instabilities which appear because
of inhomogeneities in the plasma.

8. The interaction of charged particles with a turbulent
plasma (specifically, when the plasma is viewed as an ionized
state of matter) is one of the primary objects for study in turbulent plasma physics. The excitation of plasma turbulence is usually accompanied by the formation of random electromagnetic fields.
Therefore, a charged particle incident on a turbulent plasma begins to "feel" the force of these random fields; it wanders through
the fields and in some cases may even extract energy from them.
In a stable medium charged particles will lose energy, but in
a turbulent medium a particle can pick up energy, or be accelerated.

9. Another problem which is connected with the electromagnetic nature of plasma turbulence is the propagation of electromagnetic waves through the plasma. The primary role here is occupied by the processes in which the electromagnetic fields of the turbulent fluctuations interact with the external electromagnetic wave. On one hand, such interactions can lead to strong scattering of the electromagnetic waves on the turbulent fluctuations and to changes in their intensities. But on the other hand, the electromagnetic waves can be amplified, picking up energy just like the particles do. In this sense a turbulent plasma is quite different from a stable plasma, for in a stable plasma electromagnetic waves can only be absorbed. For stationary turbulence the transmission and amplification of weak external electromagnetic waves are governed by the turbulence spectra.

10. It is very important to study the stability of stationary turbulence relative to low-frequency excitations such as magnetic fields excited by the turbulence, the skin effect, and others. These problems fall in the general category of low-frequency electromagnetic properties of a turbulent plasma. Radical changes in the low-frequency properties of a turbulent plasma are important in many applications, in particular the containment of a turbulent plasma.

11. Finally, it is of interest to study the processes leading to the dissipation of turbulence after the sources which generate turbulent fluctuations have been removed. Problems in this category may be grouped under the heading of the decay of turbulence.

1.7. Turbulent Plasma in Experiment
and Nature

In this section we shall become acquainted in general terms with the manifestations of turbulent plasma (i.e., with the phenomena rather than their physical interpretations). These phenomena are frequently called anomalous properties of plasma (an unfortunate choice of terms). This in fact means that they do not correspond to the classical concepts borrowed from the theory of pairwise collisions. Discrepancies between the experimental results and the collision theory appeared in the very first experiments on plasmas. For instance, Langmuir [13] observed the large scattering of electrons in a gaseous discharge, a phenomenon which could not be explained by pairwise collision theory. He also noticed that a Maxwellian distribution was set up for the scattered electrons in an anomalously short time.

There are a great number of experiments devoted to the interaction of a particle beam with a plasma. According to the elementary view, there are very few collisions between beam particles and plasma particles when the beam velocity is high. Indeed, if the plasma is highly ionized, Coulomb collisions are most important, and the cross section for Coulomb collisions decreases rapidly with the relative particle velocity.

One finds the mean free path to be

$$l \simeq \frac{1}{n_0 \sigma} \simeq \frac{1}{4\pi n_0 r_0^2} \left(\frac{v_0}{c} \right)^4 ;$$
$$r_0 = e^2/mc^2. \tag{1.144}$$

This equation shows that the mean free path increases very fast as the particle velocity goes up. Let us do a simple calculation using the following parameters: $n_0 \simeq 10^{13}$ cm^{-3}, $m = m_e$, $r_0 \simeq 2 \cdot 10^{-13}$ cm, and $v_0/c \sim 1/10$, which means that the energy of the electron beam is $mv_0^2/2 \sim 2.5$ keV. Equation (1.144) then gives $l \sim 10^8$ cm $\sim 10^3$ km. A beam with this energy ought to pass through the plasma unimpeded. We have made this estimate in order to show the marked difference between experiment and the predictions of the pair-collision theory.

Both Looney and Brown [22] and Merill and Webb [23] observed the energy loss suffered by a beam in plasma and the attendant production of strong plasma oscillations. Since then there have been a great many studies of the energy losses of a beam in plasma. Of these we should like to mention the work of the Russian physicists Fainberg [24, 25] and Zavoiskii [7, 26, 27] and their co-workers. The works cited explore in detail the microscopic characteristics of turbulence excited by a beam, with particular attention being given to correlation functions and the spectra of the oscillations excited, among other topics. These experimental studies demonstrated that the effectiveness of a beam interacting with a plasma could be 10^{13} times greater than predicted by the pair-collision theory. These anomalies in the transmission of a beam are also related to the separation of the beam into bunches due to the action of the oscillations excited and the coherent energy losses predicted by Veksler [28].

Other work has shown that even a very energetic beam can be totally suppressed in a plasma, and intense low-frequency oscillations of the plasma will be excited [29, 30]. The excitation of low-frequency ion-acoustic oscillations has been observed [30] as the result of an energetic particle beam and the nature of the oscillations excited in the beam—plasma interaction has been thoroughly studied [2, 25]. In all these experiments the interaction of the beam with the plasma was very strong.

A closely related problem is the turbulent resistance of plasma. Assume that we have a plasma located in an electric field. In a weak field an electron will acquire a velocity $v = \sqrt{2eEl/m_e}$. in one mean free path length. If this velocity is less than v_{Te} the scattering cross section is determined by v_{Te} and one finds that $\nu_{col} \sim v$, i.e., the force of friction due to pair collisions increases with velocity. This leads to equilibrium between the accelerating and friction forces, and this results in some average electron velocity. Another case arises in an electric field large enough so that the electron gains energy greater than T_e in one mean free path. Without collisions with the ions an electron may reach a velocity of v_{Te}, and since the friction force decreases as v increases when $v > v_{Te}$ the electron will undergo no further collisions and will experience an uninterrupted acceleration. These are the so-called runaway electrons. We emphasize that according to the pair-collision theory all electrons in a plasma must run away in a strong field, and the plasma resistance must drop sharply.

One can easily estimate the critical field in which the resistance of a plasma must dramatically drop, according to the pair-collision theory, by evaluating the equation $m_e \nu_{col} \bar{v} = eE^*$. Assuming that $\bar{v} = v_{Te}$ we have

$$E^* \simeq \frac{e\omega_{pe}^2}{v_{Te}^2} \approx \frac{e}{d_e^2}.$$

The magnitude of E* is very small, especially in high-temperature

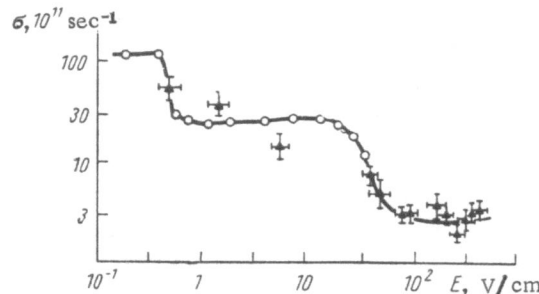

Fig. 1.9. The turbulent plasma electrical conductivity
σ as a function of the applied electric field E; ○ and
▲ are from [32] and [35] respectively.

plasmas. For instance, when n ~ 10^{13} cm^{-3} and the electron temperature is 20 eV, $E^* \sim 1$-10 V/cm.

Experiments have shown that when $E \gg E^*$ the resistance increases rapidly. This is usually accompanied by the excitation of low-frequency plasma oscillations. Turbulent resistance has been studied in [31-33]. Low-frequency vibrations in plasma when an electric field is applied have also been observed in [34, 35]. Figure 1.9 shows the experimental results, which depict the drop in plasma electrical conductivity as the electric field increases.

The problems which have been discussed here are intimately related to the turbulent heating of plasma. This heating has been observed in many experiments [27, 30]. Rapid heating does not follow from the assumption of heating due to pair-collisions because the heating occurs much faster than predicted by that theory. This type of turbulent heating has been observed by Zavoiskii and co-workers [27].

For a long time, great hopes have been entertained concerning the containment of plasma in magnetic bottles. But in many of these arrangements one observes an anomalous behavior in the plasma, which was accompanied by the development of instabilities.

Plasma diffusion is also related to these problems. According to the elementary pair-collision theory the diffusion of plasma to the walls across a strong magnetic field should fall off as $1/H^2$, that is, the diffusion ought to be relatively weak in strong magnetic fields. This prediction was the basis for the hope of thermally isolating a plasma by using strong magnetic fields. Although in a number of cases it is possible to have the so-called classical diffusion (observed in some experiments [36]), the most frequent type of diffusion encountered was much stronger. This more frequent kind of diffusion is called turbulent diffusion. Bohm [37] performed the first studies of turbulent diffusion; he found that

$$D_B \sim 1/H.$$

Turbulent diffusion has now been experimentally verified, and the diffusion coefficients found lie between those predicted by the pair-collision theory and the Bohm diffusion coefficient. The effects of turbulent diffusion are now studied in every experiment on the thermal isolation of plasma.

The following phenomena in which anomalies are observed are related to plasma radiation. Elementary considerations predict that radiation ought to come from two sources: 1) the bremsstrahlung produced by electrons colliding with ions, and 2) cyclotron radiation from the electrons, which appears in a plasma located in an external magnetic field (the electrons move in Larmor circles). While the bremsstrahlung has a continuous spectrum, the cyclotron radiation is discrete, appearing at multiples of the gyrofrequency $\nu\omega_{He}$. The first expectations for plasma heating were connected with these relatively small radiation losses due to collisions (bremsstrahlung) and cyclotron radiation. However, it has turned out that the radiant losses from plasma are significantly higher than predicted by the pair-collision theory. This is the anomalous radiation from a plasma. It was studied theoretically in [38, 39].

The first experiments on beam instabilities [24] revealed a powerful radiation emanating from a turbulent plasma. Anomalous plasma radiation was studied in detail in Zavoiskii's experiments. A large amount of work has been done by some Japanese physicists [40], who observed the anomalous plasma radiation at frequencies which were multiples of the electron gyrofrequencies. The discovery of these effects opens the door to use of plasma as a generator of electromagnetic waves at many frequencies [41].

Finally, to conclude our treatment of the various anomalies, we must dwell for a moment on the passage of charged particles through plasma. According to the pair-collision representation,

the particles ought to lose energy to the plasma. This is not verified by experiment (we are considering processes in which the
plasma is turbulent). Even the first works on plasma heating in
powerful, pinch-type, pulsed discharges carried out by Artsimovich et al. [42] showed that a plasma serves to accelerate electrons and ions. Accelerated protons led to a number of nuclear
reactions, which were at first thought to be thermonuclear in
origin. Accelerated particles were observed in nearly all apparatus which used magnetic confinement of the plasma. The acceleration effect was especially noticeable when beams were passed
through a plasma. Such experiments were performed by Berezin
[43]. Nezlin [35] also experimentally observed the acceleration
of ions. Alexeff and Neidigh [30] and Smulin and Getti [29] observed electron acceleration, and probably ion acceleration as
well.

It seems probable that accelerated particles will appear in a
plasma for any active influence on the plasma. For example,
accelerated particles have been observed in experiments on coherent radiation acceleration, as suggested by Veksler [28, 44].
The observation of electrons with energies greater than the original beam energy is in marked contrast to the predictions of the
pair-collision theory.

Let us now briefly examine the possible explanations for
these anomalous features in plasma behavior. We must emphasize once again that we are not dealing with any kind of deviation
from principle here; rather, we are concerned with the behavior
which is encountered most often in a plasma. The disagreement
arises only over the assumption which implies that pair collisions
are the decisive factors. Our brief review of the facts compels
us to the conclusion that, under turbulent conditions, the fundamental macroscopic phenomena, the basic macroscopic characteristics of plasma, do not correspond to the pair-collision concept.

According to the presently accepted view, the plasma properties that we have been discussing are due to its turbulence. In
other words, collective degrees of freedom, especially plasma
oscillations, are excited in the plasma. The decisive interaction
is therefore not the pairwise particle interactions, but rather the
interactions of particles with these collective degrees of freedom.
Thus, the effective interaction of a beam with the plasma is con-

nected with the excitation of plasma turbulence; the anomalous resistivity is related to particle scattering by the plasma oscillations, the anomalous radiation comes from transforming plasma oscillations into waves which leave the plasma, the particle acceleration arises when the collective degrees of freedom transmit energy to the particles, and so on.

Turbulence effects play a major role in the plasma encountered in nature. Here we must distinguish between studies of the ionospheric plasma, the plasma in the space around the earth, the plasma of the earth's radiation belts, and to some extent the interplanetary plasma. At present the important role of turbulent processes is clear for particle acceleration in the radiation belts, the formation of the polar aurora, the dynamics of the earth's magnetosphere, and other processes. Detailed investigations have been made of the excitation of various collective plasma oscillations in the radiation belts, the sources for the strong turbulence, and the mechanisms for turbulence in the solar wind plasma.

The study of turbulence in stellar plasma, interstellar space, galaxies, metagalaxies, and the turbulence excited by gigantic explosive processes occurring in the cosmos [45, 46] all form another group of problems. For such problems we have only the information which arrives at earth, information such as the type of electromagnetic waves and particles released, that is, the basic external manifestations of plasma turbulence.

We should point out that most of the matter in the universe is ionized and thus forms a plasma. Therefore turbulent processes in plasma are of interest not only from the viewpoint of explaining phenomena related to radiation and particle acceleration, but also on a much broader scale. At the present time there is much interest in the "superstars," the quasars. The radiation from these objects seems to correspond to the anomalous radiation connected with a turbulent plasma. This possibility has been suggested by a number of investigators [47-49]. Fast cosmic-ray particles can be generated as the result of plasma turbulence. It is known that the generation of cosmic rays carries a universal character, for whenever turbulent plasma motion is observed, cosmic rays are present. This would indicate that the production of cosmic rays is one of the fundamental properties of plasma turbulence.

Turbulence is indirectly expressed in the fact that the particles which are accelerated by the turbulence become a radiation source. The radiation from many cosmic objects can be interpreted on this basis.

Chapter 2

Basic Concepts of Turbulent Plasma Theory

2.1. Statistical Ensemble Averages

The theory of turbulent plasma can be constructed on the basis of collective excitations, such as plasmons and electron and ion excitations. In order to properly introduce the plasmon concept and show how electron and ion excitations differ from the electrons and ions proper, we begin by examining the formal methods for describing turbulence. The methods in point involve averages of physical quantities taken over a statistical ensemble. These methods will enable us to describe systems which are far from statistical equilibrium. We shall show below that under these conditions one can use the collective excitation approach. As we shall see, this result has a deep physical meaning; it is far from trivial and requires detailed proof.† For systems far from statistical equilibrium the treatment is simplified somewhat because one can use the dynamic equations of motion and the thermal fluctuations can be neglected. As we shall show, this reduces to including all the so-called induced processes. It is also important to note that, by using these processes, one can find explicit expressions for the matrix elements of the interactions between excitations. Then, once these matrix elements are known, the spontaneous processes can be calculated. The results obtained for conditions of statistical equilibrium correspond to those found in the theory of thermal fluctuations. This agreement reinforces our belief that the method of elementary excitations has a much greater value.

---

†For systems in a state of near statistical equilibrium the requisite proof has been given using Green's functions [50-52].

In order to demonstrate the possibility of describing turbulence in the language of elementary excitations, which is now widely used in solid-state theory [52], we must first discuss the statistical ensemble which will be used in the averaging process. Appropriate ensembles are the initial values of the parameters which characterize the plasma's collective degrees of freedom. For nearly linear collective motions this might be the initial values of the phases of the oscillations.

The general averaging method is valid for both weak and strong turbulence and, in the final analysis, is identical with the method used in the theory of fluid turbulence. However, the concept of collective excitations interacting weakly with one another can be used only in a weakly turbulent plasma. Therefore, most of our attention will be focused on that case in the following development. The averaging method to be described here is very convenient for describing low-frequency perturbations in a turbulent plasma (Chapter 8) and also strong turbulence (Chapter 9).

We separate the turbulent and regular components of the physical quantities L which characterize the collective degrees of freedom in the plasma. We shall write L as

$$L = L^R + L^T. \tag{2.1}$$

From the definition of the turbulent components we can write

$$\langle L^T \rangle = 0, \tag{2.2}$$

so that

$$\langle L \rangle = L^R.$$

By averaging the equations of motion over a statistical ensemble one can obtain separate equations for the turbulent and regular quantities. It will be convenient to obtain equations for the average values of certain combinations of turbulent components (such as the squares of electric fields, correlation functions, etc.), which can be studied experimentally. It will also be useful to have the equations for the regular components in a form which contains only the required correlation functions. The result must be a self-consistent set of equations for the regular quantities and the correlation functions of the random quantities.

The hydrodynamic equations can be used to describe the hydrodynamic collective degrees of freedom of a plasma, and one may introduce the turbulent components of the averaged quantities, such as the mean particle density, the mean velocity, and so on. But the more usual approach is to use the plasma kinetic equations, which enable us to describe the turbulent processes in which the detailed characteristics of the plasma particle distribution play an important part.

Let f (p, r, t) be the distribution function for the plasma particles; it gives the number of plasma particles having momenta between p and p + dp. Further,

$$\int f(\mathbf{p},\, \mathbf{r},\, t)\, \frac{d\mathbf{p}}{(2\pi)^3} = n\,(\mathbf{r},\, t), \tag{2.3}$$

where n(r, t) is the density of plasma particles. We will divide f into two parts — a regular part and a turbulent part:

$$f = f^R + f^T; \tag{2.4}$$

$$\langle f^T \rangle = 0. \tag{2.5}$$

We note that this definition does not depend on the turbulent fluctuation fields being small. The turbulence causes an additional random part to appear in the electromagnetic fields. By definition we have

$$\left.\begin{aligned} \mathbf{E} &= \mathbf{E}^R + \mathbf{E}^T; \\[4pt] \langle \mathbf{E}^T \rangle &= 0. \end{aligned}\right\} \tag{2.6}$$

One need not write an equation such as Eq. (2.6) for the magnetic field because Maxwell's equation,

$$\operatorname{curl} \mathbf{H} = -\frac{1}{c}\, \frac{\partial \mathbf{E}}{\partial t},$$

allows one to express the magnetic field in terms of the electric field.

As an example, let us consider the collisionless kinetic equation for the motion of a certain type of plasma particle:

$$\frac{\partial f}{\partial t} + \mathbf{v}\, \frac{\partial f}{\partial \mathbf{r}} + \mathbf{F}\, \frac{\partial f}{\partial \mathbf{p}} = 0, \tag{2.7}$$

where

$$F = e \left(E + \left[\frac{v}{c} \, H \right] \right) \tag{2.8}$$

is the Lorentz force. By averaging Eq. (2.7) over a statistical ensemble through Eqs. (2.4) and (2.6), we obtain an expression for f^R:

$$\frac{\partial f^R}{\partial t} + v \cdot \frac{\partial f^R}{\partial r} + F^R \frac{\partial f^R}{\partial p} = - \left\langle F^T \frac{\partial f^T}{\partial p} \right\rangle. \tag{2.9}$$

Subtracting (2.9) from (2.7), we arrive at an equation for f^T:

$$\frac{\partial f^T}{\partial t} + v \frac{\partial f^T}{\partial r} + F^T \frac{\partial f^R}{\partial p} + F^R \frac{\partial f^T}{\partial p} + F^T \frac{\partial f^T}{\partial p} - \left\langle F^T \frac{\partial f^T}{\partial p} \right\rangle = 0. \tag{2.10}$$

If we now expand f^T in powers of the turbulent field E^T (and therefore powers of F^T), such that

$$f^T = \sum_{i=1}^{\infty} f^{T\,(i)}, \tag{2.11}$$

where the index i indicates that f^T is proportional to the i-th power of E^T, then, by collecting like powers of E^T, Eq. (2.10) gives us the following system of equations:

$$\frac{\partial f^{T\,(1)}}{\partial t} + \frac{v \partial f^{T\,(1)}}{\partial r} + F^T \frac{\partial f^R}{\partial p} + F^R \frac{\partial f^{T\,(1)}}{\partial p} = 0; \tag{2.12}$$

$$\frac{\partial f^{T\,(2)}}{\partial t} + v \frac{\partial f^{T\,(2)}}{\partial r} + F^T \frac{\partial f^{T\,(1)}}{\partial p} + F^R \frac{\partial f^{T\,(2)}}{\partial p} - \left\langle F^T \frac{\partial f^{T\,(1)}}{\partial p} \right\rangle = 0; \tag{2.13}$$

$$\frac{\partial f^{T\,(3)}}{\partial t} + v \frac{\partial f^{T\,(3)}}{\partial r} + F^T \frac{\partial f^{T\,(2)}}{\partial p} + F^R \frac{\partial f^{T\,(3)}}{\partial p} - \left\langle F^T \frac{\partial f^{T\,(2)}}{\partial p} \right\rangle = 0. \tag{2.14}$$

By using the expression for the Fourier transform of a product†

$$(LM)_k = \int L_{k_1} M_{k_2} \, dk_1 \, dk_2 \, \delta (k - k_1 - k_2), \tag{2.15}$$

---

†Recall that $dk = dk \, d\omega$ and $\delta (k - k_1 - k_2) = \delta (k - k_1 - k_2) \, \delta(\omega - \omega_1 - \omega_2)$.

for

$$F^R = 0 \qquad (2.16)$$

we find that

$$i\,(\omega - \mathbf{kv})\, f_k^{T\,(1)} = \int F_{k_1}^T\, \frac{\partial f_{k_2}^R}{\partial \mathbf{p}}\, dk_1\, dk_2\, \delta\,(k - k_1 - k_2); \qquad (2.17)$$

$$i\,(\omega - \mathbf{kv})\, f_k^{T\,(2)} = \int \left[F_{k_1}^T\, \frac{\partial}{\partial \mathbf{p}}\, f_{k_2}^{T\,(1)} - \left\langle F_{k_1}^T\, \frac{\partial f_{k_2}^{T(1)}}{\partial \mathbf{p}} \right\rangle \right] dk_1\, dk_2\, \delta\,(k - k_1 - k_2); $$

$$(2.18)$$

$$i\,(\omega - \mathbf{kv})\, f_k^{T\,(3)} = \int \left[F_{k_1}^T\, \frac{\partial}{\partial \mathbf{p}}\, f_{k_2}^{T(2)} - \left\langle F_{k_1}^T\, \frac{\partial}{\partial \mathbf{p}}\, f_{k_2}^{T(2)} \right\rangle \right] dk_1\, dk_2\, \delta\,(k - k_1 - k_2). $$

$$(2.19)$$

These equations enable us to write f^T in terms of f^R and F^T.

It is now easy to find the nonlinear turbulent plasma current:

$$\mathbf{j} = \mathbf{j}^R + \mathbf{j}^T; \quad \langle \mathbf{j}^T \rangle = 0; \qquad (2.20)$$

$$\mathbf{j}^T = \sum_i \mathbf{j}^{T\,(i)}; \qquad (2.21)$$

$$\mathbf{j}^{T\,(i)} = \sum_\alpha \int \dot{e}_\alpha\, \mathbf{v} f_\alpha^{T\,(i)}\, \frac{d\mathbf{p}_\alpha}{(2\pi)^3}\,. \qquad (2.22)$$

Here the index α denotes a sum over the types of plasma particles (electrons and ions).

Let us now turn to Maxwell's equations, which relate the fields $\mathbf{B}_k$ to the current:

$$\mathrm{curl}\ \mathbf{H} = \frac{4\pi}{c}\, \mathbf{j} + \frac{1}{c}\, \frac{\partial \mathbf{E}}{\partial t}; \qquad \mathrm{curl}\ \mathbf{E} = -\frac{1}{c}\, \frac{\partial \mathbf{H}}{\partial t}; \qquad (2.23)$$

we find

$$i\,[\mathbf{kH}_k] = \frac{4\pi}{c}\, \mathbf{j}_k - \frac{i\omega}{c}\, \mathbf{E}_k; \qquad [\mathbf{kE}_k] = \frac{\omega}{c}\, \mathbf{H}_k \qquad (2.24)$$

or

$$\mathbf{k}\,(\mathbf{kE}_k) - \left(k^2 - \frac{\omega^2}{c^2} \right) \mathbf{E}_k = -4\pi\, i\, \frac{\omega}{c^2}\, \mathbf{j}_k. \qquad (2.25)$$

Because of the continuity equation

$$\frac{\partial \rho}{\partial t} + \operatorname{div} \mathbf{j} = 0 \tag{2.26}$$

ρ_k can be found from j_k:

$$\rho_k = \frac{1}{\omega} (k j_k), \tag{2.27}$$

i.e., one needs to know only the current excited in the plasma as a function of $\mathbf{E}_k$ in order to find an equation for the field $\mathbf{E}_k$. Averaging Eq.(2.25) over the statistical ensemble, we have

$$\mathbf{k}\left(\mathbf{k}\mathbf{E}_k^R\right) - \left(k^2 - \frac{\omega^2}{c^2}\right)\mathbf{E}_k^R = -\frac{4\pi i \omega}{c^2}\, \mathbf{j}_k^R, \tag{2.28}$$

and, in view of Eq. (2.25), we obtain

$$\mathbf{k}\left(\mathbf{k}\mathbf{E}_k^T\right) - \left(k^2 - \frac{\omega^2}{c^2}\right)\mathbf{E}_k^T = -\frac{4\pi i \omega}{c^2}\, \mathbf{j}_k^T. \tag{2.29}$$

If the chain of equations, (2.17)-(2.19), can be broken, then by keeping a finite number of terms in the expansion of the turbulent current in (2.29), we will be able to find a specific equation for $\langle E_{k'}^T, E_k^T \rangle$, the field correlation function, and the right side of (2.9), which describes the collision between f^R and the turbulent fluctuations, will be expressed in terms of $\langle E_{k'}^T, E_k^T \rangle$. It is valid to break off the chain of equations (2.17)-(2.19) in the theory of weak turbulence. However, the equation for the correlation function,

$$k_i k_j \langle E_{k',s}^T, E_{k,i}^T \rangle - \left(k^2 - \frac{\omega^2}{c^2}\right)\langle E_{k',s}^T, E_{k,i}^T \rangle = -\frac{4\pi i \omega}{c^2}\langle E_{k',s}^T, j_{k,i}^T \rangle, \tag{2.30}$$

obtained directly from (2.29), does not assume *a priori* a particular relationship between the frequency and the wave number for the turbulent fluctuations. Consequently, if there are physical arguments at hand for selecting certain terms in this chain, then Eq. (2.30) can describe turbulence which is not weak. We should mention that there are graphical methods [53, 54] available for writing down specific terms in the expansion of the right side of Eq. (2.30) in terms of the turbulence energy, and one can, in principle, sum the series contained in the right side of (2.30).

2.2. Linear Effects in Changes in the Distribution of Turbulent Plasma Fluctuations

Spatial and temporal variations in the fluctuation spectra are possible because of linear effects in the energy transfer of the fluctuations whose group velocities depend on characteristic lengths. In the linear approximation one can use Eqs. (2.17) and (2.22):

$$
\mathbf{j}_k^{T\,(1)} = \sum_\alpha \int \frac{e^2 \mathbf{v} \left(\mathbf{E}_{k_1}^T + \mathbf{v} \cdot \left[\dfrac{\mathbf{k}_1}{\omega_1} \mathbf{E}_{k_1}^T \right] \dfrac{\partial f_{k_1}^R}{\partial p} \right)}{i\,(\omega - \mathbf{kv} + i\delta)} \, dk_2\, dk_1\, \frac{d\mathbf{p}_\alpha}{(2\pi)^3}\, \delta\,(k - k_1 - k_2).
$$

$$(2.31)$$

We note that the turbulent current is determined by the regular part of the particle distribution, which is a function of both time and coordinates (due to the k_2 dependence of f^R in the Fourier representation). This time dependence can be slow (as compared with the frequency of the turbulent pulsations). In fact, it is clear from (2.9) that the change in f^R is proportional to the turbulence energy and can be of the order of the characteristic time for the growth of instability. If this time is much greater than the period of the excited fluctuations, we may then say that the turbulence develops in a background of slow changes in the regular distribution function.

If we neglect these changes as a first approximation, then

$$
f_{k_2}^R \approx f^R \delta\,(k_2). \tag{2.32}
$$

It is important to emphasize that Eq. (2.32) is exact for stationary turbulence. We can extract the results of the quasilinear approximation [55-57] from (2.32):

$$
j_{k,\,i}^T = \sigma_{ij}\,(k)\, E_{k,\,j}^T,
$$

where σ_{ij}, the tensor found from (2.31), is the electrical conductivity tensor,

$$
\sigma_{ij}(k) = \sum_\alpha \int \frac{v_i e^2 \left[\delta_{js} \left(1 - \dfrac{\mathbf{kv}}{\omega} \right) + \dfrac{k_s v_j}{\omega} \right]}{i\,(\omega - \mathbf{kv} + i\delta)} \cdot \frac{\partial f^R}{\partial p_s} \cdot \frac{d\mathbf{p}}{(2\pi)^3}. \tag{2.33}
$$

One frequently introduces the dielectric susceptibility tensor, which is related to $\sigma_{ij}(k)$ by the equation [19, 4, 89, 90]

$$\varepsilon_{ij}(k) = \delta_{ij} + \frac{4\pi i \sigma_{ij}(k)}{\omega}. \tag{2.34}$$

It follows immediately from (2.33) that $\varepsilon_{ij}(-k) = \varepsilon_{ij}^*(k)$. Therefore, the linear dispersion relation (2.29) reduces to the form

$$\left(k^2 \delta_{ij} - k_i k_j - \frac{1}{c^2}\omega^2 \varepsilon_{ij}(k)\right) E_{k, j}^T = 0 \tag{2.35}$$

and differs from the dispersion relation studied in the preceding chapter only in that the initial distribution function for the particles is replaced by f^R.

Let us now consider (2.35). It is convenient to summarize the solutions to this equation by introducing e_k^σ, the unit electric field polarization vector for a given oscillation [10] (σ indicates the kind of oscillation):

$$(e_k^\sigma e_k^{\sigma*}) = 1. \tag{2.36}$$

Thus, for longitudinal oscillations (indexed by l) $e_k^l = \frac{k}{k}$, for Alfven and magnetoacoustic waves in a system where $k_y = 0$, $e^A = (1, 0, 0)$; $e^M = (0, 1, 0)$. These vectors are easily found from the normalization condition of (2.36) and the relationship between the various components of the field as obtained from either the dispersion relations of the previous chapter or from (2.35). Table 1 shows the values of e_k^σ for the most important types of plasma oscillations.

It is also convenient to introduce the concept of a dielectric susceptibility for a given wave:

$$\varepsilon_k^\sigma = \varepsilon^\sigma(k) = \varepsilon_{ij}(k) e_{ik}^\sigma e_{jk}^{\sigma*} + \frac{c^2}{\omega^2}(k e_k^\sigma)(k e_k^{\sigma*}). \tag{2.37}$$

Then the dispersion relation for a particular wave σ takes the simple form $\left(E_{kj} = e_{kj}^\sigma E_k^\sigma\right)$

$$\left(k^2 - c^{-2}\omega^2 \varepsilon^\sigma(k)\right) E_k^{T\sigma} = 0. \tag{2.38}$$

One must, of course, know e_k^σ in order to find (2.37). ε^σ is an important quantity since it determines the energy of the turbulent fluctuations and their mutual interactions.

For longitudinal oscillations it is convenient to use

$$\varepsilon^l(k) = \varepsilon^\sigma(k) - \frac{k^2 c^2}{\omega^2} = \varepsilon_{ij}(k)\frac{k_i k_j}{k^2} \tag{2.39}$$

and the dispersion relation for these oscillations can be written as

$$\varepsilon^l(k) = 0. \tag{2.40}$$

One finds easily from (2.33) that

$$\varepsilon_k^l = 1 + \frac{4\pi e^2}{k^2} \int \frac{\left(k\frac{\partial f^R}{\partial p}\right)}{\omega - kv + i\delta} \cdot \frac{dp}{(2\pi)^3}. \tag{2.41}$$

Table 1 shows the values for $\varepsilon^\sigma(k)$ for the more important types of plasma fluctuations.

The dispersion equation (2.38) has a complex solution

$$\omega = \omega^\sigma(k) + i\gamma_k^\sigma, \tag{2.42}$$

where γ_k^σ is the oscillation growth rate or damping rate, as the case may be. If $\gamma_k^\sigma \ll \omega^\sigma(k)$, as was the situation considered above, an approximate expression for γ_k^σ is easily obtained from (2.38):

$$\gamma_k^\sigma = -\left.\frac{\omega^2 \operatorname{Im} \varepsilon^\sigma(k)}{\frac{\partial}{\partial\omega}\omega^2 \operatorname{Re}\varepsilon^\sigma(k)}\right|_{\omega=\omega^\sigma(k)} \tag{2.43}$$

For longitudinal oscillations

$$\operatorname{Im}\frac{1}{\omega - kv + i\delta} = -\pi\delta(\omega - kv);$$

$$\gamma_k^l = -\left.\frac{\operatorname{Im}\varepsilon^l(k)}{\frac{\partial}{\partial\omega}\operatorname{Re}\varepsilon^l(k)}\right|_{\omega=\omega^l(k)} = \frac{4\pi^2 e^2}{k^2 \frac{\partial}{\partial\omega}\operatorname{Re}\varepsilon^l(k)\Big|_{\omega=\omega^l(k)}} \times$$

$$\times \int \delta(\omega^l(k) - kv)\left(k\frac{\partial f^R}{\partial p}\right)\frac{dp}{(2\pi)^3}. \tag{2.44}$$

TABLE 1. Normalized Unit Vectors and Dielectric Susceptibility of Turbulent Plasma Functions

Type of fluctuation	Designation	Normalized unit vectors e_k^σ in a system where $k_y = 0$			Dielectric susceptibility ε^σ
		$e_{k,x}^\sigma$	$e_{k,y}^\sigma$	$e_{k,z}^\sigma$	
Langmuir	l	$\sin\theta$	0	$\cos\theta$	$\dfrac{k^2 c^2}{\omega^2} + \varepsilon^l = 1 - \dfrac{\omega_{pe}^2}{\omega^2}\cos^2\theta - \dfrac{\omega_{pe}^2 \sin^2\theta}{\omega^2 - \omega_{He}^2} + \dfrac{k^2 c^2}{\omega^2}$
Ion-acoustic oscillations	s	$\sin\theta$	0	$\cos\theta$	$1 - \dfrac{\omega_{pi}^2}{\omega^2} + \dfrac{\omega_{pe}^2}{k^2 v_{Te}^2} + \dfrac{k^2 c^2}{\omega^2} = \dfrac{k^2 c^2}{\omega^2} + \varepsilon^l$
Whistlers	w	$\dfrac{1}{\sqrt{1+\cos^2\theta}}$	$\dfrac{i\,\lvert\cos\theta\rvert}{\sqrt{1+\cos^2\theta}}$	0	$2\dfrac{\omega_{pe}^2}{\omega\omega_{He}}\cdot\dfrac{\lvert\cos\theta\rvert}{1+\cos^2\theta} + \dfrac{k^2 c^2 \sin^2\theta}{\omega^2(1+\cos^2\theta)}$
Alfven waves	A	1	0	0	$\dfrac{c^2}{v_A^2} + \dfrac{k^2 c^2}{\omega^2}\sin^2\theta$
Magnetohydrodynamic waves	M	0	1	0	$\dfrac{c^2}{v_A^2}$
Magnetized acoustic oscillations	Ms	$\sin\theta$	0	$\cos\theta$	$1 - \dfrac{\omega_{pi}^2}{\omega^2}\cos^2\theta + \dfrac{\omega_{pe}^2}{k^2 v_{Te}^2} + \dfrac{k^2 c^2}{\omega^2} = \varepsilon^l + \dfrac{k^2 c^2}{\omega^2}$

This result, like that of (2.43), has a simple interpretation and can be obtained by other, much simpler means, by using the balance equations (see below).

Let us now turn to Eq. (2.30), which, according to (2.38), can be written as

$$\left(k^2 - \frac{1}{c^2}\omega^2 \, \varepsilon_k^\sigma\right)\langle E_k^{T\sigma'} E_k^{T\sigma}\rangle = 0. \tag{2.45}$$

By taking the equation for $E_{k'}^T$ and multiplying it by $E_k^{T\sigma}$, we obtain in place of (2.45)

$$\left(k^2 - k'^2 - \frac{1}{c^2}\omega^2\varepsilon^\sigma(k) + \frac{1}{c^2}\omega'^2\varepsilon^\sigma(k')\right)\langle E_{k'}^{T\sigma} E_k^{T\sigma}\rangle = 0. \tag{2.46}$$

For a stationary and uniform turbulence [see Eq. (1.17)],

$$I_{k'k}^\sigma \equiv \langle E_{k'}^{T\sigma} E_k^{T\sigma}\rangle = I_k^\sigma \delta(k + k'). \tag{2.47}$$

Because of the slight buildup or damping described by (2.43), the dependence of $I_{kk'}^\sigma$ on k + k' is not a delta function, i.e., k' ≠ −k, but is spread out over Δk, where $|\Delta k / k| \ll 1$, because the buildup or damping process is slow.

We now define

$$\varkappa = \frac{k - k'}{2}; \qquad \Delta k = k + k'; \qquad I_{k,\,k'} = \tilde{I}_{\varkappa,\,\Delta k} \tag{2.48}$$

and the correlation function $I_k(\mathbf{r}, t) = I_k(x)$, $x = \{\mathbf{r}, t\}$, which is weakly dependent on r and t:

$$I_\varkappa(x) = \int \tilde{I}_{\varkappa,\,\Delta k}\, e^{i\Delta k x}\, d\Delta k. \tag{2.49}$$

We now multiply (2.46) by $e^{i\,\Delta k x}$ and integrate over all Δk under the stipulation that

$$\frac{|\Delta k|}{|k|} \ll 1, \quad \frac{|\Delta \omega|}{\omega} \ll 1, \quad \text{and } \gamma_k \ll \omega_k.$$

We obtain

$$\int \left\{ 2\Delta kk - \frac{1}{c^2}\Delta\omega\frac{\partial}{\partial\omega}\,\omega^2\,\mathrm{Re}\,\varepsilon_k^\sigma - \frac{1}{c^2}\omega^2\Delta k\frac{\partial}{\partial k}\,\mathrm{Re}\,\varepsilon_k^\sigma - 2i\omega^2\frac{1}{c^2}\,\mathrm{Im}\,\varepsilon_k^\sigma \right\} \times$$
$$\times\, e^{i\Delta k x}\,\tilde{I}_{\varkappa,\,\Delta k} = 0. \tag{2.50}$$

Here we have taken into account the fact that $\mathrm{Im}\,\varepsilon_{-k}^\sigma = -\mathrm{Im}\,\varepsilon_k^\sigma$.

Since one can neglect small effects of order γ_k^σ/ω_k in (2.50), one can assume that

$$I_k^\sigma(\mathbf{r}, t) \simeq I_k^\sigma(\mathbf{r}, t)\,\delta\left(\omega - \omega_k^\sigma\right); \quad \omega_k^\sigma = |\,\omega^\sigma(\mathbf{k})\,|; \quad \omega > 0. \qquad (2.51)$$

Further, noting that

$$\int \Delta\omega\, e^{i\,\Delta k x}\, I_{k,\,\Delta k}^\sigma\, d\Delta k = -\frac{1}{i}\,\frac{\partial}{\partial t}\, I_k^\sigma(\mathbf{r}, t),$$

$$\int \Delta k\, e^{i\,\Delta k x}\, I_{k,\,\Delta k}^\sigma\, d\Delta k = \frac{1}{i} \cdot \frac{\partial}{\partial \mathbf{r}}\, I_k^\sigma(\mathbf{r}, t),$$

substituting (2.51) into (2.50), and integrating over $\omega > 0$, we find

$$\frac{\partial I_k^\sigma(\mathbf{r}, t)}{\partial t} + \mathbf{v}_g^\sigma\, \frac{\partial I_k^\sigma(\mathbf{r}, t)}{\partial \mathbf{r}} = 2\gamma_k^\sigma\, I_k^\sigma(\mathbf{r}, t), \qquad (2.52)$$

where $(\mathrm{Re}\,\varepsilon_k^\sigma \approx \varepsilon_k^\sigma)$

$$\mathbf{v}_g^\sigma := \frac{\partial \omega_k^\sigma}{\partial \mathbf{k}} = -\left.\frac{\dfrac{\partial}{\partial k}\left(\omega^2\, \varepsilon_k^\sigma - k^2\right)}{\dfrac{\partial}{\partial \omega}\,\omega^2\, \varepsilon_k^\sigma}\right|_{\omega = \omega_k^\sigma} \qquad (2.53)$$

is the group velocity for the turbulent fluctuations.

It is now convenient to introduce the concept of those quasi-particles which characterize the turbulent plasma fluctuations, the plasmons; Eq. (2.52) can be treated as the kinetic equation for the plasmons. We shall use k for the plasmon momentum (strictly speaking one ought to use $\hbar k$, but the plasmon equation is classical and the $\hbar$ is cancelled out, so that we shall assume $\hbar = 1$ from the outset), and the plasmon energies will be ω_k^σ.

The distribution function for σ plasmons, which indicates the number of plasmons with a given momentum within the interval from k to k + dk, is denoted by $N_k^\sigma(\mathbf{r}, t)$. This quantity is assumed to be proportional to $I_k^\sigma(\mathbf{r}, t)$:

$$N_k^\sigma(\mathbf{r}, t) = \alpha_k\, I_k^\sigma(\mathbf{r}, t). \qquad (2.54)$$

Since Eq. (2.52) is linear in I_k^σ, we have

$$\frac{\partial N_k^\sigma}{\partial t} + \mathbf{v}_g^\sigma\, \frac{\partial N_k^\sigma}{\partial \mathbf{r}} = 2\gamma_k^\sigma\, N_k^\sigma. \qquad (2.55)$$

In order to find the constant of proportionality α_k, we write the energy of the electromagnetic field as

$$W = \int_{-\infty}^{t} E \frac{\partial D}{\partial t} \, dt + \frac{H^2}{8\pi} \, , \qquad (2.56)$$

where $D_{k,\,i} = \varepsilon_{ij}(k) E_{k,\,j}$ is the induction, and

$$H_k = \frac{c}{\omega} [k E_k].$$

After averaging (2.56) over the statistical ensemble under the assumption that $E_k = E_k^\sigma e_k^\sigma$, we obtain

$$\langle W \rangle = \frac{1}{4\pi} \sum_\sigma \int I_k^\sigma \, dk \, \frac{1}{\omega_k^\sigma} \left[\frac{\partial}{\partial \omega} \omega^2 \varepsilon_k^\sigma \right]_{\omega = \omega_k^\sigma} . \qquad (2.57)$$

Note that this energy contains both the field energy and the energy of the plasma particles participating in the turbulent motions of the specified wave types. This includes the dielectric susceptibility ε_k^σ. Equation (2.54) allows us to write the turbulence energy as a sum of the plasmon energies:

$$\langle W \rangle = \int \frac{\omega_k^\sigma N_k^\sigma}{(2\pi)^3} \, dk. \qquad (2.58)$$

We then have

$$N_k^\sigma = \frac{2\pi^2}{(\omega_k^\sigma)^2} \cdot \frac{\partial}{\partial \omega} \omega^2 \varepsilon_k^\sigma \Big|_{\omega = \omega_k^\sigma} I_k^\sigma, \qquad (2.59)$$

i.e.,

$$\alpha_k = \frac{2\pi^2}{(\omega_k^\sigma)^2} \cdot \frac{\partial}{\partial \omega} \omega^2 \varepsilon_k^\sigma \Big|_{\omega = \omega_k^\sigma} .$$

For the particular case of longitudinal fluctuations (2.59) gives

$$N_k^l = 2\pi^2 \frac{\partial}{\partial \omega} \varepsilon^l \Big|_{\omega = \omega_k^l} I_k^l. \qquad (2.60)$$

Equation (2.55) can now be considered as the balance equation for the number of plasmons. The left side describes their variation in space and time because of their motion with velocities equal to the wave's group velocity; the right side gives the changes due to the balance of plasmons emitted and absorbed by the plasma particles. According to (2.44) (or the analogous equation in the general case of nonpotential fluctuations) γ_k contains $\delta(\omega - \mathbf{kv})$, the law of conservation of energy during emission:

$$\varepsilon_{\mathbf{p}-\hbar\mathbf{k}} = \varepsilon_{\mathbf{p}} - \hbar\omega = \varepsilon_{\mathbf{p}} - \hbar\mathbf{k}\,\frac{\partial\varepsilon_{\mathbf{p}}}{\partial\mathbf{p}} = \varepsilon_{\mathbf{p}} - \hbar\mathbf{kv},$$

where $\mathbf{v} = \partial\varepsilon_{\mathbf{p}}/\partial\mathbf{p}$, and $\varepsilon_{\mathbf{p}}$ is the energy ε of a particle as a function of its momentum $\mathbf{p}$. If the emission probability is $w_{\mathbf{p}}(\mathbf{k})$, the increase in the number of plasmons due to emission is $\int w_{\mathbf{p}}(\mathbf{k}) f_{\mathbf{p}}^R \cdot (N_{\mathbf{k}} + 1)\frac{d\mathbf{p}}{(2\pi)^3}$, and the decrease due to absorption is $\int w_{\mathbf{p}} f_{\mathbf{p}-\mathbf{k}}^R N_{\mathbf{k}} \cdot \frac{d\mathbf{p}}{(2\pi)^3}$, i.e.,

$$\frac{\partial N_{\mathbf{k}}}{\partial t} = N_{\mathbf{k}} \int w_{\mathbf{p}}(\mathbf{k}) \left(\mathbf{k}\,\frac{\partial f_{\mathbf{p}}^R}{\partial\mathbf{p}} \right) \frac{d\mathbf{p}}{(2\pi)^3}.$$

Here only induced processes are left. By comparing (2.44) with (2.55) we find the probability of emission of a longitudinal plasmon by a plasma particle:

$$w_{\mathbf{p}}(\mathbf{k}) = \frac{8\pi^2\,e^2}{k^2\,\dfrac{\partial\varepsilon^l}{\partial\omega}}\Bigg|_{\omega=\omega_{\mathbf{k}}^l} \delta\,(\omega_{\mathbf{k}}^l - \mathbf{kv}). \tag{2.61}$$

It is important to emphasize that we have performed an unusual "renormalization." The function f^R describes just collective electronic excitations, and f^T describes the plasmons. We now see that a "cloud" of particles of different sign forms around a plasma particle, and f^R corresponds to this complex of particles.

2.3. Nonlinear Interaction of

Turbulent Plasma Fluctuations

We shall now treat the nonlinear effects which are due to the next terms in the turbulent field expansion of (2.17)–(2.19); the terms responsible for these effects are the currents $j^{T(2)}$ and $j^{T(3)}$ in (2.22), which are second and third order, respectively in the turbulent field. We consider both these terms (not just the single $j^{T(2)}$ term) because the current $j^{T(2)}$ on the right side of (2.30) leads to an average of three fields $\langle E^T_{k_1} E^T_{k_2} E^T_{k_3} \rangle$, which, in the approximation where the fields can be treated as statistically independent, reduces to $\langle E^T_{k_1} E^T_{k_2} E^T_{k_3} \rangle \simeq \langle E^T_{k_1} \rangle \langle E^T_{k_2} \rangle \langle E^T_{k_3} \rangle$, which is then zero because of the definition of E^T. Including the weak-field correlations gives rise to effects which are of the same order as $\langle E^T j^{T\,(3)} \rangle$.

In order to more simply clarify the essence and physical meaning of these results, we shall demonstrate our argument using the longitudinal oscillations with $e^\sigma_k = \frac{k}{k}$ as an example. Then, in place of (2.30) [and the linear version (2.46)] we obtain [see the definition of ε^l in (2.39)]

$$(\varepsilon^l_k - \varepsilon^{l*}_{k'}) \langle E^{T*}_{k'} E^T_k \rangle = -\frac{4\pi i}{k\omega} \left\langle E^{T*}_{k'} \left((k j^{T(2)}_k) + (k j^{T(3)}_k) \right) \right\rangle -$$

$$- \frac{4\pi i}{k'\omega'} \left\langle E^T_k \left((k' j^{T(2)*}_{k'}) + (k' j^{T\,(3)*}_{k'}) \right) \right\rangle . \qquad (2.62)$$

Here we have used the equalities $\varepsilon^l_{-k} = \varepsilon^{l*}_k$, $E^l_{-k} = -E^{l*}_k$, which result in

$$e^l_{-k} = -e^l_k.$$

The left linear part of (2.62) is transformed in the same manner as in §2.2.

Multiplying by $(i\Delta kx)$ and integrating over k', ω on the left side, we find for $\gamma_k \simeq 0$ that

$$i \frac{1}{2\pi^2} \left(\frac{\partial N_k}{\partial t} + v_g \frac{\partial N_k}{\partial r} \right) = i \frac{\partial}{\partial t} \int_0^\infty I_k \frac{\partial \varepsilon^l_k}{\partial \omega} \bigg|_{\omega = \omega^l_k} d\omega. \qquad (2.63)$$

Noting that the turbulent fluctuations are approximately stationary on the right side of (2.62) because of the weak nonlinearity, and $\Delta k \approx 0$, we arrive at an equation which gives the change in the number of plasmons resulting from the nonlinear interactions of the turbulent fluctuations,

$$\frac{\partial N_k}{\partial t} + \mathbf{v}_g \frac{\partial N_k}{\partial \mathbf{r}} = -8\pi^3 \int d\omega dk' \left\{ \frac{1}{k\omega} \left\langle E_{k'}^{T*} \left((\mathbf{k} \mathbf{j}_k^{T(2)}) + (\mathbf{k} \mathbf{j}_k^{T(3)}) \right) \right\rangle + \right.$$

$$\left. + \frac{1}{k'\omega'} \left\langle E_k^T \left((\mathbf{k'} \mathbf{j}_{k'}^{T(2)*}) + (\mathbf{k'} \mathbf{j}_{k'}^{T(3)*}) \right) \right\rangle \right\}. \qquad (2.64)$$

From (2.18) and (2.22) we find the expression for the turbulent current to second order in the turbulent field:

$$\frac{\mathbf{j}_k^{T(2)} \mathbf{k}}{k} = \int S_{k, k_1, k_2} \left(E_{k_1}^T E_{k_2}^T - \langle E_{k_1}^T E_{k_2}^T \rangle \right) dk_1 \, dk_2 \, \delta (k - k_1 - k_2); \qquad (2.65)$$

$$S_{k, k_1, k_2} = -\frac{e^3}{kk_1 k_2} \int \frac{(\mathbf{k}\mathbf{v})}{\omega - \mathbf{k}\mathbf{v} + i\,\delta} \times$$
$$\delta \to +0$$

$$\times \left(\mathbf{k}_1 \frac{\partial}{\partial \mathbf{p}} \right) \frac{1}{\omega_2 - \mathbf{k}_2 \mathbf{v} + i\delta} \left(\mathbf{k}_2 \frac{\partial}{\partial \mathbf{p}} \right) f_\mathbf{p}^R \, d\mathbf{p} \, (2\pi)^{-3}. \qquad (2.66)$$

In a similar manner, after incorporating $f^{T(2)}$ into the equation for $f^{T(3)}$, we obtain

$$\frac{1}{k} \mathbf{k} \mathbf{j}_k^{T(3)} = \int \delta (k - k_1 - k_2 - k_3) \, dk_1 \, dk_2 \, dk_3 \times$$

$$\times \sum_{k, k_1, k_2, k_3} \left(E_{k_1}^T E_{k_2}^T E_{k_3}^T - E_{k_3}^T \langle E_{k_1}^T E_{k_2}^T \rangle - \langle E_{k_3}^T E_{k_1}^T E_{k_2}^T \rangle \right), \qquad (2.67)$$

$$\sum_{k, k_1, k_2, k_3} = \frac{i\,e^4}{kk_1 k_2 k_3} \int \frac{(\mathbf{k}\mathbf{v})}{\omega - \mathbf{k}\mathbf{v} + i\delta} \left(\mathbf{k}_3 \frac{\partial}{\partial \mathbf{p}} \right) \times$$
$$\delta \to +0$$

$$\times \frac{1}{\omega_1 + \omega_2 - (\mathbf{k}_1 + \mathbf{k}_2)\mathbf{v} + i\delta} \left(\mathbf{k}_1 \frac{\partial}{\partial \mathbf{p}} \right) \frac{1}{\omega_2 - \mathbf{k}_2 \mathbf{v} + i\delta} \left(\mathbf{k}_2 \frac{\partial}{\partial \mathbf{p}} \right) f_\mathbf{p}^R \frac{d\mathbf{p}}{(2\pi)^3}. \qquad (2.68)$$

We may now write Eq. (2.68) as $\left(\frac{\partial}{\partial \mathbf{r}} = 0 \right)$

$$\frac{\partial N_k}{\partial t} = -8\pi^3 \int_0^\infty d\omega \int dk' \left\{ \frac{1}{\omega} \langle E_k^{T*} E_{k_3}^T E_{k_4}^T \rangle \times \right.$$

$$\times S_{k, k_3, k_4} \, dk_3 \, dk_4 \, \delta (k - k_3 - k_4) + \frac{1}{\omega'} \langle E_k^T E_{k_3}^{T*} E_{k_4}^{T*} \rangle S_{k', k_3, k_4}^{*} \times$$

$$\times dk_3 \, dk_4 \, \delta (k' - k_3 - k_4) + \frac{1}{\omega} \sum_{k, k_1, k_2, k_3} \times$$

$$\times \left(\langle E_k^{T*} E_{k_1}^T E_{k_2}^T E_{k_3}^T \rangle - \langle E_k^{T*} E_{k_3}^T \rangle \langle E_{k_1}^T E_{k_2}^T \rangle \right) \times$$

$$\times\, dk_1\, dk_2\, dk_3\, \delta\,(k-k_1-k_2-k_3) + \frac{1}{\omega'}\sum_{k'}^{*}{}_{,\,k_1,\,k_2,\,k_3}\times$$

$$\times\, (\langle E_k^T E_{k_1}^{T*} E_{k_2}^{T*} E_{k_3}^{T*}\rangle - \langle E_k^T E_{k_1}^{T*}\rangle\langle E_{k_1}^{T*} E_{k_3}^{T*}\rangle)\times$$

$$\times\, dk_1\, dk_2\, dk_3\, \delta\,(k'-k_1-k_2-k_3)\Big\}. \tag{2.69}$$

Since the nonlinearity is weak, the turbulent fields are only weakly correlated among themselves. Therefore the mean of four random quantities can be approximately split into the product of the mean values of the two fields. This procedure gives zero for the average of three fields, and one must include the slight correlation of the fields. Using (2.29) and including only terms of second order in the turbulent fluctuation fields, we obtain

$$\varepsilon_{k_4}^l E_{k_4}^T = -\frac{4\pi i}{\omega_4}\int dk_1\, dk_2\, \delta\,(k_4-k_1-k_2)\, S_{k_4,\,k_1,\,k_2}\,(E_{k_1}^T E_{k_2}^T - \langle E_{k_1}^T E_{k_2}^T\rangle). \tag{2.70}$$

Multiplying (2.70) by $E_{k'}^{T*} E_{k_2}^T$ and averaging over the ensemble, we find an approximate expression for the mean value of three turbulent fields in terms of the average of four fields. In this way we can more accurately define each of the three fields involved in $\langle E_{k'}^T E_{k_3}^T E_{k_4}^T\rangle$.

Equation (2.69) takes this simple form:

$$\frac{\partial N_k}{\partial t} = -8\pi^3\int_0^\infty d\omega\int dk'\, dk_1\, dk_2\, dk_3\Big\{\frac{\delta\,(k-k_1-k_2-k_3)}{\omega}\times$$

$$\times\sum_{k,\,k_1,\,k_2,\,k_3}^{\mathrm{eff}}\big[\langle E_{k'}^{T*} E_{k_1}^T E_{k_2}^T E_{k_3}^T\rangle - \langle E_{k'}^{T*} E_{k_3}^T\rangle\langle E_{k_1}^T E_{k_2}^T\rangle\big] +$$

$$+\frac{\delta\,(k'-k_1-k_2-k_3)}{\omega'}\sum_{k'k_1,\,k_2,\,k_3}^{\mathrm{eff}}\big[\langle E_k^T E_{k_1}^{T*} E_{k_2}^{T*} E_{k_3}^{T*}\rangle -$$

$$-\langle E_k^T E_{k_3}^{T*}\rangle\langle E_{k_1}^{T*} E_{k_2}^{T*}\rangle\big]\Big\} - 32\pi^4 i\int\frac{dk_1\, dk_2\, dk_1'\, dk_2'}{\omega\omega'}\, dk'd\omega\Big(\frac{1}{\varepsilon_{k'}^{l*}}-\frac{1}{\varepsilon_k^l}\Big)\times$$

$$\times\, \delta\,(k-k_1-k_2)\, \delta\,(k'-k_1'-k_2')\, S_{k,\,k_1,\,k_2}\times$$

$$\times\, S_{k',\,k_1',\,k_2'}^{*}\big(\big\langle E_{k_1}^T E_{k_2}^T E_{k_1'}^{T*} E_{k_2'}^{T*}\big\rangle - \big\langle E_{k_1}^T E_{k_2}^T\big\rangle\big\langle E_{k_1'}^{T*} E_{k_2'}^{T*}\big\rangle\big), \tag{2.71}$$

$$\sum_{k,\,k_1,\,k_2,\,k_3}^{\mathrm{eff}} = \sum_{k,\,k_1,\,k_2,\,k_3} - \frac{4\pi i}{(\omega_1+\omega_2)\,\varepsilon_{k_1+k_2}^l}\times$$

$$\times\, S_{k_1+k_2,\,k_1,\,k_2}\,(S_{k,\,k_3,\,k_1+k_2}+S_{k,\,k_1+k_2,\,k_3}). \tag{2.72}$$

By splitting the average of four fields into pairs we see that of the three possible pairs one drops out of the result. Moreover, allowing for the fact that the turbulent fluctuations are approximately stationary,

$$\langle E_{k_1}^{T*} E_{k_2}^{T} \rangle = I_{k_1} \delta (k_1 - k_2), \tag{2.73}$$

we obtain†

$$\frac{\partial N_{\mathbf{k}}}{\partial t} = N_{\mathbf{k}} \int v_{\mathbf{k}, \, \mathbf{k}_1} N_{\mathbf{k}_1} \, d\mathbf{k}_1 + \int w_{\mathbf{k}_1, \, \mathbf{k}_2} N_{\mathbf{k}_1} N_{\mathbf{k}_2} \, d\mathbf{k}_1 \, d\mathbf{k}_2, \tag{2.75}$$

where

$$v_{\mathbf{k}, \, \mathbf{k}_1} = \frac{4}{\pi \omega_k^l} \operatorname{Re} \left\{ \sum_{k, \, k, \, k_1, \, -k_1}^{\text{eff}} + \sum_{k, \, k_1, \, k, \, -k_1}^{\text{eff}} + \sum_{k, \, k, \, -k_1, \, k_1}^{\text{eff}} + \right.$$

$$\left. + \sum_{k, \, -k_1, \, k, \, k_1}^{\text{eff}} \right\} \left(\frac{\partial \varepsilon_k^l}{\partial \omega} \cdot \frac{\partial \varepsilon_{k_1}^l}{\partial \omega_1} \right)^{-1} \Bigg|_{\omega = \omega_k^l, \, \omega_1 = \omega_{k_1}^l}, \tag{2.76}$$

$$w_{\mathbf{k}_1, \, \mathbf{k}_2} = \frac{8\pi}{\omega^2} (G_{k, \, k_1, \, k_2} + G_{k, \, -k_1, \, k_2} + G_{k, \, k_1, \, -k_2} + G_{k, \, -k_1, \, -k_2}) \times$$

$$\times \left(\frac{\partial \varepsilon_k^l}{\partial \omega} \cdot \frac{\partial \varepsilon_{k_1}^l}{\partial \omega_1} \cdot \frac{\partial \varepsilon_{k_2}^l}{\partial \omega_2} \right)^{-1} \Bigg|_{\omega = \omega_k^l, \, \omega_1 = \omega_{k_1}^l, \, \omega_2 = \omega_{k_2}^l}, \tag{2.77}$$

$$G_{k, \, k_1, \, k_2} = |S_{k, \, k_1, \, k_2} + S_{k, \, k_2, \, k_1}|^2 \delta (k - k_1 - k_2). \tag{2.78}$$

Thus, Eq. (2.75) (first obtained in [6]) contains quadratic combinations of the number of quanta and describes the scattering processes of these quanta on others of the same type.

---

† $\langle E_{k_1}^{T} E_{k_2}^{T} \rangle = -I_{k_1} \delta (k_1 + k_2)$ because $E_{-k_1} = -E_{k_1}^*$; $\mathbf{E}_{k_1} = \dfrac{\mathbf{k}_1}{k_1} E_{k_1}$;

$$I_k = \frac{1}{2\pi^2 \left| \dfrac{\partial \varepsilon_k^l}{\partial \omega} \right|_{\omega = \omega_k^l}} \left(N_{\mathbf{k}} \delta (\omega - \omega_k^l) + N_{-\mathbf{k}} \delta (\omega + \omega_k^l) \right); \quad \omega_k^l > 0. \tag{2.74}$$

2.4. The Interpretation of Nonlinear Interactions of Turbulent Fluctuations as Induced Plasmon — Plasmon and Plasmon — Plasma Particle Scattering

Two kinds of terms can be identified in the nonlinear equations (2.75). The first of these is interpreted as the effects of induced combination scattering of waves by other waves. This group includes all the terms in the second member of (2.75). These terms are nonzero if the following conservation laws are satisfied:

$$k = k_1 + k_2; \qquad \omega_k^l = \omega_{k_1}^l + \omega_{k_2}^l. \qquad (2.79)$$

Since these conservation rules describe the "splitting" of one wave into two new waves, such processes are called decay processes. The four terms of (2.77) contain various combinations of processes which differ from (2.79) in that some of the interacting waves are absorbed rather than radiated.

The second term of (2.75) describes only spontaneous decay processes. If $w_{k_1, k_2}^{(1)} (2\pi)^3$ is the probability of the process given in (2.79) [corresponding to the first term of the four in (2.77)], the decrease in the number of quanta due to this process is

$$- w_{k_1, k_2}^{(1)} N_k (N_{k_1} + 1) (N_{k_2} + 1)(2\pi)^3,$$

and the increase resulting from the inverse process is

$$w_{k_1, k_2}^{(1)} N_{k_1} N_{k_2} (N_k + 1)(2\pi)^3,$$

i.e.,

$$\frac{\partial N_k}{\partial t} = \int w_{k_1, k_2}^{(1)} (N_{k_1} N_{k_2} - N_k N_{k_1} - N_k N_{k_2}) \, dk_1 \, dk_2. \qquad (2.80)$$

The last two terms in (2.80) represent induced decays since they are proportional to the number of original decaying quanta, N_k. These terms are quite naturally contained in v_{k, k_1} of (2.76). There are additional terms which correspond to other decay processes. Equation (2.76) also describes induced scattering processes.

In order to isolate the induced decays of interest to us in (2.80), we note that because of (2.79) $k_1 + k_2$ is a proper solution; i.e., $\varepsilon_{k_1+k_2} = 0$, and as a consequence the second term of Σ^{eff} in (2.72), which contains $1/\varepsilon_{k_1+k_2}$, plays a decisive role in this region of wave numbers. Setting $\delta\Sigma^{\text{eff}} = \Sigma^{\text{eff}} - \Sigma$, we obtain

$$\delta\Sigma^{\text{eff}}_{k, k_1, k_2, k_3} = -4\pi i \frac{1}{\varepsilon^l_{k_1+k_2}(\omega_1+\omega_2)} S_{k_1+k_2, k_1, k_2} \times$$

$$\times (S_{k, k_3, k_1+k_2} + S_{k, k_1+k_2, k_3}) = -4\pi i \int \delta(k' - k_1 - k_2) \frac{1}{\omega' \varepsilon^l_{k'}} \times$$

$$\times S_{k', k_1, k_2} \left(S_{k, k_3 k'} + S_{k, k', k_3} \right) dk'. \tag{2.81}$$

We shall now consider the contribution to $\delta\Sigma$ from the third and fourth terms of (2.76) (designated as $v^{(1)}_{k, k_1}$) †:

$$v^{(1)}_{k, k_1} = \frac{16}{\omega^l_k \frac{\partial \varepsilon^l_k}{\partial\omega} \cdot \frac{\partial \varepsilon^l_{k_1}}{\partial\omega_1}} \text{Im} \int \delta(k_2 - k + k_1) \frac{dk_2}{\omega_2 \varepsilon^l_{k_2}} \times$$

$$\times (S_{k_2, k, -k_1} + S_{k_2, -k_1, k}) (S_{k, k_1, k_2} + S_{k, k_2 k_1}) \Big|_{\omega=\omega^l_k, \ \omega_1=\omega^l_{k_1}}. \tag{2.82}$$

Using the explicit form (2.66) for S, it is easy to prove that the following identity is valid‡:

$$\frac{1}{\omega_2}(S_{k_2, k, -k_1} + S_{k_2, -k_1, k}) = -\frac{1}{\omega}(S_{-k, -k_1, -k_2} + S_{-k, -k_2, -k_1}) =$$

$$= \frac{1}{\omega}(S^*_{k, k_1, k_2} + S^*_{k, k_2, k_1}). \tag{2.83}$$

The last equality results from the reality of the nonlinear current $j^{(2)}_{-k} = -j^{(2)*}_k$ and the reality of the fields $E_{-k} = -E^*_k$. We thus ob-

---

†The first and second terms of (2.76) describe other decays for which the conservation rules differ by a change in the signs of the frequency and wave vectors in (2.79).

‡The first equality in (2.83) can also be obtained from the general theorems concerning the nondissipative character of the nonlinear current $j^{(2)}_k$ [58].

tain

$$v^{(1)}_{k, \, k_1} = -2\int w^{(1)}_{k_1, \, k_2} \, dk_2. \tag{2.84}$$

Assuming that all our expressions are symmetric under the substitution $1 \rightleftarrows 2$, we see that the last two terms of (2.80) can be rewritten as twice the second term, which then corresponds to our result in (2.84). This concludes the proof.

We can now show that the remaining terms of the nonlinear interaction in (2.76) also have a simple interpretation, for they describe the effects of induced scattering of turbulent fluctuations by plasma particles. According to (2.76) only the real part figures in the result, but because of (2.68) one must take into account only the imaginary parts present in the denominators of (2.68). For example,

$$\text{Im} \, \frac{1}{\omega - kv + i \, \delta} = -\pi\delta \, (\omega - kv).$$

$$\delta \rightarrow +0$$

The conservation $\omega = kv$ corresponds to the emission of a wave by the plasma particles and makes a small correction to the linear constant of (2.55) (see §2.8 for more detail), whereas the dispersion relation

$$\omega_1 \pm \omega_2 = (k_1 \pm k_2) \, v \tag{2.85}$$

describes the process of emitting two waves or the process of scattering turbulent fluctuations by plasma particles.

Assume the plasma particle has energy ε_p before the interaction; after emitting two waves, or emitting one and absorbing another, its energy becomes

$$\varepsilon_{p-(k_1 \pm k_2)} \approx \varepsilon_p - (k_1 \pm k_2) \, \frac{\partial \varepsilon_p}{\partial p} = \varepsilon_p - (k_1 \pm k_2) \, v$$

$$\left(v = \frac{\partial \varepsilon_p}{\partial p} = \text{particle velocity} \right).$$

But the difference between ε_p and $\varepsilon_{p-(k_1 \pm k_2)}$ is equal to $-(\omega_1 \pm \omega_2)$ from the conservation laws, and this leads to Eq. (2.85).

Because of the above, we are limited to the consideration of just the processes represented by Eq. (2.85). For the third and fourth terms† of v_{k,k_1} we find that the contribution of Σ [see (2.72)] gives

$$v^c_{k,\,k_1} = \frac{4e^4}{k_1^2\, k^2\, \dfrac{\partial \varepsilon^l_k}{\partial \omega}\bigg|_{\omega=\omega^l_k} \dfrac{\partial \varepsilon^l_{k_1}}{\partial \omega_1}\bigg|_{\omega_1=\omega^l_{k_1}}} \times$$

$$\times \int \frac{1}{(\omega^l_k - kv)}\left(k_1 \frac{\partial}{\partial p}\right)\delta\,(\omega^l_k - \omega^l_{k_1} - (k-k_1)\,v)\left\{\left(k\frac{\partial}{\partial p}\right)\frac{1}{\omega^l_{k_1} - k_1\, v}\left(k_1 \frac{\partial}{\partial p}\right) - \right.$$

$$\left. - \left(k_1 \frac{\partial}{\partial p}\right)\frac{1}{\omega^l_k - kv}\left(k\frac{\partial}{\partial p}\right)\right\}\frac{f^R\,dp}{(2\pi)^3}. \qquad (2.86)$$

It is not hard to see that the term

$$\left(\frac{1}{\omega^l_{k_1} - k_1\, v} - \frac{1}{\omega^l_k - kv}\right)\left(k\frac{\partial}{\partial p}\right)\left(k_1 \frac{\partial}{\partial p}\right)$$

in the braces in (2.86) goes to zero because of the δ function; the result is then

$$v^c_{k,\,k_1} = \frac{4e^4\,(kk_1)^2}{m_e^2\, k^2\, k_1^2\,\left|\dfrac{\partial \varepsilon^l_k}{\partial \omega}\cdot\dfrac{\partial \varepsilon^l_{k_1}}{\partial \omega_1}\right|_{\substack{\omega=\omega^l_k \\ \omega_1=\omega^l_{k_1}}}} \times$$

$$\times \int \frac{dp\delta\,(\omega^l_k - \omega^l_{k_1} - (k-k_1)\,v)}{(2\pi)^3\,(\omega^l_k - kv)^4}\left(k - k_1,\, \frac{\partial f^R}{\partial p}\right). \qquad (2.87)$$

This result can be simply interpreted as the induced Thomson scattering of the wave ω^l_k by the plasma particles. The wave ω^l_k forces the plasma particles into oscillation, and the oscillating particles emit the wave $\omega^l_{k_1}$. Assume that $w^c_p\,(k, k_1)$ is the probability of scattering. Then the decrease in the number of plasmons with frequency ω^l_k due to the induced absorption of the wave

---

†The first and second terms describe the emission of two waves, and can be treated in the same fashion.

$\omega_{\mathbf{k}}^l$ and the emission of the wave $\omega_{\mathbf{k}_1}^l$ will be given by

$$- \int w_{\mathbf{p}}^c (\mathbf{k}, \mathbf{k}_1) N_{\mathbf{k}} N_{\mathbf{k}_1} \frac{d\mathbf{k}_1 \, d\mathbf{p}}{(2\pi)^6} f_{\mathbf{p}}^R.$$

But the reverse process is proportional to the number of particles with final momentum $\mathbf{p} + \mathbf{k} - \mathbf{k}_1$, i.e.,

$$\frac{\partial N_{\mathbf{k}}}{\partial t} = \int w_{\mathbf{p}}^c (\mathbf{k}, \mathbf{k}_1) \left(N_{\mathbf{k}_1} N_{\mathbf{k}} f_{\mathbf{p}+\mathbf{k}-\mathbf{k}_1}^R - N_{\mathbf{k}} N_{\mathbf{k}_1} f_{\mathbf{p}}^R \right) d\mathbf{p} \, d\mathbf{k}_1 (2\pi)^{-6} \simeq$$

$$\simeq N_{\mathbf{k}} \int N_{\mathbf{k}_1} w_{\mathbf{p}}^c (\mathbf{k}, \mathbf{k}_1) \frac{d\mathbf{p} \, d\mathbf{k}_1}{(2\pi)^6} \left((\mathbf{k} - \mathbf{k}_1) \frac{\partial f_{\mathbf{p}}^R}{\partial \mathbf{p}} \right). \qquad (2.88)$$

By comparing with (2.87), one can now find the probability of this process:

$$w_{\mathbf{p}}^c (\mathbf{k}, \mathbf{k}_1) = \frac{4 \left| \Lambda_{\mathbf{k}, \mathbf{k}_1}^c \right|^2 \delta \left(\omega_{\mathbf{k}}^l - \omega_{\mathbf{k}_1}^l - (\mathbf{k} - \mathbf{k}_1) \mathbf{v} \right) (2\pi)^9}{\left. \dfrac{\partial \varepsilon_{\mathbf{k}}^l}{\partial \omega} \right|_{\omega = \omega_{\mathbf{k}}^l} \left. \dfrac{\partial \varepsilon_{\mathbf{k}_1}^l}{\partial \omega_1} \right|_{\omega_1 = \omega_{\mathbf{k}_1}^l} (\omega_{\mathbf{k}}^l)^2}, \qquad (2.89)$$

where

$$\Lambda_{\mathbf{k}, \mathbf{k}_1}^c = \frac{1}{m_e} \cdot \frac{(\mathbf{k}\mathbf{k}_1)}{k k_1} \cdot \frac{e^3 (2\pi)^{-3} \omega_{\mathbf{k}}^l}{\left(\omega_{\mathbf{k}}^l - \mathbf{k}\mathbf{v} \right)^2}. \qquad (2.90)$$

This same expression for the probability can be computed directly by calculating the intensity of the waves emitted with frequency $\omega_{\mathbf{k}_1}^l$ due to the vibration of a plasma charge under the influence of the wave with frequency $\omega_{\mathbf{k}}^l$ [59-62].

Of course, Compton scattering is not the only mechanism for the scattering of turbulent fluctuations by plasma particles. The other type of scattering has been named nonlinear scattering [59] and is interpreted as the transition emission[†] of the wave $\omega_{\mathbf{k}}^l$ by the inhomogeneities created in the plasma by the wave $\omega_{\mathbf{k}_1}^l$. These effects are described by the second term of Σ^{eff} in (2.72). The

---

[†]See [63-65] for more on transition emission.

third and fourth terms† of (2.76) give

$$v^N_{k,\,k_1} = \cfrac{16}{\omega^l_k \left.\cfrac{\partial \varepsilon^l_k}{\partial \omega}\right|_{\omega = \omega^l_k} \left.\cfrac{\partial \varepsilon^l_{k_1}}{\partial \omega_1}\right|_{\omega_1 = \omega^l_{k_1}}} \times$$

$$\times\; \mathrm{Im}\left\{ \frac{1}{(\omega - \omega_1)(\varepsilon^l_{k-k_1})}(S_{k-k_1,\,k,\,-k_1} + S_{k-k_1,\,-k_1,\,k}) \times \right.$$

$$\left. \times (S_{k,\,k-k_1,\,k_1} + S_{k,\,k_1,\,k-k_1}) \right\}\Bigg|_{\substack{\omega = \omega^l_k \\ \omega_1 = \omega^l_{k_1}}}. \qquad (2.91)$$

$v^N_{k,\,k_1}$ contains terms proportional to $\mathrm{Im}\,(\varepsilon^l_{k-k_1})^{-1}$. As has been shown, it is this term which leads to induced decay if $\varepsilon^l_{k-k_1}$ is nearly zero, i.e., if the frequency of the wave for $k - k_1$ is close to one of the eigenfrequencies of the turbulent fluctuations.

We shall assume that the wave $k - k_1$ is far removed from any of the eigenfrequencies, that is, it lies in the region where the plasma is opaque. This occurs when the difference between the frequencies of the interacting waves is small (this is always true for Langmuir waves):

$$\mathrm{Im}\,\frac{1}{\varepsilon^l_{k-k_1}} = -\frac{\mathrm{Im}\,\varepsilon^l_{k-k_1}}{|\varepsilon_{k-k_1}|^2} =$$

$$= \sum_\alpha \frac{4\pi^2 e^2}{|k-k_1|^2\,|\varepsilon^l_{k-k_1}|^2} \int \left((k-k_1)\frac{\partial f^R_\alpha}{\partial p_\alpha}\right)\delta\,(\omega^l_k - \omega^l_{k_1} - (k-k_1)\,v_\alpha)\,\frac{dp_\alpha}{(2\pi)^3}.$$

$$(2.92)$$

We have made use of Eq. (2.41) here.

Equation (2.92) makes it clear that, just as for v^c_{k,k_1}, the term of interest in (2.90) is proportional to a delta function which describes the scattering of the fluctuations by the plasma particles. Since $\omega - \omega_1$ lies in the opaque region, the current is dissipated. Taking into account that the imaginary part is determined by **resonances** due to scattering, we find directly from (2.65) and (2.66)

---

†Again, the first and second terms describe two-wave emission processes.

that

$$S_{k-k_1,\,k,\,-k_1} = \frac{\omega - \omega_1}{\omega}\, S_{k,\,k-k_1,\,k_1}, \tag{2.93}$$

where S_{k,k_1,k_2} is the nonlinear current, symmetrized in the last two indices:

$$S_{k,\,k_1,\,k_2} = S_{k,\,k_2,\,k_1}. \tag{2.94}$$

Therefore, (2.91) can be written as

$$v^N_{k,\,k_1}\left(\frac{64}{(\omega^l_k)^2\,\dfrac{\partial \varepsilon^l_k}{\partial \omega}\bigg|_{\omega=\omega^l_k}\,\dfrac{\partial \varepsilon^l_{k_1}}{\partial \omega_1}\bigg|_{\omega_1=\omega^l_{k_1}}}\right)^{-1} =$$

$$= \mathrm{Im}\,\frac{1}{\varepsilon^l_{k-k_1}}\,(S_{k,\,k-k_1,\,k_1})^2 = -\,\frac{\mathrm{Im}\,\varepsilon^l_{k-k_1}}{|\,\varepsilon^l_{k-k_1}\,|^2}\,|\,S_{k,\,k-k_1,\,k_1}\,|^2 +$$

$$+ 2\,(\mathrm{Im}\,S_{k,\,k-k_1,\,k_1})\,\mathrm{Re}\,\frac{1}{\varepsilon^l_{k-k_1}}\,S_{k,\,k-k_1,\,k_1}. \tag{2.95}$$

To obtain this result we have employed the identity transformations which allow the result to be expressed as the sum of two terms. The first term gives the nonlinear scattering and the second gives the interference between the nonlinear scattering and the Compton scattering. One is convinced of the first assertion by using (2.92) in the first term of (2.95). We obtain an additional term in the probability for scattering by electrons:

$$w^N_p(k,\,k_1) = \frac{4\,|\,\Lambda^N_{k,\,k_1}\,|^2\,\delta\,(\omega^l_k - \omega^l_{k_1} - (k-k_1)\,v)\,(2\pi)^9}{\left|\dfrac{\partial \varepsilon^l_k}{\partial \omega}\right|_{\omega=\omega^l_k}\left|\dfrac{\partial \varepsilon^l_{k_1}}{\partial \omega_1}\right|_{\omega_1=\omega^l_{k_1}}(\omega^l_k)^2}, \tag{2.96}$$

where in this case

$$\Lambda^N_{k,\,k_1} = \sum_a \frac{4\pi e^2}{kk_1}\int \frac{dp_a}{(2\pi)^3}\,\frac{e^2(2\pi)^{-3}\omega^l_k}{\omega^l_k - kv_a + i\,\delta}\times$$

$$\times\left\{\left(k_1\frac{\partial}{\partial p_a}\right)\frac{1}{\omega^l_k - \omega^l_{k_1} - (k-k_1)\,v_a + i\,\delta}\left((k-k_1)\frac{\partial}{\partial p_a}\right)+\right.$$

$$+ \left((\mathbf{k} - \mathbf{k}_1) \frac{\partial}{\partial p_\alpha} \right) \frac{1}{\omega_{k_1}^l - \mathbf{k}_1 \mathbf{v}_\alpha} \left(\mathbf{k}_1 \frac{\partial}{\partial p_\alpha} \right) \bigg\} f_\alpha^R \frac{1}{|\mathbf{k} - \mathbf{k}_1|^2 \, \varepsilon_{k - k_1}^l}. \tag{2.97}$$

Note that the probability in (2.97) contains just the nonlinear current. The nonlinear scattering can be found as the emission intensity of the nonlinear current

$$j = 2 \int S_{k, \, k_1, \, k_2} \, E_{k_1}^T \, E_{k_2}^l \, dk_1 \, dk_2 \, \delta \, (k - k_1 - k_2), \tag{2.98}$$

set up by the turbulent field of the incident wave and by the longitudinal field of a uniformly moving charge [62]†:

$$E_k^l = - \frac{4\pi i}{k \varepsilon_k^l (2\pi)^3} \, \delta \, (\omega - kv). \tag{2.99}$$

The fact that the second term of (2.95) describes the interference of the nonlinear and Compton scatterings is most easily seen by using

$$\operatorname{Im} S_{k, \, k - k_1, \, k_1} = - \frac{e^3 \, \omega \pi \, (\mathbf{kk}_1)}{2mkk_1 \, |\mathbf{k} - \mathbf{k}_1|} \times$$

$$\times \int \frac{\delta \, (\omega - \omega_1 - (\mathbf{k} - \mathbf{k}_1) \, \mathbf{v})}{(\omega - kv)^2} \left((\mathbf{k} - \mathbf{k}_1) \frac{\partial f^R}{\partial \mathbf{p}} \right) \frac{d\mathbf{p}}{(2\pi)^3}. \tag{2.100}$$

The total scattering probability can be written in the form of (2.96), where $\Lambda_{k, \, k_1}^N$ is replaced by

$$\Lambda_{\mathbf{k}, \, \mathbf{k}_1} = \Lambda_{\mathbf{k}, \, \mathbf{k}_1}^c + \Lambda_{\mathbf{k}, \, \mathbf{k}_1}^N. \tag{2.101}$$

Thus, $\Lambda_{\mathbf{k}, \, \mathbf{k}_1}$ plays the part of the scattering amplitude. Both this interpretation and presence of the interference effects, which can dramatically lower the total scattering cross section, were presented in [66, 67]. As pointed out in [66], it is possible to have $|\Lambda_{\mathbf{k}, \mathbf{k}_1}^c| \ll |\Lambda_{\mathbf{k}, \mathbf{k}_1}^N|$ for scattering of high-frequency fluctuations on ions since the ion masses are so large. Thus, in some cases the cross section for scattering on ions can be much greater than that for scattering by electrons.

---

†The field of a uniformly moving charge is easily found from the equation

$$ik \varepsilon_k^l E_k^l = 4\pi \rho_k = \frac{4\pi e}{(2\pi)^3} \, \delta \, (\omega - kv).$$

A graphic picture of nonlinear scattering arises by noting
that in a plasma a screening cloud of charge is formed around the
scattering charge (the screening cloud carries the opposite sign
of charge). The scattering electron is surrounded by a cloud of
positive charge. It can be created by a lack of electrons since it
is mainly electrons which move under the influence of a high-
frequency wave. The screening charge has a sign which is opposite to that of the scattering charge and the waves scattered from
this screening charge decrease the total cross section. Moreover,
the oscillations of the screening charge cloud and the scattering
charge are evidently in phase; this explains the interference effects
in the scattering.

Only the electrons in the screening cloud can participate in
scattering of high-frequency waves by ions. Thus, the ion scattering is sizable and the interference effects are now missing.

This interpretation shows that processes occurring in a turbulent plasma involve not just the individual charge, but rather a
group of charges, or better still, the collective excitations of the
plasma. As a consequence, in place of plasma electrons we have
electrons which are surrounded by a charged cloud, and instead
of ions we have ions enveloped in an electron cloud. But we shall
still call them electrons and ions of the turbulent plasma (in order
to be brief), always keeping in mind what this designation really
implies. The "real" electrons and ions participate in both the
turbulent pulsations and the average motions.

This discussion enables us to better understand the physical
sense behind our separation of the distribution function into a
regular part and a turbulent part. The regular part has a very
transparent interpretation: f^R characterizes the distribution of
collective electron and ion excitations, that is, the distribution
of charges without the "fur coat" of other charges. The turbulent
part f^T characterizes the turbulent collective degrees of freedom — the plasmons. The possibility of introducing concepts like
collective excitations has been completely validated for systems
either close to thermodynamic equilibrium or in total thermodynamic equilibrium [50, 51]. The new assertion contained in our
interpretation just presented is that the concept of collective excitations can be extended to systems which are far from a state
of thermodynamic equilibrium. The physical notion of describing
the collective interactions of turbulent fluctuations in terms of

collective excitations was first put forth in [68] and developed
further in [60-62, 69, 70]. We will develop fully the general theory
of turbulence, using the idea of collective excitations, at a later
point; therefore it is not necessary here to generalize the results
pertaining to the particular case of the interaction of longitudinal
turbulent fluctuations. We only point out now that the similarity
with the collective excitations of a system nearly in equilibrium
is only present in the balance equations obtained in §2.3. The
essential physical difference between the notions of collective
excitations in the theory of turbulent plasma and in systems which
are close to statistical equilibrium lies in the non-single-valued
relationship between the frequencies and wave numbers in a turbulent plasma. The correlation effects describe this non-single-
valuedness.

2.5. The Correlation of Turbulent

Fluctuations in Plasma

The equations obtained in §2.4 are best described as balance
equations or kinetic equations which depict the excitation and interaction of plasmons. They are also suitable for nonstationary turbulence. In formulating the balance equations it was convenient to
determine those combinations of quantities which contained the
imaginary part ε_k^i [see Eq. (2.62)], which characterizes processes
leading to a change in the number of plasmons.

The effects of correlation between the turbulent fluctuations
are given by the dependence of $I_{k,\omega}$ on ω. In the linear approximation the time correlation (which depends on ω) is uniquely
coupled to the spatial correlation (which depends on k), i.e.,
$I_{k,\omega} = I_k \delta(\omega - \omega_k)$. We emphasize that, both in this case and the
other, the question revolves around the dependence of the average
value of the fields $\langle E(r_1, t_1)E(r_2 t_2) \rangle$ on $t_2 - t_1$ and $r_2 - r_1$. But the
dependence on $(r_1 + r_2)/2$ and $(t_1 + t_2)/2$ is taken into account in the
form of a slow dependence on the average coordinates and the
average time $\frac{\partial}{\partial t} + v_g \frac{\partial}{\partial r}$.

It turns out that one can neglect correlation in the balance
equations, i.e., we set $I_{k,\omega} \approx I_k \delta(\omega - \omega_k)$. This is possible because the equation contains just small terms (of the order of
γ_k/ω_k), and because one takes the integral $\int I_{k,\omega} d\omega$ over all frequencies [see Eq. (2.64)] and it should not be very sensitive to the
detailed form of the distribution of the fluctuations over ω. Under

these conditions a more precise treatment of the correlations gives small corrections which are beyond the limits of accuracy for the equations we started with (for we have thrown away the nonlinear currents above third order in the field).

Thus, we may say that, to the same degree of precision, these equations can be written as equations for $\int I_{k,\omega} \, d\omega = I_k$, which are the quantities which describe the spectra for the turbulent fluctuations. That is, one can look for the turbulence spectrum without being concerned about the nature of the time correlations for the fluctuations.

The field correlations are measured in most experiments on plasma turbulence, and are in fact one of their most important characteristics. In turbulent plasma one loses the strict one-to-one correspondence between fluctuation frequency and wave number, i.e., for each k there is now a spectrum of frequencies occupying some region around $\omega = \omega_k$. For weak turbulence the width of the line $\Delta\omega$ is always much less than ω_k and the characteristic correlation time is of the order $\Delta\tau \sim 1/\Delta\omega$ (from the general properties of the Fourier transform).

In order to take into account the correlation, we write down the equation for I_k assuming that the turbulence is stationary $\left(\frac{\partial}{\partial t} + v_g \frac{\partial}{\partial r} = 0\right)$. The nonstationary correlation is easily included by using the results of §2.4. To maintain the greatest generality we shall not specify the dependence of the coefficients of the nonlinear turbulent currents on the regular components of the particle velocities, their distribution, and so on. This approach will provide us with general equations in which the nonlinear currents can be represented as solutions to any specific dynamic equations we like (such as the two-fluid hydrodynamic equations or the kinetic equations which involve pair-wise particle collisions). This helps to clarify the influence of certain factors, such as the pair collisions, on the nonlinear interactions and correlations of turbulent fluctuations.

Under conditions of stationary turbulence, we have

$$\langle E^T_{k,\, i} E^{T*}_{k',\, l} \rangle = I_{k,\, il}\, \delta\,(k - k').$$ (2.102)

Let us consider longitudinal turbulent fluctuations where

$$I_{k,\, ij} = I_k \frac{k_i\, k_j}{k^2}\,.$$

We can write the general expression for the nonlinear turbulent current which is second order in the turbulent field as

$$j_k^{T\,(2)} = \int S_{k,\,k_1,\,k_2} \left(E_{k_1}^T E_{k_2}^T - \langle E_{k_1}^T E_{k_2}^T \rangle \right) dk_1\, dk_2\, \delta\,(k - k_1 - k_2). \qquad (2.103)$$

Here $j_k = (\mathbf{k} j_k)/k$; $E_k = (\mathbf{k} E_k)/k$. The term $\langle E_{k_1}^T E_{k_2}^T \rangle$ is required on the right side of (2.103) in order to satisfy the requirement that $\langle j_k^{T\,(2)} \rangle = 0$.

S_{k,k_1,k_2} can always be symmetrized in the variables k_1, k_2:

$$S_{k,\,k_1,\,k_2} = S_{k,\,k_2,\,k_1}. \qquad (2.104)$$

In addition, S_{k,k_1,k_2} satisfies the obvious criterion that

$$S_{k,\,k_1,\,k_2}^{*} = - S_{-k,\,-k_1,\,-k_2}, \qquad (2.105)$$

which follows from the reality of the current j_k and the fields.

The general expression for the current in the next approximation in the turbulent field, satisfying the equation $\langle j_k^{T\,(3)} \rangle = 0$, takes the form

$$j_k^{T\,(3)} = \int \Big\{ \Sigma_{k,\,k_1,\,k_2,\,k_3} \left(E_{k_1}^T E_{k_2}^T E_{k_3}^T - \langle E_{k_1}^T E_{k_2}^T E_{k_3}^T \rangle \right) +$$
$$+ G^{(1)}_{k,\,k_1,\,k_2,\,k_3} E_{k_1}^T \langle E_{k_2}^T E_{k_3}^T \rangle + G^{(2)}_{k,\,k_1,\,k_2,\,k_3} E_{k_2}^T \langle E_{k_1}^T E_{k_3}^T \rangle +$$
$$+ G^{(3)}_{k,\,k_1,\,k_2,\,k_3} E_{k_3}^T \langle E_{k_1}^T E_{k_2}^T \rangle \Big\} \times$$
$$\times dk_1\, dk_2\, dk_3\, \delta\,(k - k_1 - k_2 - k_3). \qquad (2.106)$$

In analogy with (2.104), we can assume that the following relations are satisfied:

$$\Sigma_{k,\,k_1,\,k_2,\,k_3} = \Sigma_{k,\,k_2,\,k_1,\,k_3} = \Sigma_{k,\,k_2,\,k_3,\,k_1} =$$
$$= \Sigma_{k,\,k_1,\,k_3,\,k_2} = \Sigma_{k,\,k_3,\,k_1,\,k_2} = \Sigma_{k,\,k_3,\,k_2,\,k_1}. \qquad (2.107)$$

By changing variables in the second term of (2.106) we can quite easily reduce it to

$$- \int \Sigma'_{k,\,k_1,\,k_2,\,k_3} E_{k_1}^T \langle E_{k_2}^T E_{k_3}^T \rangle \cdot dk_1\, dk_2\, dk_3\, \delta\,(k - k_1 - k_2 - k_3), \qquad (2.108)$$

where

$$\sum{}'_{k,\,k_1,\,k_2,\,k_3} = -\,G^{(1)}_{k,\,k_1,\,k_2,\,k_3} -G^{(2)}_{k,\,k_2,\,k_1,\,k_3} -G^{(3)}_{k,\,k_3,\,k_2,\,k_1}. \tag{2.109}$$

By comparing (2.106) with (2.67) we see that $\Sigma' = \Sigma$ for a collisionless kinetic equation. Using Maxwell's equation for a turbulent field,

$$-\omega^2\,\varepsilon^l_k\,E^T_k = 4\pi i\omega\,(j^{T\,(2)}_k + j^{T\,(3)}_k), \tag{2.110}$$

multiplying it by $E^T_{k'}$, and integrating over k', we find

$$\omega^2\,\varepsilon^l_k\,I_k = 4\pi i\omega \int dk'\, S_{k,\,k_1,\,k_2} \langle E^T_{k'},$$

$$E^T_{k_1},\ E^T_{k_2} \rangle\, dk_1\,dk_2\,\delta\,(k-k_1-k_2) + 4\pi i\omega \int dk'\,dk_1\,dk_2\,dk_3\,\delta\,(k-k_1-k_2-k_3)\,\times$$

$$\times\,(\textstyle\sum_{k,\,k_1,\,k_2,\,k_3} \langle E^T_{k'}\,E^T_{k_1}\,E^T_{k_2}\,E^T_{k_3}\rangle - \langle E^T_{k'}\,E^T_{k_1}\rangle\,\langle E^T_{k_2}\,E^T_{k_3}\rangle\,\textstyle\sum'_{k,\,k_1,\,k_2,\,k_3}). \tag{2.111}$$

The average of four fields can be approximately split into three pairwise products, whereas the averages over three fields must be more carefully evaluated by using the following approximate relationship:

$$E^T_k = -\frac{4\pi i}{\omega \varepsilon^l_k} \int S_{k,\,k_1,\,k_2}\,(E^T_{k_1}\,E^T_{k_2} - \langle E^T_{k_1}\,E^T_{k_2}\rangle)\ \delta\,(k-k_1-k_2)\,dk_1\,dk_2. \tag{2.112}$$

This leads to the equation $\langle E^T_k\,E^T_{k'}\rangle = -I_k\,\delta\,(k+k')$, so that

$$\varepsilon^l_k\,I_k = +\frac{8\pi i}{\omega}\Big\{ I_k \int \textstyle\sum^{\text{eff}}_{k,\,k_1}\,I_{k_1}\,dk_1 - 4\pi i \int I_{k_1}\,I_{k_2}\,|S_{k,\,k_1,\,k_2}|^2\,\frac{dk_1\,dk_2\,\delta\,(k-k_1-k_2)}{(\omega_1+\omega_2)\,\varepsilon^l_{-k_1-k_2}} \Big\}. \tag{2.113}$$

Here we have included the fact that [from (2.67)]

$$\textstyle\sum_{k,\,k_1,\,k_2,\,k_3} = \textstyle\sum'_{k,\,k_1,\,k_2,\,k_3}; \tag{2.114}$$

$$\textstyle\sum^{\text{eff}}_{k,\,k_1} = \frac{1}{2}\,(\textstyle\sum^{\text{eff}}_{k,\,k_1,\,k,\,-k_1} + \textstyle\sum^{\text{eff}}_{k,\,k_1,\,-k_1,\,k}) - \frac{8\pi i S_{k,\,k-k_1,\,k_1}\,S_{k-k_1,\,k,\,-k_1}}{(\omega-\omega_1)\,\varepsilon^l_{k-k_1}}; \tag{2.115}$$

$$S_{k-k_1,\,k,\,-k_1} = \frac{1}{2}\,(S_{k-k_1,\,k,\,-k_1} + S_{k-k_1,\,-k_1,\,k}). \tag{2.116}$$

Equation (2.113) is more general than (2.75)[†] and is integrated. Specific expressions for the correlation functions of the turbulent fluctuations can be found from (2.113) if one knows S and Σ for the particular types of collective motions.

We must mention that (2.113) is a convenient expression for finding the balance equation and studying the correlation effects in the tails of the correlation curves, that is, for ω much different from the ω_k^l in the linear theory. But near a resonance such that $\omega - \omega_k^l$ is almost zero it must be computed in a more precise manner. For because $\mathrm{Re}\,\varepsilon^l(\omega_k^l, \mathbf{k}) = 0$ near a resonance, the left side of (2.114) now takes the form

$$\omega \left.\frac{\partial \varepsilon_k^l}{\partial \omega}\right|_{\omega=\omega_k^l} \left((\omega - \omega_k^l)\,I_k + i\gamma_k^l\,I_k\right).$$

Thus, $\omega - \omega_k^l$ is proportional to I [according to the right side of (2.113)]. I (or more exactly the dimensionless quantity W/nT) is a small parameter so that the correlation width is much smaller than ω_k^l, which corresponds to the weak-turbulence approximation. However, the last term on the right side of (2.113) contains $1/\varepsilon_{-k_1-k_2}^l$, which is equal to $1/\varepsilon_{-k}^l$ because $k_1 + k_2 = k$. For $\omega \approx \omega_k^l$, $1/\varepsilon_{-k}^l \sim 1/\omega - \omega_k^l \sim W^{-1}$. This simply indicates that the last term in (2.113) was not computed with sufficient accuracy.

Let us now return to Eqs. (2.111) and (2.112). We shall use the following physical argument.[‡] Let the mean frequency width which characterizes the correlations be $\Delta\omega$.[§] We shall employ the small parameter $\Delta\omega/\omega_k^l \ll 1$. We separate the fields for the low frequency ω_L, which have frequencies of order $\Delta\omega$, from the

---

[†]In order to get the right side of (2.75) from (2.113), one must take the imaginary part of (2.114), divide by $\omega \dfrac{\partial \varepsilon_k^l}{\partial \omega}$ and integrate the result over ω. By using the fact that

$$I_k = \frac{1}{2\pi^2 \left.\dfrac{\partial \varepsilon_k^l}{\partial \omega}\right|_{\omega = \omega_k^l}} \cdot \left(N_k \delta\left(\omega - \omega_k^l\right) + N_{-k}\,\delta\left(\omega + \omega_k^l\right)\right),$$

one is easily convinced that the equation thus obtained is identical with (2.75).

[‡]Justification may also be obtained through use of graphical techniques [54].

[§]$\Delta\omega$ can be the characteristic half-width of the correlation lines, i.e., it indicates the frequency shift needed to drop the intensity by a factor of 2.

high-frequency fields which have frequencies of the same order as the turbulent fluctuations under investigation. We will assume that the high-frequency fields decay, so that for the three high-frequency fields we have the condition $\mathbf{k} = \mathbf{k_1} + \mathbf{k_2}$, $\omega = \omega_1 + \omega_2$. Since the interaction, which is proportional to $\delta(k - k_1 - k_2)$, contains both positive and negative frequencies, it must correspond to both the sum and difference of the frequencies ω_1 and ω_2. The low-frequency fields can arise only when the frequencies for the high-frequency fields are of opposite signs. Therefore, in the first approximation, we can write for the low-frequency fields that

$$E_k^{TL} = -\frac{4\pi i}{\omega \varepsilon_k^l} \int S_{k,\,k_1,\,k_2} \left(E_{k_1}^{TH} E_{k_2}^{TH} - \langle E_{k_1}^{TH} E_{k_2}^{TH} \rangle \right) dk_1 \, dk_2 \, \delta(k - k_1 - k_2).$$

(2.117)

The problem in the theory is to produce an equation which contains just E_k^{TH} (or more precisely, the average of its quadratic combinations). We can therefore exclude the fields E_k^{TL}.

The mean value of $\langle E_{k'}^T E_{k_1}^T E_{k_2}^T \rangle$, which appears in (2.111), can be rewritten in the following form by assuming that the field $E_{k'}^T$ can be only a high-frequency field:

$$\langle E_{k'}^T E_{k_1}^T E_{k_2}^T \rangle = \langle E_{k'}^{TH} E_{k_1}^{TH} E_{k_2}^{TH} \rangle + \langle E_{k'}^{TH} E_{k_1}^{TL} E_{k_2}^{TH} \rangle + \langle E_{k'}^{TH} E_{k_1}^{TH} E_{k_2}^{TL} \rangle. \quad (2.118)$$

The last two terms of (2.118), transformed with the help of (2.117), give terms which depend on S in Σ^{eff} [see Eq. (2.115)]. The first term of (2.118) immediately gives that feature of $1/\varepsilon_{-k}^l$ which we were concerned about above. We now write the equation for $E_{k'}^{TH}$ more exactly:

$$\varepsilon_{k'}^l E_{k'}^{TH} = -\frac{4\pi i}{\omega'} \int S_{k',\,k_1',\,k_2'} \left(E_{k_1'}^{TH} E_{k_2'}^{TH} + E_{k_1'}^{TH} E_{k_2'}^{TL} + E_{k_1'}^{TL} E_{k_2'}^{TH} - \right.$$
$$\left. - \langle E_{k_1'}^{TH} E_{k_2'}^{TH} + E_{k_1'}^{TH} E_{k_2'}^{TL} + E_{k_1'}^{TL} E_{k_2'}^{TH} \rangle \right) \delta(k' - k_1' - k_2') dk_1' dk_2' -$$
$$- \frac{4\pi i}{\omega'} \int dk_1' \, dk_2' \, dk_3' \, \Sigma_{k',\,k_1',\,k_2',\,k_3'} \left(E_{k_1'}^{TH} E_{k_2'}^{TH} E_{k_3'}^{TH} - \right.$$
$$\left. - E_{k_1'}^{TH} \langle E_{k_2'}^{TH} E_{k_3'}^{TH} \rangle - \langle E_{k_1'}^{TH} E_{k_2'}^{TH} E_{k_3'}^{TH} \rangle \right) \times$$
$$\times \delta(k' - k_1' - k_2' - k_3'). \quad (2.119)$$

In the last term we have neglected the low-frequency fields, which, according to (2.117), give higher powers of the high-frequency fields. Multiplying (2.119) by $E_{k_1}^{TH} E_{k_2}^{TH}$ and calculating the average value of $\langle E_{k'}^{TH} E_{k_1}^{TH} E_{k_2}^{TH} \rangle$, we obtain

$$\epsilon_{k'}^l \langle E_{k'}^{TH} E_{k_1}^{TH} E_{k_2}^{TH} \rangle = -\frac{4\pi i}{\omega'} \int S_{k', \, k_1', \, k_2'} \, dk_1' dk_2' \times$$

$$\times \left(\left\langle E_{k_1}^{TH} E_{k_2}^{TH} E_{k_1'}^{TH} E_{k_2'}^{TH} \right\rangle - \langle E_{k_1}^{TH} E_{k_2}^{TH} \rangle \left\langle E_{k_1'}^{TH} E_{k_2'}^{TH} \right\rangle \right) \times$$

$$\times \delta \left(k' - k_1' - k_2' \right) - \frac{8\pi i}{\omega'} \int S_{k', \, k_1', \, k_2'} \, dk_1', dk_2' \left(\left\langle E_{k_1}^{TH} E_{k_2}^{TH} E_{k'1}^{TH} E_{k_2'}^{TL} \right\rangle - \right.$$

$$\left. - \langle E_{k_1}^{TH} E_{k_2}^{TH} \rangle \left\langle E_{k_1'}^{TH} E_{k_2'}^{TL} \right\rangle \right) \delta \left(k' - k_1' - k_2' \right) -$$

$$- \frac{4\pi i}{\omega'} \int \mathbf{\Sigma}_{k', \, k_1', \, k_2', \, k_3'} \left(\left\langle E_{k_1}^{TH} E_{k_2}^{TH} E_{k_1'}^{TH} E_{k_2'}^{TH} E_{k_3'}^{TH} \right\rangle - \right.$$

$$- \langle E_{k_1}^{TH} E_{k_2}^{TH} \rangle \left\langle E_{k_1'}^{TH} E_{k_2'}^{TH} E_{k_3'}^{TH} \right\rangle - \left\langle E_{k_2'}^{TH} E_{k_3'}^{TH} \right\rangle \left\langle E_{k_1}^{TH} E_{k_2}^{TH} E_{k_1'}^{TH} \right\rangle \right) \times$$

$$\times \delta \left(k' - k_1' - k_2' - k_3' \right) dk_1' dk_2' dk_3'.$$

We have used the symmetry property of S_{k, k_1, k_2}. Splitting the average of four fields into pairs and substituting (2.117), we find that

$$\epsilon_{k'}^l \langle E_{k'}^{TH} E_{k_1}^{TH} E_{k_2}^{TH} \rangle = -\frac{8\pi i}{\omega'} S_{k', \, -k_1, \, -k_2} \, \delta \left(k' + k_1 + k_2 \right) I_{k_1} I_{k_2} -$$

$$- \frac{4\pi i}{\omega'} \int \mathbf{\Sigma}_{k', \, k_1', \, k_2', \, k_3'}^{\text{eff}} \left(\left\langle E_{k_1}^{TH} E_{k_2}^{TH} E_{k_1'}^{TH} E_{k_2'}^{TH} E_{k_3'}^{TH} \right\rangle - \right.$$

$$- \langle E_{k_1}^{TH} E_{k_2}^{TH} \rangle \left\langle E_{k_1'}^{TH} E_{k_2'}^{TH} E_{k_3'}^{TH} \right\rangle - \left\langle E_{k_1}^{TH} E_{k_2}^{TH} E_{k_1'}^{TH} \right\rangle \times$$

$$\times \left\langle E_{k_2'}^{TH} E_{k_3'}^{TH} \right\rangle \right) \delta \left(k' - k_1' - k_2' - k_3' \right) dk_1' dk_2' dk_3'. \qquad (2.120)$$

Here $\mathbf{\Sigma}_{k, \, k_1, \, k_2, \, k_3}^{\text{eff}}$ corresponds to a symmetrized version of (2.72) [see (2.115)].

One must now approximate the average of the field products by the possible products of averages of three fields and two fields (the remaining divisions contain $\langle E^T \rangle$, which is zero by definition). Thus, for those terms containing Σ^{eff} in the right side of (2.120) we find

$$\frac{8\pi i}{\omega'} \langle E_{k'}^{TH} E_{k_1}^{TH} E_{k_2}^{TH} \rangle \int \mathbf{\Sigma}_{k' \, k_1'}^{\text{eff}} I_{k_1'} \, dk_1' +$$

$$+ \frac{4\pi i}{\omega'} I_{k_1} \int \widetilde{\mathbf{\Sigma}}_{k', \, -k_1, \, k_2', \, k_3'}^{\text{eff}} \left\langle E_{k_2}^{TH} E_{k_2'}^{TH} E_{k_3'}^{TH} \right\rangle \delta \left(k' + \right.$$

$$+ k_1 - k_2' - k_3') \, dk_2' \, dk_3' + \frac{4\pi i}{\omega'} I_{k_2} \int \widetilde{\Sigma}^{\text{eff}}_{k', -k_2, k_2', k_3'} \times$$

$$\times \left\langle E^{TH}_{k_1} E^{TH}_{k_2'} E^{TH}_{k_3'} \right\rangle \delta \left(k' + k_2 - k_2' - k_3' \right) dk_2' \, dk_3'. \qquad (2.121)$$

Here $\Sigma^{\text{eff}}_{kk_1}$ corresponds to (2.115) and

$$\widetilde{\Sigma}^{\text{eff}}_{k', k_1, k_2, k_3} = \Sigma^{\text{eff}}_{k', k_1, k_2, k_3} + \Sigma^{\text{eff}}_{k', k_2, k_1, k_3} + \Sigma^{\text{eff}}_{k', k_3, k_2, k_1}.$$

We now introduce the nonlinear permeability as defined by the equation

$$\varepsilon^N_k = - \frac{8\pi i}{\omega} \int \Sigma^{\text{eff}}_{kk_1} I_{k_1} \, dk_1. \qquad (2.121')$$

It is clear from (2.120) and (2.121) that

$$\langle E^{TH}_{k'} E^{TH}_{k_1} E^{TH}_{k_2} \rangle = I_{k_1, k_2} \, \delta \left(k' + k_1 + k_2 \right),$$

where I_{k_1, k_2} satisfies the equation

$$I_{k_1, k_2} + \frac{4\pi i}{(\omega_1 + \omega_2) \left(\varepsilon^l_{-k_1 - k_2} + \varepsilon^N_{-k_1 - k_2} \right)} \times$$

$$\times \left\{ I_{k_1} \int dk_1' \, \widetilde{\Sigma}^{\text{eff}}_{-k_1 - k_2, -k_1, k_1', -k_2 - k_1'} I_{k_1', -k_2 - k_1'} + \right.$$

$$\left. + I_{k_2} \int dk_1' \, \widetilde{\Sigma}^{\text{eff}}_{-k_1 - k_2, -k_2, k_1', -k_1 - k_1'} I_{k_1', -k_1 - k_1'} \right\} =$$

$$= \frac{8\pi i \, S_{-k_1 - k_2, -k_1, -k_2}}{(\omega_1 + \omega_2)} I_{k_1} I_{k_2} \frac{1}{\left(\varepsilon^l_{-k_1 - k_2} + \varepsilon^N_{-k_1 - k_2} \right)}. \qquad (2.122)$$

This is an integral equation. Its solution is easily found with accuracy sufficient for our purposes. The fluctuation intensity is a small parameter (W/nT ≪ 1). It is necessary to include the terms ~I_k in the denominators of (2.122) because $\varepsilon^l_{-k_1 - k_2}$ is nearly zero. Thus, if the denominator contains $1/\varepsilon^l_{k''}$, where k'' is not equal to $k_1 + k_2$, then such a term does not have a singularity and one need not retain the corrections of order I . We can solve (2.122) by iteration. Neglecting the integrated term on the left as a first approximation, we have

$$I_{k_1, k_2} = \frac{8\pi i \, I_{k_1} I_{k_2} \, S_{-k_1, -k_2, -k_1, -k_2}}{(\omega_1 + \omega_2) \left(\varepsilon^l_{-k_1 - k_2} + \varepsilon^N_{-k_1 - k_2} \right)}. \qquad (2.123)$$

Substituting this expression into the integral term of (2.122), we are assured that only denominators like $1/\varepsilon_{k_2}$ arise, where k_2 is integrated out according to (2.122).[†] By induction we see that this result holds for the approximation that the solution is of the form of (2.123), where the numerator is a series in I_k. It is thus sufficient to keep just the first term in this series, which means that we shall use (2.123) as is. From the definition of ε_k^N (2.121') we have in place of (2.113) [71]

$$\omega\left(\varepsilon_k^l + \varepsilon_k^N\right) I_k = 32\pi^2 \int \frac{|S_{k,\,k_1,\,k_2}|^2 I_{k_1} I_{k_2}\, \delta\,(k - k_1 - k_2)}{\omega\left(\varepsilon_{-k}^l + \varepsilon_{-k}^N\right)}\, dk_1\, dk_2 \equiv R_k^N. \qquad (2.124)$$

This equation is the desired equation for the fluctuation correlations in the plasma. The balance equation is obtained from this expression in the same manner as it is derived from (2.113). Of course we now are faced with certain peculiarities, about which we shall have more to say later. Equation (2.124) is a complex integral equation for the correlation functions whose solution can only be found in special cases. We shall limit ourselves here to a few general comments. First, since $\varepsilon_{-k} = \varepsilon^*_k$, we can write (2.124) as

$$I_k = \frac{32\pi^2}{\omega^2} \int \frac{|S_{k,\,k_1,\,k_2}|^2 I_{k_1} I_{k_2}\, \delta\,(k - k_1 - k_2)\, dk_1\, dk_2}{|\varepsilon_k^l + \varepsilon_k^N|^2}. \qquad (2.125)$$

It is clear from this form of the equation that I_k is always positive, as it should be from its definition. Secondly, the numerator of (2.125) is a rather weak function of ω and $\mathbf{k}$ when $\omega \to \omega_k^l$. This is also true of ε_k^N. We may therefore state that the correlation function I_k has the form

$$I_k = \frac{\alpha_k}{(\omega - \omega_k^N) + (\gamma_k^N)^2}, \qquad (2.126)$$

where α_k, ω_k^N, γ_k^N are weak functions of ω and $\mathbf{k}$ when ω is nearly equal to ω_k^l. Using the expansions near resonance we can write

$$\omega_k^N = \omega_k^l - \frac{\operatorname{Re} \varepsilon_k^N}{\dfrac{\partial \varepsilon_k}{\partial \omega}\bigg|_{\omega = \omega_k^l}}; \qquad (2.127)$$

[†]As in the balance equation, the quantity $1/\varepsilon$ under the integral sign gives no singularity.

$$\gamma_k^N = -\frac{\text{Im}\left(\varepsilon_k^l + \varepsilon_k^N\right)}{\dfrac{\partial \varepsilon_k}{\partial \omega}\bigg|_{\omega = \omega_k^l}}.$$

$$(2.127')$$

Thus, there arises a nonlinear shift in the frequency and the spectrum width. The correlation width is characterized by γ_k^N. When estimating the magnitude of this quantity one should be aware of the fact that the balance equation indicates the compensation of the positive and negative γ_k, and the magnitude of γ_k^N can be much smaller than any of the individual terms which enter into it (both linear and nonlinear).

From the form of (2.126) one can obtain an important result concerning the effect of correlation on the interactions of turbulent fluctuations. Equation (2.125) describes a decay process [this is clear from the function $\delta(k - k_1 - k_2)$]. There is a wide class of fluctuations for which decays are weakly forbidden; small changes in the fluctuation frequencies can result in such interaction processes being allowed. For instance, almost all fluctuations with a nearly linear dispersion relation,

$$\omega_k \approx \text{const}\, k,$$

fall in this category.

If three waves are propagating in the same direction, the restriction that $k_1 + k_2 = k_3$ will necessitate that $\omega_{k_1} + \omega_{k_2} = \omega_{k_3}$. Such fluctuations include ion-acoustic, Alfven, and magnetohydrodynamic waves, among others. But since their spectra are not strictly linear, the decays are forbidden. But broadening the spectra of these fluctuations because of their correlations can make the decays allowed. Since $\Delta\omega \sim \gamma_k^N$, the requisite conditions for decay in (2.126) must be fulfilled to within an accuracy of $\sim \gamma_k^N$. In addition to the importance of the correlation effects, the frequency shifts can also play a major role [see (2.127)].

For turbulence which is strictly nondecaying one must either take into account decays of higher order in the turbulence energy, or one must include the correlation effects which are related to the pronounced dependence of ε_N^l on $\omega - \omega_k^l$ (see §4.6 for more detail).

In a number of applications it is very important to examine the possibility of turbulent fluctuations arising at low frequencies.

The general equation for the correlations, (2.124), was used to analyze the shape of the resonance curve close to the resonance frequency, where $\Delta\omega/\omega_k^l \ll 1$. But portions of the resonance curve can also be found at frequencies much less than ω_k^l. Such turbulent fluctuations will not, of course, have any unique connection between the frequencies and wave numbers. Their relative amplitudes are small. As a rough estimate, they are

$$\sim \frac{(\gamma_k^N)^2}{(\omega-\omega_k^N)^2+(\gamma_k^N)^2} \sim \frac{(\gamma_k^N)^2}{(\omega_k^N)^2} \quad \text{when } \omega \ll \omega_k^N.$$

However, the absolute value can be significantly greater than the thermal noise level if the oscillation intensity in the resonance region is large enough. In the resonance region the left side of (2.124) is small because $\varepsilon_k^l \approx 0$; the right side is small because it contains an extra power of the turbulence energy, which is small when the turbulence is weak. Outside the resonance ε_k^l is rather large, but the left side of (2.124) remains small because I_k is small. Thus, at low frequencies one can neglect ε_k^N and write

$$I_k^L = 32\pi^2 \int dk_1 \, dk_2 \cdot \frac{I_{k_1} I_{k_2} \, \delta(k-k_1-k_2)|S_{k,\,k_1,\,k_2}|^2}{(\omega_1+\omega_2)^2 |\varepsilon_{k_1+k_2}^l|^2}. \tag{2.128}$$

We must point out that the resonance curve need not be continuous, which means that there may be regions in which there are no fluctuations. This is the case for a strictly nondecaying spectrum such as the spectrum for Langmuir oscillations. If we consider frequencies ω which are either close to, or on the order of, ω_{pe}, then the integral in (2.128) is nearly zero. For frequencies equal to the difference between Langmuir oscillation frequencies, I_k^L becomes quite large because the maxima of the spectra for I_{k_1} and I_{k_2} now make contributions to the right side of (2.128). This effect of exciting fluctuations at frequencies equal to or less than the difference in the fluctuation frequencies has been observed experimentally [72]. This should be kept in mind when treating the problem of turbulent heating. The vast majority of the plasma particles may not be resonant with the high-frequency fluctuations near the center of the correlation curve of (2.126), but they can be resonant with the low-frequency fluctuations of (2.128). Thus, high-frequency fluctuations can heat the plasma by

exciting low-frequency oscillations. We shall show below that
the final result of this interaction in the balance equations for
particles, which also describe heating, is heating due to the induced scattering of plasmons by the plasma "particles" (§2.7)
when all terms of the same order are included.

Finally, for a strictly nondecaying spectrum the presence of
low-frequency fluctuations in (2.128) can effectively mean that the
right side of (2.125) is not equal to zero. For, if one of the I_k in (2.125)
is assumed to be low-frequency (2.128), and is substituted into
(2.125), we obtain a right side which is proportional to the third
power of the high-frequency oscillation energy. Evidently, including the nonlinear currents which were omitted will lead to effects
of the same order. It turns out that the sum of all the terms describes a four-plasmon interaction, that is, the process of plasmon scattering on plasmons (see §2.9 and §4.3).

2.6. The Quasilinear Approximation

Let us now study the equations for the regular part of the
distribution function, f^R. We assume that $E^R = 0$; Eq. (2.9) then
takes the form

$$\frac{\partial f^R}{\partial t} + \mathbf{v}\frac{\partial f^R}{\partial \mathbf{r}} = -\left\langle \mathbf{F}^T \frac{\partial f^T}{\partial \mathbf{p}}\right\rangle. \tag{2.129}$$

Now expand f^T in powers of the turbulent field $\mathbf{F}^T$. From
(2.17) the first approximation gives

$$i\,(\omega - \mathbf{k}\mathbf{v})f_k^{T(1)} = \int \left(\mathbf{F}_{k_1}^T \frac{\partial}{\partial \mathbf{p}}\right) f_{k_2}^R \, dk_1\, dk_2\, \delta\,(k - k_1 - k_2)$$

and in place of (2.129) we have

$$\frac{\partial f^R}{\partial t} + \mathbf{v}\frac{\partial f^R}{\partial \mathbf{r}} = -\left\langle \mathbf{F}^T \frac{\partial f^{T(1)}}{\partial \mathbf{p}}\right\rangle = -\left\langle \int \mathbf{F}_{k'}^T \cdot \frac{\partial f_k^{T(1)}}{\partial \mathbf{p}}\, e^{i\,(k+k')\,x}\, dk\, dk'\right\rangle =$$

$$= i \int\limits_{\delta \to +0} dk\, dk'\, \langle F_{k'}^T, {}_i F_{k_1}^T, {}_j\rangle \frac{\partial}{\partial p_i}\, \frac{1}{(\omega - \mathbf{k}\mathbf{v} + i\,\delta)} \cdot \frac{\partial f_{k_2}^R}{\partial p_j}\, \times$$

$$\times e^{i\,(k+k')\,x}\, \delta\,(k - k_1 - k_2)\, dk_1\, dk_2. \tag{2.130}$$

This equation, which includes the effect of turbulent fluctuations on the regular part of the distribution function in the first

approximation, is called a quasilinear equation [56]. It can be simplified if one assumes that f^R changes much more slowly than the turbulent fields, i.e., $f^R_{k_1}$ on the right of (2.130) can be approximately changed to $f^R \delta(k_2)$. Integrating over k_1 gives

$$\frac{\partial f^R}{\partial t} + \mathbf{v}\, \frac{\partial f^R}{\partial \mathbf{r}} = i\, \frac{\partial}{\partial p_i} \int \langle F^{T*}_{k',\,i}\, F^T_{k,\,i} \rangle \frac{e^{i\,(k-k')\,x}}{\omega - kv + i\delta}\, dk\, dk'\, \frac{\partial f^R}{\partial p_j}. \tag{2.131}$$

Equation (2.131) is already a differential equation. The right side describes the "collision" of the collective plasma excitations (overgrown with electrons and ions) with the turbulent fluctuations.

For the sake of simplicity we shall consider longitudinal turbulent fluctuations where

$$\langle F^{T*}_{k',\,i}\, F^T_{k,\,j} \rangle = \frac{k'_i\, k_j}{kk'}\, e^2 \langle E^{T*}_{k'}\, E^T_k \rangle.$$

Let us examine the quasilinear equation when $\gamma^l_k \ll \omega^l_k$. In the first approximation, assuming that the amplitudes of the turbulent fields are slowly varying, one can assume that

$$\langle E^{T*}_{k'}\, E^T_k \rangle \simeq I_k\, \delta(k-k').$$

Equation (2.131) then takes the new form

$$\frac{\partial f^R}{\partial t} + \mathbf{v}\, \frac{\partial f^R}{\partial \mathbf{r}} = i\, e^2\, \frac{\partial}{\partial p_i} \int \frac{k_i\, k_j}{k^2}\, dk\, I_k\, \frac{1}{\omega - kv + i\delta} \cdot \frac{\partial f^R}{\partial p_j}. \tag{2.132}$$

Because the right side is real, one must take into account just the imaginary part of the denominator:

$$\text{Im}\, \frac{1}{\omega - kv + i\delta} = -\pi\delta\,(\omega - kv).$$

We find that

$$\frac{\partial f^R}{\partial t} + \mathbf{v}\, \frac{\partial f^R}{\partial \mathbf{r}} = e^2\pi\, \frac{\partial}{\partial p_i} \int \frac{dk}{k^2}\, k_i\, k_j I_k\, \delta(\omega - \mathbf{kv})\, \frac{\partial f^R}{\partial p_j}. \tag{2.133}$$

This equation has a simple interpretation. We express I_k in terms of the number of plasmons, neglecting the correlation widths of the frequencies as a first approximation,

$$I_k = \frac{1}{2\pi^2 \left.\dfrac{\partial \varepsilon_k^l}{\partial \omega}\right|_{\omega = \omega_k^l}} \left(N_k \delta(\omega - \omega_k^l) + N_{-k} \delta(\omega + \omega_k^l) \right). \qquad (2.134)$$

We have

$$\frac{\partial f^R}{\partial t} + \mathbf{v}\, \frac{\partial f^R}{\partial \mathbf{r}} = \frac{\partial}{\partial p_i}\, D_{ij}\, \frac{\partial f^R}{\partial p_j}, \qquad (2.135)$$

where

$$D_{ij} = \int k_i\, k_j\, w_{\mathbf{p}}(\mathbf{k})\, N_k\, \frac{d\mathbf{k}}{(2\pi)^3},$$

and

$$w_{\mathbf{p}}(\mathbf{k}) = \frac{e^2}{\pi k^2} \cdot \frac{(2\pi)^3}{\left.\dfrac{\partial \varepsilon_k^l}{\partial \omega}\right|_{\omega = \omega_k^l}} \delta(\omega_k^l - \mathbf{k}\mathbf{v}) \qquad (2.136)$$

is the probability of plasmon emission introduced above [see (2.61)].†
Equation (2.134) can be obtained by balancing the induced emission
and absorption of plasmons by the particles in the turbulent plasma. By emitting a plasmon a particle loses momentum $\mathbf{k}$ and
makes a transition from the state $\mathbf{p}$ into the state $\mathbf{p} - \mathbf{k}$. Using
the principle of detailed balance, one finds that the probabilities of
plasmon absorption by particles in the state $\mathbf{p} - \mathbf{k}$ is $w_{\mathbf{p}}(\mathbf{k})$; for particles in state $\mathbf{p}$ the probability is $w_{\mathbf{p}+\mathbf{k}}(\mathbf{k})$. Balancing absorption with
emission we have

$$\frac{\partial f_{\mathbf{p}}^R}{\partial t} + \mathbf{v}\, \frac{\partial f_{\mathbf{p}}^R}{\partial \mathbf{r}} = -\int \{ w_{\mathbf{p}}(\mathbf{k}) \left[(N_k + 1) f_{\mathbf{p}}^R - N_k f_{\mathbf{p}-\mathbf{k}}^R \right] +$$

$$+ w_{\mathbf{p}+\mathbf{k}} \left[N_k f_{\mathbf{p}} - (N_k + 1) f_{\mathbf{p}+\mathbf{k}}^R \right] \} \, d\mathbf{k}\, (2\pi)^{-3}. \qquad (2.137)$$

Expanding this in terms of $\mathbf{k}$, and including just the induced
processes, we arrive at Eqs. (2.136) and (2.135).

---

†It is not hard to see that (2.136) is the probability for plasmon emission by a uniformly moving charge (Cerenkov radiation — see [73]). One readily obtains Eq.
(2.136) by using the field for a charge in uniform motion, $E_k = -\dfrac{4\pi i e}{k \varepsilon_k^l}\, \delta(\omega - \mathbf{k}\mathbf{v})(2\pi)^{-3}$,
in the expression for the radiated power $\int (\mathbf{j}\mathbf{E})\, d\mathbf{r} = \int \omega_k^l\, w_{\mathbf{p}}(\mathbf{k})\, d\mathbf{k}\,(2\pi)^{-3}$.

It is also important that only particles which satisfy the absorption and emission conservation laws can interact in our approximation. From (2.136) this gives

$$\omega_k^l = \mathbf{kv}.$$

Particles satisfying this constraint are called resonant particles. Most often the number of resonant particles is small when compared with the total number of plasma particles. It is therefore of interest to examine how the nonresonant particles interact with turbulent fluctuations. In reality the resonance is never infinitely sharp (where it might be described by a δ function); thus, the "wings" of the resonance can contain particles with velocities somewhat different from the velocities of the resonant particles, and these particles are still able to participate in the interaction. Both the correlation of turbulent fluctuations (see §2.5) and instabilities in the turbulence can produce such effects.

Let us write the basic quasilinear equation (2.130) for longitudinal turbulent fluctuations, replacing k' with −k' and integrating over k:

$$\frac{\partial f^R}{\partial t} + \mathbf{v}\frac{\partial f^R}{\partial \mathbf{r}} = \frac{\partial}{\partial p_i}\, i\, e^2 \int dk'\, dk_1\, dk_2\, \frac{k_i'\, k_{1j}}{k'\, k_1}\, \langle E_{k'}^{T*} E_{k_1}^T \rangle \times$$

$$\times\; \frac{1}{\omega_1 + \omega_2 - (k_1 + k_2)\mathbf{v} + i\,\delta}\, \frac{\partial f_{k_2}^R}{\partial p_j}\, \exp\left[i\, x\,(k_1 + k_2 - k')\right]. \qquad (2.138)$$

The width of the resonance can be related to the inhomogeneity and instability of f^R and the average of the fields of the turbulent fluctuations $\langle E_k^{T*} E_{k_1}^T \rangle$. To simplify the explanation of the physics, we shall treat only those effects related to instability, which assumes therefore that both f^R and the turbulence are homogeneous. If both ω_2 and $\mathbf{k}_2$ are neglected in the denominator of (2.138), we obtain $\int f_{k_2}^R e^{ik_2 x}\, dk_2 = f^R(x)$, which is the localized distribution function computed above. We shall keep the next term of the expansion in ω_2 and k_2. Because f^R is homogeneous, $f_{k_2}^R = f_{\omega_2}^R \delta(\mathbf{k}_2)$ $\int \omega_2 e^{i k_2 x} f_{k_2}^R\, dk_2 = i\frac{\partial f^R}{\partial t}$. By comparing the term thus obtained on the right side of (2.138) with $\partial f^R/\partial t$ on the left, we see that it is small if the turbulence is weak (W/nT ≪ 1).

Let us find the change in the quasilinear interaction induced by the instability of the turbulent fluctuations. Assuming that

$\omega_2 \sim k_2 \sim 0$ in the denominator of (2.138), we have

$$\frac{\partial f^R}{\partial t} + \mathbf{v} \frac{\partial f^R}{\partial \mathbf{r}} = \frac{\partial}{\partial p_i} D_{ij} \frac{\partial f^R}{\partial p_j}, \qquad (2.139)$$

where

$$D_{ij} = i e^2 \int \frac{k_i' k_{1j}}{k' k_1} \frac{\langle E_k^{T*} E_{k_1}^T \rangle e^{i(k_1 - k')x}}{\omega_1 - k_1 \mathbf{v} + i \delta} \, dk' \, dk_1. \qquad (2.140)$$

We now write $\omega_1 = (\omega_1 + \omega')/2 + (\omega_1 - \omega')/2$ and $k_1 = (k_1 + k')/2 + (k_1 - k')/2$. In terms of the notation used in §2.2, $(k' + k_1)/2 = \varkappa$; $k_1 - k' = \Delta k$; $\varkappa = \{\varkappa, \omega\}$; $\omega_1 = \omega + \Delta\omega/2$; $k_1 = \varkappa + \Delta k/2$. We allow for the fact that $\Delta k = 0$ for stable fluctuations. $\Delta\omega$ appears only in the denominator of (2.140). By expanding the denominator of (2.140) in terms of Δk and setting $\langle E_k^{T*} E_{k_1}^T \rangle = |E_\varkappa(\Delta k)|^2$, we easily find the correction to D_{ij}:

$$\delta D_{ij} = -i e^2 \int \frac{\varkappa_i \varkappa_j}{\varkappa^2} d\varkappa \frac{(\Delta\omega - \mathbf{v}\,\Delta\varkappa) |E_\varkappa(\Delta k)|^2}{2(\omega - \varkappa\mathbf{v} + i \delta)^2} e^{i\,\Delta kx} \, d\Delta k. \qquad (2.141)$$

We now replace $\varkappa$ with k; then because of the homogeneity of the turbulence we find

$$\delta D_{ij} = \frac{e^2}{2} \operatorname{Re} \int \frac{k_i k_j}{k^2} \cdot \frac{dk}{(\omega - k\mathbf{v} + i \delta)^2} \cdot \frac{\partial I_k}{\partial t}. \qquad (2.142)$$

After integrating by parts we arrive at the final expression for δD_{ij}:

$$\delta D_{ij} = \frac{e^2}{2} \operatorname{Re} \int \frac{dk}{\omega - k\mathbf{v} + i \delta} \cdot \frac{k_i k_j}{k^2} \cdot \frac{\partial}{\partial t} \cdot \frac{\partial}{\partial \omega} I_k. \qquad (2.143)$$

Using this correction we can now find the time rate of change of the total momentum $\mathbf{P}^R = \int \mathbf{p} f^R \frac{d\mathbf{p}}{(2\pi)^3}$ of the plasma particles:

$$\frac{\partial \mathbf{P}^R}{\partial t} = -e^2 \pi \int \frac{\mathbf{k}}{k^2} I_k \, \delta(\omega - \mathbf{k}\mathbf{v}) \, dk \left(\mathbf{k} \frac{\partial f^R}{\partial \mathbf{p}} \right) \frac{d\mathbf{p}}{(2\pi)^3} -$$

$$- \frac{e^2}{2} \operatorname{Re} \int \frac{\mathbf{k}}{k^2} \frac{d\mathbf{p}\,dk}{\omega - \mathbf{k}\mathbf{v} + i \delta} \left(\mathbf{k} \frac{\partial f^R}{\partial \mathbf{p}} \right) \frac{\partial}{\partial t} \cdot \frac{\partial}{\partial \omega} I_k \frac{1}{(2\pi)^3}. \qquad (2.144)$$

The coefficients present in (2.144) can be expressed in terms of

$$\frac{\partial}{\partial \omega} \operatorname{Re} \varepsilon_k^{lR} = \frac{4\pi e^2}{k^2} \operatorname{Re} \int \frac{\partial}{\partial \omega} \frac{\left(k \frac{\partial f^R}{\partial p}\right)}{\omega - kv + i \delta} \cdot \frac{dp}{(2\pi)^3} ; \qquad (2.145)$$

$$\operatorname{Im} \varepsilon_k^{lR} = -\frac{4\pi^2 e^2}{k^2} \int \delta (\omega - kv) \left(k \frac{\partial f^R}{\partial p}\right) \frac{dp}{(2\pi)^3} . \qquad (2.146)$$

We then obtain

$$\frac{\partial \mathbf{P}^R}{\partial t} = \frac{1}{4\pi} \int k \frac{\partial}{\partial \omega} \operatorname{Re} \varepsilon_k^{lR} \left\{ \frac{\operatorname{Im} \varepsilon_k^{lR}}{\frac{\partial}{\partial \omega} \operatorname{Re} \varepsilon_k^{lR}} I_k + \frac{1}{2} \cdot \frac{\partial}{\partial t} I_k \right\} dk. \qquad (2.147)$$

If one takes into account the terms linear in I_k in an approximate way, then $\mathbf{P}^R$ is conserved. In fact, the first term in the braces of (2.147) is $-\gamma_k I_k$ and the second is $\gamma_k I_k$. But when the nonlinear effects in the equation for $\frac{\partial}{\partial t} I_k$ are taken into account we obtain

$$\frac{\partial \mathbf{P}^R}{\partial t} = \frac{1}{8\pi^3} \int k \left(\frac{\partial N_k}{\partial t}\right)_{nl} dk = \frac{1}{8\pi^3} \int k v_{\mathbf{k}, \, \mathbf{k}_1} \times$$

$$\times N_{\mathbf{k}} N_{\mathbf{k}_1} dk \, dk_1 + \int \frac{1}{8\pi^3} k w_{\mathbf{k}_1 \mathbf{k}_2} N_{\mathbf{k}_1} N_{\mathbf{k}_2} dk \, dk_1 \, dk_2. \qquad (2.148)$$

Now in order to know whether the current is conserved when the nonlinear effects are included, one must calculate the interaction between the plasma particles and the turbulent fluctuations to the next order in N_k and combine that result with (2.148). One notes immediately that the decay processes make no contribution to (2.148). In fact, Eq. (2.84) is still valid for these decay processes, and as a consequence the first term in (2.148) cancels the second term in that equation. Hence, (2.148) contains only terms appropriate to effects due to induced scattering. And it is just these effects which figure in the next approximation for the interaction of particles and turbulent fluctuations. They compensate for the contribution from scattering effects in the right side of (2.148) (see below). Thus, $\partial \mathbf{p}^R/\partial t = 0$.

This last result is rather easily understood. Indeed, since the momentum of the electromagnetic field is zero for potential oscillations, we have

$$\frac{d}{dt} \int \mathbf{p} f \frac{d\mathbf{p}}{(2\pi)^3} = 0, \quad \left\langle \int \mathbf{p} \frac{\partial f}{\partial t} \cdot \frac{d\mathbf{p}}{(2\pi)^3} \right\rangle = 0,$$

or

$$\frac{\partial}{\partial t} \int \mathbf{p} f^R \frac{d\mathbf{p}}{(2\pi)^3} = 0. \tag{2.149}$$

For the particle energy $W^R = \int \frac{p^2}{2m} f^R \frac{d\mathbf{p}}{(2\pi)^3}$ the corresponding equation differs from (2.144) in that, instead of k being under the integral, we have the quantity $kv = k \frac{d}{dp} \cdot \frac{p^2}{2m}$, which, in turn, can be replaced by ω in both the first term (because of the δ function) and in the second (because the integral containing the factor $\omega - \mathbf{kv}$ instead of $\mathbf{kv}$ goes to zero). Equation (2.147) will then differ in that the factor ω will appear in place of k; an integration by parts will give a linear term $\frac{1}{8\pi} \int (\operatorname{Re} \varepsilon_k^{lR} - 1) \frac{\partial}{\partial t} I_k \, dk$. Noting that $\operatorname{Re} \varepsilon_k^l = 0$, we find

$$\frac{\partial}{\partial t} \left(W^R + \frac{1}{8\pi} \int I_k \, dk \right) = 0, \tag{2.150}$$

that is, the sum of the particle energy and the electric field energy of the turbulent fluctuations is conserved. Equation (2.150) can also be obtained directly by averaging the exact expression for the conservation of energy.

The growth rate is small in comparison with the frequency because the number of resonant particles is small compared with the total number of particles. The first of the two terms in (2.144) contains contributions from just the resonant particles, but the second term includes a small factor which describes the weak instability and which is, in a linear approximation, of order γ_k/ω_k, but the basic contribution comes from the nonresonant particles, which are ω_k/γ_k times more plentiful than the resonant particles [6].

If we isolate just the resonant particles (that is, integrate only over that region of $\mathbf{p}$ in which the resonance condition is fulfilled) we are quite easily satisfied that the momentum of the resonant particles is not conserved. We will denote the total momentum of the resonant particles as

$$\mathbf{P}_{\text{res}}^R = \int_{\text{res}} \mathbf{p}\, f^R\, \frac{d\mathbf{p}}{(2\pi)^3}\,.$$

We then have

$$\frac{\partial \mathbf{P}_{\text{res}}^R}{\partial t} = \frac{1}{4\pi} \int \mathbf{k}\, \mathrm{Im}\, \varepsilon_k^{lR}\, I_k\, dk = -\int \frac{\mathbf{k}\gamma_k^l}{4\pi} \cdot \frac{\partial}{\partial \omega}\, \mathrm{Re}\, \varepsilon_k^{lR}\, I_k\, dk. \quad (2.151)$$

One can introduce the notion of plasmon momentum $\mathbf{P}_{\text{pl}}$ such that

$$-\frac{\partial \mathbf{P}_{\text{res}}^R}{\partial t} = -\frac{\partial}{\partial t}\, \mathbf{P}_{\text{pl}}\,. \quad (2.152)$$

Since only resonant particles are being included,

$$2\gamma_k^l\, I_k = \frac{\partial}{\partial t}\, I_k, \quad (2.153)$$

and we have

$$\mathbf{P}_{\text{pl}} = \int \frac{1}{8\pi}\, \mathbf{k}\, \left(\frac{\partial}{\partial \omega}\, \mathrm{Re}\, \varepsilon_k^{lR} \right) I_k\, dk. \quad (2.154)$$

This expression corresponds to a model of a plasmon gas in which each plasmon has momentum $\mathbf{k}$. From the equation

$$\frac{\partial}{\partial \omega}\, \mathrm{Re}\, \varepsilon_k^{lR}\, I_k = \frac{1}{2\pi^2}\, \{ N_k\, \delta\, (\omega - \omega_k^l) + N_{-k}\, \delta\, (\omega + \omega_k^l) \} \quad (2.155)$$

we obtain

$$\mathbf{P}_{\text{pl}} = \int \mathbf{k} N_k\, \frac{d\mathbf{k}}{(2\pi)^3}\,. \quad (2.156)$$

From our discussion above, this momentum is identical with that of the nonresonant particles which vibrate adiabatically in the field of the turbulent fluctuations.

In this same spirit we can also introduce the notion of the plasmon energy:

$$\frac{\partial W_{\text{res}}^R}{\partial t} = \int \frac{\omega}{4\pi}\, \mathrm{Im}\, \varepsilon_k^{lR}\, I_k\, dk = -\frac{\partial W_{\text{pl}}}{\partial t}\,, \quad (2.157)$$

where

$$W_{pl} = \frac{1}{8\pi} \int \omega \frac{\partial}{\partial \omega} \operatorname{Re} \varepsilon_k'^R I_k \, dk = \int \frac{\omega_k^l N_k \, dk}{(2\pi)^3} . \qquad (2.158)$$

The plasmon energy is equal to the electric field energy plus the kinetic energy of the nonresonant particles in the field of the turbulent fluctuations. The latter is easily found if one simply subtracts the electric field energy $\int (I_k/8\pi) dk$ from Eq. (2.158).

If it is required to determine just the total energy and momentum transmitted to the nonresonant particles, we can limit our study to those effects which are connected with the resonant particles. For the quasilinear equations (which in the first approximation include just effects which are first order in the turbulent fluctuation energy or only resonant particles) are in fact a closed, self-consistent system of equations. In principle, this system of equations allows one to use the initial values of f^R and N_k to find f^R and N_k at any other time. But, when N_k is known, one can then determine the energy and momentum of the plasmons and, therefore, the energy and momentum of the nonresonant particles as well.

2.7.   Correlation of Turbulent Fluctuations and Its Effect on the Interaction with Plasma Particles

It has been shown that the correlation of turbulent fluctuations smears the resonance curve and destroys the strictly one-to-one correspondence between frequency and wave number for turbulent fluctuations. These effects appear for both external (stochastic) fields and fields excited in the plasma because of instability.

We shall attempt to determine if the broadening of the resonance leads to more effective interaction between the plasma particles and the turbulent fluctuations (more precisely, to a more effective energy exchange). This problem is very important to the stochastic heating of plasma by external random fields; it has been explored extensively in various experiments [75, 76]. Since the role of turbulent nonstationarity is identical in this problem to the case discussed earlier, we shall assume at the outset that the turbulence is stationary.

The general quasilinear equation for stationary turbulence (2.132) already includes correlation effects. Only the specific form, Eq. (2.134), which does not include correlations, leads to Eqs. (2.135) and (2.136), in which only resonant particles are included in the interaction. For the resonant particles, inclusion of the correlations has little effect on their interactions since the correlation broadening $\Delta\omega$ of the resonance is always much smaller than ω_k. Assume that the phase velocity ω/k of the fluctuations is much greater than the average particle velocity. Only a small number of resonant particles, those which satisfy the condition $\omega = \mathbf{kv}$, i.e., have velocities greater than ω/k, interact with such fluctuations.

Correlation broadening is very important for the particles which are strictly nonresonant with the fluctuations; these are the particles whose velocities are much less than the phase velocity of the fluctuations. For, although the interaction takes place far off resonance and is much reduced in magnitude, the total number of particles involved can be much greater than the number of resonant particles. This leads us to expect effects similar to those discussed above for nonstationary fluctuations. However, it is easy to show in this case that in addition to the correlation effects (treated in [74]), there are other effects of the same order of magnitude which strongly compensate for the correlation effects in a number of situations. The results for this case have a simple physical interpretation (see below).

In our case the fluctuations can be resonant with a large number of particles if their frequencies are sufficiently low (the tail of the resonance). We can therefore use Eq. (2.128). Substituting this expression into (2.133), we obtain the following equation for the nonresonant particle diffusion coefficient (for the diffusion due to fluctuation correlations):

$$D_{ij} = 32\pi^3 e^2 \int dk_1 dk_2 \times$$

$$\times \frac{(k_{1i}+k_{2i})(k_{1j}+k_{2j})\, I_{k_1} I_{k_2} |S_{k_1+k_2, k_1, k_2}|^2 \delta(\omega_1+\omega_2-(k_1+k_2)\,\mathbf{v})}{(k_1+k_2)^2 (\omega_1+\omega_2)^2 |\varepsilon^l_{k_1+k_2}|^2}. \tag{2.159}$$

Keeping just those terms containing the frequency difference $\omega_{k_1}^l - \omega_{k_2}^l$ and using the approximation of (2.134), we obtain

$$D_{ij} = \frac{16\,e^2}{\pi} \int d\mathbf{k}_1 d\mathbf{k}_2 \times$$

$$\times \frac{(k_{1i}-k_{2i})(k_{1j}-k_{2j})\,N_{\mathbf{k}_1}N_{\mathbf{k}_2}\left|S_{\mathbf{k}_1-\mathbf{k}_2,\,\mathbf{k}_1,-\mathbf{k}_2}\right|^2 \delta\left(\omega_{k_1}^l - \omega_{k_2}^l - (k_1-k_2)\,\mathbf{v}\right)}{\left.\dfrac{\partial \varepsilon_{k_1}^l}{\partial \omega_1}\right|_{\omega_1 = \omega_k^l}\left.\dfrac{\partial \varepsilon_{k_2}^l}{\partial \omega_2}\right|_{\omega_2 = \omega_{k_2}^l}(\mathbf{k}_1-\mathbf{k}_2)^2\,(\omega_1-\omega_2)^2\left|\varepsilon_{\mathbf{k}_1-\mathbf{k}_2}^l\right|^2}. \qquad (2.160)$$

Note that (2.160) accounts for the scattering of plasma particles on turbulent fluctuations; this is just the nonlinear part of the scattering. Equation (2.160) can be written as

$$D_{ij} = \int w_{\mathbf{p}}^N(\mathbf{k}_1,\,\mathbf{k}_2)\,N_{\mathbf{k}_1}N_{\mathbf{k}_2}(k_{1i}-k_{2i})(k_{1j}-k_{2j})\,\frac{d\mathbf{k}_1\,d\mathbf{k}_2}{(2\pi)^6}, \qquad (2.161)$$

where $w_{\mathbf{p}}^N(\mathbf{k}_1,\,\mathbf{k}_2)$ is the probability of nonlinear scattering introduced above in (2.96). It was shown earlier that Compton scattering takes place in addition to the nonlinear scattering. Since (2.160) is quadratic in the turbulence energy, one must include all effects of the same order.

The general expression for the collision integral for plasma particle scattering on turbulent fluctuations takes the form

$$\frac{\partial f^R}{\partial t} + \mathbf{v}\,\frac{\partial f^R}{\partial \mathbf{r}} = I^T,$$

where

$$I^T = -\frac{\partial}{\partial p_i}\,e\langle E_i^T\,f^T\rangle = -\frac{\partial}{\partial p_i}\int dk\,dk'\,\frac{k_i'}{k'}\,e\langle E_{k'}^{T*}\,f_k^T\rangle\,e^{i\,(k-k')\,x}. \qquad (2.162)$$

Above only the first term of the expansion of f^T in terms of the turbulent field E^T was taken into account. In the approximation that f^R is stationary the next terms in the expansion are of the form [see (2.18) and (2.19)]

$$f_k^{T\,(2)} = e^2 \int \frac{dk_1\,dk_2\,\delta(k-k_1-k_2)}{i\,(\omega-\mathbf{k}\mathbf{v}+i\,\delta)}\left(\frac{\mathbf{k}_1}{k_1}\cdot\frac{\partial}{\partial \mathbf{p}}\right)\times$$

$$\times \frac{1}{i\,(\omega_2-\mathbf{k}_2\,\mathbf{v}+i\,\delta)}\left(\frac{\mathbf{k}_2}{k_2}\cdot\frac{\partial f^R}{\partial \mathbf{p}}\right)(E_{k_1}^T E_{k_2}^T - \langle E_{k_1}^T E_{k_2}^T\rangle); \qquad (2.163)$$

$$f_k^{T(3)} = e^3 \int \frac{dk_1\, dk_2\, dk_3\, \delta\,(k - k_1 - k_2 - k_3)}{i\,(\omega - \mathbf{kv} + i\,\delta)} \times$$

$$\times \left(\frac{k_1}{k_1} \cdot \frac{\partial}{\partial \mathbf{p}}\right) \frac{1}{i\,(\omega_2 + \omega_3 - (k_2 + k_3)\,\mathbf{v} + i\,\delta)} \times$$

$$\times \left(\frac{k_2}{k_2} \cdot \frac{\partial}{\partial \mathbf{p}}\right) \frac{1}{i\,(\omega_3 - k_3\,\mathbf{v} + i\,\delta)} \left(\frac{k_3}{k_3} \cdot \frac{\partial f^R}{\partial \mathbf{p}}\right) \times$$

$$\times \left(E_{k_1}^T E_{k_2}^T E_{k_3}^T - E_{k_1}^T \langle E_{k_2}^T E_{k_3}^T\rangle - \langle E_{k_1}^T E_{k_2}^T F_{k_3}^T\rangle\right). \qquad (2.164)$$

As before, we shall use as an example the simplest case of longitudinal turbulent fluctuations. The procedure in the following calculation repeats the computations used in the nonlinear interaction of turbulent fluctuations. Averages over three turbulent fields can be expressed in terms of averages over four, which in turn are divided into the possible pairs of products. Denoting the contributions to the collision integral I^T from $f^{T(2)}$ and $f^{T(3)}$ respectively as $f^{T(2)}$, $I^{T(3)}$, we have

$$I^{T(3)} = e^4 \int \frac{dk_1\, dk}{k_1^2\, k^2}\, \pi I_{k_1} I_k \left(k\, \frac{\partial}{\partial \mathbf{p}}\right) \frac{1}{\omega - \mathbf{kv} + i\,\delta} \left(k_1\, \frac{\partial}{\partial \mathbf{p}}\right) \delta\left(\omega - \omega_1 - (k - k_1)\,\mathbf{v}\right) \times$$

$$\times \left\{\left(k\, \frac{\partial}{\partial \mathbf{p}}\right) \frac{1}{\omega_1 - k_1\,\mathbf{v}} \left(k_1\, \frac{\partial f^R}{\partial \mathbf{p}}\right) - \left(k_1\, \frac{\partial}{\partial \mathbf{p}}\right) \frac{1}{\omega - \mathbf{kv}} \left(k\, \frac{\partial f^R}{\partial \mathbf{p}}\right)\right\}. \qquad (2.165)$$

Here we have taken into account the nonresonant character of the fluctuation $\omega \neq \mathbf{kv}$; thus, an imaginary part can arise because of the denominator

$$\text{Im}\, \frac{1}{\omega - \omega_1 - (k - k_1)\,\mathbf{v} + i\,\delta} = -\pi\delta\,(\omega - \omega_1 - (k - k_1)\,\mathbf{v}).$$

As a consequence of the terms proportional to $(\omega - \omega_1 - (k - k_1)\mathbf{v}$ in the δ-functions in (2.165), one obtains

$$I^{T(3)} = \frac{e^4\, \pi}{m} \int \frac{dk_1\, dk\, (kk_1)}{k^2\, k_1^2}\, I_{k_1} I_k \times$$

$$\times \left(k\, \frac{\partial}{\partial \mathbf{p}}\right) \frac{1}{\omega - \mathbf{kv}} \left(k_1\, \frac{\partial}{\partial \mathbf{p}}\right) \delta\,(\omega - \omega_1 - (k - k_1)\,\mathbf{v}) \times$$

$$\times \left\{\frac{1}{(\omega_1 - k_1\,\mathbf{v})^2} \left(k_1\, \frac{\partial f^R}{\partial \mathbf{p}}\right) - \frac{1}{(\omega - \mathbf{kv})^2} \left(k\, \frac{\partial f^R}{\partial \mathbf{p}}\right)\right\}. \qquad (2.166)$$

As a further result of the δ-function, the expression in braces in

(2.166) is easily transformed into

$$\frac{1}{2}\left[\frac{1}{(\omega_1-\mathbf{k}_1\,\mathbf{v})^2}+\frac{1}{(\omega-\mathbf{k}\mathbf{v})^2}\right]\left((\mathbf{k}_1-\mathbf{k})\,\frac{\partial f^R}{\partial \mathbf{p}}\right).$$

Making the substitution $k_1 \rightleftarrows k$ in (2.166) and taking half the sum of the resulting expressions, we obtain

$$I^{T\,(3)} = \frac{e^4\pi}{4m}\int\frac{d\mathbf{k}_1\,d\mathbf{k}}{k^2\,k_1^2}\,(\mathbf{k}\mathbf{k}_1)\,I_{k_1}\,I_k\times$$

$$\times\left\{\left(\mathbf{k}\,\frac{\partial}{\partial\mathbf{p}}\right)\frac{1}{\omega-\mathbf{k}\mathbf{v}}\left(\mathbf{k}_1\,\frac{\partial}{\partial\mathbf{p}}\right)-\left(\mathbf{k}_1\,\frac{\partial}{\partial\mathbf{p}}\right)\frac{1}{\omega_1-\mathbf{k}_1\,\mathbf{v}}\left(\mathbf{k}\,\frac{\partial}{\partial\mathbf{p}}\right)\right\}\times$$

$$\times\,\delta\left(\omega-\omega_1-(\mathbf{k}-\mathbf{k}_1)\,\mathbf{v}\right)\left[\frac{1}{(\omega_1-\mathbf{k}_1\,\mathbf{v})^2}+\frac{1}{(\omega-\mathbf{k}\mathbf{v})^2}\right]\left((\mathbf{k}_1-\mathbf{k})\,\frac{\partial f^R}{\partial\mathbf{p}}\right).\qquad(2.167)$$

The expression in braces in (2.167) can now be transformed into

$$\left(\mathbf{k}\,\frac{\partial}{\partial\mathbf{p}}\right)\left(\mathbf{k}_1\,\frac{\partial}{\partial\mathbf{p}}\right)\left(\frac{1}{\omega-\mathbf{k}\mathbf{v}}-\frac{1}{\omega_1-\mathbf{k}_1\,\mathbf{v}}\right)-\frac{(\mathbf{k}\mathbf{k}_1)}{m}\left(\mathbf{k}\,\frac{\partial}{\partial\mathbf{p}}\right)\frac{1}{(\omega-\mathbf{k}\mathbf{v})^2}+$$

$$+\frac{(\mathbf{k}\mathbf{k}_1)}{m}\left(\mathbf{k}_1\,\frac{\partial}{\partial\mathbf{p}}\right)\frac{1}{(\omega_1-\mathbf{k}_1\,\mathbf{v})^2}.\qquad(2.168)$$

Since the first term in (2.168) is zero by virtue of the δ-function, we have

$$I^{T\,(3)} = \frac{e^4\pi}{2m^2}\int\frac{d\mathbf{k}\,d\mathbf{k}_1}{k^2\,k_1^2}\,(\mathbf{k}\mathbf{k}_1)^2\,I_{k_1}\,I_k\left((\mathbf{k}-\mathbf{k}_1)\,\frac{\partial}{\partial\mathbf{p}}\right)\frac{1}{(\omega-\mathbf{k}\mathbf{v})^4}\times$$

$$\times\,\delta\left(\omega-\omega_1-(\mathbf{k}-\mathbf{k}_1)\,\mathbf{v}\right)\left((\mathbf{k}-\mathbf{k}_1)\,\frac{\partial f^R}{\partial\mathbf{p}}\right).\qquad(2.169)$$

Now, using the approximate expressions in (2.134) in order to express the turbulence energy in terms of the number of quanta, we find

$$I^{T\,(3)} = \frac{\partial}{\partial p_i}\,D_{ij}\,\frac{\partial f^R}{\partial p_j}\,;$$

$$D_{ij}=\int(k_i-k_{1i})\,(k_j-k_{1j})\,w_p^C\,(\mathbf{k},\,\mathbf{k}_1)\,N_{k_1}\,N_{k_2}\,d\mathbf{k}_1\,d\mathbf{k}_2\,(2\pi)^{-6};\qquad(2.170)$$

$$w_p^C(\mathbf{k},\,\mathbf{k}_1) = \frac{4e^4\,(2\pi)^3}{m^2\,\dfrac{\partial \varepsilon_k^l}{\partial \omega}\bigg|_{\omega\,=\,\omega_k^l}\,\dfrac{\partial \varepsilon_{k_1}^l}{\partial \omega_1}\bigg|_{\omega_1\,=\,\omega_{k_1}^l}\,(\omega - \mathbf{k}\mathbf{v})^4} \times$$

$$\times \frac{(\mathbf{k}\mathbf{k}_1)^2}{k^2\,k_1^2}\,\delta\,(\omega_k^l - \omega_{k_1}^l - (\mathbf{k} - \mathbf{k}_1)\,\mathbf{v}). \qquad (2.171)$$

Equation (2.171) agrees exactly with the probability for Compton scattering obtained earlier, (2.89). By using much the same calculation it is easy to see that $I^{T(2)}$ describes the interference of the nonlinear and Compton scatterings.

This analysis has important consequences. First, the correlation of fluctuations and the interaction of particles with the tails of the resonance broadening introduce quantities which are of the same order of magnitude as the effects, which are of next higher order in the turbulence energy in those equations which describe the particle—turbulent fluctuation interactions. Second, the results obtained here show that a complete description of these interactions is possible without including the tails of the correlation interactions, by instead taking into account the induced-scattering effects. Induced scattering is also a resonance interaction, but this resonance is of higher order in $\omega - \omega_1 = (\mathbf{k}_1 - \mathbf{k}_1)\mathbf{v}$ and those particles selected by this condition are different from those satisfying the condition that $\omega = \mathbf{k}\mathbf{v}$. Thus, one need only include the resonance interactions, and the correlation effects can be neglected in a first approximation (see also Chapter 4). Third, the strong compensation of the nonlinear and Compton scatterings has a significant impact on the conclusions pertaining to the effectiveness of stochastic heating of plasma electrons. For instance, for Langmuir fluctuations (see below) the effectiveness of the electron heating is reduced by the factor $(v_{Te}/v_p)^2$, where v_p is the phase velocity of the Langmuir fluctuations. For scattering of high-frequency oscillations on ions, the Compton scattering is small (due to the large ion mass) and the nonlinear scattering becomes dominant; all heating effects are due solely to the correlation of turbulent fluctuations. This is not the case for heating by low-frequency fluctuations, and Compton scattering can turn out to be the decisive factor for the ions.

The equations for f^R can be quite easily obtained from simple balance considerations if one notes that in the scattering process a particle goes from a state of momentum $\mathbf{p}$ into a state of momentum $p - k + k_1$. The balance equation then takes the simple form

$$\frac{\partial f^R}{\partial t} + v \frac{\partial f^R}{\partial r} = - \int [w_p(k, k_1)(f_p - f_{p-k+k_1}) +$$

$$+ w_{p+k-k_1}(f_p - f_{p+k-k_1})] N_{k_1} N_k \frac{dk_1\, dk}{(2\pi)^6}. \qquad (2.172)$$

Here we have included only the effects of induced scattering; $w_p(k, k_1)$ is the total scattering probability (both nonlinear and Compton). Expanding (2.172) in terms of k and k_1 we find

$$\frac{\partial f^R}{\partial t} + v \frac{\partial f^R}{\partial r} = \frac{\partial}{\partial p_i} D_{ij} \frac{\partial f^R}{\partial v_j};$$

$$D_{ij} = \int w_p(k, k_1) N_{k_1} N_k (k_i - k_{1i})(k_j - k_{1j}) \frac{dk_1\, dk}{(2\pi)^6}, \qquad (2.173)$$

which agrees with the result obtained above.

From Eq. (2.173) and the equation for the wave it is easily seen that the sum of the wave and particle energies is conserved during induced scattering. Thus, it is as if our entire treatment method is a closed system, in which f^R describes the plasma electron and ion overgrown with an electromagnetic fur coat (electron and ion excitations), and N_k describes the plasma excitations — the plasmons. Their interactions are given by the probabilities, and the total energy of all the quasiparticles is conserved.

Incidentally, this picture of the system rather simply provides us with the physical consequences pertaining to the plasma heating. During induced scattering the number of plasmons is conserved since for each plasmon absorbed there is one emitted. Thus, from energy conservation, heating must only take place when the total plasmon energy is changed. Therefore, if there is heating, the transfer of turbulence energy is accomplished by reducing the average fluctuation frequency. For a series of fluctuations the frequencies in the entire spectrum change very little (for example, this is the case for Langmuir fluctuations, whose frequencies are close to ω_{pe}). As a consequence, a substantial transfer of such fluctuations along the spectrum cannot result in much heating. The fluctuations which are more conducive to heating are those whose frequencies decrease rapidly as k drops.

The heating effectiveness can be evaluated if one knows the characteristic time for transfer along the spectrum. If the transfer times τ were known in some frequency interval $\delta\omega$ and for the turbulence energy W, then the heating rate is of order $\frac{\delta\omega}{\omega} \, W \, \frac{1}{\tau}$.[†]

2.8. Turbulent Resonance Broadening
of Particle – Plasmon Interactions

It was assumed in §2.7 that the particles and plasmons were nonresonant with the majority of the plasma particles (i.e., $\omega \neq \mathbf{kv}$). When there is resonance, with $\omega = \mathbf{kv}$, a condition frequently satisfied for the low-frequency fluctuations, the linear damping and quasilinear effects become very pronounced. Calculation of the nonlinear corrections to both γ and D_{ij} demands that accurate account be taken of the turbulence broadening of the resonances at $\omega = \mathbf{kv}$. The correlations considered above reflect effects integrated over all the particles; they can be quite insensitive to this type of broadening, which, in some degree, is similar to the correlation effects themselves, even though it is an independent effect. We need to include both the effects of this broadening and the correlation effects only for fluctuations which are resonant with particles; that is, they should be included in any calculation of corrections to the quasilinear effects.

The problem of broadening of wave–particle resonances was approached by Dupree [77]. The general form of the equations which include this broadening was presented by Kadomtsev [6], but he did not find the solutions to them, nor did he discuss the physical consequences to be derived from them. This problem was treated in more detail in [78], where it was shown that under the resonance condition $\omega = \mathbf{kv}$ the turbulent processes are adequately described by the usual quasilinear equations, and the nonlinear effects in both γ and D_{ij} are always small corrections to the quasilinear equations if $W/nT \ll 1$, which indicates that they can be dropped. It is no longer possible to expand in terms of the turbulence energy when treating those corrections.

The discussion which follows is taken from [78]. We must note here that the broadening of the resonances where $\omega = \mathbf{kv}$ can

---

[†]In addition to this heating there is also heating from the transfer of energy in the region of intense absorption. We shall consider only the heating coming from the transfer of turbulent energy across the spectrum.

only affect the correlation effects through changes in the nonlinear currents S and Σ. The effect of resonance broadening is
most easily understood if we examine the effect of high-frequency
short-wavelength fluctuations with frequency $\omega(\mathbf{k})$ and wave number $\mathbf{k}$ on the low-frequency (Ω) vibrations, where (q), $\Omega \ll \omega$,
$q \ll k$. Both the high- and low-frequency oscillations are assumed
to be in resonance with the particles, i.e., $\Omega = \mathbf{qv}$ and $\omega = \mathbf{kv}$, and
they both belong to the same branch of the turbulent fluctuations
(such as the ion-acoustic branch). When discussing the action of
high-frequency oscillations on the low-frequency oscillations, it
will be assumed that the energy of the high-frequency fluctuations
is so great, and the low-frequency energy so small, that the reverse effect of the low-frequency oscillations on the high-frequency
modes can be neglected. Without pretending to be rigorous
(the exact theory will be presented later), we can illustrate the
effect which high-frequency oscillations have on the low-frequency
modes by including their average effect in the quasilinear equations. We must emphasize strongly that this quasilinear equation
is written only for the turbulent component of the distribution
function, and it makes sense only in that the low-frequency fluctuations are "sensitive" to the average effect of the high-frequency
oscillations since $\Omega \ll \omega$ and $q \ll k$:

$$- i \Omega f_q^T + i \mathbf{q} \mathbf{v} f_q^T + i \frac{e}{m} E_q^T \left(\frac{\mathbf{q}}{q} \cdot \frac{\partial f^R}{\partial \mathbf{v}} \right) = \frac{\partial}{\partial v_i} D_{ij} \frac{\partial f_q^T}{\partial v_j} ;$$

$$D_{ij} = \pi \int \frac{k_i k_j}{k^2} \cdot \frac{e^2}{m^2} I_k \, \delta \, (\omega - \mathbf{kv}) \, dk. \tag{2.174}$$

The coefficient D_{ij} describes the action of the low-frequency
fluctuations; this action can be considered as a kind of effective
"turbulent collision" for the low-frequency fluctuations. These
turbulent collisions also broaden the resonance. Let us consider
the case of greatest interest to our applications, in which the
condition $v \gg \Omega/q$ is fulfilled for most of the plasma particles
(i.e., the mean thermal velocity of the particles is much larger
than the fluctuation phase velocity). Then the condition $\Omega = \mathbf{qv}$
is fulfilled for just those particles whose velocities are nearly
perpendicular to $\mathbf{q}$. By introducing the angle θ, which is the angle
between $\mathbf{v}$ and the perpendicular to $\mathbf{q}$, we can write the condition
for approximate resonance as $\Omega = qv\theta$. Thus one must have
$\Omega/qv = \theta \ll 1$. Under these conditions the diffusion term of (2.174)

is written as $D \frac{\partial^2}{\partial \theta^2} f_q^T$; then, when θ changes, this term moves the particle off resonance in a manner such that the smaller θ is, the faster the particle is moved off the resonance. Then, in this situation, (2.174) is transformed into

$$
\left.
\begin{aligned}
f_q^T &= g \frac{e}{m} E_q^T \left(\frac{\mathbf{q}}{q} \cdot \frac{\partial f^R}{\partial v} \right) ; \\
g &= -i \int_0^\infty e^{-iqv\theta\tau + i\Omega\tau - \frac{D\tau^3}{3} q^2 v^2} d\tau .
\end{aligned}
\right\}
\tag{2.175}
$$

When far from resonance, $g = 1/(\Omega - qv\theta)$, but for $\Omega \to qv\theta$, $g \to -i(q^2 v^2 D/3)^{-1/3} \Gamma(1/3)/3$, that is, if one assumes that when $\Omega \to kv\theta$, $g \approx 1/i\nu_{eff}$, then $\nu_{eff} \sim D^{1/3} \sim (W/nT)^{1/3}$. This analysis makes it clear that the nonlinear effects cannot be expanded in terms of the oscillational energy near the resonance $\omega = \mathbf{kv}$. The general theory of the resonance broadening effects [78] can be constructed using the general equation (2.10) for the turbulent components of the distribution function. As before we set $E^R = 0$ for simplicity. For longitudinal fields, such that $E_k = -ik\varphi$, (2.10) can be written as

$$
-(\omega - \mathbf{kv}) f_k^T = e\varphi_k \left(\mathbf{k} \frac{\partial f^R}{\partial \mathbf{p}} \right) + e \frac{\partial}{\partial \mathbf{p}} \int \mathbf{k}_1 \, dk_1 \, dk_2 \times
$$

$$
\times \delta(k - k_1 - k_2) (\varphi_{k_1}^T f_{k_2}^T - \langle \varphi_{k_1}^T f_{k_2}^T \rangle);
$$

$$
dk = d\mathbf{k} \, d\omega \quad k = \{\mathbf{k}, \omega\};
$$

$$
\delta(k - k_1 - k_2) = \delta(\mathbf{k} - \mathbf{k}_1 - \mathbf{k}_2) \delta(\omega - \omega_1 - \omega_2).
\tag{2.176}
$$

From the right side of this equation we isolate the term diagonal in f_k^T and designate it as $i\,\hat{v}_k(\mathbf{p}) f_k^T$, where $\hat{v}_k(\mathbf{p})$ is in general an operator [see Eq. (2.174)]. We introduce the operator $\hat{g}_k(\mathbf{p})$, which satisfies the equation

$$
\hat{g}_k(\mathbf{p}) \left(\omega - \mathbf{kv} + i\,\hat{v}_k(\mathbf{p}) \right) f_k^T(\mathbf{p}) = f_k^T(\mathbf{p}).
\tag{2.177}
$$

The formal solution to (2.176) will be

$$
f_k^T = -e\varphi_k^T \hat{g}_k(\mathbf{p}) \left(\mathbf{k} \frac{\partial f^R}{\partial \mathbf{p}} \right) + i\hat{g}_k(\mathbf{p}) \hat{v}_k(\mathbf{p}) f_k^T -
$$

$$
-\hat{g}_k(\mathbf{p}) \frac{\partial}{\partial \mathbf{p}} e \int \mathbf{k}_1 \, dk_1 \, dk_2 \, \delta(k - k_1 - k_2) \times (\varphi_{k_1}^T f_{k_2}^T - \langle \varphi_{k_1}^T f_{k_2}^T \rangle).
\tag{2.178}
$$

Upon substituting this solution into the nonlinear term of the original equation we find

$$(\omega - \mathbf{k}\mathbf{v} + i\,\hat{v}_k(\mathbf{p}))\,f_k^T + e\varphi_k\left(\mathbf{k}\,\frac{\partial f^R}{\partial \mathbf{p}}\right) =$$

$$= i\,\hat{v}_k(\mathbf{p})f_k^T(\mathbf{p}) + \frac{\partial}{\partial p_i}B_{ij}\frac{\partial f^R}{\partial p_j} - i\,e\,\frac{\partial}{\partial p_i}\times$$

$$\times \int k_{1i}\,dk_1\,dk_2\,\delta\,(k - k_1 - k_2)\,(\varphi_{k_1}^T\,\hat{g}_{k_2}(\mathbf{p})\,\hat{v}_{k_2}(\mathbf{p})\,f_{k_2}^T -$$

$$- \langle\varphi_{k_1}^T\,\hat{g}_{k_2}(\mathbf{p})\,\hat{v}_{k_2}(\mathbf{p})\,f_{k_2}^T\rangle) + \frac{\partial}{\partial p_i}e^2\int dk_1\,dk_2\,dk_3\,\delta\,(k - k_1 - k_2 - k_3)\times$$

$$\times k_{1i}\,k_{2j}\,\hat{g}_{k-k_1}(\mathbf{p})\frac{\partial}{\partial p_j}(\varphi_{k_1}^T\,\varphi_{k_2}^T\,f_{k_3}^T - \varphi_{k_1}^T\langle\varphi_{k_2}^T\,f_{k_3}^T\rangle - \langle\varphi_{k_1}^T\,\varphi_{k_2}^T\,f_{k_3}^T\rangle);$$

$$(2.179)$$

$$B_{ij} = e^2\int k_{1i}\,k_{2j}\,dk_1\,dk_2\times$$

$$\times\,\delta\,(k - k_1 - k_2)\,\hat{g}_{k_2}(\mathbf{p})\,(\varphi_{k_1}^T\,\varphi_{k_2}^T - \langle\varphi_{k_1}^T\,\varphi_{k_2}^T\rangle).$$

The right side of (2.179) can contain no terms diagonal in f_k^T assigned to $\hat{v}$, at least in the main terms. A main term is a term containing $\varphi_{k_1}^T\,\varphi_{k_2}^T\,f_{k_2}^T$ where $k_2 = -k_1$. This gives an expression for the operator $I_{k_1} = |\varphi_{k_1}|^{2k_1^2}$:

$$i\,\hat{v}_k\,f_k^T = e^2\frac{\partial}{\partial p_i}\int k_{1i}\,k_{1j}\,|\varphi_{k_1}|^2\,\hat{g}_{k-k_1}(\mathbf{p})\frac{\partial f_k^T}{\partial p_j}\,dk_1. \qquad (2.180)$$

In order to describe the resonance broadening we shall consider a new kind of perturbation theory, the basis of which is the summing of the most important terms near resonance in the general expansion in terms of the turbulence energy. This approximation corresponds to keeping just the first term on the right side of (2.179), i.e.,

$$f_k^{T(0)}(\mathbf{p}) = -\hat{g}_k(\mathbf{p})\,e\varphi_k^T\left(\mathbf{k}\,\frac{\partial f^R}{\partial \mathbf{p}}\right). \qquad (2.181)$$

Noting that $\hat{g}$ operates only on the particle momenta,

$$\hat{g}_k(\mathbf{p})\,f_k^T(\mathbf{p}) = \int g_k(\mathbf{p},\,\mathbf{p}')\,f_k^T(\mathbf{p}')\,dp',$$

we can rewrite (2.180) and (2.177) as

$$\int d\mathbf{p}' \left[(\omega - \mathbf{k}\mathbf{v})\, \delta(\mathbf{p} - \mathbf{p}') - e^2 \frac{\partial}{\partial p_i} \times \right.$$

$$\left. \times \int k_{1i}\, k_{1j} |\varphi_{k_1}|^2\, dk_1\, g_{k-k_1}(\mathbf{p}, \mathbf{p}') \frac{\partial}{\partial p_j} \right] g_k(\mathbf{p}', \mathbf{p}'') = \delta(\mathbf{p} - \mathbf{p}''). \qquad (2.182)$$

When $W/nT \to 0$, $g \to 1/(\omega - \mathbf{k}\mathbf{v})$ and the approximation of (2.181) corresponds to the usual linear approximation. Thus, the new perturbation theory for nonresonant fluctuations leads to the same results obtained earlier by the more conventional method.

One can also use the new perturbation theory to describe non-resonant fluctuations by noting that part of the nonlinear effects are already included in (2.181). Using the approximation of (2.181) it is rather easy to construct a perturbation theory which will lead to a nonlinear equation for $|\varphi|^2$:

$$\left. \begin{aligned} \varepsilon_k |\varphi_k|^2 &= |\varphi_k|^2 \sum_\alpha \int \Lambda^\alpha_{kk_1} |\varphi_{k_1}|^2\, dk_1 + \\[6pt] &+ 2\sum_\alpha \frac{\omega_{p\alpha}^4\, m_\alpha^2}{k^4\, n^2\, \varepsilon_k} \int |\lambda^\alpha_{k,\, k_1,\, k_2}|^2 |\varphi_{k_1}|^2 \times \\[4pt] &\qquad \times |\varphi_{k_2}|^2\, dk_1\, dk_2\, \delta(k - k_1 - k_2), \\[6pt] \varepsilon_k &= \sum_\alpha \frac{\omega_{p\alpha}^2 m_\alpha}{nk^2} \int \hat{g}_k(\mathbf{p}) \left(\mathbf{k}\, \frac{\partial f_\alpha^R}{\partial \mathbf{p}} \right) \frac{d\mathbf{p}}{(2\pi)^3} + 1, \\[6pt] \Lambda^\alpha_{k,\, k_1} &= \frac{e^2\, \omega_{p\alpha}^2\, m_\alpha}{nk^2} \int \hat{g}_k(\mathbf{p}) \left(k_1\, \frac{\partial}{\partial \mathbf{p}} \right) \times \\[4pt] &\quad \times \hat{g}_{k-k_1}(\mathbf{p}) \left(\mathbf{k}\, \frac{\partial}{\partial \mathbf{p}} \right) \hat{g}_{-k_1}(\mathbf{p}) \left(k_1\, \frac{\partial f^R}{\partial \mathbf{p}} \right) \frac{d\mathbf{p}}{(2\pi)^3} + \\[6pt] &+ 2\, \frac{\omega_{p\alpha}^2\, m_\alpha}{k^2\, n\, \varepsilon_{k-k_1}} \lambda_{k,\, k_1,\, k-k_1}\, \lambda_{k-k_1,\, k,\, -k_1}, \end{aligned} \right\} \qquad (2.183)$$

$$\lambda_{k,\, k_1,\, k_2} = \frac{1}{2} \int \hat{g}_k(\mathbf{p}) \left[\left(k_1\, \frac{\partial}{\partial \mathbf{p}} \right) \hat{g}_{k_2}(\mathbf{p}) \left(k_2\, \frac{\partial}{\partial \mathbf{p}} \right) + 1 \rightleftarrows 2 \right] f^R\, \frac{d\mathbf{p}}{(2\pi)^3}. \qquad (2.184)$$

Here ε_k is a modified linear permeability, which differs from the ordinary permeability in that $1/(\omega - \mathbf{k}\mathbf{v})$ is replaced by $\hat{g}(\omega - \mathbf{k}\mathbf{v})$, which allows for the turbulence broadening of the resonance. Those terms of (2.183) which are nonlinear in relation to this ε_k are of order $(W/nT)^{1/3}$ under the resonance condition $\omega = \mathbf{k}\mathbf{v}$.

But when $\omega \neq \mathbf{kv}$, one can expand (2.183) in terms of W, and the sum of the nonlinear terms from g and from (2.183), which are of the same order of magnitude in this instance, gives the nonlinear interaction discussed in detail earlier. In the neighborhood of the resonance $\nu_{\text{eff}}/kv \ll 1$ one can obtain an explicit expression for $\hat{g}$ for the resonant fluctuations. Let us introduce a new variable $\eta = \omega - \mathbf{kv}$ into (2.182). Since η is small near the resonance, the largest contribution comes from the terms containing $\partial/\partial\eta$. Then

$$g_k(\mathbf{p}, \mathbf{p}') = g^{(0)}_{\eta k}(\mathbf{p})\, \delta\,(\mathbf{p} - \mathbf{p}') + \eta g^{(1)} + \cdots,$$

$$\left(\eta + \frac{\partial}{\partial\eta} D_\eta \frac{\partial}{\partial\eta}\right) g^{(0)}_{\eta k}(\mathbf{p}) = 1, \tag{2.185}$$

$$D_\eta = e^2 \int (\mathbf{kk}_1)^2\, dk_1\, |\varphi_{k_1}|^2\, g^{(0)}_{-\eta_1,\, \mathbf{k}-\mathbf{k}_1}(\mathbf{p}). \tag{2.186}$$

These equations differ from (2.174) only in their notation; their solution has the form

$$g_{\eta,\, k}(\mathbf{p}) = -i \int\limits_0^\infty \exp\left[i\,\eta\tau - D_0(\mathbf{k}, \mathbf{p})\, \frac{\tau^3}{3}\right] d\tau, \tag{2.187}$$

where $D_0(\mathbf{k}, \mathbf{p})$ is defined by the equation

$$D_0(\mathbf{k}, \mathbf{p}) = \frac{e^2}{m^2} \int dk_1\, (\mathbf{kk}_1)^2\, |\varphi_{k_1}|^2 \int\limits_0^\infty d\tau\, e^{\,i\,\eta\tau - \frac{D_0(\mathbf{k}_1,\, \mathbf{p})\,\tau^3}{3}}. \tag{2.188}$$

This solution enables us to study the effect of turbulence broadening on the linear (actually, the quasilinear) effects of damping and buildup of oscillations. These corrections always are small for the resonance fluctuations. This is completely natural in a sequential perturbation theory. It is easily seen that for the resonance fluctuations the expansion is in terms of

$$\left(\frac{eE}{mv_T\, \nu_{\text{eff}}}\right)^2 \sim \left(\frac{W}{nT}\right)^{1/3}.$$

Thus, one can in fact neglect the nonlinear effects for the resonance fluctuations; only the quasilinear effects need be included and one can use the usual expressions for the growth rates and diffusion coefficients. This latter statement is valid because the resonance broadening only weakly influences the integral quantities γ and D. The important physical result is that the actual interaction is not strictly resonant and the conservation laws need be satisfied only up to an accuracy which is of the same magnitude as the broadening. However, inclusion of this effect when the interactions of higher order in the oscillation energy are neglected is the most frequent type of excessive accuracy. One may use the concept of elementary excitations interacting among themselves through radiation and plasmon—particle scattering.

However, one must be aware that to consider the scattering of plasmons by particles, which must simultaneously emit the very same plasmon, makes little sense because this effect is always a small correction. Thus, the interaction of ion-acoustic fluctuations, which undergo heavy damping and amplification by electrons, is not due to the scattering of these fluctuations on resonant electrons. In the balance equations the interaction of particles and plasmons contains an integral over all fluctuation frequencies, so that frequency broadening due to correlations is of little importance. In the same manner, the balance equation also contains an integral over the particle velocities, so that broadening of the resonance $\omega = \mathbf{kv}$ from plasmon—particle interactions is not important either.

2.9. General Balance Equations

for the Interactions of Particles

and Turbulent Fluctuations

It has been shown earlier that the detailed microscopic description of turbulent processes in plasma can be replaced by a description based on certain average quantities — the averaged distribution function f^R and the average squares of the turbulent fluctuation amplitudes I_k. f^R characterizes the distribution of the electron and ion quasiparticles, which differ from the ordinary electrons and ions in that they are clothed in a "fur coat" because of their interactions with turbulent fluctuations. This coat is a cloud of charges of the opposite sign. N_k characterizes the distribution over $\mathbf{k}$ of the plasmons in the turbulent plasma. One may therefore create a turbulence based on the notion of plasmons and plasma quasiparticles interacting among themselves [59]. In order to mathematically formulate such a theory it is necessary to write the balance equation for the quasiparticles and plasmons, and to

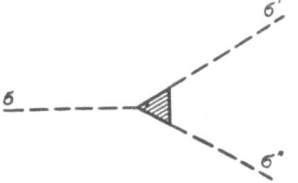

Fig. 2.1. Schematic representation of a three-plasmon decay process involving the three waves σ, σ', σ''.

provide by independent means the probabilities for various inter-action processes.

Graphical methods are very convenient for setting up the balance equation [60]. Interactions of plasmons with each other and with both electrons and ions (electron and ion excitations) are included. Such interactions can be classified according to the number of plasmons involved in a particular interaction. The simplest plasmon—plasmon interaction is the decay of one plas-mon σ into two others, σ' and σ'' (Fig. 2.1). A particular case of this process is the emission of one plasmon by another (σ emits σ' in Fig. 2.2). A more complex event is the emission of three plasmons by a simple plasmon (Fig. 2.3a) or the transformation of two plasmons into two new ones (Fig. 2.3b). A particular case of the latter type is the scattering of one plasmon by another (Fig. 2.4).

The simplest instance of an interaction between a particle and a plasmon is emission of a plasmon by a plasma particle (Fig. 2.5). A somewhat more complicated interaction is the ab-sorption of one plasmon σ' and the emission of another σ (Fig. 2.6). A specific case for this process is the scattering of a plasmon σ by a particle (Fig. 2.7).

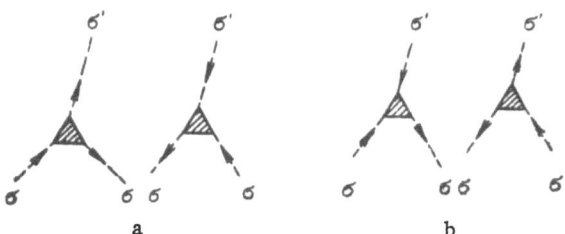

a   b

Fig. 2.2. Possible processes for emitting σ' and σ by waves in the state k^σ: a) emission and its inverse process; b) absorp-tion and its inverse.

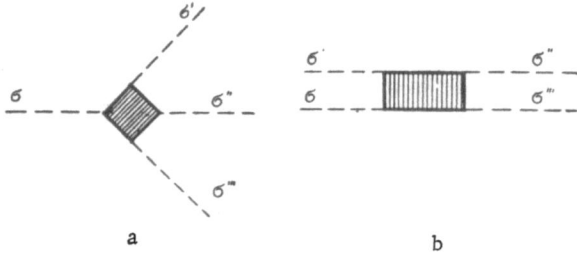

Fig. 2.3. Possible four-plasmon decays involving the waves
σ, σ', σ'', σ'''.

Fig. 2.4. Plasmon–plasmon scattering.

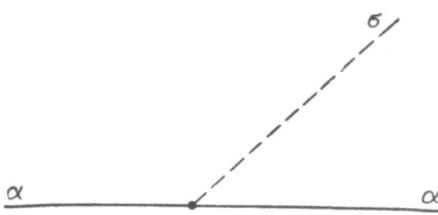

Fig. 2.5. Emission of a plasmon by a plasma particle.

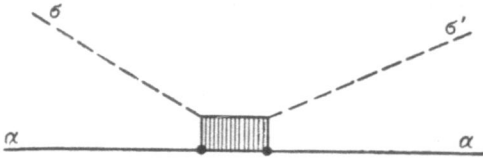

Fig. 2.6. Scattering of the plasmon σ by a particle and
its subsequent transformation into a new plasmon σ'.

In principle the vertex parts of these diagrams ought to be
broken down also. Thus, in addition to the true interaction of four
plasmons one can also have the interaction composed of a repeti-
tion of a three-plasmon interaction (Fig. 2.8). Further, along
with the "true" (that is to say, Compton) scattering, one can also
have a contribution involving a three-plasmon interaction (Fig. 2.9,
nonlinear scattering).

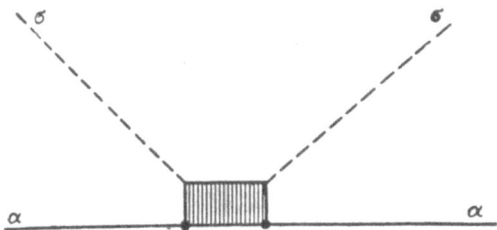

Fig. 2.7. Scattering of the plasmon σ without its conversion into a new wave.

Fig. 2.8. Vertex parts for a four-plasmon decay process.

The general balance equation for the plasmon σ takes into account changes in the number of plasmons due to their interactions among themselves (decay process) and because of interactions with plasma quasiparticles (scattering processes):

$$\frac{dN_{\mathbf{k}}^{\sigma}}{dt} = \frac{dN_{\mathbf{k}}^{\sigma}}{\partial t} + \mathbf{v}_g^{\sigma} \frac{\partial N_{\mathbf{k}}^{\sigma}}{\partial r} = \left(\frac{dN_{\mathbf{k}}^{\sigma}}{dt}\right)_{\text{decay}} + \left(\frac{dN_{\mathbf{k}}^{\sigma}}{dt}\right)_{\text{scat}}.$$

We shall begin by considering the decay processes. We have already introduced the idea of the number $N_{\mathbf{k}}^{\sigma}$ of quanta of a specific type σ. We shall now examine the induced decay of these quanta into others, σ' and σ''. Let $\mathbf{k}$, $\mathbf{k}'$, $\mathbf{k}''$ be the momenta for the appropriate plasmons and $w_{\sigma}^{\sigma' \sigma''}(\mathbf{k}, \mathbf{k}', \mathbf{k}'')$ be the probability of a given type of decay. The change in the number of plasmons due to the decay of a σ plasmon and the increase in number brought about

Fig. 2.9. Vertex parts of a wave—particle interaction.

by the reverse process is given by

$$\frac{dN_{\mathbf{k}}^{\sigma}}{dt} = -\int w_{\sigma}^{\sigma'\sigma''} (\mathbf{k},\, \mathbf{k}',\, \mathbf{k}'') \{ N_{\mathbf{k}}^{\sigma} (N_{\mathbf{k}'}^{\sigma'} + 1) \times$$

$$\times (N_{\mathbf{k}''}^{\sigma''} + 1) - (N_{\mathbf{k}}^{\sigma} + 1)\, N_{\mathbf{k}'}^{\sigma'} N_{\mathbf{k}''}^{\sigma''} \} \, \frac{dk'\,dk''}{(2\pi)^{6}} =$$

$$= \int w_{\sigma}^{\sigma'\sigma''} (\mathbf{k},\, \mathbf{k}'\, \mathbf{k}') (N_{\mathbf{k}'}^{\sigma'} N_{\mathbf{k}''}^{\sigma''} - N_{\mathbf{k}}^{\sigma} N_{\mathbf{k}'}^{\sigma'} - N_{\mathbf{k}}^{\sigma} N_{\mathbf{k}''}^{\sigma''}) \, \frac{dk'\,dk''}{(2\pi)^{6}}. \qquad (2.189)$$

We have dropped the terms linear in $N_{\mathbf{k}}^{\sigma}$. The more complex processes of decay into three plasmons or the transformation of two plasmons into two new ones gives

$$\frac{dN_{\mathbf{k}}^{\sigma}}{dt} = \int w_{\sigma}^{\sigma'\sigma''\sigma'''} (\mathbf{k},\, \mathbf{k}',\, \mathbf{k}'',\, \mathbf{k}''') (N_{\mathbf{k}'}^{\sigma'} N_{\mathbf{k}''}^{\sigma''} N_{\mathbf{k}'''}^{\sigma'''} - N_{\mathbf{k}}^{\sigma} N_{\mathbf{k}'}^{\sigma'} N_{\mathbf{k}''}^{\sigma''} -$$

$$- N_{\mathbf{k}}^{\sigma} N_{\mathbf{k}'}^{\sigma'} N_{\mathbf{k}'''}^{\sigma'''} - N_{\mathbf{k}}^{\sigma} N_{\mathbf{k}'}^{\sigma'} N_{\mathbf{k}'''}^{\sigma'''}) \, \frac{dk'\,dk''\,dk'''}{(2\pi)^{9}}. \qquad (2.190)$$

In a similar manner we obtain equations for the processes shown in Fig. 2.3b which differ from (2.190) only in the sign of the second term on the right hand side.

The plasma particles do not participate in the processes we have been describing. Therefore the sum of the energies and momenta of all the plasmons is conserved. Because of the conservation laws the probabilities for these processes must take the form

$$w_{\sigma}^{\sigma'\sigma''} (\mathbf{k},\, \mathbf{k}',\, \mathbf{k}'') = |\Lambda_{\sigma}^{\sigma'\sigma''}|^{2}\, \delta\, (\mathbf{k} - \mathbf{k}' - \mathbf{k}'')\, \delta\, (\omega_{\mathbf{k}}^{\sigma} - \omega_{\mathbf{k}'}^{\sigma'} - \omega_{\mathbf{k}''}^{\sigma''}); \qquad (2.191)$$

$$w_{\sigma}^{\sigma'\sigma''\sigma'''} (\mathbf{k},\, \mathbf{k}',\, \mathbf{k}'',\, \mathbf{k}''') = |\Lambda_{\sigma}^{\sigma'\sigma''\sigma'''}|^{2}\, \delta\, (\mathbf{k} - \mathbf{k}' - \mathbf{k}'' - \mathbf{k}''') \times$$

$$\times \delta\, (\omega_{\mathbf{k}}^{\sigma} - \omega_{\mathbf{k}'}^{\sigma'} - \omega_{\mathbf{k}''}^{\sigma''} - \omega_{\mathbf{k}'''}^{\sigma'''}), \qquad (2.192)$$

and so on. We can extract the conservation laws for the total energy and momentum from (2.191) and (2.192):

$$\frac{d}{dt} \sum_{\sigma} W^{\sigma} = 0, \qquad \frac{d}{dt} \sum_{\sigma} \mathbf{P}^{\sigma} = 0,$$

$$W^{\sigma} = \int \omega_{\mathbf{k}}^{\sigma} N_{\mathbf{k}}^{\sigma} \frac{dk}{(2\pi)^{3}}, \qquad \mathbf{P}^{\sigma} = \int \mathbf{k} N_{\mathbf{k}}^{\sigma} \frac{dk}{(2\pi)^{3}}.$$

In the emission of one wave σ by the others ($\sigma'' = \sigma$ for Fig. 2.2a and $\sigma''' = \sigma$ for Fig. 2.3) the total number of waves is conserved:

$$\frac{d}{dt} N^\sigma = 0, \qquad N^\sigma = \int \frac{N_k^\sigma\, dk}{(2\pi)^3}.$$

If the momentum of the emitted wave is small compared with the momentum of the emitting wave, $k' \ll k$, and if $N_k^\sigma \ll N_{k'}^{\sigma'}$ the equation for $N_{k'}^{\sigma'}$ can then be written in differential form:

$$\frac{dN_{k'}^{\sigma'}}{dt} = N_{k'}^{\sigma'} \int w_\sigma^{\sigma'\sigma}(\mathbf{k}, \mathbf{k}', \mathbf{k}'') (N_k^\sigma - N_{k''}^\sigma) \frac{dk''\, dk}{(2\pi)^6} = N_{k'}^{\sigma'} \int w_\sigma^{\sigma'}(\mathbf{k}, \mathbf{k}') \left(\mathbf{k}' \frac{\partial N_k^\sigma}{\partial \mathbf{k}} \right) \frac{dk}{(2\pi)^3},$$

$$w_\sigma^{\sigma'}(\mathbf{k}, \mathbf{k}') = \int w_\sigma^{\sigma'\sigma}(\mathbf{k}, \mathbf{k}', \mathbf{k}'') \frac{dk''}{(2\pi)^3}.$$

Here we have set $\mathbf{k}'' = \mathbf{k} - \mathbf{k}'$ and expanded in terms of k'/k.

In order to write an analogous equation for the wave σ, one must take into account the fact that when $\sigma'' = \sigma$, (2.189) does not describe all the possible processes; it gives just those shown in Fig. 2.2a; it does not give those of Fig. 2.2b. The probabilities for these other processes can be expressed in terms of $w_\sigma^{\sigma'}$ only when $\sigma'' = \sigma$. The first graph of Fig. 2.2a differs from the first one in Fig. 2.2b in that the initial state of the plasmon σ is replaced by the final state, which is $\mathbf{k} - \mathbf{k}'$ for Fig. 2.2a, i.e., the probability of the process shown in the first graph of Fig. 2.2a is $w_\sigma^{\sigma'}(\mathbf{k} + \mathbf{k}', \mathbf{k}')$. Thus we find

$$\frac{dN_k^\sigma}{dt} = -\int \{ w_\sigma^{\sigma'}(\mathbf{k}, \mathbf{k}') [(N_{k'}^{\sigma'} + 1) N_k^\sigma - N_{k-k'}^\sigma N_{k'}^{\sigma'}] +$$
$$+ w_\sigma^{\sigma'}(\mathbf{k} + \mathbf{k}', \mathbf{k}') [N_k^\sigma N_{k'}^{\sigma'} - (N_k^\sigma + 1) N_{k+k'}^\sigma] \} \frac{dk'}{(2\pi)^3}.$$

By expanding in terms of k' we obtain a diffusion equation:

$$\frac{dN_k^\sigma}{dt} = \frac{\partial}{\partial k_i} D_{ij}^{\sigma\sigma'} \frac{\partial N_k^\sigma}{\partial k_j}, \qquad D_{ij}^{\sigma\sigma'} = \int w_\sigma^{\sigma'}(\mathbf{k}, \mathbf{k}') N_{k'}^{\sigma'} k_i' k_j' \frac{dk'}{(2\pi)^3}.$$

In order to continue it will be important to know the intensity of the spontaneous emission of a σ plasmon due to the decay process. We shall use the example of a three-plasmon decay process as described by (2.189). Going to the limit $N_k^\sigma \to 0$ in the exact equation (2.189), we have

$$\frac{dN_k^\sigma}{dt}\bigg|_{N_k^\sigma \to 0} = \int w_\sigma^{\sigma'\sigma''}(k, k', k'')\, N_{k'}^{\sigma'}\, N_{k''}^{\sigma'''}\, \frac{dk'\, dk''}{(2\pi)^6}.$$

From this we can find the radiated power in the σ wave emitted by the waves σ' and σ'':

$$\frac{dW^\sigma}{dt} = \frac{d}{dt}\int \frac{\omega_k^\sigma N_k^\sigma\, dk}{(2\pi)^3} = \int \omega_k^\sigma\, N_{k''}^{\sigma''}\, N_{k'}^{\sigma'}\, w_\sigma^{\sigma'\sigma''}(k, k', k'')\, \frac{dk\, dk'\, dk''}{(2\pi)^9}. \qquad (2.193)$$

This radiated power can be found from the nonlinear current in the plasma.

Let us now consider further the motion of the plasma particles. We shall characterize the motion of a single quasiparticle in the presence of external magnetic fields by the projection of its momentum along the magnetic field p_z, its energy ε, and the coordinates of the center of the Larmor circle. In the quantum description for the general case of relativistic particles we have (see [79], for example)

$$\varepsilon^2 = m^2 c^4 + c^2\left(p_z^2 + p_\perp^2\right),$$

where $p_\perp^2 = 2|eH|\hbar\,(n + s)$, in which n is an integer characterizing the particle energy level in a magnetic field and s is the quantum number corresponding to the spin; generally, one can neglect any change in this quantum number during the interaction processes. If the vector potential in the external magnetic field is chosen to be in the gauge $A_x = -H_y$, $A_z = A_y = 0$, then p_x is a conserved quantity. The coordinate of the center of the Larmor circle is related to p_x by the equation

$$y = -\frac{cp_x}{eH}.$$

The particle distribution function depends on three quantities, p_x, p_z, and n:

$$f = f_{p_z,\, p_x,\, n}.$$

We will now consider a process in which the particle goes from the state p_z, p_x, n into the state p'_z, p'_x, n'. We shall use k_z, k_x, and ν to denote the following quantities (as before we set $\hbar = 1$):

$$k_z = p_z - p'_z, \quad k_x = p_x - p'_x, \quad \nu = n - n'.$$

It will be assumed that the change in energy $\Delta\varepsilon$ is small:

$$\varepsilon\Delta\varepsilon = p_z k_z + \nu |eH|.$$

As an example let us assume that the change in the particle's state is connected with the emission of the wave ω_k^σ. The probability for this emission depends on the parameters of the original state p_z, p_x, n and on changes in the energy $\Delta\varepsilon$ and momentum k: $w_{p_z, p_x, n}(k, \Delta\varepsilon)$. Since for a given k (that is, given k_z) $\Delta\varepsilon$ is determined by the integer ν, we can just as well replace $\Delta\varepsilon$ with ν, so that $w = w_{p_z, p_x, n}(k, \nu)$. From energy conservation $\Delta\varepsilon = \omega_k^\sigma$ and

$$w_{p_z, p_x, n}^\sigma (k, \nu) \sim \delta (\omega_k^\sigma - k_z v_z - \omega_H \nu),$$

where $\omega_H = |eH| c/\varepsilon$ is the cyclotron frequency of the plasma particle.

We can now write down the balance equation for the distribution function f:

$$\frac{df_{p_z, p_x, n}}{dt} = - \sum_{\nu=-\infty}^{\infty} \int w_{p_z, n}^\sigma (k, \nu) \left[f_{p_z, p_x, n} (N_k^\sigma + 1) - \right.$$

$$\left. - N_k^\sigma f_{p_z - k_z, p_x - k_x, n-\nu} \right] + w_{p_z + k_z, p_x + k_x, n+\nu} (k, \nu) \left[N_k^\sigma f_{p_z, p_x, n} - \right.$$

$$\left. - (N_k^\sigma + 1) f_{p_z + k_z, p_x + k_x, n+\nu} \right] dk (2\pi)^{-3}. \tag{2.194}$$

The first two terms of (2.194) describe the emission of a plasmon ω_k^σ in the state p_z, p_x, n together with the reverse process of absorbing a plasmon from the state $p_z - k_z$, $p_x - k_x$, $n - \nu$. The last two terms correspond to plasmon absorption in the state p_z, p_x, n and the emission of a plasmon from the state $p_z + k_z$, $p_x + k_x$, $n + \nu$.

Let us now assume that the particle motion is quasiclassical, which means that the quantum number n is very large and $\nu \ll n$.

It is convenient to use $p_\perp$ instead of n to characterize the particle motion:

$$p_\perp^2 = 2\varepsilon\omega_H n, \quad p_\perp \Delta p_\perp = \varepsilon\omega_H \nu.$$

It will also be assumed that the distribution function depends on p_z, $p_\perp$, y:

$$f_{p_z, p_x, n} \to f_{p_z, p_\perp, y}, \quad f_{p_z-k_z, p_x-k_x, n-\nu} \to f_{p_z, p_\perp, y} - k_z \frac{\partial}{\partial p_z} f_{p_z, p_\perp, y} -$$

$$- \frac{1}{v_1} \omega_H \nu \frac{\partial}{\partial p_\perp} f_{p_z, p_\perp, y} + \frac{k_x c}{eH} \cdot \frac{\partial}{\partial y} f_{p_z, p_\perp, y},$$

and so on.

Expanding f up to the second derivatives, we obtain an equation of the form

$$\frac{df}{dt} = \frac{\partial}{\partial \lambda_i} \sum_{\nu=\infty}^{\infty} \int \frac{dk}{(2\pi)^3} N_k^\sigma \Delta\lambda_i \Delta\lambda_j w_{p_z, p_\perp, y}^\sigma (k, \nu) \frac{\partial f}{\partial \lambda_j} +$$

$$+ \frac{\partial}{\partial \lambda_i} \sum_{\nu=-\infty}^{\infty} \int \frac{dk}{(2\pi)^3} \Delta\lambda_i w_{p_z, p_\perp, y}^\sigma (k, \nu) f, \qquad (2.195)$$

$$\lambda_1 = p_z, \quad \Delta\lambda_1 = k_z, \quad \lambda_2 = p_\perp, \quad \Delta\lambda_2 = \frac{\omega_H \nu}{v_\perp},$$

$$\lambda_3 = y, \quad \Delta\lambda_3 = - \frac{k_x c}{|eH_0|}.$$

The first term in (2.195) describes the induced emission and absorption effects and the second accounts for spontaneous effects.

Let us now consider processes which are of higher order in the number of emitted or absorbed waves. Such processes will include the scattering of waves by plasma particles. We denote the scattering probability as $w_{p_z, p_\perp y}^{\sigma\sigma'} (k, k', \nu)$. In this case the momentum change is $k - k'$ (the wave for k is absorbed and that for k' is emitted). The energy change will then be

$$\varepsilon\Delta\varepsilon = p_z (k_z - k_z') + \omega_H \varepsilon\nu.$$

From energy conservation $\Delta\varepsilon = \omega_k^\sigma - \omega_{k'}^\sigma$, so that $w_{p_z, p_\perp, y}^{\sigma\sigma_1}$ is proportional to

$$\delta (\omega_k^\sigma - \omega_{k'}^{\sigma'} - v_z (k_z - k_z') - \omega_H \nu).$$

The balance equation can be written in analogy with (2.194):

$$\frac{df}{dt} = -\sum_{\nu=-\infty}^{\infty} \int \frac{dk\,dk'}{(2\pi)^6} \left\{ w_{p_z,\,p_x,\,n}^{\sigma\sigma'}(k,\,k',\,\nu) \left[f_{p_z,\,p_x,\,n}\, N_k^{\sigma}(N_{k'}^{\sigma'}+1) - \right. \right.$$

$$- f_{p_z-k_z+k_z',\,p_x-k_x+k_x',\,n-\nu}\,(N_k^{\sigma}+1)\,N_{k'}^{\sigma'} \Big] +$$

$$+ w_{p_z+k_z-k_z',\,p_x+k_x-k_{x'},\,n+\nu}^{\sigma\sigma'}(k,\,k',\,\nu) \left[f_{p_z,\,p_x,\,n}\,(N_k^{\sigma}+1)\,N_{k'}^{\sigma'} - \right.$$

$$- f_{p_z+k_z-k_z',\,p_x+k_x-k_x',\,n+\nu}\,(N_{k'}^{\sigma'}+1)\,N_k^{\sigma} \Big] \Big\}. \qquad (2.196)$$

The expansion in terms of $\Delta p_\perp$, Δp_z, Δy produces an equation similar to (2.195):

$$\frac{df}{dt} = \frac{\partial}{\partial\lambda_i} \sum_{\nu=-\infty}^{\infty} \int \frac{dk\,dk'}{(2\pi)^6}\, N_k^{\sigma}\,N_{k'}^{\sigma'}\,\Delta\lambda_i'\,\Delta\lambda_j'\,w_{p_z,\,p_\perp,\,y}^{\sigma\sigma'}(k,\,k',\,\nu)\,\frac{\partial f}{\partial\lambda_j},$$

$$\Delta\lambda_1' = k_z - k_z';\quad \Delta\lambda_2' = \frac{\omega_H\,\nu}{v_\perp},\quad \Delta\lambda_3' = -\frac{(k_x - k_x')\,c}{|eH_0|}. \qquad (2.197)$$

Only the induced processes have been included in (2.197).

It remains to take into account the processes of plasmon emission and scattering by plasma particles in those equations which describe the changes in the number of plasmons.

As an example, let us consider the changes in the plasmon population due to their emission and absorption by plasma particles

$$\frac{dN_k^{\sigma}}{dt} = \sum_{n,\,\nu} \int dp_z \left\{ (N_k^{\sigma}+1)\,w_{p_z,\,p_x,\,n}^{\sigma}(k,\,\nu)\,f_{p_z,\,p_x,\,n} - \right.$$

$$- w_{p_z,\,p_x,\,n}^{\sigma}(k,\,\nu)\,N_k^{\sigma}\,f_{p_z-k_z,\,p_x-k_x,\,n-\nu} \Big\} = \sum_{\nu=-\infty}^{\infty} \int \frac{dp}{(2\pi)^3}\,N_k^{\sigma} \times$$

$$\times w_{p_z,\,p_\perp,\,y}^{\sigma}(k,\,\nu)\,\Delta\lambda_i\,\frac{\partial f}{\partial\lambda_i} + \sum_{\nu=-\infty}^{\infty} \int dp\,(2\pi)^{-3}\,w_{p_z,\,p_\perp,\,y}^{\sigma}(k,\,\nu)\,f_{p_z,\,p_\perp,\,y}.$$

$$(2.198)$$

The first term of (2.198) gives the induced processes, while the second accounts for the spontaneous events.

Let us calculate the radiated power for the quanta σ as $N_k^\sigma \to 0$:

$$Q^\sigma = \int \frac{\omega_k^\sigma}{(2\pi)^3} \, dk \, \frac{dN_k^\sigma}{dt} \bigg|_{N_k^\sigma \to 0} = \sum_{\nu=-\infty}^{\infty} \int \frac{\omega_k^\sigma}{(2\pi)^3} \, w_{p_z, \, p_\perp}^\sigma (k, \nu) \times$$

$$\times \frac{dk \, dp}{(2\pi)^3} f_{p_z, \, p_\perp} = \int f_{p_z, \, p_\perp} \frac{dp}{(2\pi)^3} \frac{d}{dt} W_{p_z, \, p_\perp}^\sigma, \qquad (2.199)$$

where $\dfrac{d}{dt} W_{p_z, \, p_\perp}^\sigma$ is the power radiated by a single plasma particle.

We now turn to the particle scattering effects. In analogy with (2.197) we obtain

$$\frac{dN_k^\sigma}{dt} = \sum_{n, \, \nu} \int w_{p_z, \, p_x, \, n}^{\sigma\sigma'} (k, \, k', \, \nu) \left\{ N_{k'}^{\sigma'} (N_k^\sigma + 1) f_{p_z, \, p_x, \, n} - N_k^\sigma (N_{k'}^{\sigma'} + 1) \times \right.$$

$$\left. \times f_{p_z - k_z + k_z', \, p_x - k_x + k_x', \, n - \nu} \right\} \frac{dk' \, dp_z}{(2\pi)^3} = \sum_{\nu=-\infty}^{\infty} \int \frac{dp \, dk'}{(2\pi)^6} N_k^\sigma N_{k'}^{\sigma'} \, \Delta\lambda_i' \frac{\partial f}{\partial\lambda_i} \times$$

$$\times w_{p_z, \, p_\perp}^{\sigma\sigma'} (k, \, k', \, \nu) + \sum_{\nu=-\infty}^{\infty} \int \frac{dp \, dk'}{(2\pi)^6} N_{k'}^{\sigma'} w_{p_z, \, p_\perp}^{\sigma\sigma'} (k, \, k', \, \nu) f_{p_z, \, p_\perp} -$$

$$- \sum_{\nu=-\infty}^{\infty} N_k^\sigma \int \frac{dp \, dk'}{(2\pi)^6} w_{p_z, \, p_\perp}^{\sigma\sigma'} (k, \, k', \, \nu) f_{p_z, \, p_\perp}. \qquad (2.200)$$

The first term in (2.200) corresponds to the nonlinear effects arising from the interactions of the waves N_k^σ and $N_{k'}^\sigma$, and the last two terms are spontaneous scattering terms. In the limit $N_k^\sigma \to 0$ we have

$$Q^\sigma = \int \frac{\omega_k^\sigma}{(2\pi)^3} \, dk \, \frac{dN_k^\sigma}{dt} \bigg|_{N_k^\sigma \to 0} = \sum_{\nu=-\infty}^{\infty} \int \frac{dp \, dk \, dk'}{(2\pi)^9} N_{k'}^{\sigma'} f_{p_z, \, p_\perp} \times$$

$$\times w_{p_z, \, p_\perp}^{\sigma, \, \sigma'}(k, \, k', \, \nu) = \int f_{p_z, \, p_\perp} \frac{dp}{(2\pi)^3} \cdot \frac{d}{dt} W_{p_z, \, p_\perp}^{\sigma, \, \sigma'}, \qquad (2.201)$$

where $\dfrac{d}{dt} W_{p_z, \, p_\perp}^{\sigma, \, \sigma'}$ is the power scattered by a single plasma charge.

2.10. Probability Calculations

We will show a simple method of calculating the probabilities for the processes of wave decay, emission, and scattering which is a generalization of the method presented by Landau (see [4, 60]). We shall use the correspondence principle; that is, we shall find from the nonlinear equations the power emitted by the wave σ in the limit where its intensity is small. To this end we will sum up the results of (2.193), (2.199), and (2.201).

The radiated power of the wave σ can be computed from the work done by the field of the wave σ (generated in some process) on the current which is the source of the radiation. From Maxwell's equation

$$(k^2 \delta_{ij} - k_i k_j - \omega^2 \varepsilon_{ij}) E_{k, j} = 4\pi i \omega j_{k, i} \qquad (2.202)$$

one can find the field strength for the polarization e_k^σ $(E_{k, j} = e_{k, j}^\sigma E_k^\sigma)$:

$$(k^2 - \omega^2 \varepsilon_k^\sigma) E_k^\sigma = 4\pi i \omega (e_k^{\sigma*} j_k). \qquad (2.203)$$

For a decay process the current in (2.203) is the nonlinear current

$$j_{k, i}^{(2)} = 2 \int S_{ij, l}(k, k_1, k_2) E_{k_1, j}^{\sigma'} E_{k_2, l}^{\sigma''} \delta(k - k_1 - k_2) dk_1 dk_2,$$

set up by the fields of the waves σ' and σ'', $E_{kl}^{\sigma'} = e_{kl}^{\sigma'} E_k^{\sigma'}$, and the radiated power can be found from the equation

$$Q^\sigma = -\langle j^{(2)}(r, t) E^{(2)}(r, t) \rangle = -4\pi i \int \frac{\omega dk' dk e^{i(k-k')x} \langle (e_k^\sigma j_k^{(2)*})(e_k^\sigma j_k^{(2)}) \rangle}{(k^2 - \omega^2 \varepsilon_k^\sigma)}.$$

$$(2.204)$$

Since the waves σ and σ' are not correlated, the average over the turbulent fluctuations gives k = k' in (2.204). Since (2.204) is real only the imaginary part of $\frac{1}{k^2 - \omega^2 \varepsilon_k^\sigma}$, which is $i\pi\delta(k^2 - \omega^2 \varepsilon_k^\sigma)$ in the transparency region, enters into the calculation. Using (2.134) to write the expressions containing $I_k^{\sigma'}$ and $I_k^{\sigma''}$ in terms of the number of quanta, we obtain the result shown in (2.193). By comparison, we then obtain an expression for the probability of three-plasmon decay in terms of the nonlinear plasma current:

$$w_\sigma^{\sigma'\sigma''}(k, k', k'') = 32\rho (2\pi)^7 \delta(k - k' - k'') \delta(\omega_k^\sigma - \omega_{k'}^{\sigma'} - \omega_{k''}^{\sigma''}) (\omega_{k'}^{\sigma'})^2 \times$$

$$\times (\omega_{k''}^{\sigma''})^2 |S_{\sigma\sigma'\sigma''}(\omega_k^\sigma, k, \omega_{k'}^{\sigma'}, k', \omega_{k''}^{\sigma''}, k'')|^2 \left[\frac{\partial}{\partial \omega} \omega^2 \varepsilon_k^\sigma \bigg|_{\omega = \omega_k^\sigma} \times \right.$$

$$\times \frac{\partial}{\partial \omega} \omega'^2 \varepsilon_{k'}^{\sigma'} \Big|_{\omega'=\omega_{k'}^{\sigma'}} \frac{\partial}{\partial \omega} \omega''^2 \varepsilon_{k''}^{\sigma''} \Big|_{\omega''=\omega_{k''}^{\sigma''}} \Big]^{-1} ; \tag{2.205}$$

$$S_{\sigma\sigma', \sigma''}(k, k', k'') = e_{k, l}^{\sigma*} e_{k', j}^{\sigma'} e_{k'', l}^{\sigma''} S_{ijl}(k, k', k''), \tag{2.206}$$

where e_k^σ are normalized unit vectors introduced in §2 of this chapter $(\rho = 1, \sigma' \neq \sigma''; \rho = 1/2, \sigma' = \sigma'')$.

In addition to the current $j^{(3)}$, which describes the process shown in Fig. 2.3a, for four-plasmon decay one must also include the effects which arise from iterating the current $j^{(2)}$. According to Ref. 80 the matrix elements for the two graphs cited commute with each other very strongly in the case of wave scattering on plasma particles, as an example. The general formula for the probability of a four-plasmon decay takes the form

$$w_\sigma^{\sigma'\sigma''\sigma'''}(k, k', k'', k''') = 288(2\pi)^8 \delta(k - k' - k'' - k''') \delta(\omega_k^\sigma - \omega_{k'}^{\sigma'} -$$

$$- \omega_{k''}^{\sigma''} - \omega_{k'''}^{\sigma'''}) (\omega_{k'}^{\sigma'})^2 (\omega_{k''}^{\sigma''})^2 (\omega_{k'''}^{\sigma'''})^2 |\Sigma_{\sigma\sigma'\sigma''\sigma'''}^{\text{eff}} (\omega_k^\sigma, k; \omega_{k'}^{\sigma'}, k';$$

$$\omega_{k''}^{\sigma''}, k''; \omega_{k'''}^{\sigma'''}, k''')|^2 \Big[\frac{\partial}{\partial \omega} \omega^2 \varepsilon_k^\sigma \frac{\partial}{\partial \omega'} \omega'^2 \varepsilon_{k'}^{\sigma'} \times$$

$$\times \frac{\partial}{\partial \omega''} \omega''^2 \varepsilon_{k''}^{\sigma''} \frac{\partial}{\partial \omega'''} \omega'''^2 \varepsilon_{k'''}^{\sigma'''} \Big]^{-1}_{\substack{\omega=\omega_k^\sigma, \omega'=\omega_{k'}^{\sigma'} \\ \omega''=\omega_{k''}^{\sigma''}, \omega'''=\omega_{k'''}^{\sigma'''}}} ; \tag{2.207}$$

$$\Sigma_{\sigma\sigma'\sigma''\sigma'''}^{\text{eff}}(k, k', k'', k''') = e_{k, l}^{\sigma*} e_{k', j}^{\sigma'} e_{k'', l}^{\sigma''} e_{k''', s}^{\sigma'''} \Sigma_{i, j, l, s}^{\text{eff}}(k, k', k'', k''');$$

$$\Sigma_{ijls}^{\text{eff}}(k, k' k'', k''') = \Sigma_{ijls}(k, k', k'', k''') +$$

$$+ 8\pi i (\omega - \omega') S_{ijm}(k, k', k - k') \Pi_{mn}(k - k') S_{nls}(k - k', k'', k''');$$

$$(k^2 \delta_{ij} - k_i k_j - \frac{1}{c^2}\omega^2 \varepsilon_{ij}) \Pi_{j'} = \delta_{il}.$$

We can use the expression for the power emitted by a single particle to compute the emission and scattering probabilities. The current density thus created is

$$j = ev(t) \delta(r - r(t)). \tag{2.208}$$

The magnetized and unmagnetized particles must be treated separately. The magnetization criterion is the ratio of the emitted wave's wavelength to the particle's Larmor radius (the particle is magnetized if its Larmor radius is much smaller than the wavelength).

We will now consider the general case of a particle moving in a magnetic field. Since the speed is constant in a magnetic field, the equation of motion for the charge can be written as

$$\frac{dv_x}{dt} = -\omega_H v_y, \qquad \frac{dv_y}{dt} = \omega_H v_x, \qquad \omega_H = \frac{eH_0 c}{\varepsilon},$$

from which we find

$$v_x = v_\perp \cos(\omega_H t + \varphi_0), \qquad v_y = -v_\perp \sin(\omega_H t + \varphi_0), \qquad v_z = \text{const},$$

$$x = x_0 + \frac{v_\perp}{\omega_H} \sin(\omega_H t + \varphi_0), \qquad y = y_0 + \frac{v_\perp}{\omega_H} \cos(\omega_H t + \varphi_0),$$

$$z = v_z t + z_0. \tag{2.209}$$

The Fourier transform of the current in (2.208) is

$$j_k = \frac{1}{(2\pi)^4} \int j_i(\mathbf{r},\, t) e^{-i\mathbf{kr}+i\omega t}\, d\mathbf{r}\, dt = \frac{e}{(2\pi)^4} \int \mathbf{v}\,(t) e^{-i\mathbf{kr}\,(t)+i\omega t} dt. \tag{2.210}$$

If the explicit forms for $\mathbf{v}(t)$ and $\mathbf{r}(t)$ are known in (2.209), then (2.210) gives a rather simple means of computing j_k. For example we have for j_{kz} (when x is in the $\mathbf{k}$, $\mathbf{H_0}$ plane)

$$j_{k,\,z} = \frac{ev_z}{(2\pi)^4} \int e^{i\omega t - i\zeta \sin(\omega_H t + \varphi_0) - ik_z v_z t - i\mathbf{kr}_0}\, dt, \tag{2.211}$$

$$\zeta = \frac{k_\perp v_\perp}{\omega_H}. \tag{2.212}$$

By using the well-known expansion

$$e^{-i\zeta \sin \varphi} = \sum_{\nu=-\infty}^{\infty} e^{-i\nu\varphi} J_\nu(\zeta), \tag{2.213}$$

where j_ν is a Bessel function, we can calculate the integral in (2.211) without great difficulty:

$$j_{k,[z} = \frac{ev_z}{(2\pi)^3} \sum_{\nu=-\infty}^{\infty} e^{-i\nu\varphi_0 - i\mathbf{kr}_0} J_\nu(\zeta)\, \delta(\omega - k_z v_z - \nu\omega_H). \tag{2.214}$$

The remaining components are computed in the same manner, and the general expression for the current takes the form

$$j_{k, i}^{a} = \frac{e_a}{(2\pi)^3} \sum_{\nu=-\infty}^{\infty} \Gamma_{k, \nu, i}^{a} e^{-i\nu\varphi_\bullet - i k r_\circ} \delta(\omega - k_z v_z - \nu\omega_{Ha}); \quad (2.215)$$

$$
\begin{aligned}
\Gamma_{k, \nu, 1} &= \frac{\nu J_\nu(\zeta_a)}{\zeta_a} v_\perp; \\
\Gamma_{k, \nu, 2} &= -i v_\perp J_\nu'(\zeta_a); \\
\Gamma_{k, \nu, 3} &= v_z J_\nu(\zeta_a).
\end{aligned}
\right\} \quad (2.216)
$$

By using (2.203) we can find the average over the initial phase φ_0 of the work done by the field E^σ, created by the current density j_k, on this very same current density. This work is identical with the radiated power:

$$\frac{dW_{p_z, p_\perp}^{\sigma}}{dt} = -\int j(r, t) E(r, t) dr = -(2\pi)^3 \int \frac{dk d\omega d\omega'}{k^2 - \omega^2 \varepsilon^\sigma} 4\pi i \omega (j_{k, \omega}^* e_k^\sigma) \times$$

$$\times (e_k^{\sigma*} j_k) \exp[-i(\omega - \omega')t] = -\frac{4\pi i}{(2\pi)^3} \sum_{\nu, \nu'=-\infty}^{\infty} \int \exp[-i(\omega - \omega')t -$$

$$-i(\nu - \nu')\varphi_0] (\Gamma_{k\nu'}^* e_k^\sigma)(\Gamma_{k, \nu} e_k^{\sigma*}) \delta(\omega - k_z v_z - \nu\omega_H) \times$$

$$\times \delta(\omega' - k_z v_z - \nu'\omega_H) \frac{dk d\omega d\omega'}{k^2 - \omega^2 \varepsilon^\sigma}. \quad (2.217)$$

The average over the initial phase is

$$\frac{1}{2\pi} \int e^{-i(\nu - \nu')\varphi_0} d\varphi_0 = \delta_{\nu, \nu'}. \quad (2.218)$$

We then have

$$Q_{p_z, p_\perp}^\sigma = \frac{dW_{p_z, p_\perp}^\sigma}{dt} = \frac{e^2}{\pi} \sum_{\nu=-\infty}^{\infty} \int \frac{\omega_k^\sigma dk}{\left| \frac{\partial}{\partial \omega} \omega^2 \varepsilon_k^\sigma \right|_{\omega=\omega_k^\sigma}} \times$$

$$\times |\Gamma_{k, \nu} e_k^{\sigma*}|^2 \delta(\omega_k^\sigma - k_z v_z - \nu\omega_H). \quad (2.219)$$

By comparing this result with (2.199) we find

$$w^\sigma_{p_z, p_\perp}(\mathbf{k}, v) = 8\pi^2 e^2 \frac{\left|\Gamma_{\mathbf{k}, v}\, e^{\sigma*}_{\mathbf{k}}\right|^2}{\left|\dfrac{\partial}{\partial\omega}\, \omega^2 \varepsilon^\sigma_k\right|_{\omega=\omega^\sigma_{\mathbf{k}}}} \delta\left(\omega^\sigma_{\mathbf{k}} - k_z v_z - v\omega_H\right). \tag{2.220}$$

This equation solves the general problem of computing the probabilities of plasma particle radiation. When $v = 0$ the radiation is called Cerenkov radiation and when $v \neq 0$ it is cyclotron radiation. This notation is selected because in a frame of reference which moves along the magnetic field with the charge's velocity v_z, the frequency ω^σ_k is transformed into $(\omega^\sigma_k)' = \omega^\sigma_k - k_z v_z$ due to the Doppler effect. Thus, when $v = 0, (\omega^\sigma_k)' = 0$, that is, a constant field acts on the particle and the particle is in resonance with the wave. When $v \neq 0$ we have $(\omega^\sigma_k)' = v\omega_H$, so that the frequency is a multiple of the gyrofrequency of the electron; we now have cyclotron resonance. Note that for longitudinal waves

$$\left(\Gamma_{\mathbf{k}, v}\, e^{\sigma*}_{\mathbf{k}}\right) = \frac{\Gamma_{\mathbf{k}, v}\, k}{k} = \frac{\omega^\sigma_{\mathbf{k}}}{k} J_v(\zeta)$$

and

$$\frac{\partial}{\partial\omega}\, \omega^2 \varepsilon^\sigma_k\Big|_{\omega=\omega^\sigma_{\mathbf{k}}} = (\omega^l_{\mathbf{k}})^2 \frac{\partial \varepsilon^l_k}{\partial\omega}\Big|_{\omega=\omega^l_{\mathbf{k}}}$$

so that we obtain

$$w^l_{p_z, p_\perp}(\mathbf{k}, v) = 8\pi^2 \frac{e^2}{k^2} \frac{J^2_v(\zeta)\, \delta\left(\omega^l_{\mathbf{k}} - k_z v_z - v\omega_H\right)}{\dfrac{\partial\varepsilon^l_k}{\partial\omega}\Big|_{\omega=\omega^l_{\mathbf{k}}}}. \tag{2.221}$$

One frequently encounters the situation in which either one of the particles (electrons or ions) or both of them can, for all practical purposes, be considered as nonmagnetized; that is, the average Larmor radius of the particle is much larger than $1/k_\perp = \lambda_\perp$, which is the wavelength of the plasmon taken transverse to the magnetic field. In this case $\zeta \gg 1$ and all the v up to some very large number make a contribution, which makes use of (2.221) inconvenient.

It is simplest to assume from the outset that the charge moves almost in a straight line during the time the wave σ is emitted, and to use the following expression for the current density:

$$j_k = \frac{ev}{(2\pi)^3}\, e^{-ikr_0}\, \delta\,(\omega - kv).\qquad(2.222)$$

This gives the following expression for the probabilities:

$$w_p^\sigma\,(k) = 8\pi^2\, e^2\, \frac{|ve_k^\sigma|^2\, \delta\,(\omega_k^\sigma - kv)}{\left.\dfrac{\partial}{\partial\omega}\,\omega^2\, \varepsilon_k^\sigma\right|_{\omega=\omega_k^\sigma}},\qquad(2.223)$$

and for longitudinal waves,

$$w_p^l\,(k) = 8\pi^2\, \frac{e^2}{k^2}\, \frac{\delta\,(\omega^l - kv)}{\left.\dfrac{\partial\varepsilon_k^l}{\partial\omega}\right|_{\omega=\omega_k^l}}.\qquad(2.224)$$

The latter expressions can be used for a plasma in which $H_0 = 0$.

We now turn to the scattering probability. As pointed out in the first approximation, the scattering amplitude is described by two diagrams (see Figs. 2.6 and 2.9). Physically, the first diagram describes a process in which the wave σ' sets the charge into oscillation and the charge, as a consequence of these oscillations, emits the wave σ (the radiation results in all kinds of waves, but we are now interested in just this process). In the first approximation the current excited by the vibrating charge is proportional to the field $E_k^{\sigma'}$ of the wave:

$$j_k^{ac} = \sum_{v=-\infty}^{\infty} \int \Lambda^{ac}\,(k,\,k',\,v)\,E_k^{\sigma'}\,dk'\,\delta\,(\omega - \omega' - (k_z - k_z')\,v_z - v\omega_{Ha}).\qquad(2.225)$$

Usually the vector Λ^c can be easily computed from perturbation theory if one assumes in the first approximation that the charge moves in a helical path; the field $E_k^{\sigma'}$ then weakly perturbs this motion.

The second kind of scattering, shown in Fig. 2.9, can be quite graphically interpreted as the radiation from a charge which moves along a helix (without perturbation) in a plasma whose density and other parameters are perturbed by the wave $E_k^{\sigma'}$. The resulting radiation is similar to the transition radiation from a charge moving in a medium having a density which varies periodically in both space and time [63–65].

The current describing such radiation is nonlinear, created by the field $E_{k'}^{\sigma'}$ and the field E_k^a of the charge moving along its helical path:

$$j_{k,i}^{aN} = \int S_{ijl}(k, k_1, k_2)\, dk_1\, dk_2 \times$$

$$\times\, \delta(k - k_1 - k_2)\,(E_{k_1,j}^a E_{k_2,l}^{\sigma'} + E_{k_1,j}^{\sigma'} E_{k_2,l}^a). \tag{2.226}$$

Assuming that S_{ijl} has the following symmetry property:

$$S_{ijl}(k, k_1, k_2) = S_{i,l,j}(k, k_2, k_1),$$

we find

$$j_{k,i}^{aN} = 2\int S_{ijl}(k, k', k - k')\, E_{k',j}^{\sigma}\, E_{k-k',l}^a\, dk'. \tag{2.227}$$

The field $E_{k,l}^a$ can be found from the current $j_{k,l}^a$ of (2.215) and Maxwell's equation:

$$\left(k^2\, \delta_{ij} - k_i\, k_j - \frac{1}{c^2}\omega^2\, \varepsilon_{ij}\right) E_{k,j}^a = 4\pi i\omega j_{k,i}^a\, c^{-2}. \tag{2.228}$$

We define the inverse Maxwell operator $\Pi_{ij}(k)$ as

$$\Pi_{si}(k)\left(k^2\, \delta_{ij} - k_i\, k_j - \frac{1}{c^2}\omega^2\, \varepsilon_{ij}\right) = \delta_{sj}. \tag{2.229}$$

Then

$$E_{k-k',l}^a = 4\pi i\,(\omega - \omega')\,\Pi_{ls}(k - k')\sum_{\nu=-\infty}^{\infty}\frac{e}{(2\pi)^3}\,\Gamma_{k-k',s}^{a'} \times$$

$$\times\, \exp\left[-i\,(\mathbf{k} - \mathbf{k}')\,\mathbf{r}_0 - i\nu\varphi_0\right]\,\delta\left(\omega - \omega' - (k_z - k_z')\,v_z - \nu\omega_{Ha}\right). \tag{2.230}$$

Here Γ' is the vector of (2.216) in an arbitrary coordinate system rather than the frame of reference in which $k_y = 0$, where (2.216) is valid; it differs from Γ in that Γ_1' and Γ_2' are linear combinations of Γ_1 and Γ_2. The current which describes the nonlinear scattering has a form similar to (2.225):

$$j_k^{aN} = \sum_{v=-\infty}^{\infty} \int \Lambda^N (k, k', v) E_k^{\sigma'} dk' \times$$
$$\times \exp \left[-i(k-k')r_0 - iv\varphi_0 \right] \delta (\omega - \omega' - (k_z - k_z') v_z - v\omega_{Ha}), \qquad (2.231)$$

where

$$\Lambda_i^N (k, k', v) = \frac{8\pi ei (\omega - \omega')}{(2\pi)^3} e_{k', j}^{\sigma'} S_{ijl} (k, k', k-k') \Pi_{ls} (k-k') \Gamma_{k-k', s}^{'a}. \qquad (2.232)$$

The total current describing the scattering is then the sum of j_k^{aN} and j_k^{ac}. This means that the amplitudes corresponding to the two graphs add, while the probabilities do not. In the classical limit this means that these two mechanisms for emitting σ waves interfere with one another. In the case of unmagnetized particles, the sum over δ is absent in both (2.225) and (2.231), while the argument of the ν function becomes $\omega - \omega' - (k - k')\mathbf{v}$.

The calculation of the scattering probability by using the current $j_k^{aN} + j_k^{ac}$ is very much like our previous calculations. We will therefore just cite the result [81]:

$$w_{p_z, p_\perp}^{\sigma' \sigma} (k, k', v) = \frac{4 (2\pi)^9 \left| \Lambda_{\sigma\sigma'} (\omega_k^\sigma, k, \omega_{k'}^{\sigma'}, k', v) \right|^2}{\frac{\partial}{\partial\omega} \omega^2 \varepsilon_k^\sigma \Big|_{\omega = \omega_k^\sigma} \frac{\partial}{\partial\omega} \omega^2 \varepsilon_{k'}^{\sigma'} \Big|_{\omega = \omega_{k'}^{\sigma'}}} \times$$
$$\times (\omega_{k'}^{\sigma'})^2 \delta (\omega_k^\sigma - \omega_{k'}^{\sigma'} - (k_z - k_z') v_z - v\omega_{Ha}), \qquad (2.233)$$

where

$$\Lambda_{\sigma\sigma'} = e_{k, i}^{\sigma*} (\Lambda_i^{ac} + \Lambda_i^{a, N}). \qquad (2.234)$$

There is a range of the parameters in which the nonlinear scattering can be the most important (for scattering on thermal ions and so on). Thus, one can in a number of instances obtain some simpler and more explicit equations for the scattering probabilities. We note that the operator Π_{ij} is the Green's function for the virtual wave in the nonlinear scattering process shown in Fig. 2.9. Of special importance is the case where the virtual wave can be considered to be longitudinal, at least to a good approximation. It frequently turns out that such scattering is the most probable. In this case

$$\Pi_{ij} \approx - \frac{1}{\omega^2 \, \varepsilon_k^l} \cdot \frac{k_i \, k_j}{k^2}, \tag{2.235}$$

where ε^l is determined by (2.39). Using the equality

$$\frac{(\Gamma_{k-k'}^{'a}, k-k')}{|k-k'|} = \frac{\omega_k^\sigma - \omega_{k'}^{\sigma'}}{|k-k'|} J_\nu(\zeta_-^a), \qquad \zeta_-^a = \frac{|k-k'|_\perp \, v_\perp}{\omega_{Ha}},$$

we find the expression

$$w_{p_z, \, p_\perp}^{l', \, l}(k, \, k', \, v) = \frac{32 \, \left| S^l \left(\omega_k^l, \, k, \, \omega_{k'}^{l'}, \, k', \, \omega_k^l - \omega_{k'}^{l'}, \, k-k' \right) \right|^2}{\pi \, |k-k'|^2 \, |\varepsilon^l \left(\omega_k^l - \omega_{k'}^{l'}, \, k-k' \right)|^2} \times$$

$$\times \, \frac{(2\pi)^6 \, e^2 J_\nu^2 \left(\zeta_-^a \right) \delta \left(\omega_k^l - \omega_{k'}^{l'} - \left(k_z - k_z' \right) v_z - \nu \omega_{Ha} \right)}{\left(\omega_k^l \right)^2 \frac{\partial}{\partial \omega} \varepsilon_k^l \Big|_{\omega = \omega_k^l} \frac{\partial}{\partial \omega'} \varepsilon_{k'}^{l'} \Big|_{\omega' = \omega_{k'}^{l'}}} \tag{2.236}$$

in the case where the nonlinear scattering is predominant for scattering of longitudinal waves into longitudinal waves [58]; here

$$S^l(k, \, k_1, \, k_2) = S_{ijl}(k, \, k_1, \, k_2) \frac{k_i \, k_{1j} \, k_{2l}}{|k| \, |k_1| \, |k_2|}. \tag{2.237}$$

When the particle motion can be treated as unmagnetized one can, in analogy with (2.224), replace (2.236) with [58]

$$w_p^{l, \, l'}(k, \, k') = \frac{32}{\pi} \, \frac{\left| S^l \left(\omega_k^l, \, k, \, \omega_{k'}^{l'}, \, k', \, \omega_k^l - \omega_{k'}^{l'}, \, k-k' \right) \right|^2}{|k-k'|^2 \, |\varepsilon^l \left(\omega_k^l - \omega_{k'}^{l'}, \, k-k' \right)|^2} \times$$

$$\times \, \delta \left(\omega_k^l - \omega_{k'}^{l'} - (k-k') v \right) e^2 \, (2\pi)^6 \left(\omega_k^l \right)^{-2} \times$$

$$\times \left(\frac{\partial}{\partial \omega} \varepsilon_k^l \right)_{\omega = \omega_k^l}^{-1} \left(\frac{\partial}{\partial \omega'} \varepsilon_{k'}^{l'} \right)_{\omega' = \omega_{k'}^{l'}}^{-1}. \tag{2.238}$$

This equation can also be conveniently used for an isotropic plasma. These equations enable us to make explicit calculations of the coefficients in the nonlinear equations for both longitudinal oscillations and arbitrary electromagnetic oscillations.

When considering the effects of induced scattering of high-frequency turbulent fluctuations in a magnetically active plasma on ions, one must know the nonlinear current [according to (2.237)] for which one of the frequencies is much smaller than the other two. This nonlinear current will be determined by the electrons. We can write an equation for the corrections to the electron distribution function $f_k^{(2)}$:

$$
\left[i\,(\omega - \mathbf{kv}) + \left(\omega_{He}\,\frac{\partial}{\partial\varphi} \right) \right] f_k^{(2)} = -e \int F_{k_1} \times
$$

$$
\times\,\frac{\partial f_{k_2}^{(1)}}{\partial \mathbf{p}}\,\delta\,(k - k_1 - k_2)\,dk_1\,dk_2;
$$

$$
F_k = E_k \left(1 - \frac{\mathbf{kv}}{\omega} \right) + \frac{\mathbf{k}\,(\mathbf{vE}_k)}{\omega}\;.
$$

(2.239)

The function $f_k^{(2)}$ is a correction of second order in the field E_k when $H_0 \neq 0$. We shall assume that the fields for k_1 and k are high-frequency fields. Then, neglecting the terms of order $\mathbf{kv}/\omega$, one can write

$$
\left(i\omega + \omega_{He}\frac{\partial}{\partial\varphi} \right) f_k^{(2)} = -e \int E_{k_1}\,\frac{\partial f_{k_2}}{\partial \mathbf{p}}\,\delta\,(k - k_1 - k_2)\,dk_1\,dk_2.
$$

(2.240)

Multiplying (2.239) by v_z, v_x, v_y and integrating over $\mathbf{p}$, we find a system of equations for the current components:

$$
j_{k,\,x}^{(2)} = \frac{\omega_{pe}^2\,i\omega}{4\pi n_0\,(\omega^2 - \omega_{He}^2)} \int \left(E_{k_1\,x} + i\,\frac{\omega_{He}}{\omega}\,E_{k_1\,y} \right) \times
$$

$$
\times\,dk_1\,dk_2\,\delta\,(k - k_1 - k_2) \int f_{k_2}^{(1)}\,\frac{d\mathbf{p}}{(2\pi)^3}\;;
$$

$$
j_{k,\,y}^{(2)} = \frac{i\omega_{pe}^2\,\omega}{4\pi n_0\,(\omega^2 - \omega_{He}^2)} \int \left(E_{k_1\,y} - i\,\frac{\omega_{He}}{\omega}\,E_{k_1\,x} \right) \times
$$

$$
\times\,dk_1\,dk_2\,\delta\,(k - k_1 - k_2) \int f_{k_2}^{(1)}\,\frac{d\mathbf{p}}{(2\pi)^3}\;;
$$

$$
j_{k,\,z}^{(2)} = \frac{i\omega_{pe}^2}{4\pi\omega n_0} \int dk_1\,dk_2\,\delta\,(k - k_1 - k_2)\,E_{k_1,\,z} \int f_{k_2}^{(1)}\,\frac{d\mathbf{p}}{(2\pi)^3}\;.
$$

Using the fact that

$$\int f_{k_1}^{(1)} \frac{dp}{(2\pi)^3} = -\frac{1}{e}\rho_{k_1}^{(1)} = -\frac{\left(kj_{k_1}^{(1)}\right)}{e\omega_2} =$$

$$= -\frac{k_{2i}\sigma_{ij}(k_2)E_{k_1 j}}{e\omega_2} = -\frac{k_{2i}}{4\pi ei}\left(\varepsilon_{ij}(k_2) - \delta_{ij}\right)E_{k_1, i}, \qquad (2.241)$$

we obtain relations which express the nonlinear currents in terms of the linear electronic permeability [58]:

$$S_{1, ij} - i\frac{\omega_{He}}{\omega}S_{2, ij} = \frac{\omega_{pe}^2 k_{2s}}{(4\pi)^2 n_0 e\omega}\delta_{i1}\left(\varepsilon_{sj}^{(e)}(k_2) - \delta_{sj}\right), \qquad (2.242)$$

$$S_{2, ij} + i\frac{\omega_{He}}{\omega}S_{1, ij} = \frac{\omega_{pe}^2 k_{2s}\delta_{i2}}{(4\pi)^2 e\omega n_0}\left(\varepsilon_{sj}^{(e)}(k_2) - \delta_{sj}\right), \qquad (2.243)$$

$$S_{3, ij} = \frac{\omega_{pe}^2 \delta_{i3} k_{2s}}{(4\pi)^2 e\omega n_0}\left(\varepsilon_{sj}^{(e)}(k_2) - \delta_{sj}\right). \qquad (2.244)$$

In addition to the S_{ijl} one sometimes needs to know the nonlinear currents [see (2.226)] in which $k_1 \rightleftarrows k_2$ and $i \rightleftarrows j$. This then solves the general problem of finding the possible probabilities. This procedure is easily extended to processes involving a larger number of plasmons.

Thus, we see that it is possible to construct a closed system of equations which include the interactions of turbulent fluctuations among themselves and with the plasma particles. Further, we have presented a method for computing the matrix elements for these processes without involving ourselves in the complicated process of deriving the complete equations. The basic physical premises which are at the heart of this method were presented in [59] and further developed in [60, 61, 68-70] among others (see the review [62]). It has been assumed throughout that only a homogeneous external magnetic field is present. In the more interesting case in which the plasma is inhomogeneous or the external fields are weak, the inhomogeneity can be neglected in the nonlinear interaction.

Chapter 3

Collective Dissipation and Excitation of Turbulent Fluctuations

3.1. Collective Dissipation of Turbulent Fluctuations in Plasma

A necessary element for establishing a stationary turbulence in plasma is the presence of a region for the dissipation and excitation of fluctuations. As has already been mentioned, a very effective collective absorption of fluctuations can take place in plasma in addition to the usual absorption from pairwise collisions, emission of electromagnetic waves, and particle acceleration. We shall study in detail the collective mechanisms for absorption and excitation of fluctuations. The theory of turbulence requires that one know not only the location of the region producing intense dissipation of the fluctuations, but in addition one needs to know how the absorption changes as the fluctuation frequencies change. It was pointed out in the previous chapter that the particle heating during spectral transfer reduces the frequency of the fluctuations. Further, the direction of this transfer can be altered if the thermal particles have a nonequilibrium distribution. A magnetic field will increase the number of intense-dissipation regions, thereby promoting turbulent heating, because one can have the situation where a transfer of turbulent energy in any spectral direction will encounter an absorption region.

Collective dissipation effects can be separated into linear and nonlinear effects. Our first concern will be with the linear effects.

The most effective dissipation takes place for the low-frequency fluctuations. As a rule, low-frequency fluctuations have small phase velocities, which means that they are efficiently absorbed by plasma particles. For the simplest case of Cerenkov resonance with $H_0 = 0$ we have $\omega = \mathbf{k}\mathbf{v} = kv \cos\theta$, $\omega/k = v_{ph} = v\cos\theta$, i.e., the wave phase velocity must be less than the particle velocity for an effective interaction. Therefore, fluctuations having phase velocities smaller than the mean thermal velocities of the electrons can be efficiently absorbed by the electrons.

To begin, we shall consider the absorption of fluctuations in a uniform plasma with $H_0 = 0$. From (2.61) the absorption coefficient for electrons having a Maxwellian distribution is

$$\gamma_k^l = \frac{1}{2} \int w_{\mathbf{p}}^l \left(\mathbf{k} \, \frac{\partial f}{\partial \mathbf{p}} \right) \frac{d\mathbf{p}}{(2\pi)^3} = - \int \frac{w_{\mathbf{p}}^l \, \omega_k^l \, f}{2 \, (2\pi)^3 \, m_e \, v_{Te}^2} \, d\mathbf{p}. \tag{3.1}$$

Substituting $\omega_k^l = \omega^s$ for ion-acoustic fluctuations from (1.42), and

$$\varepsilon_k^s = - \frac{\omega_{pi}^2}{\omega^2} + \frac{1}{k^2 \, d_e^2}, \qquad \frac{\partial \varepsilon_k^s}{\partial \omega} = \frac{2\omega_{pi}^2}{\omega^3},$$

$$w_{\mathbf{p}}^s = \frac{e^2 \, (\omega_k^s)^3}{\omega_{pi}^2 \, k^2} \, \delta(\omega_k^s - \mathbf{k}\mathbf{v}) \, (2\pi)^2,$$

$$f = \frac{n \, (2\pi)^{3/2}}{m_e^3 \, v_{Te}^3} \, \exp\left\{ \frac{-v^2}{2v_{Te}^2} \right\},$$

we find

$$\gamma_k = - \sqrt{\frac{\pi}{8}} \sqrt{\frac{m_e}{m_i}} \, \frac{\omega_k^s}{(1 + k^2 \, d_e^2)^{3/2}}. \tag{3.2}$$

The damping of waves due to Cerenkov absorption and emission by plasma particles was first studied by Landau [15] for Langmuir oscillations. This damping is called Landau damping. We reemphasize that (3.2) holds for electrons in a Maxwell distribution.

In analogy to (3.1), we have for nonresonant Langmuir fluctuations [15] that

$$\left.\begin{aligned} \frac{\partial \varepsilon_k^l}{\partial \omega} &= \frac{2\omega_{pe}^2}{\omega^3} = \frac{2}{\omega_{pe}}, \\ \gamma_k &= -\sqrt{\frac{\pi}{8}} \frac{\omega_{pe}^4}{k^3 d_e^3} \exp\left\{-\frac{\omega_{pe}^2}{2k^2 v_{Te}^2} - \frac{3}{2}\right\}. \end{aligned}\right\} \qquad (3.3)$$

When effects due to turbulence broadening of the resonances are included in γ, a correction $\delta\gamma$ appears, which, according to §2.8 for a Maxwell distribution, is [78]

$$\frac{\delta\gamma}{\gamma} = \frac{1}{\pi} \int_0^\infty e^{-v^2} dy^2 \int_0^\infty \frac{\sin\tau\, d\tau}{\tau} \times$$

$$\times \left\{ \left[\cos\frac{\omega}{kv_{Te}} \cdot \frac{\tau}{y} - \frac{\sin\left(\frac{\omega}{kv_{Te}} \cdot \frac{\tau}{y}\right)}{\left(\frac{\omega}{kv_{Te}} \cdot \frac{\tau}{y}\right)} \cdot \frac{\tau^3 \alpha}{y^4} \right] \exp\left(-\frac{\alpha\tau^3}{3y^4}\right) - 1 \right\};$$

$$\alpha = \pi \frac{e^2}{m_e^2} \int \frac{|\varphi_{k_1}|^2 k_1\, dk_1}{kv_{Te}^4} = \frac{\nu_{eff}}{kv_{Te}}.$$

The integral contains cutoff factors with the following characteristics: $\tau_1 = y\,(kv_{Te}/\omega)$ and $\tau_2 = \alpha^{-1/3}y^{4/3}$. If the level of turbulence is low, $\tau_2 \gg \tau_1$, then $\delta\gamma/\gamma \sim (W/nT)(m_i/m_e)$. This limit holds for $W/nT_e \ll (m_e/m_i)^2$ and the maximum $\delta\gamma/\gamma \sim m_e/m_i \ll 1$. When $W/nT_e \gg (m_e/m_i)^2$ we have $\tau_1 \gg \tau_2$ and $\delta\gamma/\gamma \sim (W/nT_e)^{1/2} \ll 1$. It is clear from (3.3) that the Landau damping is appreciable if the phase velocity is not very large compared with the thermal velocity; however, when $W/nT_e \ll 1$ the damping is exponentially small.

The phase velocity of ion–acoustic waves is greater than v and the ion Landau damping is very much like that of Eq. (3.3):

$$\gamma_k = -\sqrt{\frac{\pi}{8}} \frac{\omega_s^4}{k^3 v_{Ti}^3} \exp\left\{-\frac{\omega_s^2}{2k^2 v_{Ti}^2}\right\}. \qquad (3.4)$$

In order to have turbulent absorption of Langmuir waves in a homogeneous plasma, one must have either an effective transfer into the Landau absorption region (that is, a decrease in the oscillation phase velocities), or a transfer of the Langmuir oscillations into ion-acoustic waves. In the first instance the frequency of the turbulent oscillations increases, while in the second it decreases.

If the phase velocities of the Langmuir waves increase until they are on the order of the speed of light, then, as a rule, the electromagnetic radiation effects become significant and this type of transfer can produce another dissipative process for radiating energy from the plasma.

Since the ion-acoustic oscillations undergo relatively strong absorption, their transfer is possible if the transfer growth rate is greater than (3.2). If the condition for transfer is satisfied in a relatively narrow range Δk, transfer in any direction away from this interval leads to absorption [Eq. (3.2)].

We now turn to a consideration of the fluctuations in a magneto-active homogeneous plasma. Landau damping, for which $\nu = 0$, will be treated first. According to (2.50), for a Maxwellian distribution the damping due to electrons is

$$
\begin{aligned}
\gamma_{\mathbf{k},\,0}^{\sigma} &= \frac{1}{2} \int w_{\mathbf{p}}^{\sigma}(\mathbf{k},\,0)\,\Delta\lambda_i\,\frac{\partial f}{\partial\lambda_i}\cdot\frac{d\mathbf{p}}{(2\pi)^3} = \\
&= \frac{1}{2}\int w_{\mathbf{p}}^{\sigma}(\mathbf{k},\,0)\left(k_z\frac{\partial f}{\partial p_z}+\frac{\omega_H\,\nu}{\upsilon_\perp}\cdot\frac{\partial f}{\partial p_\perp}\right)\frac{d\mathbf{p}}{(2\pi)^3} = -\frac{1}{2}\int w_{\mathbf{p}}^{\sigma}(\mathbf{k},\,0)\times \\
&\times\frac{\omega_{\mathbf{k}}^{\sigma}}{m_e\,\upsilon_{Te}^2}\,f\,\frac{d\mathbf{p}}{(2\pi)^3} = -\frac{e^2\,\omega_{\mathbf{k}}^{\sigma}}{2\pi}\int\frac{\left|e_{\mathbf{k}}^{\sigma*}\,\Gamma_{\mathbf{k},\,0}\right|^2\,\delta\left(\omega_{\mathbf{k}}^{\sigma}-k_z\upsilon_z\right)f}{\left.\dfrac{\partial\omega^2\,\varepsilon_{\mathbf{k}}^{\sigma}}{\partial\omega}\right|_{\omega=\omega_{\mathbf{k}}^{\sigma}}\,m_e\,\upsilon_{Te}^2}\,d\mathbf{p}.
\end{aligned}
\tag{3.5}
$$

Integrating over v_z we find

$$
\gamma_{\mathbf{k},\,0}^{\sigma} = -(2\pi)^{3/2}\,\frac{e^2\,n\,\exp\left\{-\dfrac{(\omega_{\mathbf{k}}^{\sigma})^2}{2k_z^2\,\upsilon_{Te}^2}\right\}\omega_{\mathbf{k}}^{\sigma}}{m_e\left.\dfrac{\partial}{\partial\omega}\,\omega^2\,\varepsilon_{\mathbf{k}}^{\sigma}\right|_{\omega=\omega_{\mathbf{k}}^{\sigma}}\left|k_z\right|\upsilon_{Te}^5}\times
$$

$$
\times\int_0^\infty\left|e_z^{\sigma*}\,\frac{\omega_{\mathbf{k}}^{\sigma}}{k_z}\,J_0(\zeta)-ie_y^{\sigma*}\,\upsilon_\perp\,J_0'(\zeta)\right|^2 e^{-\dfrac{\upsilon_\perp^2}{2\upsilon_{Te}^2}}\upsilon_\perp\,d\upsilon_\perp.
\tag{3.6}
$$

By including the assumption that $k_y = 0$, which was made in our derivation of the emission probabilities, we obtain the following expression for longitudinal oscillations in which $e_k^l = k/k$:

$$\frac{\partial \omega^2 \, \varepsilon_k^l}{\partial \omega} \Bigg|_{\omega = \omega_k^l} = (\omega_k^l)^2 \frac{\partial \varepsilon_k^l}{\partial \omega} \Bigg|_{\omega = \omega_k^l} \, ,$$

$$\gamma_{k,\,0}^l = -\sqrt{\frac{\pi}{2}} \cdot \frac{\omega_k^l \exp\left\{ -\dfrac{(\omega_k^l)^2}{2k_z^2 \, v_{Te}^2} \right\}}{\dfrac{\partial \varepsilon_k^l}{\partial \omega} \Bigg|_{\omega = \omega_k^l}} \, e^{-\mu} \, I_0(\mu) \, \frac{\omega_{pe}^2}{k^2 \, k_z \, v_{Te}^3} \, , \tag{3.7}$$

where I_0 is a Bessel function with the following imaginary argument:

$$\mu = \frac{k_\perp^2 \, v_{Te}^2}{\omega_{He}^2} \, . \tag{3.8}$$

Equation (3.7) was obtained by using the following value for the integral:

$$\int_0^\infty e^{-\frac{v_\perp^2}{2v_{Te}^2}} J_\nu^2 \left(\frac{k_\perp \, v_\perp}{\omega_{He}} \right) v_\perp \, dv_\perp = v_{Te}^2 e^{-\mu} \, I_\nu(\mu). \tag{3.9}$$

Equation (3.7) can be used for many different types of spectra. If the electrons are magnetized, which means that the fluctuation wavelength (more precisely, the value of λ, which is related to the projection of k on a direction perpendicular to H_0 by the equation $\lambda = 1/k_\perp$) is much larger than the mean Larmor radius for the electrons, then $\mu \ll 1$ and we have

$$\gamma_{k,\,0}^l = -\sqrt{\frac{\pi}{2}} \, \frac{\omega_{pe}^2 \, \omega_k^l \exp\left\{ -\dfrac{(\omega_k^l)^2}{2k_z^2 \, v_{Te}^2} \right\}}{k^2 \, k_z \, v_{Te}^3 \dfrac{\partial \varepsilon^l}{\partial \omega} \Bigg|_{\omega = \omega_k^l}} \, . \tag{3.10}$$

We note that the damping is exponentially weak for those waves which are propagating strictly transverse to the magnetic field, so that $k_z \to 0$. Therefore, the spectral transfer which decreases the angle between k and H_0 increases the absorption of the waves. According to (1.39) there are two branches for high–

frequency fluctuations in a magnetic field. The frequency of the ω_+ branch increases with the angle, so a drop in the frequencies of the turbulent fluctuations produces a decrease in the angle and a concomitant increase in absorption. The situation is just the opposite for the ω_- branch.

In a weak field $\omega_{He} \ll \omega_{pe}$ and

$$\omega_+^2 = \omega_{pe}^2 + \omega_{He}^2 \sin^2 \theta, \quad (\partial\varepsilon/\partial\omega) \simeq \frac{2}{\omega_{pe}}.$$

so that, from (3.10)

$$\gamma_{k,\,0}^l = -\sqrt{\frac{\pi}{8}}\, \omega_{pe}\, \frac{\omega_{pe}^3}{k^2 k_z v_{Te}^3} \exp\left\{-\frac{\omega_{pe}^2}{2k_z^2 v_{Te}^2} - \frac{\omega_{He}^2 \sin^2 \theta}{2k_z^2 v_{Te}^2}\right\}. \qquad (3.11)$$

The other branch $\omega_- = \omega_{He}|\cos\theta|$ is weakly damped:

$$\gamma_{k,\,0}^l = -\sqrt{\frac{\pi}{8}}\, \omega_{He}|\cos\theta|\frac{\omega_{He}^3}{k^3 v_{Te}^3} \sin^2\theta \exp\left\{-\frac{\omega_{He}^2}{2k^2 v_{Te}^2}\right\}, \qquad (3.12)$$

if its wavelength is much greater than the electron Larmor radius.

In a strong magnetic field $\omega_{He} \gg \omega_{pe}$ for the wave

$$\omega_+^2 \simeq \omega_{He}^2 + \omega_{pe}^2 \sin^2 \theta$$

we have

$$\gamma_{k,\,0}^l = -\sqrt{\frac{\pi}{8}}\, \omega_{pe}\, \frac{\sin^2 \theta}{|\cos\theta|} \cdot \frac{1}{k^3 d_e^3} \times$$

$$\times \exp\left\{-\frac{\omega_{He}^2}{2k_z^2 v_{Te}^2} - \frac{\omega_{pe}^2 \sin^2 \theta}{2k_z^2 v_{Te}^2}\right\}, \qquad (3.13)$$

which indicates that the damping becomes important if the wavelength is comparable to either the Debye radius or the Larmor radius, but the wave $\omega_- = \omega_{pe}\cos\theta$ is damped with a damping constant given in (3.3) (as if the field were absent).

Turning to the low-frequency fluctuations we note that if the electrons are magnetized, but $\omega^\sigma \ll k_z v_{Te}$, then according to (3.10) the damping rate will have the form

$$\gamma^l_{k,0} = -\sqrt{\frac{\pi}{2}} \cdot \frac{\omega^2_{pe}\,\omega^l_k}{k^2\,k_z\,v^3_{Te}\,\partial\varepsilon^l/\partial\omega\big|_{\omega=\omega^l_k}}. \tag{3.14}$$

For example, for magnetized ion-acoustic oscillations

$$\frac{\partial\varepsilon}{\partial\omega} = \frac{2\omega^2_{pi}}{(\omega)^3}\cos^2\theta, \quad \omega^{Ms}_k = k_z v_s.$$

We then obtain

$$\gamma^l_{k,0} = -\sqrt{\frac{\pi}{8}}\sqrt{\frac{m_e}{m_i}}\,\omega^{Ms}_k. \tag{3.15}$$

For longitudinal ion waves $\omega = \omega_{Hi}\cos\theta$, and it is to this limit that magnetic ion-acoustic waves tend when k increases in a dense plasma; we then have

$$\left.\begin{aligned}
&\frac{\partial\varepsilon}{\partial\omega} = \frac{2\omega^2_{pi}}{\omega^3_{Hi}\sin^2\theta\cos\theta};\\[2mm]
&\gamma^l_{k,0} = -\sqrt{\frac{\pi}{8}}\left(\frac{m_e}{m_i}\right)^2\frac{\omega^3_{He}\sin^2\theta\,|\cos\theta|}{k^3\,v^3_{Te}}\,\omega_{Hi}.
\end{aligned}\right\} \tag{3.16}$$

Nonmagnetic ion-acoustic waves, whose frequencies are much higher than ω_{Hi}, are damped quite heavily, and

$$\gamma_{k,0} = -\sqrt{\frac{\pi}{8}}\sqrt{\frac{m_e}{m_i}}\cdot\frac{kv_s}{|\cos\theta|}; \quad k_\perp v_{Te} < \omega_{He}. \tag{3.17}$$

One can use (3.6) for waves which are not longitudinal. The unit vector e^σ_k for an Alfven wave is directed along x and, according to (3.6), its Landau damping goes to zero.† For a fast magnetoacoustic wave in the case where its velocity $v_A = \dfrac{H_0}{\sqrt{4\pi nm_i}} \gg v_s$

---

† If one includes the next term in the expansion in terms of ω/ω_{Hi}, the damping of the Alven wave is no longer zero; instead it turns out to be given bv

$$\gamma^A_{k,0} = -\sqrt{\frac{\pi}{2}}\,kv_A\,|\cos\theta|\,\frac{m_e}{m_i}\cot^2\theta\,\frac{v_{Te}}{v_A}\cdot\frac{\omega^2_A}{\omega^2_{Hi}}.$$

the polarization vector is along y, so that (3.6) gives

$$\gamma_{k,0}^{M} = -\sqrt{\frac{\pi}{2}\frac{m_e}{m_i}}\; k v_A \frac{v_{Te}}{v_A}\frac{\sin^2\theta}{|\cos\theta|}. \tag{3.18}$$

For whistlers $\omega_{Hi} \ll \omega \ll \omega_{He}$ and

$$
\left.
\begin{aligned}
&e_x^w = \frac{1}{\sqrt{1+\cos^2\theta}}; \quad e_y^w = \frac{i\,|\cos\theta|}{\sqrt{1+\cos^2\theta}}; \\
&\gamma_{k,0} = -\sqrt{\frac{\pi}{2}}\; k v_{Te}\frac{\sin^2\theta}{\omega_{He}}\,\omega^w; \\
&\omega^w = \frac{\omega_{He}\,k^2\,|\cos\theta|\,c^2}{\omega_{pe}^2}; \\
&\frac{\partial}{\partial\omega}\,\omega^2 \varepsilon_k^\sigma = \frac{2\omega_{pe}^2\,|\cos\theta|}{\omega_{He}\,(1+\cos^2\theta)}.
\end{aligned}
\right\} \tag{3.19}
$$

In a magnetoactive plasma there can also be cyclotron absorption of turbulent fluctuations in addition to Landau damping. It corresponds to processes for which $\nu \neq 0$.

For a Maxwellian electron distribution we find, using (2.220), that

$$
\gamma_\nu = -\frac{e^2\,\omega_k^\sigma\,(2\pi)^{3/2}\exp\left\{-\left(\dfrac{\omega^\sigma - \nu\omega_{He}}{\sqrt{2}\,k_z v_{Te}}\right)^2\right\}}{k_z m_e v_{Te}^5\,\dfrac{\partial\omega^2\varepsilon_k^\sigma}{\partial\omega}\bigg|_{\omega=\omega^\sigma}} \times
$$

$$
\times \int |e_k^{\sigma *}\cdot\Gamma_{k\nu}|^2 \exp\left\{-\frac{v_\perp^2}{2v_{Te}^2}\right\} v_\perp\,dv_\perp. \tag{3.20}
$$

For quasilongitudinal oscillations we have

$$|e_k^{\sigma *}\cdot\Gamma_{k\nu}| = \frac{1}{k}\left(k_z v_z + \frac{k_\perp v_\perp}{z}\,\nu\right)J_\nu = \frac{\omega^\sigma}{k}\,J_\nu$$

and

$$\gamma_\nu = -\sqrt{\frac{\pi}{2}}\cdot\frac{\omega_{pe}^2\,\omega^\sigma}{k^2\,k_z\,v_{Te}^3\,\dfrac{\partial\varepsilon_k^\sigma}{\partial\omega^\sigma}}\exp\left\{-\frac{(\omega^\sigma-\nu\omega_{He})^2}{2k_z^2\,v_{Te}^2}\right\}e^{-\mu}\,I_\nu(\mu). \tag{3.21}$$

Here we have used (3.9). Cyclotron absorption by electrons can be significant when the frequencies of the turbulent fluctuations are nearly equal to multiples of the cyclotron frequency ω_{He}. In a weak magnetic field $\omega_{pe} \gg \omega_{He}$, the damping of the high–frequency oscillations in the branch where $\omega_- = \omega_{He} \cos \theta$ is formally given by ($\nu = 1$)

$$\gamma_1 = -\sqrt{\frac{\pi}{8}}\, \omega_{He} |\cos \theta| \frac{\omega^3_{He} \sin^2 \theta}{(k v_{Te})^3} \exp\left\{ -\frac{2\omega^2_{He} \sin^4 \frac{\theta}{2}}{k^2 v^2_{Te} \cos^2 \theta} \right\}. \quad (3.22)$$

In a strong magnetic field such that $\omega_{He} \gg \omega_{pe}$, the branch $\omega^2_+ = \omega^2_{He} + \omega^2_{pe} \sin^2\theta$ has a frequency close to the cyclotron frequency. For this wave

$$\gamma_1 = -\sqrt{\frac{\pi}{8}}\, \omega_{pe} \frac{\omega^3_{pe} \sin^2 \theta}{k^3 v^3_{Te} |\cos \theta|} \exp\left\{ -\frac{\omega^4_{pe} \sin^4 \theta}{8\omega^2_{He} \cos^2 \theta k^2 v^2_{Te}} \right\}. \quad (3.23)$$

In the absence of Landau damping the cyclotron damping does not go to zero for transverse waves if their frequencies are close to the cyclotron frequencies of the plasma. We resort to the general formula (3.20) in order to determine the absorption coefficient. It is evident for this case that

$$\left| e^{\sigma *}_k \cdot \Gamma_{kv} \right|^2 = \left| e^{\sigma *}_k (\Gamma_{kv})_\perp \right|^2,$$

where $(\Gamma_{kv})_\perp$ is the projection of the vector Γ_{kv} on the plane perpendicular to k. In order to find $(\Gamma_{kv})_\perp$, we must rotate the coordinate system around the y axis so that the z axis lies along k. Then, we obtain from (2.16) that

$$(\Gamma_{kv})_{x'} = (\Gamma_{vk})_x \frac{k_z}{k} - (\Gamma_{kv})_z \frac{k_x}{k} = \frac{\omega^\sigma k_z - k^2 v_z}{k_\perp k} J_v; \quad (\Gamma_{kv})_{y'} = (\Gamma_{kv})_y. \quad (3.24)$$

Now x' and y' are perpendicular to k, so that the two components in (3.24) describe two possible polarizations of the transverse waves. When $\omega^\sigma \to \nu\omega_{He}$ it follows from (3.20) that, for the polarization along x',

$$\gamma_v = -\sqrt{\frac{\pi}{8}}\, \omega^2_{He} v^2 \frac{\omega^2_{pe}}{k^3 v^3_{Te}} \cdot \frac{|\cos \theta|}{\sin^2 \theta} \exp\left\{ -\frac{(\omega - \nu\omega_{He})^2}{2k^2_z v^2_{Te}} \right\} e^{-\mu} I_v(\mu). \quad (3.25)$$

For $\nu = 1$ and

$$I_1 \simeq \frac{\mu}{2} = \frac{k^2 \sin^2 \theta v_{Te}^2}{\omega_{He}^2} \qquad (3.26)$$

we obtain

$$\gamma_1 = -\sqrt{\frac{\pi}{8}} \cdot \frac{\omega_{pe}^2}{2k v_{Te}} |\cos\theta| \exp \cdot \left\{ -\frac{(\omega - \omega_{He})^2}{2k_z^2 v_{Te}^2} \right\}. \qquad (3.27)$$

If ω is very nearly equal to ω_{He}, γ_1 can become much larger than ω_{He}, which contradicts the original assumption used in deriving the equation for the absorption.

Thus, near the frequency ω_{He} the transverse waves can be heavily absorbed by the plasma electrons. Therefore, any spectral transfer of high-frequency fluctuations into transverse waves with frequencies close to the cyclotron frequency can lead to significant electron heating. It evidently only needs the possibility of transfer into transverse waves of those frequencies for which the spectral transitions go toward ω_{He}.

In some cases it is difficult to achieve cyclotron heating of the plasma by external electromagnetic waves because the electromagnetic waves do not penetrate to any great depth due to the strong absorption in the surface layer of the plasma; thus, only the thin surface layer is heated. As a rule, during the turbulent heating, fluctuations appear outside the region of intense absorption, in particular, outside the cyclotron resonance region. This is because the transfer of oscillations is possible into those regions where their absorption is small. From our previous discussion such turbulent fluctuations can be spectrally transferred into the region of intense cyclotron absorption, in particular. Then the entire plasma is heated. This possibility for turbulent heating was examined in [81].

Cyclotron absorption on ions, that is, when the frequencies of the turbulent fluctuations are transferred into the region of ion cyclotron absorption, can produce turbulent heating of the ions. The appropriate equation for cyclotron absorption by ions differs from (3.20) in that all the subscripts indicating electrons (e) are replaced by the ion subscript (i).

This, in the absence of collisions, there are very strong absorption mechanisms for the turbulent fluctuations; they are Landau absorption and cyclotron absorption, and these two methods can produce absorption coefficients which are much larger than those appropriate to absorption due to pairwise collisions.

The important result here is that the low-frequency turbulent fluctuations are damped much more strongly, as a rule, than the high-frequency fluctuations. As a result, the transfer processes which tend to lower the turbulent fluctuation frequencies are the most important for turbulent heating. Finally, absorption of low-frequency oscillations by electrons is always effective, so that we must expect the electrons to be heated more strongly by the low-frequency fluctuations.

A number of conditions must be satisfied to obtain effective ion heating. The ions must succeed in absorbing the low-frequency fluctuations before they heat the electrons. The resonant character of the absorption effects has a special effect on the ions; instead of heating all the ions only a small fraction of them are accelerated.

In a bounded plasma it is important that the plasma dimensions be larger than the mean free path over which the turbulent fluctuations can be absorbed:

$$L > \frac{v_g}{\gamma}. \tag{3.28}$$

From this point of view the optimum condition for absorption is to have waves with small group velocities and large damping coefficients. If the spectral transfer leads to turbulent fluctuations in the region where the group velocities are small, this will aid the plasma heating.

Transverse waves having large velocities v_g comparable to the speed of light cannot satisfy (3.28) even when the wave frequency is the first cyclotron harmonic. Then a plasma layer of thickness L is transparent to the cyclotron radiation (the optical thickness is less than unity). In this case it is essential to take into account just the spontaneous cyclotron radiation as described by (2.199). The energy losses of the plasma due to spontaneous cyclotron emission were treated in [82, 83].

Finally, the essential assumption in our derivation of the absorption coefficient is that the particle distribution is nearly Maxwellian. In the general case, for an arbitrary distribution, the sign of γ_k can vary. When $\gamma_k > 0$ the oscillations are amplified. It is possible to show that, no matter what the plasma particle distribution is, the longitudinal turbulent fluctuations are always absorbed if the distribution is isotropic. Any process which makes the particle distribution isotropic increases the absorption of turbulent fluctuations.

We will dwell briefly on the nonlinear mechanisms for dissipating turbulence. One such mechanism is the change in the oscillation frequency during the spectral transfers and the conversion of a given type of fluctuation into other types. In these instances the total oscillation energy is either not altered (a decay process) or a portion of the energy is given to the plasma particles (a scattering process). In this last case it makes sense to speak of the dissipation of turbulence energy only if the frequency of the turbulent fluctuations is lowered as the result of some nonlinear interaction. Another effect which has a significant effect on the absorption is a change in the distribution function for the resonant particles because of either nonlinear or quasilinear effects. In fact, the mechanisms for linear absorption are related to Cerenkov or cyclotron resonance. Usually, only a small number of particles satisfy the resonance condition. Because of the turbulence the distribution for these particles can depart from a Maxwell form. In particular, because of the diffusion of the resonant particles in velocity space, df/dp, which determines the linear damping, can decrease. This type of drop in absorption is possible when the resonant particles are only a small part of the whole, and the energy of the turbulent fluctuations is large enough to significantly alter their state.

3.2. Turbulence Excitation

Mechanisms

The excitation of turbulence is one of the most complicated problems in turbulent plasma physics. Therefore, we shall give here just a short resume of some of the methods for exciting turbulence. This problem is complex because, first, the initial stages of the excitation may not be turbulent in many cases, since only a small number of vibrational modes are excited and the system becomes turbulent only gradually. Evidently such an initial stage occurs for aperiodic plasma instabilities, when the excitation growth rate is either on the order of, or greater than, the oscillation frequencies. Under such conditions the system can pass through a number of states, in which certain parameters undergo radical changes, and finally stochastization of oscillations takes place. Problems of randomization are closely related to problems of ergodicity in systems [84-86], and there is no general answer to the question of a randomization time. Second, the very stage of transporting the system into a state of stationary turbulence (if such a transition does occur) is a complicated process in itself, in which both linear and nonlinear phenomena play important roles. Third, the theory of stability of a turbulent plasma can produce conclusions which differ markedly from those produced by the theory for a nonturbulent plasma. The electromagnetic properties of a turbulent plasma for low frequencies are significantly different from their counterparts in a nonturbulent plasma. For example, a turbulent plasma may well be stable relative to the appearance of a certain type of oscillation while the nonturbulent plasma is not. In laboratory conditions, of course, one would be interested in posing such a question for the low-frequency instabilities, which are most dangerous in the problem of magnetic confinement of the plasma. A problem closely related to the instability of a turbulent plasma is the aftereffect in the excitation of the turbulence.

As an example, let us assume that some agent excites turbulence in the plasma for a time τ_0 and this turbulence decays over a time $\tau'_0 \gg \tau_0$. The agent will be assumed to be a beam of charged particles which penetrates the plasma and acts for some finite period τ_0. If this agent once again acts on the plasma after exciting the turbulence, and the time between successive applications of the agent is much smaller than τ'_0, it cannot excite oscillations in the plasma. Because of spectral transfer the turbulent fluctuations excited in the first application of the exciting mechanism can move into a region of high phase velocities. Then, in the second application of the interaction the beam instabilities will be stabilized, pumping all the fluctuations produced into a region where they do not interact with the beam, for all practical purposes. This example shows how important even a relatively weak prior plasma turbulence is in the excitation theory.

Under laboratory conditions the plasma is bounded, which means that it is inhomogeneous, and this causes a certain instability related to that inhomogeneity which is difficult to remove; it can therefore lead to some prior turbulence. Finally, contact with the walls of the chamber containing the plasma can play a role in the removal of heat and destruction of the turbulent fluctuations which is by no means a secondary role.

It should therefore be clear that, under laboratory conditions, the initial state of the plasma (before the exciting agent acts; the beam of charged particles for example) is frequently not precisely known. In a number of cases this fact makes a detailed comparison of experiment and theory rather difficult, although for the most part the effects predicted by turbulence theory agree well with the observations.

The mechanisms for generating plasma turbulence can be classified after the initial state of the system (composed of a plasma and beam, for example) has been specified, or after specifying what it interacts with. The initial state determines the type of linear or nonlinear plasma instability; as things progress it can be modified significantly through the development of many channels along which the process of developing turbulence progresses.

The simplest excitation mechanisms are those which are described by growth rates $\gamma \ll \omega_k^g$.

Now, having completed our outline of the general difficulties in describing the processes involved in turbulence excitation, we shall list those initial states frequently encountered, and from which plasma turbulence arises [6].

1. The interaction of a beam and a plasma. A beam or group of particles can be created outside the plasma and injected into it from the outside, or it can appear by some other means in the plasma. The instability appears when the beam particle velocities are sufficiently large [87, 88]. Aperiodic instabilities can develop for low beam velocities.

2. There is an instability related to an anisotropic velocity distribution for the particles. Such an anisotropy can be generated by a temperature difference among the particles in different directions which is a consequence of some nonuniformity in their heating, perhaps due to the system geometry. The following kinds of instabilities fall into this class.

3. The loss-cone instability, which arises in magnetic traps because the traps are not able to hold those particles which are moving inside a certain cone of angles along the external magnetic field.

4. The cyclotron instability is a combination of the beam and anisotropy instabilities. It arises when a beam of particles is injected across the external magnetic field.

5. When an external electric field is acting on the plasma the beam instability and the anisotropy and cyclotron instabilities appear. These are related to the effects of the turbulent electrical conductivity of the plasma.

6. The drift instability appears in the plasma when the plasma is nonuniform and the nonuniformity is maintained by a magnetic field. It arises because the plasma inhomogeneity causes currents to appear on the plasma surface, and these currents are unstable just as the currents created by an external electric field are unstable. This instability can lead to turbulent diffusion and turbulent thermal conductivity [6].

7. Similar mechanisms appear when axial nonlinear waves and shock waves act on the plasma [21]. Currents and fields leading to turbulence appear at the fronts of such waves.

8. Similar turbulence mechanisms come into play when the plasma interacts with strong high-frequency fields.

9. Among the nonlinear mechanisms for creating turbulence are turbulence excitation by intense radiation, particularly high-frequency fields and the radiation from optical quantum generators.

3.3. The Interaction of a Plasma and a Beam of Charged Particles

3.3.1. Linear Effects

The linear effects of beam instability in plasma have been treated by Bohm and Gross [87] and Akhiezer and Fainberg [88]. In the linear theory the excitation of longitudinal waves is described by the general dispersion equation [89, 90]

$$\varepsilon_k^l = \frac{k_i \, \varepsilon_{ij} \, k_j}{k^2} = 0. \qquad (3.29)$$

If we are considering a monoenergetic beam, the dielectric constant for the plasma and beam can be found if one knows the dielectric constant for the plasma ε_0^l and beam ε_1^l separately:

$$\varepsilon^l = \varepsilon_0^l + \varepsilon_1^l - 1, \tag{3.30}$$

and the beam dielectric constant can be found by studying a system in which the beam is at rest [89]. For nonrelativistic beams one need include only the Doppler effect [in ε_1^l, ω is replaced by $(\omega - ku)$, where u is the beam velocity]. If the thermal motion is negligible then $\varepsilon_0^l = 1 - \omega_{pe}^2/\omega^2$, i.e.,

$$\varepsilon_k^l = 1 - \frac{\omega_{p1}^2}{(\omega - ku)^2} - \frac{\omega_{pe}^2}{\omega^2}, \quad \omega_{p1}^2 = \frac{4\pi n_1 e^2}{m_e}. \tag{3.31}$$

When $n_1/n_0 \ll 1$ the solution to (3.31) is

$$\omega = ku \pm \omega_{p1} \left(1 - \frac{\omega_{pe}^2}{(ku)^2} \right)^{-\frac{1}{2}}. \tag{3.32}$$

This solution is written for values of **ku** which are not close to ω_{pe}. An instability appears when **ku** $< \omega_{pe}$. The maximum growth rate, however, occurs for ω_{pe} nearly equal to **ku** when

$$\omega = \omega_{pe} \left[1 - \frac{1 \pm i\sqrt{3}}{2^{4/3}} \left(\frac{n_1}{n_0} \right)^{1/3} \right]. \tag{3.33}$$

The region in which these equations are applicable is rather narrow, but when the beam is not quite monoenergetic, such that $\Delta u/u \sim (n_1/n_0)^{1/3}$, one can still use the equations for a weak excitation and have $\gamma \sim \omega_{pe}(n_1/n_0)(u/\Delta u)^2$ (see Chapter 2). This criterion follows from comparing this γ with the growth rate in (3.33). The oscillation in (3.33) is hydrodynamic and, strictly speaking, statistical methods cannot be used to describe it. However, it has been shown [91] that there can be large increases in Δu in the hydrodynamic stages, and the beam moves into a region of weak excitation after undergoing small changes in some of its parameters.

According to [59, 94] the instability related to anisotropies in the beam−plasma distribution function [92, 93] is of little importance when $n_1/n_0 \ll 1$ for the beam. This is because a small spread in either the transverse or longitudinal velocities can eliminate this type of instability.

We shall treat in somewhat greater depth the instabilities which arise for beams with large velocity spreads. For low-density beams one can neglect changes in the dispersion properties of the plasma fluctuations when the beam is present and one need take into account only the growth of these fluctuations, which is described by the growth rate of (2.198):

$$\gamma_k^\sigma = \frac{e^2}{2\pi} \sum_{\nu,\alpha} \int \frac{|\Gamma_\nu^\alpha e_k^{\sigma*}|^2}{\frac{\partial}{\partial\omega} \omega^2 \varepsilon_k^\sigma} \delta\left(\omega_k^\sigma - k_z v_z - \nu\omega_{H\alpha}\right) \times$$

$$\times \left(k_z \frac{\partial f_p^\alpha}{\partial p_z} + \frac{\omega_{H\alpha}\nu}{v_\perp} \cdot \frac{\partial f_p^\alpha}{\partial p_\perp} - \frac{k_x c}{eH} \cdot \frac{\partial f_p^\alpha}{\partial y} \right) d\mathbf{p}. \qquad (3.34)$$

Let f_i^z be the distribution function for the beam particles; the beam can be nonuniform in the y direction, which is perpendicular to the external magnetic field.

If the beam moves along the magnetic field, then

$$f_1^\alpha = \frac{(2\pi)^{3/2} n_1(y)}{v_{T_\perp}^2 v_{T_\parallel} m_\alpha^3} \exp\left\{ -\frac{v_\perp^2}{2v_{T_\perp}^2} - \frac{(v_z - u_\alpha)^2}{2v_{T_\parallel}^2} \right\}.$$

Then (3.34) can be written in the following form for longitudinal waves:

$$\gamma_k^l = \frac{\sqrt{\frac{\pi}{2}}}{v_{T_\parallel}^2 k^2 \frac{\partial}{\partial\omega} \varepsilon_k^l} \sum_{\nu,\alpha} I_\nu(\mu_\alpha) e^{-\mu_\alpha} \frac{\omega_{1\alpha}^2}{|k_z| v_{T_\parallel}} \times$$

$$\times \exp\left\{ -\frac{(\omega_k^l - \nu\omega_{H\alpha} - k_z u_\alpha)^2}{2k_x^2 v_{T_\parallel}^2} \right\} \left[(k_z u_\alpha - \right.$$

$$\left. -\omega_k^l + \nu\omega_{H\alpha}) \left(1 + \frac{k_x}{k_z \omega_{H\alpha}} \cdot \frac{\partial u_\alpha}{\partial y} \right) - \right.$$

$$\left. -\frac{v_{T_\parallel}^2}{v_{T_\perp}^2} \left(\omega_{H\alpha}\nu + \frac{k_x}{\omega_{H\alpha}} v_{T_\perp}^2 \frac{\partial}{\partial y} \ln n_1 \right) \right]. \qquad (3.35)$$

Here we have used the relations

$$\mu_a = \frac{k_x^2 v_{T_\perp}^2}{\omega_{Ha}}, \quad (k_y = 0), \quad \omega_{1a}^2 = \frac{4\pi n_{1a} e^2}{m_a}.$$

It is quite clear from (3.34) and (3.35) that two mechanisms exist for excitations: Cerenkov when $\nu = 0$ and cyclotron when $\nu \neq 0$. Moreover, with a nonuniformity along the y axis in both the density distribution and the particle velocities in the beam, new possibilities for cyclotron and Cerenkov instability appear.

If $\mu_{\alpha\,max} \ll 1$ one can have an effective excitation only in the first harmonics $\nu = 0, \pm 1$; thus, with $\nu = 0$ and an electron beam present,

$$\gamma_{k,\,\nu=0}^l = \sqrt{\frac{\pi}{2}} \; \frac{\omega_{1e}^2}{\dfrac{\partial}{\partial\omega}\varepsilon_k^l} \; \frac{1}{k^2 v_{T_\parallel}^2 \, |k_z|\, v_{T_\parallel}} \times$$

$$\times \left[(k_z u_e - \omega_k^l)\left(1 + \frac{k_x}{k_z \omega_{He}}\cdot\frac{\partial u_e}{\partial y}\right) - \frac{k_x}{|k_x|}\,\omega_1 \right], \qquad (3.36)$$

where

$$\omega_1 = \frac{|k_x|\, v_{T_\parallel}^2}{\omega_{He}} \cdot \frac{\partial}{\partial y} \ln n_1$$

is the drift frequency of the beam particles. It is now clear that the effect of beam–density nonuniformity is observable only in the excitation of frequencies which are of the order of the beam drift frequencies. If the nonuniformity is absent, the buildup takes place for waves in which $k_z u_e > \omega_k^l$ (the Cerenkov condition). The effect related to a nonuniformity in the beam's velocity profile du_e/dy can also cause an instability — this is the so-called slipping instability, which has been studied in the hydrodynamic limit [95–97]. It can appear even for $\omega_k^l < k_z u_e$ if

$$\frac{k_x}{k_z \omega_{He}} \cdot \frac{\partial u_e}{\partial y} > 0. \qquad (3.37)$$

In a homogeneous plasma we can estimate the maximum growth rate for exciting some branch of the high-frequency plasma waves in a strong magnetic field ($\omega_{pe} \ll \omega_{He}$):

$$\gamma^\sigma_{k, \nu=0} = \sqrt{\frac{\pi}{8} \cdot \frac{n_1}{n_0} \cdot \frac{\omega^4_{pe} \sin^2\theta}{\omega_{He}|\cos\theta|}} \frac{(k_z u_e - \omega_{He})}{k^3 v^3_{T\parallel}} e^{\frac{-(\omega_{He}-k_z u_e)^2}{2k^2_z v^2_{T\parallel}}} ; \qquad (3.38)$$

$$\omega \sim \omega_{He} ;$$

$$\gamma^\sigma_{k, \nu=0} = \sqrt{\frac{\pi}{8} \cdot \frac{n_1}{n_0} \cdot \frac{\omega^3_{pe}}{k^3 v^3_{T\parallel}}} (k_z u_e - \omega_{pe}) e^{\frac{-(\omega_{pe}-k_z u_e)^2}{2k^2_z v^2_{T\parallel}}} ; \qquad (3.39)$$

$$\omega = \omega_{pe}|\cos\theta|.$$

An estimate of the growth rate due to a cyclotron instability for the branch ω_{He} is

$$\gamma_k \simeq \frac{1}{2}\sqrt{\frac{\pi}{2} \cdot \frac{n_1}{n_0} \cdot \frac{\omega^2_{pe}}{\omega_{He}}} \sin^2\theta \frac{u_e}{v_{T\parallel}} ; \qquad (3.40)$$

for the branch $\omega_{pe}\cos\theta$ it is

$$\gamma_k \approx \frac{1}{4}\sqrt{\frac{\pi}{2} \cdot \frac{n_1}{n_0}} \frac{\sin^2\theta |\cos\theta|\omega^3_{pe}}{\omega^2_{He}} \cdot \frac{u_e}{v_{T\parallel}} . \qquad (3.41)$$

Thus, in the cyclotron instability of the beams there is an increase in the growth rate as the beam velocity increases in a direction perpendicular to the magnetic field. The strong increase in the buildup of frequencies $\omega \sim \omega_{He}$ as $v_{T\perp}$ increases has been experimentally observed in [98].

3.3.2. Quasilinear Effects in the Beam—

Plasma Interaction

The effects of quasilinear relaxation can significantly alter the distribution of electrons in a beam, which can lose a significant fraction of its energy [55, 56, 99, 100]. This was shown very clearly in [91]. It was found there [91] that even beams with a small spread in velocities ["monoenergetic" beams; their instabil-

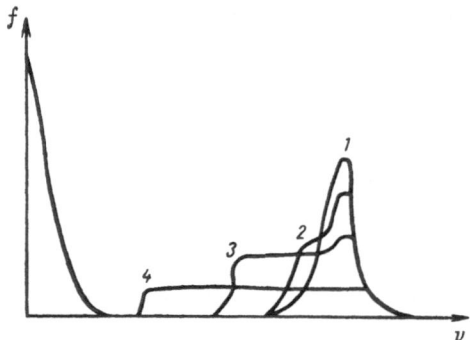

Fig. 3.1. The formation of a shock wave in velocity space during the development of a beam instability. Curves 1-4 are various stages in the relaxation of the beam's distribution function as a function of time.

ity is described by (3.33)] quickly go into a state of weak excitation under the action of the generated oscillations when

$$\gamma \sim \omega_{pe} \frac{n_1}{n_0} \left(\frac{u}{\Delta u} \right)^2, \tag{3.42}$$

and when the velocity spread increases up to[†]

$$\frac{\Delta u}{u} \gtrsim \left(\frac{n_1}{n_0} \right)^{1/3}. \tag{3.43}$$

Below we shall encounter quasilinear relaxation, which, in the case of a one-dimensional process (which is possible only in a strong magnetic field such that $\omega_{He} \gg \omega_{pe}$), leads to the formation of a plateau in the distribution function (Fig. 3.1). This plateau has been observed experimentally [101].

The spectrum for the Langmuir fluctuations which arise at the plateau stage can be obtained from a one-dimensional quasilinear equation.

A one-dimensional theory is sufficient if all the beam particles move in the same direction at both the initial moment of time and throughout the whole quasilinear relaxation process. This condition is possible if there is a strong external magnetic field

[†]The theory in [91] is based on equations for the moments of the distribution function and permits one to find only the mean value of $(\Delta u)^2$.

in the direction of the beam motion, which will then limit the possibility of the beam particles developing any velocity component perpendicular to the original direction of motion. In addition, we shall assume that the beam velocity is greater than v_{Te}, and we shall treat just the generation of Langmuir waves in the beam direction. Equations (2.139) and (2.52) then take the form

$$\frac{\partial f_p^R}{\partial t} = \frac{\partial}{\partial v} D \frac{\partial f_p^R}{\partial v}, \tag{3.44}$$

$$D = \frac{\omega_{pe}^3}{8\pi^3 m_e n_0 v} N_{\frac{\omega_{pe}}{v}}, \tag{3.45}$$

$$\frac{\partial N_k}{\partial t} = 2\gamma_k N_k; \quad \gamma_k = \frac{e^2 \omega_{pe}}{4\pi k^2} \cdot \frac{\partial f_p^R}{\partial v}\Big|_{v = \frac{\omega_{pe}}{k}}. \tag{3.46}$$

Setting $N(v) = N_k\big|_{k = \frac{\omega_{pe}}{v}}$ we have

$$\frac{\partial N(v)}{\partial t} = \frac{\omega_{pe} v^2 m_e}{8\pi^3 n_0} \cdot \frac{\partial f^R(v)}{\partial v} N(v). \tag{3.47}$$

Upon substituting $\frac{\partial f^R(v)}{\partial v} N(v)$ from (3.49) into (3.47), we obtain from

$$\frac{\partial f^R(v)}{\partial t} = \frac{\partial}{\partial t} \cdot \frac{\partial}{\partial v} \cdot \frac{\omega_{pe}^2}{m_e^2 v^3} N(v) \tag{3.48}$$

the integral

$$f^R(v, t) - f^R(v, 0) = \frac{\partial}{\partial v} \cdot \frac{\omega_{pe}^2}{m_e^2 v^3} \{N(v, t) - N(v, 0)\}. \tag{3.49}$$

By considering those values of N which fall in the region outside the initial beam distribution $f^R(v, 0) = 0$, and assuming that the energy density of the fluctuations is significantly greater than its initial value $N(v, 0)$, we have

$$N(v, \infty) = \frac{v^3 m_e}{\omega_{pe}^2} \int^v dp\, f^R(v, \infty) \simeq (2\pi)^3 \frac{v^4 n_1 m_e}{\omega_{pe}^2 u}, \quad v = \frac{\omega_{pe}}{k},$$

where u is the initial beam velocity and $v_{Te} \ll v < u$. The spectral energy density of the turbulent fluctuations takes the form

$$W_k(\infty) = \frac{\omega_{pe}}{(2\pi)^3} N\left(\frac{\omega_{pe}}{k}, \; \infty\right) = \frac{n_1 m_e \omega_{pe}^3}{k^4 u}, \qquad \frac{\omega_{pe}}{u} < k < \frac{1}{d_e}. \qquad (3.50)$$

The total energy imparted to the turbulent fluctuations is

$$W = \int W_k dk = \frac{n_1 m_e u^2}{3}. \qquad (3.51)$$

Half of this energy is in the electric field of the turbulent fluctuations, and the other half is contained in the nonresonant plasma particles, which adiabatically vibrate in the field of the turbulent fluctuations. Note that the spectrum of the turbulent fluctuations is a power function of k. However, unlike fluids, the entire region of the spectrum in (3.50) corresponds to the region of excitation (an energy-containing region).

The dynamics involved in forming the plateau have been studied theoretically in [102]. According to this work [102] the plateau forms when a nonlinear self-similar wave propagates in velocity space. The work of Levitskii and Shashurin [103] shows that this effect is observable. The quasilinear integral of (3.49) enables us to eliminate $f^R(v, t)$ from (3.46) and obtain a single equation for $N(v)$:

$$\frac{\partial N(v)}{\partial t} = \frac{\omega_{pe} v^2 m_e}{8\pi^2 n_0} N(v, \; t) \frac{\partial f^R(v, \; 0)}{\partial v} + \frac{\omega_{pe}^3 v^2}{8\pi^2 n_0 m_e} N(v, \; t) \times$$

$$\times \frac{\partial^2}{\partial v^2} \cdot \frac{1}{v^3} (N(v, \; t) - N(v, \; 0)). \qquad (3.52)$$

The self-similar solution, as obtained below from (3.52), is valid if the nonlinear interaction does not become too strong. This is possible only for sufficiently low beam-particle velocities, for then the effects of induced scattering on ions will be unimportant [104].

Assume that the initial beam has a narrow spread of velocities such that $\Delta u = \Delta v_1 \ll v_1 = u$.[†] We shall consider those values

---

[†]Recall that $\dfrac{\Delta v_1}{v_1} \gg \left(\dfrac{n_1}{n_0}\right)^{1/3}$ in order to use the statistical description.

of v which are outside Δv_1 [for which $f^R(v, 0) = 0$], and the stage where the oscillations have intensities significantly in excess of their initial values, $N(v, t) \gg N(v, 0)$. Then Eq. (3.52) takes the form

$$\frac{\partial W(v, t)}{\partial t} = \frac{\omega_{pe}^3}{8\pi^2 n_0 m_e} v^2 W(v, t) \frac{\partial^2 W(v, t)}{\partial v^2}, \qquad W(v, t) = \frac{N(v, t)}{v^3}. \qquad (3.53)$$

According to the assumption used in [102] the dynamics of the beam instability development leads to the production of waves in velocity space (either thermal waves or shock waves). At the front of this wave the quantity W changes quite rapidly, reaching some plateau behind the front. As the wave front moves toward lower values of v the magnitude of this plateau slowly drops (see Fig. 3.1). Near the wave front we can seek solutions to (3.53) which are of the form

$$W = W(v - v_*(t), t) = W(\xi, t), \quad \xi = v - v_*(t), \qquad (3.54)$$

which assumes that W and dv_*/dt are slowly varying functions of t. Neglecting the time derivative of W and assuming that $\xi \ll v_*(t)$, we find

$$-\frac{dv_*}{dt} \cdot \frac{dW}{d\xi} = \frac{\omega_{pe}^3}{8\pi^2 n_0 m_e} v_*^2 W \frac{\partial^2 W}{\partial \xi^2}. \qquad (3.55)$$

Integrating (3.55) over ξ we have

$$-\frac{8\pi^2 n_0 m_e}{\omega_{pe}^3} \cdot \frac{dv_*}{dt} \cdot \frac{1}{v_*^2} \ln \frac{W(\xi, t)}{W(0, t)} = \frac{\partial W(\xi, t)}{\partial \xi} - \frac{\partial W(\xi, t)}{\partial \xi}\bigg|_{\xi=0} \sim \frac{\partial W(\xi, t)}{\partial \xi}. \qquad (3.56)$$

The point $\xi = 0$ is selected so that $\dfrac{\partial W}{\partial \xi}\bigg|_{\xi=0}$ is small compared with $\partial W/\partial \xi$ when $\xi \neq 0$. By treating those ξ for which $W(\xi, t)$ is sufficiently large one can replace the ln term in (3.56) with $\ln[W(\infty, t)/W(0, t)]$, where $W(\infty, t)$ is the value of W when $\xi \to \infty$, i.e., behind the wave front. Then

$$W(\xi, t) = -\xi \frac{8\pi^2 n_0 m_e}{\omega_{pe}^3} \cdot \frac{1}{v_*^2} \cdot \frac{dv_*}{dt} \ln \frac{W(\infty, t)}{W(0, t)}. \qquad (3.57)$$

$W(\infty, t)$ can be found from the matching conditions. By integrating over the region of the plateau behind the wave front we find from (3.49)

$$\int_{v_*}^{v_1} f^R(v, t)\, dv = \int_{\Delta v_1} f^R(v, 0)\, dv = n_1 m_e^{-1} = f(\infty, t)(v_1 - v_*), \qquad (3.58)$$

which expresses the conservation of particle number rule. On the other hand, for v outside Δv_1 but near the wave front, (3.49) gives

$$f(v, t) \simeq f(\infty, t) = \frac{\partial}{\partial v}\, \omega_{pe}^2\, W/m_e^2 = \omega_{pe}^2\, \frac{\partial W/m_e^2}{\partial \xi} = -\frac{1}{v_*^2}\, \frac{dv_*}{dt}\, \ln \frac{W(\infty, t)}{W(0, t)}\, \frac{8\pi^2 n_0}{\omega_{pe} m_e}. \tag{3.59}$$

We have used (3.57) to obtain the last equation. By comparing (3.58) with (3.59) an equation for the motion of the wave front can be obtained

$$\frac{n_1}{v_* - v_1} = \frac{1}{v_*^2} \cdot \frac{dv_*}{dt}\, \frac{8\pi^2 n_0}{\omega_{pe}} \cdot \ln\left(\frac{W(\infty, t)}{W(0, t)}\right) \tag{3.60}$$

For our purposes it is sufficiently accurate to neglect the time dependence of the expression in the large parentheses. When $v_* - v_1 \ll v_1$ ($v_* - v_1 \gg \Delta v_1$) we have

$$\frac{v_1 - v_*}{v_1} = \sqrt{\frac{\omega_{pe}\, t}{4\pi^2 \ln \dfrac{W(\infty, t)}{W(0, t)}} \cdot \frac{n_1}{n_0}}. \tag{3.61}$$

If $v_* \ll v_1$ then

$$\frac{v_*}{v_1} = \frac{8\pi^2}{t\omega_{pe}}\, \frac{n_0}{n_1} \cdot \ln \frac{W(\infty, t)}{W(0, t)}. \tag{3.62}$$

Equation (3.61) can be used to estimate the characteristic time for the beam energy to change when the particle velocities in the beam change by an amount on the order of Δv_1:

$$t = \frac{1}{\omega_{pe}}\, 4\pi^2 \ln \frac{W(\infty)}{W(0)} \cdot \frac{n_0}{n_1} \left(\frac{\Delta v_1}{v_1}\right)^2. \tag{3.63}$$

The change in beam energy $mv_1 \Delta v_1$ when $\Delta v_1 \sim v_1$ is about $\sim mv_1^2$, i.e., the beam loses energy in an amount comparable to its initial energy. From energy conservation this goes into the oscillations, which means that $W(\infty) \sim n_1 m v_1 \Delta v_1$. But the initial oscillation energy is $\sim T\omega_{pe}^3 \Delta v_1/v_1^4$, i.e.,

$$\ln \frac{W(\infty)}{W(0)} \simeq \ln \left(\frac{v_1^2}{v_{Te}^2} n_1 \frac{v_1^3}{\omega_{pe}^3} \right).$$

The expression in the large parentheses is usually a rather large number.

The question of when the quasilinear relaxation proceeds up to the formation of the plateau depends on the boundary conditions, the nonlinear effects (see below), and the excitation conditions (whether the beam injection is continuous or pulsed).

If a particle beam is continuously injected into a half-space occupied by a plasma, the oscillation buildup effect [105, 106] is possible.

In this effect the group velocities of the Langmuir oscillations are much smaller than the velocity of the beam particles. Because of this effect the accumulation of high-energy fluctuations is concentrated only near the plasma boundary. This nonuniformity in the fluctuation distribution can produce an instability which is connected with the Miller force, and it leads to a significant role for the nonlinear effects.

A local increase in the fluctuation intensity can mean that the nonlinear effects will be very important in this region even when such nonlinear effects are weak in other regions. In turn, this leads to turbulent fluctuations which are locally isotropic, and there will be intense scattering of the beam particles which enter this region.

We must also point out that a change in the beam distribution function as a result of the quasilinear relaxation can, in general, lead to new instabilities, which either did not appear or showed up only weakly for the original distribution, which held sway before the quasilinear relaxation [107, 108]. Physically this comes about because the formation of the plateau can stabilize one type of oscillation, but such a state will not be stable for other oscillations.

This effect can be simply illustrated using (3.34). For example, let us consider a Cerenkov instability ($\nu = 0$) for two vibrations, one high-frequency vibration ω_k^σ and one low-frequency vibration $\omega_k^{\sigma'}$. The growth rate for each of them is proportional to

$$k_z \frac{\partial f_1^\alpha}{\partial v_z} - \frac{k_x cm}{eH} \frac{\partial f_1^\alpha}{\partial y}.$$

Assume that for the high-frequency wave the effect of beam non-uniformity, given by the second term, is of little importance, that is, for these waves the growth rate is proportional to $\partial f_1^\alpha / \partial v_z$. Neglecting the beam nonuniformity in this case is valid if

$$k_z \frac{\partial f_1^\alpha}{\partial v_z} \approx \frac{\omega^\sigma n_1}{m (\Delta v)^2 u}$$

is much larger than

$$\frac{c k_x n_1}{eHL (\Delta v)} \simeq \frac{n_1}{m \omega_H LL_1 (\Delta v)} \left(k_x \sim \frac{1}{L_1} \right).$$

The development of an instability in the high-frequency fluctuations produces the plateau and $\Delta v \sim u$. Thus, the nonuniformity for high-frequency vibrations is unimportant when

$$\omega^\sigma \gg \omega_{D1} = \frac{u^2}{L \omega_H L_1}. \qquad (3.64)$$

Here ω_{D1} is the beam drift frequency ($k_x \sim 1/L$ for the fundamental mode).

Let us now study the low-frequency oscillations $\omega_k^{\sigma'}$. The nonuniform term in the growth rate can be very important for these waves, and if $\omega_k^{\sigma'}$ is resonant with respect to those same particles a decrease in $\partial f_1/\partial v_z$ will mean that the growth rate for the low-frequency oscillations increases. The characteristic value of the growth rate is easily estimated from (3.34) [108]:

$$\gamma \simeq \frac{n_1}{n_0} \omega_{D1}. \qquad (3.65)$$

Generally speaking, (3.65) is very small. However, in order to eliminate it completely, it is necessary that $\partial f_1/\partial v_z$ be decreased further until it is negative, but when this occurs the high-frequency fluctuations start to be absorbed. Because of energy conservation, the total energy in the beam and the low- and high-frequency fluctuations is conserved. However, the beam loses energy, and as a consequence the absorption of the high-frequency fluctuations is connected with the conversion of their energy into the low-frequency fluctuations. This process is rather slow. If the beam is maintained in the plasma for a period of time $\sim 1/\gamma$, for example by injection from the outside, the nonlinear processes can play a more important part in the spectral conversion.

The quasilinear effects of the cyclotron excitation of low-frequency fluctuations (Alfven waves and whistlers in particular) by broadened ionic beams were treated in [109].

3.3.3. Nonlinear Stabilization of Beam

Instabilities

The very important role of nonlinear effects in the development of a beam instability was first demonstrated in [104]. According to this paper [104] the effects of spectral conversion in scattering of turbulent fluctuations by ions can transport the fluctuations into a region where the fluctuation is not resonant with respect to the beam particles; this results in stabilization. When $T_i > T_e (v_{Te}^2/u^2)(m_i/m_e)$ the condition for stabilization takes the form $\frac{v_{Ti}}{u}\left(\frac{u}{\Delta u}\right)^3 \frac{T_i}{T_e}\left(1 + \frac{T_e}{T_i}\right)^2 \ll 1$. This criterion contains just beam and plasma parameters; the turbulent fluctuation energy does not enter. When $T_i < T_e (v_{Te}^2/u^2)(m_i/m_e)$ the spectral conversion for scattering on ions can be accomplished in a sort of relay-race manner, and will be accompanied by rather extensive trends toward isotropy. In this case stabilization[†] is also possible when the condition $(m_i/m_e)(v_{Te}/u)^3(v_{Te}/\Delta u) \ll 1$ is fulfilled.

It is not possible to have stabilization for small beam velocities in a homogeneous plasma. Thus, if the conversion is accomplished through scattering on electrons, and $u/v_{Te} < (m_i/m_e)^{1/5}$, the quasilinear relaxation will take place faster than the spectral conversion. As the plasma density increases, the role of the Coulomb collisions in spectral conversion will increase (see §4.5). An increase in the intensity of spectral conversion in this case can mean that even beams with relatively low velocities will be

---

[†] In some sense this stabilization is not complete. The nonlinear effects simply substantially increase the time taken to broaden the beam, and this turns out to be much larger than the reciprocal of the linear growth rate.

stabilized. The dependence of the intensity with which turbulent fluctuations in the plasma are excited by the beams has been experimentally studied in [98] as a function of plasma density.

Evidently, one can assign both the decrease in intensity of high-frequency oscillations, as the beam current increases, and their spectrum broadening [98, 110], to the nonlinear effects. A significant factor in the appearance of nonlinear effects can be nonuniformities in the fluctuations' distribution, which leads to accumulation effects. The theoretical study of such effects within the models used in [105, 106], which take into account the nonlinearities, has shown that there are solutions which give a periodic cutoff in the buildup of the oscillations which alternates with the quasilinear relaxation effects. Such relaxation-type low-frequency oscillations have also been obtained in [111] in studies of stabilizing a beam instability through decay processes: Two quasistationary states are then possible and these are split off one after the other, taking turns. These theoretical deductions are in qualitative agreement with the experimental results of [98], in which the low-frequency relaxation oscillations have a frequency nearly equal to the frequency of the ion-acoustic oscillations; they are also in agreement with the cutoff effects observed in [112].

3.4. Excitation of Turbulence by Anisotropy and Loss-Cone Instabilities

3.4.1. Linear Effects

We shall consider the instabilities of the plasma particle distributions in a magnetic field, for which there are no directional velocities along the magnetic field. We will assume that the distribution function takes the form

$$f_p = f\left(v_z^2, v_\perp\right). \tag{3.66}$$

An appropriate example is a particle beam injected into the plasma across the magnetic field, which is usually the case when filling magnetic traps. In this case both the theory and the general picture are compatible in a certain respect with the theory described above and the phenomena present when a beam travels through

the plasma; the exception is, in this case, that the cyclotron instability is the fundamental one. Another example appropriate
to Eq. (3.66) is the particle distribution in a magnetic trap, where
because of the loss cone there are no particles having small $v_\perp$.
The indicated anisotropy leads to an instability. As our third and
final example, we have a distribution for which the temperatures
are different along and transverse to the magnetic field. Such a
distribution can arise during cyclotron heating of a plasma, when,
primarily because of the heating, the particle velocities perpendicular to the external magnetic field are increased. All these examples fall into the same general category – there is an anisotropy in the plasma particle distribution in the presence of magnetic fields. As with the beam instabilities, when the degree of
anisotropy is large there is strong excitation and the instability
has an aperiodic nature. It can be assumed that the strong instability also leads to the elimination of the strong anisotropy,
and quite rapidly leads to a weak instability that develops more
slowly, which then controls the time scale for the process. In the
latter case one may use (3.34).

Assume that the particle distribution function is of the form

$$f_{p\alpha} = \frac{(2\pi)\, n_\alpha(y)}{m_\alpha^3 v_\perp\, v_{T\alpha}^2(y)}\, \exp\left\{ -\frac{(v_\perp - u_\alpha)^2 + v_z^2}{2 v_{T\alpha}^2(y)} \right\}. \tag{3.67}$$

This distribution function describes the first of our examples
above. Substituting (3.67) into (3.34) for the excitation of longitudinal waves we easily find that

$$\gamma_k^l = \sum_\alpha \frac{\omega_{p\alpha}^2\, \sqrt{\pi/2}}{\sqrt{2\pi}\, \left| \dfrac{\partial \varepsilon_k^l}{\partial \omega} \right|_{\omega=\omega_k^l}} k^2 \times$$

$$\times \sum_{\nu=-\infty}^{\infty} \exp\left\{ -\frac{(\omega_k^l - \nu\omega_{H\alpha})^2}{2 k_z^2\, v_{T\alpha}^2} \right\} \frac{1}{v_{T\alpha}^2} \int_0^\infty \frac{dv_\perp}{v_{T\alpha}} \left\{ J_\nu^2(z_\alpha) \frac{\nu\omega_{H\alpha}\left(1 + \dfrac{v_{T\,\alpha}^2}{v_\perp^2}\right) - \omega_k^l}{k_z\, v_{T\alpha}} \right. -$$

$$\left. - \nu J_\nu(z_\alpha)\, J_\nu'(z_\alpha) \frac{k_\perp\, v_{T\alpha}}{k_z\, v_\perp} - \frac{k_\perp c v_{T\alpha}}{e H\, k_z} \cdot \frac{\partial}{\partial y} \ln f_{p\alpha} \right\} \exp\left[-\frac{(v_\perp - u_\alpha)^2}{2 v_{T\alpha}^2} \right] ;$$

$$\tag{3.68}$$

$$z_\alpha = \frac{k_\perp\, v_\perp}{\omega_{H\alpha}}.$$

If the nonuniformity is small, then when $\nu = 0$ the instability is absent, according to (3.68). In other words, the instability can only be of a cyclotron nature. Various limiting cases could be obtained from (3.68) (such as hot and cold electrons and ions, anisotropic ions and isotropic electrons, or the reverse, and so on). These cases have been treated in [113-116, 109] or, in the case of a strong excitation, in [117-121].

Let us examine a number of examples. Assume that the electrons are magnetic, isotropic, and the ion velocities u_i are much greater than the electron thermal velocities. Then, in a strong magnetic field ($\omega_{He} \gg \omega_{pe}$), neglecting the nonuniformity effects, we shall examine the excitation of longitudinal Langmuir waves $\omega_{pe} \cos \theta$. The instability occurs because of the anisotropy in the ion distribution. From (3.68) we have

$$\gamma_{k,\nu} = \frac{1}{4} \cdot \frac{\omega_{pi}^2}{k^2 v_{Ti}^2} \omega_{pe} J_\nu^2 \left(\frac{k u_i \sin \theta}{\omega_{Hi}} \right) \frac{\nu \omega_{Hi} - \omega_{pe} \cos \theta}{k v_{Ti}} \times$$

$$\times \exp \left\{ -\frac{(\omega_{pe} \cos \theta - \nu \omega_{Hi})^2}{2 k^2 v_{Ti}^2 \cos^2 \theta} \right\}. \tag{3.69}$$

An instability arises when $\nu \omega_{Hi} > \omega_{pe} \cos \theta$. In order to effectively remove harmonics it is required that $k u / \omega_{Hi} \sim 1$, i.e., $\gamma \sim (\omega_{pi}^2 / \omega_{Hi}^2)(u/v_{Ti})^3 \omega_{pe}$.

Let us now consider those waves which propagate along the magnetic field with $k_\perp = 0$. According to (3.68) we now have $\nu = 0$, so that the longitudinal waves are damped. However, different types of waves, such as the transverse waves, can build up [114]. Waves for which the thermal motion is insignificant, i.e., for which

$$k v_{Ta} \ll |\omega_k^\sigma - \omega_{Ha}|, \quad k v_{Ta} \ll \omega_k^\sigma,$$

will appear as the superposition of two waves which are circularly polarized, that is,

$$e_k^\pm = \left\{ \frac{1}{\sqrt{2}}, \frac{\mp i}{\sqrt{2}}, 0 \right\}.$$

We then have

$$\gamma_{\mathbf{k}}^{\pm} = \frac{e^2}{4\pi} \sum_{\nu=-\infty}^{\infty} \sum_{a} \int \frac{v_{\perp}^2}{\frac{\partial}{\partial\omega}\,\omega^2\,\varepsilon_k^{\pm}} \left(\frac{\nu J_\nu(z_a)}{z_a} \pm J_\nu'(z_a) \right)^2 dp_a \ \times$$

$$\times\ \delta\left(\omega_{\mathbf{k}}^{\pm} - k_z v_z - \nu\omega_{Ha}\right) \left(k_z \frac{\partial f_{p_a}}{\partial p_z} + \frac{\omega_{Ha}\,\nu}{v_{\perp}} \cdot \frac{\partial f_{p_a}}{\partial p_{\perp}} - \frac{k_{\perp}\,c}{eH} \cdot \frac{\partial f_{p_a}}{\partial y} \right). \qquad (3.70)$$

When $k_{\perp} \to 0$, $z_\alpha \to 0$ so that for the polarization $e_{\mathbf{k}}^{+}$ all that remains is $\nu = 1$, while for $e_{\mathbf{k}}^{-}$ we have just $\nu = -1$. Then

$$\gamma_{\mathbf{k}}^{\pm} = \frac{e^2}{2} \sum_{a} \int \frac{v_{\perp}^3\,dv_{\perp}}{\frac{\partial}{\partial\omega}\,\omega^2\,\varepsilon_k^{\pm}}\, \delta\left(\omega_{\mathbf{k}}^{\pm} - k v_z \mp \omega_{Ha}\right)\ \times$$

$$\times \left(k_z \frac{\partial f_{p_a}}{\partial p_z} \pm \frac{\omega_{Ha}}{v_{\perp}} \cdot \frac{\partial f_{p_a}}{\partial p_{\perp}} - \frac{k_{\perp}\,c\partial f_{p_a}}{eH\,\partial y} \right) m_a^2 dp_z. \qquad (3.71)$$

In the region where $\omega \ll \omega_{Hi}$ we find for $\omega = k v_A$† and the distribution of (3.3) that

$$\gamma_{\mathbf{k}} = \sqrt{\frac{\pi}{2}} \cdot \frac{1}{4} \cdot \frac{v_A\,u_i^2}{v_{Ti}c^2} \cdot \frac{\omega_{pi}^2}{k^2\,v_{Ti}^2}\,(\pm 1)\,\omega_{Hi}\,e^{\frac{-\omega_{Hi}^2}{2k^2\,v_{Ti}^2}}. \qquad (3.72)$$

For a distribution appropriate to a temperature anisotropy,

$$f_{\mathbf{p}} = \frac{(2\pi)^{3/2}\,n}{v_{T_{\parallel}}\,v_{T_{\perp}}^2\,m^3}\,\exp\left(-\frac{v_{\perp}^2}{2v_{T_{\perp}}^2} - \frac{v_z^2}{2v_{T_{\parallel}}^2} \right), \qquad (3.73)$$

we find [114] that the maximum growth rate is obtained when

$$k \sim \frac{\omega_{Hi}}{v_A}\left| 1 - \frac{T_{\perp}}{T_{\parallel}} \right|$$

and

$$\gamma_{k\,\max} \simeq \omega_{pi} \frac{v_{T_{\parallel}}}{c}\,\exp\left\{ -\frac{v_A^2}{2v_{T_{\parallel}}^2} \cdot \frac{T_{\parallel}^2}{(T_{\perp} - T_{\parallel})^2} \right\}. \qquad (3.74)$$

---

†This wave is a superposition of an Alfven wave and a magnetoacoustic wave.

In similar fashion, it is not difficult to find the growth rate for whistlers $\omega^\sigma = k^2 c^2 \omega_{He} / \omega_{pe}^2$ from (3.68) and the distribution of (3.74):

$$\gamma_{k\ max} = \frac{\sqrt{\pi}}{2\sqrt{2}} \omega_{pe} \frac{v_{T_{||}}}{c} \left(\frac{T_\perp - T_{||}}{T_\perp} \right)^{3/2} \exp\left\{ -\frac{v_A^2}{2v_{T_{||}}^2} \cdot \frac{T_\perp}{T_\perp - T_{||}} \right\}. \tag{3.75}$$

In the same way we can treat the loss-cone instability which corresponds to that particle distribution in which f goes to zero in a specific cone of angles such that $v_z > (1/\alpha) v_\perp$, where $\alpha = H_{max}/H_{min}$ is the aspect ratio. It is evident that in reality the drop in the distribution function as the angle increases is quite smooth. Rosenbluth and Post [122] have studied the loss-cone instability for hot ions for which $T_i \gg T_e$ with respect to the excitation of a longitudinal electron wave with $\omega = \omega_{pe} \cos \theta$. ($\omega_{pe} \ll \omega_{He}$). If in addition $\omega_{Hi} \ll \omega_{pe}$, the ions may be treated as unmagnetized with $kv_{Ti}/\omega_{Hi} \gg 1$, i.e., they move along straight lines, and the probability can be written as

$$w_p^l(k) = \frac{e^2 (2\pi)^3}{\pi k^2 \left. \frac{\partial \varepsilon_k^l}{\partial \omega} \right|_{\omega=\omega_k^l}} \delta(\omega_k^l - kv). \tag{3.76}$$

Then when $k_\perp \gg k_z$

$$\gamma_k^l = \frac{e^2}{2\pi k^2 \left. \frac{\partial \varepsilon^l}{\partial \omega} \right|_{\omega=\omega_k^l}} \int \delta(\omega_k^l - kv) \left(\frac{k_\perp v_\perp}{v_\perp} \frac{\partial f_{p_i}}{\partial p_\perp} + k_z \frac{\partial f_{p_i}}{\partial p_z} \right) dp =$$

$$= \frac{\omega_{pi}^2 \omega_k^l m_i^{3/2}}{\sqrt{2} k^2 \left. \frac{\partial \varepsilon_k^l}{\partial \omega} \right|_{\omega=\omega_k^l}} \int_{\varepsilon_{\perp 0}}^{\infty} \frac{d\varepsilon_\perp}{k_\perp \sqrt{\varepsilon_\perp - \varepsilon_{\perp 0}}} \cdot \frac{\partial \varphi(\varepsilon_\perp)}{\partial \varepsilon_\perp}, \tag{3.77}$$

where

$$\varphi(\varepsilon_\perp) = \int f_{p_i} \frac{dp_z}{8\pi^3} m_i, \quad \varepsilon_{\perp 0} = \frac{m_i (\omega_k^l)^2}{2k_\perp^2}.$$

The function $\varphi(\varepsilon_\perp)$ is the particle distribution function in terms of the energy $\varepsilon_\perp = m_i v_\perp^2/2$ and satisfies the normalization condition that $\int_0^\infty \varphi(\varepsilon_\perp) d\varepsilon_\perp = 1$.

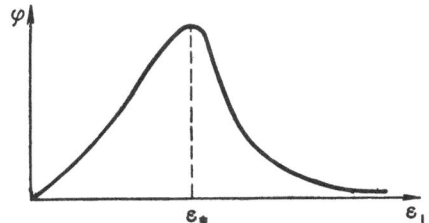

Fig. 3.2. The distribution function for plasma particles in terms of the energy $\varepsilon_\perp = m v_\perp^2 /2$ for a loss-cone instability: $v_\perp$ is the particle velocity perpendicular to the external magnetic field H_0.

The expression for the growth rate in (3.77) was obtained by Rosenbluth and Post [122]. The distribution function $\varphi(\varepsilon_\perp)$ must go to zero when $\varepsilon_\perp \to 0$ since such particles under this condition are not in bottlenecks because of their loss, and naturally $\varphi(\infty) = 0$. If $\varphi(\varepsilon_\perp)$ has a single maximum (Fig. 3.2) and ε_* is that value of $\varepsilon_\perp$ at the maximum, then it is evident that for $\varepsilon_{\perp 0} > \varepsilon_*$, $\partial\varphi/\partial\varepsilon_\perp < 0$, which means that (3.77) is negative and we have damping. But if $\varepsilon_{\perp 0} < \varepsilon_*$ then as a rule there will be some instability. As an example, consider the distribution

$$\varphi(\varepsilon_\perp) = \frac{\varepsilon_\perp}{\varepsilon_*^2} \exp\left(-\frac{\varepsilon_\perp}{\varepsilon_*}\right), \tag{3.78}$$

which satisfies all the requirements stipulated. We find

$$\gamma_k^l = \omega_{pe} \frac{\cos^2\theta}{\sin\theta} \sqrt{\pi \frac{m_i}{m_e}} \cdot \frac{\omega_{pi}^3}{2k^3 v_*^3} e^{-\frac{\varepsilon_{\perp 0}}{\varepsilon_*}} \left(1 - 2\frac{\varepsilon_{\perp 0}}{\varepsilon_*}\right), \quad \varepsilon_* = \frac{m_i v_*^2}{2}. \tag{3.79}$$

Buildup occurs when $\varepsilon_{\perp 0} < \varepsilon_*/2$. When $\varepsilon_{\perp 0} \ll \varepsilon_*$ the growth rate is independent of $\varepsilon_{\perp 0}$. This is because the primary contribution then comes from particles with energies greater than $\varepsilon_{\perp 0}$.

Under these conditions the growth rate of (3.77) takes the following form in the general case:

$$\gamma_k^l = \frac{\omega_{pi}^2 \omega_k^l m_i^{3/2}}{k^2 \frac{\partial\varepsilon}{\partial\omega}\Big|_{\omega=\omega_k^l} \sqrt{2k_\perp}} \int_0^\infty \frac{d\varepsilon_\perp}{\sqrt{\varepsilon_\perp}} \cdot \frac{\partial\varphi}{\partial\varepsilon_\perp}. \tag{3.80}$$

If $\partial\varphi/\partial\varepsilon_\perp \to 0$ when $\varepsilon_\perp \to 0$ faster than $\varepsilon_\perp^{1/2}$ then, upon integrating (3.80) by parts, we see that there will always be an instability:

$$\gamma_k^l = \frac{1}{2} \cdot \frac{\omega_{pi}^2 \, \omega_k^l \, m_i^{3/2}}{k^2 \dfrac{\partial\varepsilon}{\partial\omega}\bigg|_{\omega=\omega_k^l} \sqrt{2k_\perp}} \int_0^\infty \frac{\varphi(\varepsilon_\perp)}{\varepsilon_\perp^{3/2}} \, d\varepsilon_\perp. \tag{3.81}$$

It is clear that the instability can be eliminated only by including the diffusion of particles into the region of small $\varepsilon_\perp$.

3.4.2. Quasilinear and Nonlinear Effects

Quasilinear effects for aperiodic instabilities due to anisotropies in the distribution when $\gamma_k \ll kv_T$ were treated in [109]. It was shown there that the total energy given to the oscillations in a condition of weak supercriticality is proportional to the small parameter γ_k/kv_T, which characterizes the degree of supercriticality. A quasilinear theory of instability such as given by (3.77) was studied in [123]. The corresponding system of quasilinear equations which describe the diffusion process can be obtained without difficulty by using the quasilinear equations of Chapter 2.

We will treat in somewhat greater detail the quasilinear effects connected with the loss-cone instability when $\varepsilon_\perp \gg \varepsilon_{\perp 0}$. By using (3.77) we can obtain an equation for $\varphi(\varepsilon_\perp)$:

$$\frac{\partial\varphi(\varepsilon_\perp)}{\partial\tau} = \frac{\partial}{\partial\varepsilon_\perp} \cdot \frac{1}{\sqrt{\varepsilon_\perp}} \cdot \frac{\partial\varphi(\varepsilon_\perp)}{\partial\varepsilon_\perp}, \qquad d\tau = dt \int \frac{(\omega_k^l)^2 \, e^2 \, N_k \, dk \, \sqrt{m_i/2}}{2\pi^2 k^2 k_\perp \dfrac{\partial\varepsilon}{\partial\omega}\bigg|_{\omega=\omega_k^l}}. \tag{3.82}$$

Thus, an increase in the energy of turbulent fluctuations decreases the time scale for diffusion as given in (3.82), but a change in $\varphi(\varepsilon_\perp)$ appears in the time scale for buildup or damping of turbulent fluctuations. This situation is characteristic of the general case of three-dimensional processes in quasilinear relaxation [107].

From (3.82) we see that the final stationary state corresponds to $(1/\sqrt{\varepsilon_\perp})(\partial\varphi/\partial\varepsilon_\perp) = \text{const}$. But from (3.80) there is always either a buildup or damping. In order that the state be truly stationary one must have $N_k = 0$ or $\partial\varphi/\partial\varepsilon_\perp = 0$. Moreover, the plateau in $\varphi(\varepsilon_\perp)$

cannot extend to ∞ because $\varphi(\varepsilon_\perp)$ decreases. Therefore, $d\varphi/d\varepsilon_\perp$ becomes negative, which means that $N_k = 0$. Thus, the quasilinear relaxation process takes place so that initially the ions excite the turbulent fluctuations; then, after changing the distribution, they absorb all the fluctuations excited.

For the case of beam injection across the magnetic field, an indication of the development of nonlinear effects can be obtained from the successive processes of relaxation cutoffs of the instability [124].

3.5. The Excitation of Turbulence by a Constant Electric Field

3.5.1. Linear Effects

If the force acting on the plasma electrons from an external field is greater than the maximum friction force due to collisions between the electrons and ions,

$$eE > (\nu_{\text{col max}} m_e v_{T_e}) = m_e n_0 4\pi v_{Te}^2 \left(\frac{e^2}{m_e v_{Te}^2}\right)^2 \Lambda = \frac{e^2 \omega_{pe}^2}{v_{Te}^2} \Lambda = eE_D \qquad (3.83)$$

(Λ is the Coulomb logarithm), the electrons will be accelerated without any obstruction [125]. Here E_D is the Drieser field [126].

When the directional velocity $u = (eE/m_e)t$ of the electrons is greater than v_{Te}, the hydrodynamic Buneman instability (first treated in [127]) appears. An expression for the appropriate growth rate is easily obtained from (3.31) and (3.33) by going to the reference system where the electrons are at rest. Then ω_{pi}^2 is replaced with ω_{pi}^2, or the parameter n_1/n_0 is replaced by m_e/m_i:

$$\gamma = \text{Im } \omega_k = \frac{\sqrt{3}}{2^{4/3}} \left(\frac{m_e}{m_i}\right)^{1/3} \omega_{pe}. \qquad (3.84)$$

This value for the growth rate can be attained if u is much greater than v_{Te}. As shown in [128] by numerical methods, and in [129] analytically, the development of the Buneman instability produces a rapid heating of the electrons so that the condition $v_{Te} < u$ is destroyed.

In a plasma for which $T_e \gg T_i$ it is possible to excite ion-acoustic fluctuations as well as other low-frequency oscillations. Unlike the Buneman instability this excitation is not aperiodic; therefore the basic turbulent effects can be determined by this slower stage. It should be mentioned that the Buneman and ion-acoustic fluctuations belong to the same vibrational branch. If $T_e \gg T_i$ at the very outset, the ion-acoustic oscillations will start to appear at much lower drift velocities u: $u > 3v_{Ti}$. We can obtain the growth rate if we substitute the following into (3.34):

$$f_p^e = (2\pi)^{3/2} m_e^{-3} v_{Te}^{-3} \left(\exp\left[-(\mathbf{v} - \mathbf{u})^2 / 2v_{Te}^2 \right] \right) n_0,$$

$$\gamma_k^l = \frac{e^2}{2\pi \left. \dfrac{\partial e_k^l}{\partial \omega} \right|_{\omega = \omega_k^l} k^2} \int \delta(\omega_k^l - \mathbf{k}\mathbf{v}) \left(\mathbf{k} \frac{\partial f_p^e}{\partial \mathbf{p}} \right) d\mathbf{p} =$$

$$= \sqrt{\frac{\pi}{8}} \cdot \frac{k v_s}{(1 + k^2 d_e^2)^{3/2}} e^{-\dfrac{(\omega_k^l - ku)^2}{2k^2 v_{Te}^2}} \left(\frac{uk}{k v_{Te}} - \sqrt{\frac{m_e/m_i}{1 + k^2 d_e^2}} \right). \qquad (3.85)$$

If $u/v_{Te} \gg (m_e/m_i)^{1/2}$ then (3.85) will always describe the generation of oscillations. When $k \gg 1/d_e$ the magnitude of γ_k drops as $1/k^2$; if $k \ll 1/d_e$ we have $\gamma_e \sim k$. The maximum growth rate is obtained for $k \sim 1/d_e$. In this region the Landau damping on the ions is

$$\gamma_{ki} = -\omega_{pi} \sqrt{\frac{\pi}{8}} \left(\frac{T_e}{T_i} \right)^{3/2} e^{-\dfrac{T_e}{2T_i}}.$$

As the directional velocity increases the nonisothermal requirement is weakened until when $u \sim v_{Te}$ [the limit of applicability for (3.85)] the nonisothermal nature of the plasma plays no role. We note that when $T_e \gg T_i$ and the Landau damping on ions is exponentially small, the growth rate of (3.85) must be compared with the damping rate of ion-acoustic waves due to ion—ion collisions [130].

A detailed numerical calculation of the growth rates in the linear theory has been presented in [131]. The results of those calculations are shown in Fig. 3.3, which gives the maximum growth rates as functions of the nonisothermicity and the electrons' directional velocities. One can now follow the transition from an ion-acoustic instability to a Buneman instability. These results show that when u = v_{Te}, $\gamma_{max} \approx 0.2\,\omega$, i.e., one can still speak of a weak excitation.

Equation (3.34) gives the growth rate for exciting the various low-frequency oscillations in a magnetoactive plasma. In fact, one can use a previously obtained result, Eq. (3.35), if ω_{ia}^2 is replaced by ω_{pi}^2. In a low-pressure plasma ($v_A^2 \gg v_s^2$) the buildup of magnetized acoustic oscillations is described by the equation

$$\gamma_k = \sqrt{\frac{\pi}{8}} |\cos\theta| k v_s \left(\frac{u}{v_{Te}} - \sqrt{\frac{m_e}{m_i}} \right). \qquad (3.86)$$

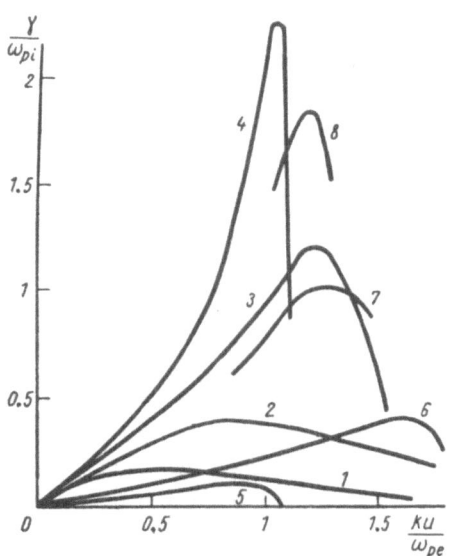

Fig. 3.3. Plots of γ/ω_{pi}, which is the growth rate for the buildup of oscillations divided by the ion frequency ω_{pi} in the presence of electronic directed motion relative to the ions with velocity u, as a function of ku/ω_{pe} (k is the wave number for the excited fluctuations). These graphs were obtained in [131] by numerical techniques.

Curve	T_i/T_e	u/v_{Te}
1	0	0.707
2	0	1.414
3	0	2.83
4	0	5.66
5	1	1.7
6	1	2.12
7	1	2.83
8	1	3.85

When $\cos \theta \sim 1$ this growth rate is of the same order as (3.85). With this same condition, but for longitudinal ion waves with $\omega_{Hi}\cos \theta$, (3.35) gives

$$\gamma_k = \sqrt{\frac{\pi}{8}} \, \omega_{Hi} \, |\cos \theta| \, \frac{v_A^2}{v_{Te}^2} \cdot \frac{\omega_{pe}^2}{k^2} \sin^2 \theta \left(\frac{u}{v_{Te}} - \frac{\omega_{Hi}}{k v_{Te}} \right). \qquad (3.87)$$

In the same way, when $v_A \gg v_s$, we can find the growth rate for exciting longitudinal ion-cyclotron waves from (3.35) [132, 133]. To find the growth rate for magnetoacoustic and Alfven waves, it is sufficient to use the following unit vectors in (3.34):

$$e_k^A = \{1, 0, 0\}, \quad e_k^M = \{0, 1, 0\}. \qquad (3.88)$$

For example,

$$\gamma_k^M = k v_A \sin^2 \theta \frac{v_s^2}{v_A^2} \left(\frac{u}{v_{Te}} - \frac{v_A}{v_{Te}|\cos \theta|} \right) \sqrt{\frac{\pi}{2}}. \qquad (3.89)$$

Buildup of these waves occurs when $u > v_A$.

Equation (3.71) can describe the buildup of circularly polarized waves which propagate along H_0. The possibility for such buildup depends rather heavily on the electron temperature anisotropy along the electric field and perpendicular to it. These instabilities are much like those considered in §3.4.[†] Those instabilities connected with the temperature nonuniformity and plasma density, together with the nonuniformity in the profile of the electron's directed velocity, are quite important [95-97, 135]. For a weak excitation these instabilities can be described by (3.35) for longitudinal waves, or by (3.34) in the more general case. When $\omega \gg \omega_{Hi}$ the ion-acoustic oscillations are not magnetized and the growth rates stay at the same values as in the absence of a magnetic field. The maximum is found at $k \sim 1/d_e$. When compared with this growth rate all those of Eqs. (3.86) to (3.89) turn out to be much smaller and they need be taken into account if the condition $T_e \gg T_i$ is not satisfied.

The above maximum growth rate can increase when there is a nonuniformity in the profile of the electrons' directed velocities.

[†] Hydrodynamic instabilities of this type were treated in [134].

It follows from (3.35) that the slipping instability can arise even when $u < \omega^\sigma/k_z$ if $k_x u/k_z \omega_{He} L > 1$. Similarly, the temperature anisotropy, which enters into (3.35) through the factor

$$-\nu\omega_{He}\left(1-\frac{T_{\|}}{T_{\perp}}\right), \tag{3.90}$$

so that it is absent when $T_{\|} = T_{\perp}$, can turn out to be very strong in large magnetic fields. If $k \sim 1/d_e$ corresponds to unmagnetized electrons, for which $k \gg \omega_{He}/v_{Te}$ and whose motion can be treated as being rectilinear, then the growth rate is described by (3.85) and the dependence on ω_{He} is now missing.

3.5.2. Quasilinear Effects

The initial stage of quasilinear relaxation due to development of ion-acoustic instability in an electric field has been treated in [136] using numerical integration of the quasilinear equations under the assumption that the electric field is so weak that

$$E \ll \frac{n_0 v_{Te}^3}{\omega_{pe}^3} E_D \sqrt{\frac{m_e}{m_i}} \quad \text{or} \quad \frac{E^2}{8\pi n_0 T_e} \ll \frac{m_e}{m_i}.$$

In satisfying this inequality one can, in the first approximation, neglect the effect of the electric field on the quasilinear interactions of the particles with the turbulent fluctuations, and one can then use the following:

$$\frac{\partial f_p^R}{\partial t} + \frac{e}{m_e}\mathbf{E}\,\frac{\partial f_p^R}{\partial v} = \frac{\partial}{\partial p_i}\,D_{ij}\,\frac{\partial f_p^R}{\partial v_j}, \tag{3.91}$$

$$D_{ij} = \int \frac{k_i k_j}{\pi k^2 \left.\dfrac{\partial \varepsilon_k^\sigma}{\partial \omega}\right|_{\omega=\omega_k^\sigma}} \delta\,(\omega_k^\sigma - \mathbf{k}\mathbf{v})\,N_k^\sigma\,\frac{dk}{(2\pi)^3}. \tag{3.92}$$

Field and Fried [136] have stated that the non-one-dimensional nature of the diffusion is very important. This is easy to understand if one notes that the electron velocities are far greater than the wave phase velocities; this means that the waves interact primarily with the electrons which are moving at nearly right angles to the waves.

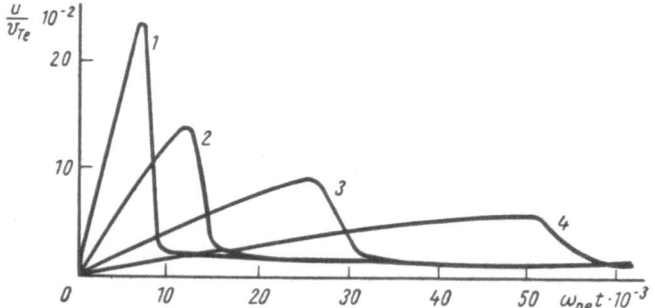

Fig. 3.4. Average directed velocity u of plasma electrons in an electric field as a function of time t in the initial phase of instability development (moments approximation). These curves were obtained by numerical calculation in [136].

Curve	E_0/E_D	ω_{pe}/ν_e
1	60	$2 \cdot 10^6$
2	20	$2 \cdot 10^6$
3	60	$2 \cdot 10^7$
4	20	$2 \cdot 10^7$

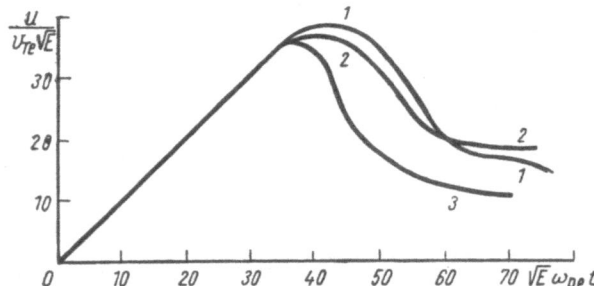

Fig. 3.5. Directed velocity of electrons in a strong electric field E_0 as a function of time ($E = eE_0/m_e v_{Te}\omega_{pe}$).

Curve	E_0/E_D	ω_{pe}/ν_e
1	50	10^6
2	10	10^6
3	10	$2 \cdot 10^6$

Figures 3.4 and 3.5 show the time dependence of the average electron velocities; these graphs show that initially the electron velocity increases linearly with time, but because of the development of intense turbulence the velocity begins to drop off, finally reaching a nearly constant value.

A solution to the quasilinear equations under quasistationary conditions, which means that the acceleration due to the electric field is compensated by the frictional forces of the ion-acoustic oscillations, has been examined in [137]. The solutions were extended to pair collisions with neutral particles in a weakly ionized plasma in [138, 6, 139], and to Coulomb collisions in a completely ionized plasma in [140]. Without dwelling on the details of the analysis, we would simply point out that according to [137] the work done by the electric field is dispersed primarily by increasing the vibrational energy and heating the electrons; however, the current remains essentially constant, independent of the electric field. The current is $j = en_0 u_*$, where u_* is the critical value of the electron velocity at which the ion-acoustic instability begins to appear. It was also shown in [137] that the energy of the ion-acoustic oscillations increases linearly with time. This growth naturally goes on only until the nonlinear interactions come into play. It was shown in [140] that the inclusion of collisions leads to a limitation on the increase in energy of the turbulent fluctuations as a function of time. However, in the majority of cases the nonlinear interactions become important at an energy of the turbulent fluctuations which is lower than the energy at which saturation sets in. In this connection, according to [139], in strong magnetic fields limited by angular diffusion the formation of a plateau in the distribution for electrons is important; it brings to a halt the diffusion process, but does not completely eliminate the quasilinear growth rate.

3.5.3. Nonlinear Effects

The nonlinear effects were first discussed in [138], in which the effects of induced scattering of ion-acoustic oscillations by ions were examined. The characteristic nonlinear growth rate of spectral transfer is of order $\gamma \sim \omega_{pi} (W/nT_i)\theta_0^2$, where θ_0 is the aperture angle of the interacting vibrations. Because of this spectral transfer the acoustic oscillations are moved into a region of small k, where they are then absorbed because of pairwise colli-

sions. This liquidates the small growth rate which was not eliminated by the quasilinear relaxation. A more rigorous solution to
the nonlinear equations which includes the effects of nonuniform
spectral transfer has been considered in [141]. The decay of the
ion-acoustic waves will produce a more intense spectral transfer
[142] (see Chapter 4 for more detail).

Many workers have observed a large turbulent resistivity
[26, 31-33]. One indication that the resistivity is high in the plasma is the uniform current distribution over the cross section,
and the external field will not experience any skin effect. According to [143-146], up to 30% of the energy is injected into the plasma and the electrons can be heated to 1.5-10 keV, with the ions
reaching energies of up to 200 eV. Anomalous plasma resistivity
has also been observed in attempts to create a high-current plasma betatron [147] and in work on plasma sources [148]. It was
also observed in [31]. It resisted the acceleration of particles.
The excitation of ion-cyclotron waves in plasma with a current
was observed in [149]. The spectra of ion-acoustic instabilities
in the presence of a current were studied in detail experimentally
in [150, 151].

3.6. Excitation of Turbulence in an Inhomogeneous Plasma. Drift Instabilities

The problems of instability in an inhomogeneous plasma and the
related problem of turbulent diffusion are quite important in discussions of plasma confinement. Diffusion effects cannot be
treated in detail without an in-depth study of the structure and
topology of the confining magnetic fields appropriate to the specific apparatus used to confine the plasma. Such problems are
rather technical and require the analysis of many details pertaining to the construction of the apparatus.

Leaving this kind of detail to the reviews [152, 6, 153], we
shall here consider briefly some of the basic ideas which will enable us to understand the physics of instabilities in an inhomogeneous plasma.

3.6.1. Linear Effects

Since we are not concerned here with the problems connected with the aperiodic hydrodynamic instabilities of an inhomogeneous plasma, we shall consider as before just those effects which lead to weak excitation of turbulent fluctuations. Returning to (3.34), it is assumed that the distribution function is of the form

$$f_p^\alpha = \frac{n_\alpha(y)}{(2\pi)^{-3/2}\, v_{T_{\perp,\alpha}}^2\, v_{T_{\parallel,\alpha}}}\, \exp\left\{-\frac{v_\perp^2}{2v_{T_{\perp,\alpha}}^2(y)} - \frac{v_z^2}{2v_{T_{\parallel,\alpha}}^2}\right\}, \qquad (3.93)$$

which takes into account the nonuniformity in the particle density and the nonuniformity and anisotropy of the temperature. By using (3.9) we easily find the following equation for longitudinal waves, which will replace (3.35):

$$\gamma_k^l = \sqrt{\frac{\pi}{2}} \sum_\alpha \frac{\omega_{p\alpha}^2}{v_{T_{\parallel,\alpha}}^3\, k^2 k_z \frac{\partial \varepsilon^l}{\partial \omega}\Big|_{\omega=\omega_k^l}} \sum_{\nu=-\infty}^{\infty} \left\{\exp\left(-\frac{(\omega_k^l - \nu\omega_{H\alpha})^2}{2k_z^2 v_{T_{\parallel,\alpha}}^2}\right)\right\} \times$$

$$\times \left\{\left[\nu\omega_{H\alpha}\left(1 - \frac{T_{\parallel,\alpha}}{T_{\perp,\alpha}}\right) - \omega_k^l - \frac{k_\perp v_{T_{\parallel,\alpha}}^2}{eH}\left(\frac{1}{n_\alpha}\cdot\frac{dn_\alpha}{dy} - \frac{1}{2T_{\parallel,\alpha}}\times\right.\right.\right.$$

$$\left.\left.\times\frac{dT_{\parallel,\alpha}}{dy}\left(1 - \frac{(\omega_k^l - \nu\omega_{H\alpha})^2}{k_z^2 v_{T_{\parallel,\alpha}}^2}\right)\right)\right] e^{-\mu_\alpha} I_\nu(\mu_\alpha) - \frac{k_\perp v_{T_{\parallel,\alpha}}^2}{eH}\cdot\frac{1}{T_{\perp,\alpha}}\times$$

$$\times\frac{dT_{\perp,\alpha}}{dy}\mu_\alpha \frac{d}{d\mu_\alpha} I_\nu(\mu_\alpha) e^{-\mu_\alpha}\right\}. \qquad (3.94)$$

We shall consider frequencies in the range

$$k_z v_{Ti} \ll \omega_k^l \ll k_z v_{Te} \quad (\omega_k \ll \omega_{Hi}).$$

Then, because of the factor

$$\exp\left\{-\frac{(\omega_k^l - \nu\omega_{Hi})^2}{2k_z^2 v_{Ti}^2}\right\}$$

the buildup or damping on ions will be exponentially small only if ω_k^l is not very close to $\nu\omega_{Hi}$. Excluding this case, consider the effect which appears due to the electrons. Assuming that the electrons are magnetized, $\mu_e \ll 1$, we see from (3.94) that the term describing the nonuniformity of the transverse temperatures is small and will be important only in noncyclotron buildup, when $\nu = 0$:

$$\gamma_k^l = \sqrt{\frac{\pi}{2}} \cdot \frac{\omega_{pe}^2}{v_{T_\|}^3 \, k^2 \, k_z \, \dfrac{\partial \varepsilon^l}{\partial \omega}\bigg|_{\omega=\omega_k^l}} \left(\omega_D - \omega_k^l - \omega_D \frac{1}{2} \cdot \frac{d \ln T_\|}{d \ln n} \right), \quad (3.95)$$

where

$$\omega_D = -\frac{k_x v_{T_\|}^2}{eH} \cdot \frac{1}{n} \cdot \frac{dn}{dy};$$

$|\omega_D|$ is the drift frequency.

If we are considering the buildup of drift waves, i.e., if $\omega^l \approx \omega^D$, then (3.95) will give the criterion for instability [154, 155] (the instability of an inhomogeneous plasma was first considered in [156]):

$$\frac{d \ln T_\|}{d \ln n} < 0. \qquad (3.96)$$

Noting that $\dfrac{\partial \varepsilon^l}{\partial \omega} = \dfrac{\omega_{pi}^2}{k^2 v_s^2} \, \omega_D^{-1}$, we obtain

$$\gamma_k^\sigma = -\sqrt{\frac{\pi}{8}} \cdot \frac{\omega_D^2}{k_z v_{T_\|}} \cdot \frac{d \ln T_{\| e}}{d \ln n}. \qquad (3.97)$$

Here the growth rate is much smaller than the frequency since $\omega_D \ll k_z v_{T_{\| e}}$. In addition $\omega_D \gg k_z v_s$, which is a necessary condition in order that ω_k^l be close to the drift frequency. If the small difference between ω_k^l and ω_D is taken into account, where

$$\omega_k^l = \omega_D + \frac{k_z^2 v_s^2}{\omega_D},$$

we then obtain the following instability criterion [157]:

$$\frac{1}{2} \frac{d \ln T_{\|, e}}{d \ln n} < -\frac{k_z^2 v_s^2}{\omega_D^2}. \qquad (3.98)$$

As shown in [157], when $\dfrac{d \ln T_{\parallel,e}}{d \ln n} = 0$ the plasma can become

unstable for small k_z since when $\dfrac{d \ln T_{\parallel,e}}{d \ln n_e} = 0$ in (3.95), even small differences between ω_k^g and ω_D become very significant. Such a difference exists even if $k_z \to 0$, if one takes into account the small corrections connected with the possible displacement of the ions in a wave transverse to the strong magnetic field. The vibrational frequency is found from the equation

$$\varepsilon^l = \varepsilon_{ij} \frac{k_i\, k_j}{k^2} = \varepsilon_{zz} \frac{k_z^2}{k^2} + \varepsilon_{xx} \frac{k_x^2}{k^2} = 0,$$

which gives, when ε_{xx} is neglected, the equation

$$\varepsilon_{zz} = \frac{\omega_{pl}^2}{k_z^2 v_s^2} \left(1 - \frac{\omega_D}{\omega}\right) = 0,$$

$$\omega_k = \omega_D \qquad (k_z v_s \ll \omega_D).$$

When $\omega_k \ll \omega_{Hi}$ we have $\varepsilon_{xx} = 1 + \omega_{pi}^2 / \omega_{Hi}^2$, i.e.,

$$\omega_k = \frac{\omega_D}{1 + \dfrac{k_x^2 v_s^2}{\omega_{Hi}^2}}, \qquad \omega_D - \omega_k = \omega_D \frac{k_x^2 v_s^2}{\omega_{Hi}^2}. \tag{3.99}$$

When $\dfrac{d \ln T_{\parallel,e}}{d \ln n} = 0$ this gives the following expression for the growth rate:

$$\gamma = \sqrt{\frac{\pi}{2} \cdot \frac{\omega_D^2}{k_z v_{Te}} \cdot \frac{k_\perp^2 v_s^2}{\omega_{Hi}^2}}. \tag{3.100}$$

Clearly this buildup occurs when

$$k_\perp > k_z \frac{\omega_{Hi}}{\omega_D}.$$

We note that when $T_e \gg T_i$, according to (1.124)–(1.127) one branch goes into ω_D while the other tends toward $k_z^2 v_s^2/\omega_D$ as $k_z \to 0$. Thus, in the limit we find the following instability condition from (3.95):

$$\frac{d \ln T_{\parallel,e}}{d \ln n} > 2. \tag{3.101}$$

Strictly speaking, since $\omega_k^\sigma > k_z v_{Ti}$, (3.101) is appropriate to the case of cold ions; $T_i \to 0$. However, even in an isothermal plasma, where $T_e = T_i$, we obtain the condition given in (3.101) (as shown in [154]) if we consider the frequency region where $\omega \ll k_z v_{Ti}$ and take into account the fact that the excitation can be strong. In fact, (3.100) already contains the effects due to the finiteness of the ion Larmor radius, which is of the order $k_\perp^2 v_{Ti}^2 / \omega_{Hi}^2$. It has already been illustrated why it is important to take such effects into account. These questions are of interest because of [158], where it was shown that the finiteness of the ion Larmor radius can stabilize certain low-frequency plasma fluctuations. The studies reported in [156, 155] showed that when the finite Larmor radius is taken into account the plasma becomes unstable for any value of $d \ln T / d \ln n$.

In the absence of a temperature gradient, but with a large density gradient, ω_D can be larger than ω_{Hi}. Then, when $T_e \gg T_i$ one can excite the usual nonmagnetic acoustic oscillations with $\omega = k v_s$, $\partial \varepsilon / \partial \omega = 2 (\omega_{pi}^2 / \omega_s^3)$; Eq. (3.95) then gives [159]

$$\gamma_k^\sigma = \sqrt{\frac{\pi}{8}} \sqrt{\frac{m_e}{m_i}} \cdot \frac{k}{k_z} (\omega_D - k v_s). \qquad (3.102)$$

Thus in this case an instability arises when the drift velocity exceeds the sound velocity.

In order to consider the excitation of nonlongitudinal waves it is sufficient to take into account just the difference in the unit polarization vectors for these waves in (3.34); this then leads to an equation similar to (3.95). The nonpotentiality must be taken into account if $\omega_k^\sigma / k_z = v_p > v_A$ [160]. The buildup of ion-cyclotron waves in inhomogeneous plasma was treated in [161]. When $k_z \neq 0$ there is a buildup of drift waves in the electrons when

$$\omega_D > \omega_{Hi}. \qquad (3.103)$$

A general condition for noncyclotron buildup in magnetized electrons is easily obtained from (3.34) (see [162]):

$$k_z \frac{\partial f_p^e}{\partial v_z} - \frac{k_\perp}{\omega_{He}} \cdot \frac{\partial f_p^e}{\partial y} > 0. \qquad (3.104)$$

3.6.2. Quasilinear and Nonlinear Effects

One cannot assume the distribution function to be Maxwellian, as was done in (3.93), when the reverse effect of the generated waves is taken into account. When $\nu = 0$, (3.34) gives, for magnetized electrons,

$$\frac{\partial N_k^\sigma}{\partial t} = 2\gamma_k^\sigma N_k^\sigma = \frac{e^2 N_k^\sigma}{k^2 m_e \pi \left.\frac{\partial \varepsilon^\sigma}{\partial \omega}\right|_{\omega = \omega_k^\sigma}} \int \left(k_z \frac{\partial f_{p_z}^e}{\partial v_z} - \frac{k_x}{\omega_{He}} \cdot \frac{\partial f_{p_z}^e}{\partial y} \right) \times$$

$$\times \delta\left(\omega_k^\sigma - k_z v_z\right) dp_z. \tag{3.105}$$

Here $f_{p_z} = \int f_p \, dp_x \, dp_y$ is a one-dimensional distribution function, which depends on p_z and y. In similar fashion, the general diffusion equation gives

$$\frac{df_{p_z}}{dt} = \int d\mathbf{k} \left(k_z \frac{\partial}{\partial v_z} - \frac{k_x}{\omega_{He}} \cdot \frac{\partial}{\partial y} \right) \frac{e^2 \delta\left(\omega_k^\sigma - k_z v_z\right) N_k^\sigma}{k^2 m_e^2 \pi \left.\frac{\partial \varepsilon^\sigma}{\partial \omega}\right|_{\omega = \omega_k^\sigma}} \times$$

$$\times \left(k_z \frac{\partial}{\partial v_z} - \frac{k_x}{\omega_{He}} \cdot \frac{\partial}{\partial y} \right) f_{p_z}. \tag{3.106}$$

Equations (3.105) and (3.106) reveal that the particles diffuse with respect to both velocities and the y coordinate, which is perpendicular to the confining external magnetic field. We note that (3.106) describes not only the effects of diffusion under the influence of turbulent fluctuations connected with the plasma inhomogeneity, but those effects which appear as the result of other longitudinal oscillations whose excitation and dissipation energy regimes (and therefore energy levels as well) are determined by factors unrelated to the plasma inhomogeneity.

Note that Eq. (3.106) describes the diffusion of resonance particles. As shown in Chapter 2, one need consider just the effects connected with resonance particles, assuming that the momentum of nonresonance particles is uniquely related to the momentum of the resonance particles.

The change in average momentum of the resonance particles in interactions with longitudinal oscillations is exactly equal to the change in momentum of the nonresonance particles. The momentum of the nonresonance particles is the plasmon momentum. If the turbulent fluctuations are high in frequency, the resonance and nonresonance particles are electrons. Thus, such processes cannot change the plasma density; that is, they cannot produce turbulent diffusion. Only low-frequency oscillations are connected with turbulent diffusion effects.

The quasilinear equations (3.105) and (3.106) were considered in [162-164]. According to [162] the quasilinear buildup of the spectrum given in (3.99) means that a very narrow packet of waves is formed with $k_x = \bar{k}_x$ satisfying the equation

$$\omega_{He} \frac{\omega_k^\sigma}{v_z} \cdot \frac{\partial f_{p_z}}{\partial v_z} = \bar{k}_x \frac{\partial f_{p_z}}{\partial y}. \qquad (3.107)$$

As in [107] a change of variable in Eqs. (3.105) and (3.106) changes them into one-dimensional equations, and the diffusion proceeds along the direction given by

$$\xi = \frac{v_z^2 k_z}{2\omega_k^\sigma} + \frac{\omega_{He} y k_z}{\bar{k}_x} = \text{const.}$$

From this we find that $\delta y \sim \frac{v_z \delta v_z \bar{k}_x}{\omega_{He} k_z \omega_D} \sim L \frac{v_A}{v_{Te}} \ll L$, i.e., the particle displacement is small compared with the size of the nonuniformity. However, it was shown in [164] that the nonuniformity of the diffusion process turns out to be very important (see also [107]).

Nonlinear decay effects were treated in [162, 165]. They cause a broadening of the spectrum for the wave packet mentioned above and an effective spectral transfer of the turbulent drift fluctuations into a region of strong Landau absorption by ions. Nonlinear effects in the interaction of drift oscillations in particle scattering have been considered in [166].

3.6.3. Turbulent Diffusion in Plasma

After integrating (3.106) over the particle momentum p_z and noting that $n = \int f_{p_z} \frac{dp_z}{(2\pi)^3}$, we obtain

$$\frac{\partial n}{\partial t} + \frac{\partial}{\partial y} (\langle n v_y \rangle + \langle n v_y \rangle^T) = 0, \qquad (3.108)$$

where

$$\langle nv_y \rangle^T = \int dk \frac{e^2 k_x N_k^\sigma \delta (\omega_k^\sigma - k_z v_z)}{k^2 \omega_{He} \pi \frac{\partial \varepsilon^\sigma}{\partial \omega}\Big|_{\omega = \omega_k^\sigma}} \left(k_z \frac{\partial}{\partial v_z} - \frac{k_x}{\omega_{He}} \cdot \frac{\partial}{\partial \mu} \right) \frac{f_{p_z} dp_z}{(2\pi)^3} . \tag{3.109}$$

Thus, in addition to the usual flow of matter across the magnetic field connected with the macroscopic motion of the plasma, there is also the turbulent flow given in (3.109). It determines the effects of turbulent plasma diffusion. Equation (3.109) describes the total flow connected with more than just the gradient in the plasma concentration. The coefficient of turbulent diffusion is the coefficient of proportionality between the turbulent flows that are proportional to dn/dy and dn/dy:

$$\langle nv_y \rangle^T = \langle nv_y \rangle' - D^T \frac{dn}{dy} . \tag{3.110}$$

If $f = n(y) f'(v_z)$, where $f'(v_z)$ is independent of y and characterizes the particle's velocity distribution, then

$$D^T = \int \frac{e^2 k_x^2 N_k^\sigma \delta (\omega_k^\sigma - k_z v_z) f'(v_z)}{k^2 \omega_{He}^2 \pi \frac{\partial \varepsilon^\sigma}{\partial \omega}\Big|_{\omega = \omega_k^\sigma}} \cdot \frac{dp_z}{(2\pi)^3} . \tag{3.111}$$

Note that (3.109) contains the total turbulent plasma flow across the magnetic field; this is of interest from the experimental point of view. This flow can be related to both a density gradient and a temperature gradient. Moreover, the first term of (3.109), which is proportional to $\partial/\partial v_z$, can make a contribution when the fluctuation distribution is asymmetric with respect to k_x. (If N_k^σ depends on $|k_x|$, this term drops out because the integrand is odd with respect to k_x.) Since a large number of different quantities, such as the drift velocities among others, depend on the density gradient, it is sometimes convenient to introduce the effective coefficient for turbulent diffusion, which is defined in terms of the total turbulent flow:

$$\langle nv_y \rangle^T = -D_{\text{eff}}^T \frac{dn}{dy} . \tag{3.112}$$

It is not difficult to write a general expression for D_{eff}^T. From the general equations of Chapter 2 we obtain

$$D_{eff}^T = \sum_{\alpha,\,v=-\infty}^{\infty} \int \frac{dp e^2 \left| \Gamma_v e_k^{\sigma*} \right|^2 dk N_k^\sigma k_x^2 v_{T\alpha}^2}{\pi \frac{\partial}{\partial\omega} \omega^2 \varepsilon^\sigma \Big|_{\omega=\omega_k^\sigma} \omega_{H\alpha}^2 \omega_D n (2\pi)^3} \times$$

$$\times \delta \left(\omega_k^\sigma - k_z v_z - v\omega_{H\alpha} \right) \left(k_z \frac{\partial}{\partial v_z} - \frac{k_x}{\omega_{H\alpha}} \cdot \frac{\partial}{\partial y} + \frac{v\omega_{H\alpha}}{v_\perp} \cdot \frac{\partial}{\partial v_\perp} \right) f_p, \qquad (3.113)$$

which then gives us a general expression for the total turbulent flow.

Thus, for the cyclotron instability of the ion-cyclotron waves in plasma with a current (see §3.4) we estimate the effective diffusion coefficient [132] (using the quasilinear theory) to be

$$D_{eff}^T \simeq \frac{cT_e}{eH} \left(\frac{T_e}{T_i} \right)^2 \left(\frac{u}{v_{Te}} \right)^2,$$

where $u \ll v_{Te}$ is the directed velocity of the electrons relative to the ions. Inclusion of the pairwise collisions [167] increases the diffusion coefficient by a factor of $\alpha = (\nu_{col}/\omega_{Hi})^{1/2}(v_{Te}/u)^{7/2}$ with $\alpha \gg 1$. Estimates for the drift instability give [162] $D_{eff}^T \simeq \frac{cT}{eH} \cdot \frac{c}{\omega_D n_0} \cdot \frac{dn_0}{dy}$. According to [6], if $8\pi nT/H^2 > m_i/m_e$, then

$$D_{eff}^T \simeq \frac{cT}{eH} \cdot \frac{c^2}{\omega_{pe}^2} \cdot \frac{\omega_{He}}{v_{Te}} \cdot \frac{1}{n_0} \cdot \frac{dn_0}{dy}.$$

Estimates of the diffusion coefficients are also available in [6, 166, 168, 169]. In all the available equations the diffusion coefficient is proportional to the Bohm factor cT_e/eH. The turbulent diffusion of plasma has been studied experimentally in many works [6, 170].

We must mention that the calculations of the diffusion coefficients are very rough in some cases (this is because neither the quasilinear nor the nonlinear equations can be solved exactly). It is therefore difficult to obtain exact quantitative agreement with the theory. An exception is the work of Kadomtsev and Nedospasov [171], which reports excellent agreement with experiment.

3.6.4. Turbulent Thermal Conductivity

of Plasma

In order to find the turbulent thermal conductivity one can use (3.106) to compute the change in the plasma electrons' energy perpendicular to H_0:

$$\frac{\partial}{\partial t}\int f_p \frac{v_\perp^2 m_e}{2}\frac{d\mathbf{p}}{(2\pi)^3} = \frac{\partial}{\partial y}\int \frac{k_x^2 N_\mathbf{k}^\sigma dk e^2\, \delta\left(\omega_\mathbf{k}^\sigma - k_z v_z\right)}{\omega_{He}^2\, k^2\, m_e^2\, \pi\, \dfrac{\partial \varepsilon^\sigma}{\partial \omega}\Big|_{\omega=\omega_\mathbf{k}^\sigma}}\frac{\partial}{\partial y}\ \frac{mv_\perp^2}{2}\cdot\frac{f_p\, d\mathbf{p}}{(2\pi)^3}\ .$$

$$(3.114)$$

Here, in the interest of simplicity, we have neglected the heat flow, which is independent of $\partial f/\partial y$ [assuming, as in (3.111), that $N_\mathbf{k}^\sigma$ is even in k_x]. The result from (3.114) can be written as $[f = f(\mathbf{v}_\perp)\, f'(v_z)]$

$$\frac{\partial T_\perp}{\partial t}+\frac{\partial}{\partial y}q=0;\qquad q=-\varkappa\frac{\partial T_\perp}{\partial y}\,,\qquad \varkappa=\int \frac{k_x^2 N_\mathbf{k}^\sigma e^2\, \delta\left(\omega_\mathbf{k}^\sigma - k_z v_z\right)}{\omega_{He}^2\, k^2\, m_e^2\, \pi\, \dfrac{\partial \varepsilon^\sigma}{\partial \omega}\Big|_{\omega=\omega_\mathbf{k}^\sigma}}f'(v_z)\frac{dp_z}{(2\pi)^3}\ .$$

$$(3.115)$$

Here $T_\perp = \int f(\mathbf{v}_\perp)\dfrac{mv_\perp^2}{2}dp_\perp$, where q is the heat flow. The coefficient $\varkappa$ is called the turbulent thermal conductivity of the plasma transverse to the magnetic field. It is rather easy to write a general expression for $\varkappa_{eff}$ which is analogous to that for D_{eff}, (3.113).

Turbulent thermal conductivity effects which include pairwise collisions were studied in [172].

3.7.   The Excitation of Turbulence

by Electromagnetic Waves and

Lasers

3.7.1. The Interactions of Highly Nonlinear

Waves and Shock Waves with Plasma

Assume that a strong, nonlinear wave is propagating through plasma. Quite strong electric fields are created at its front and these set up large, directional electron currents. Furthermore, the plasma distribution becomes nonuniform and anisotropic; this creates the conditions necessary for the beam-type instabilities (considered earlier), which are related to current, anisotropy, and drift instabilities.

These effects should be quite distinct, especially for low-frequency waves, which are long-wavelength, creating relatively weak nonuniformities. Thus, at the front of a nonlinear low-frequency wave the necessary conditions may prevail for the appearance of turbulent electrical conductivity, turbulent diffusion, turbulent thermal conductivity, and so on.

In a fluid strong low-frequency waves can turn into shock waves [1]. In a plasma the turbulent processes determine the structure of the collisionless shock waves [21]. With the nonlinear effects included in the study, this structure has been examined theoretically in [173-175]. The structure of nonlinear waves excluding the effects of turbulent dissipation was studied in [176-180]. It should be mentioned that the turbulent structure and width of the shock wave fronts can be obtained from the general phenomenological theory [181] if one uses an estimate of the turbulent transfer coefficients (see §3.6). The turbulent structure of shock waves has been studied experimentally in many works (see [182]).

3.7.2. Interaction of Strong High-Frequency

Waves with Plasma

The problem of turbulent processes in the interactions of strong high-frequency fields with plasma was addressed by Veksler [28] in connection with the new methods of acceleration which he had suggested. An intense high-frequency wave, which will not penetrate the plasma when $\omega < \omega_{pe}$, can exert a strong pressure on the plasma and cause it to accelerate. Since it is only those particles in the skin layer which experience the action of the electromagnetic wave, and the distribution of the electromagnetic field is highly nonuniform at the plasma boundary, one should find the low-frequency instabilities in the plasma to be those which are preferentially excited. They could be excited by turbulent transfer of momentum in the plasma. In addition to those instabilities connected with the nonuniformity of the field due to the electromagnetic wave, there can also be specific instabilities related to the anisotropy of the distribution function. In this case the anisotropy creates an external high-frequency field which vibrates the plasma charges only in the direction of the wave's electric field. The theory for this instability was constructed in [183]. If the external high-frequency field is strong enough that $eE_0/m\omega > v_{Te}$, there appears an instability which is similar to the Buneman instability [184]. It evidently leads to rapid heating of the plasma so that the condition $eE_0/m\omega > v_{Te}$ is no longer satisfied, and only weak instabilities remain [185].

Similar effects appear when lasers interact with plasmas [185-192].

We now return to Eqs. (2.113) and (2.124), which take into account the correlation effects in the interactions between fluctuations. We are interested in the excitation of longitudinal fields under the assumption that they are produced by strong high-frequency fields. The difference between the equations which take into account the transverse high-frequency fields and Eq. (2.124) is that ε_k^N and R_k^N depend on the energy density of the high-frequency fields W_k^t. The coefficients in these terms differ from those presented in that the nonlinear plasma currents must be multiplied by the transverse e_k^t rather than the longitudinal version (for example, $\sum_{k, k_1, k_2, k_3}$ in ε_k^N will be $\sum_{i j n l} \frac{k_i}{k} \cdot \frac{k_{1j}}{k_1} e_{k_2, n}^t e_{k_3, l}^t$). If the energy level of the longitudinal fields is not too high, the interaction between longitudinal fluctuations can be neglected (this is the linear stage of the excitation), while the energy density of the high-frequency fields is assumed to be given:

$$\begin{aligned} \omega \left(\varepsilon_k^l + \varepsilon_k^N \right) I_k^l &= R_k^N; \\ \varepsilon_k^N &= -\frac{8\pi i}{\omega} \int I_{k_1}^t \, dk_1 \sum_{k, k_1}^{\text{eff}}; \end{aligned} \right\} \tag{3.116}$$

$$R_k^N = 32\pi^2 \int \frac{|S_{k, k_1, k_2}|^2}{\omega \left(\varepsilon_{-k}^l + \varepsilon_{-k}^N \right)} I_{k_1}^t I_{k_2}^t \, dk_1 \, dk_2 \, \delta \left(k - k_1 - k_2 \right). \tag{3.117}$$

Here R_k^N acts as the external source of the excitation and ε_k^N describes the nonlinear growth rates of the excitation.

If the frequency is so large that $\omega \gg \omega_{pe}$, then, generally speaking, R_k^N is very small (see [185-187]). Then, for a relatively low energy level W_k^l, where one can neglect the nonlinear effects in the interactions of longitudinal waves, the left side of (3.116) is much greater than the right side. This means that one can use the following dispersion equation to analyze the excitation:

$$\varepsilon_k^l + \varepsilon_k^N = 0.$$

Since scattering on the thermal plasma particles is so small, the decay processes are most important for the high-frequency waves. Thus, one takes into account just that expression, similar to the second term in (2.117), which contains the resonant denominator for frequencies close to the linear eigenfrequencies [$1/\varepsilon_k^l$ in (2.117)].

Now consider the excitation of ion-acoustic oscillations. Let ω_k^s be the linear spectrum for the ion-acoustic oscillations, with

$$\left| \frac{\omega - \omega_k^s}{\omega_k^s} \right| \ll 1,$$

and let the spectrum of the transverse waves have an average width $\Delta\omega_*$, average frequency ω_*, and average angular spread $\Delta\theta$. By designating the difference between ω and $\omega^s(k)$ as Ω we obtain the following expression for $\varepsilon_k^N + \varepsilon_k^l$:

$$\varepsilon_k^N + \varepsilon_k^l = \frac{2\Omega\,\omega_{pi}^2}{(\omega^s(k))^3} + \frac{\omega_{pe}^2}{8k^2 v_{Te}^2} \int \frac{\omega_{pe}^4}{\omega_1^2} \frac{k \frac{\partial}{\partial k_1}}{n_0 m_e v_{Te}^2} \cdot \frac{\frac{I_{k_1}^t}{4\pi\omega_1} dk_1}{\Omega + \omega^s(k) - kv_g} = 0. \qquad (3.118)$$

When $\Omega \ll \Delta(k\mathbf{v}_g)$ one can neglect Ω in the denominator of (3.118). Then, by setting $1/(\omega - k\mathbf{v}_g) = -i\pi\delta(\omega - k\mathbf{v}_g)$ we obtain the effect of the decay instability of the transverse waves (into acoustic waves):

$$i\gamma = \Omega = \frac{i\pi(\omega^s(k))^3}{16k^2 v_{Te}^4 \omega_{pi}^2} \int \frac{\omega_{pe}^6}{n_0 m_e \omega_1^2} \left(k \frac{\partial}{\partial k_1} \right) \frac{I_{k_1}^t}{4\pi} \cdot \frac{1}{\omega_1} \delta(\omega^s(k) - kv_g) dk_1. \qquad (3.119)$$

If the frequencies of the excited oscillations are close to the Langmuir frequencies of the plasma, the dispersion relation which describes the excitation process takes the form $(\Omega = \omega - \omega^t_{k_1})$

$$\varepsilon^N_k + \varepsilon^l_k = \frac{2\Omega}{\omega_{pe}} + \int dk_1 w_{kk_1} \frac{I^t_{k_1} \frac{1}{\omega^t_{k_1}} - I^t_{k_1-k} \frac{1}{\omega^t_{k_1-k}}}{\Omega + \omega^l_k - \omega^t_{k_1} + \omega^t_{k_1-k}} = 0; \quad (3.120)$$

$$w_{k,k_1} = \frac{1}{16\pi} \cdot \frac{\omega^2_{pe} k^2}{n_0 m_e \omega^t_{k_1} \omega^t_{k-k_1}} \left(1 + \frac{(k_1 (k-k_1))^2}{k^2_1 (k-k_1)^2}\right). \quad (3.120')$$

If $\Omega \ll \Delta\omega_{max}(k)$, where $\Delta\omega_{max}(k)$ is the largest possible value of $\omega^l_k - \omega^t_{k_1} + \omega^t_{k_1-k} = \Delta\omega(k)$ when k_1 in $I^t_{k_1}$ runs over the possible values corresponding to the spectrum of the transverse waves, then the following decay instability appears:

$$i\gamma = \Omega = \frac{\pi\omega_{pe}}{2} i \int dk w_{kk_1} \delta(\Delta\omega(k)) \left(I^t_{k_1} \frac{1}{\omega^t_{k_1}} - I^t_{k_1-k} \frac{1}{\omega^t_{k_1-k}}\right). \quad (3.121)$$

A detailed study of the growth rates of the decay instability was conducted in [185-189]. If the momentum change of the transverse wave is small in the decay process, i.e., $k \ll k_1$, then the conservation of energy in the decay, which is given by a δ-function, gives us $\omega = k\mathbf{v}_g$. Plasma waves whose phase velocities are equal to the group velocity of the transverse waves are generated in the direction of a beam of transverse waves. The relativity factor for the waves $\gamma_p = 1/\sqrt{1-v^2_p}$ ($v_p = \omega/kc$) can become quite large, i.e., $\gamma_p \simeq \omega^t/\omega_{pe}$.

If the spectrum width for the transverse waves is so large that $\Delta\omega_* \gg \omega_{pe}$ (i.e., the spectrum is broad), one can use an expansion in terms of the momentum transfer k to find the growth rate. In the beam direction we have

$$\gamma_k \simeq \int w \left(k \frac{\partial}{\partial k_1}\right) N^t(k_1) dk_1, \quad (3.122)$$

where

$$w \simeq \delta(kv_g - \omega_{pe}) \frac{e^2 \omega_{pe}}{4\pi m^2_e} \left(\frac{k}{k_1}\right)^2. \quad (3.123)$$

Thus

$$\gamma_k = \omega_{pe} \frac{e^2}{4\pi m_e^2} \, \omega_1 \frac{dN(\omega_1)}{d\omega_1} \Bigg|_{\omega_1 = \frac{\omega_{pe} \, v_p}{\sqrt{1-v_p^2}}} \quad , \qquad v_p = \frac{\omega}{kc} \, . \tag{3.124}$$

The growth rate for generating waves along the beam is thus proportional to the frequency derivative of the distribution function for transverse waves. Introducing the spectral energy density $W^t(\omega)$ for the transverse waves by means of the equation

$$W^t = \frac{1}{(2\pi)^3} \int 2 \sqrt{k_1^2 c^2 + \omega_{pe}^2} \; N_{k_1}^t \, dk_1 = \int_0^\infty W^t(\omega_1) \, d\omega_1,$$

one can, when $\Delta\omega_* \ll \omega_*$, write γ_k in (3.124) in the form [185]

$$\gamma_k \simeq \omega_{pe} \frac{\pi}{4} \cdot \frac{\omega_{pe}^2}{nm_e c^2} \cdot \frac{dW^t(\omega)}{d\omega} \, . \tag{3.125}$$

Assuming that $W \sim W^t(\omega)\Delta\omega_*$ and $dW^t(\omega)/d\omega \sim W^t/(\Delta\omega_*)^2$, we have

$$\gamma_k \simeq \omega_{pe} \frac{\pi}{4} \left(\frac{\omega_{pe}}{\Delta\omega_*}\right)^2 \frac{W^t}{nm_e c^2} \, . \tag{3.126}$$

Even larger growth rates are obtained for "monochromatic" beams in which $\Delta\omega_* \ll \omega_{pe}$. In the beam direction

$$\gamma_k \simeq \omega_{pe} \frac{\pi}{2} \cdot \frac{\omega_{pe}}{\Delta\omega_*} \cdot \frac{W^t}{nm_e c^2} \, . \tag{3.127}$$

Only the decay effect has been included in obtaining (3.127) since the coalescence effects are possible only for other k when $\Delta\omega_* \ll \omega_{pe}$. The interval Δv_p of phase velocities of the generated waves around γ_p is smaller for (3.127) than for (3.126). For a monochromatic beam of transverse waves the growth rate is given by (3.133) up to angles θ of order $\sqrt{\omega_{pe}/\omega_*}$. As a result of the instability satellite frequencies appear for monochromatic beams. These results are for the case of plasma waves generated by a beam of transverse waves with a small aperture:

$$\Delta\theta \ll \left(\frac{\omega_{pe}}{\omega_*}\right)^{3/2} \, . \tag{3.128}$$

When $\Delta\theta \gg (\omega_{pe}/\omega_*)^{3/2}$ the growth rate for the same total energy in the beam of electromagnetic waves turns out to be smaller [189]:

$$\gamma = \omega_{pe}\frac{\omega_*}{\Delta\omega_*}\cdot\frac{W^t}{nm_e c^2}\cdot\frac{1}{(\Delta\theta)^2}\left(\frac{\omega_{pe}}{\omega_*}\right)^3 . \qquad (3.129)$$

Let us consider the limit $\Omega \gg \Delta\omega_{max}$ in (3.120) when the imaginary part of the denominator is insignificant. For a packet of transverse waves having a broad angular spread such that $\theta \sim 1$ when $|\Omega| \ll kc$ and $\theta \ll (\omega_{pe}/\omega_*)^2$ if $\lceil\Omega\rceil \gg (\omega_{pe}/\omega_*)^2kc$, we obtain

$$\Omega + i\gamma_k^l = \frac{1}{2\Omega}\int dk_1 I_{k_1}^t\left(k\,\frac{\partial w_{kk_1}}{\partial k_1}\right)\frac{\omega_{pe}}{\omega_*} . \qquad (3.130)$$

For $\theta \sim 1$ and $|\Omega| \ll kc$ in the case where $\theta \ll (\omega_{pe}/\omega_*)^2$ and $|\Omega| \ll (\omega_{pe}/\omega_*)^2$, we have

$$\Omega + i\gamma_k^l = -\frac{\omega_{pe}}{2\Omega^2}\int dk_1 I_{k_1}^t\, w_{kk_1}\,\frac{1}{(\omega_*)^2}\left(k^2 -\frac{(kk_1)^2}{(\omega_*)^2}\right). \qquad (3.131)$$

By substituting (3.120') into (3.130) we find

$$\Omega\left(\Omega + i\gamma_k^l\right) = -\frac{e^2k^2\omega_{pe}}{4m_e^2}\int dk_1 I_{k_1}^t\,\frac{(kk_1)}{(\omega_*)^5} . \qquad (3.132)$$

If, in the linear approximation, the waves are not damped and do not build up, i.e., if $\gamma_k^l = 0$, then for waves propagating in the direction of the beam of transverse waves there is a buildup of oscillations. The nonlinear growth rate can be evaluated from (3.129):

$$\gamma_k^N \sim \omega_{pe}\left(\frac{\omega_{pe}}{\omega_*}\right)^2\left(\frac{W^t}{nm_e v_p^2}\right)^{1/2} . \qquad (3.133)$$

The dynamics of the decay generation of Langmuir waves when $\Delta\omega_* \gg \omega_{pe}$ is, in many respects, similar to the dynamics of the development of beam instability considered above.

The energy lost by the beam of transverse waves can be of the same order of magnitude as the original energy; further, processes similar to the beam instability (a spatial instability in particular) can develop which produce a cutoff on the beam of transverse waves. The dynamics of decay generation for the narrow laser lines correspond to the appearance of a satellite. This diffusion through satellites corresponds to a shock wave in the space of the wave numbers k^t.

The experiments of [193] attest qualitatively to the excitation of turbulence due to the interactions of high-frequency waves. Evidence relative to the interactions of lasers with plasma is found in the anomalous absorption of energy in the experiments reported in [194, 195].

3.8. The Generation of Turbulence in Cosmic Plasmas

In the realm of astrophysics there is a large and varied group of physical conditions which can result in instabilities and plasma turbulence [46].

But in spite of the many different mechanisms responsible for turbulence, there are a number of characteristic features which indicate that turbulent processes play a very important role in cosmic plasmas.

First, most of the matter in the universe is found in an ionized state; that is, it is a plasma.

Second, there is no doubt that many of the processes observed under astrophysical conditions have an unstable, and perhaps explosive, nature. For example, there are frequent explosions on the surface of the sun, and there are explosions of supernovae. Furthermore, objects such as quasars are unstable [196], there are explosions in the cores of galaxies [46], and so on. Finally, it appears that metagalaxies have been formed through explosions [197]. It is apparent that the plasma is turbulent in all of these processes.

Third, there are continuously active energy sources which can maintain the plasma in a turbulent state for long periods. Among these sources are nuclear energy in stars and gravitational energy.

Fourth, the production of cosmic rays is usually related to the acceleration of particles in turbulent plasma [46, 45]. Cosmic rays are very important to the energy balance of the galaxies because their energy is comparable to the energy in the magnetic fields and the thermal energy of the interstellar plasma. This makes it easy to understand the great interest recently shown in the problem of turbulence in the cosmic plasma. There are also ways [198] of estimating the energy in the low-frequency turbulence from these observations.

Three mechanisms can be cited as specific methods for exciting turbulence in astrophysical conditions. One leads to an influx of energy from the low-frequency excitations — this is gravitational instability [199]. Another mechanism is connected with an energy influx from very high frequencies — this is the nonlinear radiation instability in plasma. The third mechanism is related to the generation of turbulence by the cosmic rays. In a gravitational instability hydrodynamic excitations appear; these correspond to collisionless shock waves, magnetohydrodynamic waves, or large-amplitude Alfven waves.

These low-frequency waves can be a source of high-frequency turbulence if the particle velocity at the wave front is greater than the average thermal velocities and a beam instability appears [200, 201]. At the front of nonlinear hydromagnetic waves there can also appear the specific instability related to the appearance of counter electron flow [202]. Very large radiation flows arise in explosive processes. This type of radiation can also be directional in character. In such a situation one must use the equations of §3.7. They lead to larger growth rates. The generation of nonrelativistic plasma waves is especially effective in this case. The generation of ion-acoustic waves is also quite effective.

We note that the electromagnetic radiation is frequently quite isotropic locally under cosmic conditions. The growth rates then found from $\varepsilon\,^{N}_{k}$ are negative, which means that turbulence is not generated. But according to (3.116) there is another possible generation mode, connected with the fact that the electromagnetic radiation can be treated as an external source which excites the turbulence. The amount of energy generated by this source in 1 second [203] is determined from the square of the spectral intensity of transverse waves: $1/4\pi = (1/4\pi)\int I_\omega d\omega n$ is the wave energy in 1 cm^3. In most situations the generation of Langmuir fluctuations is many times more effective than the generation of other types of fluctuations, and it increases rapidly as the frequency of the electromagnetic waves decreases. If ω_* is the smallest frequency in their spectrum then [203]

$$Q \approx \omega_{pe} \left(\frac{\omega_{pe}}{\omega_*} \right)^3 I. \tag{3.134}$$

We turn now to the generation of turbulence by cosmic rays. If, for example, the cosmic rays are formed through some explosion, their penetration into the surrounding plasma produces a local cosmic ray anisotropy and the appearance of a beam instability [201]. The corresponding growth rate for generating Langmuir fluctuations is greatest for phase velocities of order c [199]:

$$\gamma_k = \int w_e\,(\mathbf{k}) \left(\mathbf{k}\, \frac{\partial f_1}{\partial \mathbf{p}} \right) \frac{d\mathbf{p}}{(2\pi)^3} \sim \omega_{pe}\, \frac{n_1}{n_0} \cdot \frac{m_e}{m}\, \frac{1}{\Delta\theta}\,. \tag{3.135}$$

Here n_1 and n_0 are the concentrations of cosmic rays and the plasma, respectively, m is the mass of the cosmic ray particles, and $1/\Delta\theta$ is the degree of anisotropy in the cosmic ray distribution. It is significant that the instability is easily stabilized by the nonlinear effects.

However, these conclusions are just for the case where there is no acceleration of the low-energy particles, such as sub-cosmic rays, i.e., particles with nonrelativistic energies [46], whose spectra are a continuation of the spectrum of the cosmic rays. The peak energy can be found in these particles, and they will be primarily responsible for generating turbulence. In this case we can speak of turbulence generated by sub-cosmic rays. Moreover, for the nonrelativistic particles the nonlinear effects cannot remove the instability, and the energy of the turbulence which they have generated is on the order of the energy of the fast particles. Thus, a turbulent region ought to appear at the front of the expanding cloud formed, for example, when a supernova explodes. This region should move through the plasma at the velocity of the slow particles. Cosmic rays which scatter from this region can be trapped inside the cloud [201].

We should also point out that the plasma instability in the solar wind [205] causes turbulence in the interplanetary plasma.

There is strong turbulence in the chromospheric flares [206].
In the plasma found in the space near earth there is evidently both
a cyclotron instability and an anisotropic loss-cone instability for
particles trapped in the magnetic traps of the radiation belts [207],
formed during the streamlining of the earth's magnetosphere by
the solar wind [208].

There has been little attention devoted to the problems of
turbulent heating of the cosmic plasmas of late, in spite of the
fact that these heating processes are quite widespread. The heating of interstellar and intergalactic plasmas by cosmic rays was
discussed in [209]. Turbulent heating seems to be quite effective
in all processes in which low-frequency fluctuations are excited
directly, by gravitational instabilities or shock waves, for example.

Turbulent heating processes can be responsible for the x-ray
bremsstrahlung of the heated gas [144]. According to [144] such
processes could explain the powerful x-ray radiation coming from
x-ray sources.

Chapter 4

The Spectra of Stationary Plasma Turbulence

4.1. Types of Stationary Plasma Turbulence

The results of the previous chapters now enable us to address one of the most important problems in the theory of turbulence — the determination of the spectra for stationary plasma turbulence. Stationary turbulence is achieved when there is a balance between the generation, absorption, and transformation of the fluctuation energy across the spectrum. Some transformation is necessary because there can be no absorption in those wave number regions in which generation takes place (the balance between buildup and damping is such that the buildup dominates), and, of course, in the absorption region there is no generation. Stationarity is maintained in the generation region because the inflow of oscillation energy resulting from excitation is balanced with the energy outflow due to nonlinear transfer along the spectrum. In the absorption region balance is achieved between the energy inflow from the spectral transfer and the outflow due to absorption. Finally, in regions where there is neither absorption nor excitation, balance for a given k is maintained by the energy inflow and outflow due to nonlinear energy transformation. In this region the fluctuation spectrum does not depend on the type of instability leading to the excitation of oscillations; it is therefore universal, at least in this sense.

However, the dimensional considerations employed in fluid turbulence are not applicable in this case, for when the fluctuations have completely defined eigenfrequencies, one can always introduce lengths which are unique for the oscillations considered. The spectra for fluctuations of different collective plasma motions are quite different. In the region where the fluctuation energy is transferred from the sources of turbulence, there is a single characteristic — the turbulent power Q generated (which is the fluctuation energy generated in 1 cm^3 in 1 sec) — which determines the flow of energy along the spectrum. Furthermore, the possible absorption mechanism for the fluctuations can have an increased effect on the turbulence spectrum. This is true for at least two reasons. First, if the absorption rate is not very high, or there exist certain bands of absorption in the spectrum indicated by the nonuniform nature of the absorption, the transfer process can lead to a concentration of the turbulent fluctuations in certain regions. Then in these regions other types of nonlinear transfer can come into play, where these new types are small in regions of lower-level turbulent fluctuations. The presence of these kinds of transfer significantly alters the spectra of turbulent fluctuations. An example is the strong Langmuir turbulence. Second, there can be nonlinear energy loss mechanisms which operate on equal footing with the nonlinear spectral transfers.

It is useful to classify the types of plasma turbulence according to the absorption mechanism, which determines the final stage in the process of transforming the fluctuations for a given type of turbulence [58]. Such a classification is most convenient from a practical standpoint because it indicates the final result of instability development in a system (such as heating, radiation, and so on).

We shall use the following classes for stationary plasma turbulence:

1. Turbulence which is dissipated by absorbing the fluctuations in pair collisions of particles. This kind of turbulence is possible in a weakly ionized plasma, or when the vibrational energy is pumped away from the region of collective absorption, and only absorption due to pair collisions is possible (this is the case for isotropic weak Langmuir turbulence). This kind of turbulence is similar to fluid turbulence, where the dissipation results from viscosity due to the effects of pair collisions of particles.

2. Turbulence dissipated through the collective absorption of the fluctuations. This kind of turbulence produces an effective turbulent heating of the plasma.

3. Turbulence dissipated by the acceleration of fast plasma particles. As a rule, low-frequency fluctuations are either effectively absorbed by electrons, or they effectively interact with fluctuations, which are then absorbed by the electrons. Thus, the high-frequency fluctuations are most effective for acceleration. It can be assumed that the type of turbulence dissipated through particle acceleration is most probable for high-frequency fluctuations.

4. Fluctuations dissipated by the anomalous radiation of electromagnetic waves from the turbulent plasma. As a rule, the radiation processes increase as the magnetic field increases, just as they do when the number of fast particles accelerated by the turbulence increases. Thus, in strong magnetic fields when the particles are strongly accelerated one can have turbulence which is dissipated by radiation.

It is then clear that for practical purposes it is desirable to create the second kind of turbulence to obtain turbulent heating, the third kind for particle acceleration, and the fourth when one wishes to use the plasma as a radiation generator. In reality it is not always possible to strictly divide the turbulence into the classes given, for all the dissipation modes are present to some degree. However, the classification is convenient and it should be understood in the sense of defining the most important process for dissipating a given type of turbulence.

4.2. Nonlinear Interactions and the Spectral Transfer of Energy of Langmuir Fluctuations in an Isotropic Plasma

The most important interactions for Langmuir fluctuations are the induced scattering of plasmons l by electrons e and ions i in the plasma, and the interactions of plasmons l among themselves (called plasmon—plasmon scattering). Symbolically, these processes can be written as follows:

$$l + i \rightleftarrows l' + i', \tag{4.1}$$

$$l + e \rightleftarrows l' + e', \tag{4.2}$$

$$l + l_1 \rightleftarrows l' + l_1'. \tag{4.3}$$

In a nonisothermal plasma ($T_e \gg T_i$) one can also have the excitation of ion-acoustic oscillations:

$$l \rightleftarrows l' + s. \tag{4.4}$$

We shall treat the interactions of fluctuations in an isothermal plasma where the process described in (4.4) is not possible. The processes of (4.1) and (4.2), involving interactions with thermal electrons and ions, transform the fluctuation energy toward the lower frequencies. In fact it follows from (2.88) that

$$\frac{\partial N_k^l}{\partial t} = N_k^l \int w_p^{ll} \quad (\mathbf{k},\ \mathbf{k}_1)\left((\mathbf{k} - \mathbf{k}_1)\,\frac{\partial f_p^{(e,i)}}{\partial p}\right) N_{k_1}^l \frac{dp\,dk_1}{(2\pi)^6} =$$

$$= -N_k^l \int (\omega_k^l - \omega_{k_1}^l)\, N_{k_1}^l \frac{f_p^{(e,\,i)}}{(2\pi)^6\, mv_T^2}\, w_p^{ll}\,(\mathbf{k},\ \mathbf{k}_1)\, dp dk_1, \tag{4.5}$$

i.e., there is an increase in the waves N_k^l when $\omega_k^l < \omega_{k_1}^l$. Conservation of the number of waves follows from (4.5), and since the frequency (or energy) of each wave is only slightly dependent on $\mathbf{k}$,

$$\omega_k^l \simeq \omega_{pe} + \frac{3}{2}\,\frac{k^2 v^2_{Te}}{\omega_{pe}},$$

the energy is approximately conserved also. Therefore, the processes in (4.1) and (4.2) transform the turbulent energy toward small k.

According to (2.90), (2.97), and (2.101) an approximate expression for the probability of scattering by the plasma electrons is

$$w_p^{ll}\,(\mathbf{k},\ \mathbf{k}_1) = \frac{4e^4\,\left|\Lambda_{\mathbf{k},\,\mathbf{k}_1}^{(e)}\right|^2 \delta\left(\omega_k^l - \omega_{k_1}^l - (\mathbf{k} - \mathbf{k}_1)\,\mathbf{v}\right)(2\pi)^3}{\left.\dfrac{\partial \varepsilon_k^l}{\partial \omega}\right|_{\omega = \omega_k^l} \left.\dfrac{\partial \varepsilon_{k_1}^l}{\partial \omega_1}\right|_{\omega_1 = \omega_{k_1}^l}},$$

$$\Lambda_{\mathbf{k},\,\mathbf{k}_1}^{(e)} = \frac{1}{m_e\left(\omega_k^l\right)^2} \cdot \frac{(\mathbf{k}\mathbf{k}_1)}{k k_1}\left(\frac{2\mathbf{k}\mathbf{v}}{\omega_k^l} + \frac{\varepsilon_i^l\,(\omega_-,\ \mathbf{k}_-)}{\varepsilon^l\,(\omega_-,\ \mathbf{k}_-)}\right), \tag{4.6}$$

$$\omega_- = \omega_k^l - \omega_{k_1}^l, \quad \mathbf{k}_- = \mathbf{k} - \mathbf{k}_1. \tag{4.7}$$

Then, when the phase velocities $v_p = \omega_{pe}/k$ satisfy the condition

$$\frac{v_p}{v_{Te}} \ll \left(\frac{3m_i}{m_e}\right)^{1/3},\tag{4.8}$$

and the term ε_i/k in (4.7) is much smaller than kv/ω_k^l, Eq. (4.6) can be substituted into (4.5) to give the following expression for the nonlinear interaction:

$$\frac{\partial N_k^l}{\partial t} = \frac{3v_{Te}}{2m_e\, n_0\, \omega_{pe}}\, N_k^l \int \frac{N_{k_1}^l\, dk_1}{(2\pi)^{5/2}} \cdot \frac{(kk_1)^2}{k^2\, k_1^2} \cdot \frac{[kk_1]^2}{|k-k_1|}\left(k_1^2 - k^2\right).\tag{4.9}$$

We can use (4.9) to estimate the characteristic time τ for the spectral transfer of energy over $\Delta k \sim k$:

$$\frac{1}{\tau} \approx \omega_{pe}\frac{W}{n_0\, T_e} \cdot \frac{v_{Te}^3}{v_p^3}.\tag{4.10}$$

Since the transfer process excites waves with all possible directions with the same characteristic times, the transfer is accompanied by a rapid process of making the oscillations isotropic. This "isotropization" process is especially fast for ion scattering, which is effective if $\omega_- < |k_-|v_{Ti}$ (for electrons the analogous condition $\omega_- < |k_-|v_{Te}$ is always satisfied because $v_p > v_{Te}$). This condition indicates that the magnitude of the vector changes little ($\omega_- = 0$ when $k = k_1$). But the direction in which the fluctuations propagate can be altered by an angle of order unity. For ions Λ_{k,k_1} is large since there is no compensation between the nonlinear and Compton scattering (the Compton scattering is small since the ion masses are large):

$$\Lambda_{k,k_1}^{(i)} = \frac{(kk_1)}{kk_1} \cdot \frac{1}{(1+T_e/T_i)\, m_e\left(\omega_k^l\right)}.\tag{4.11}$$

The nonlinear interaction is described by the equation

$$\frac{\partial N_k^l}{\partial t} = N_k^l\frac{3\omega_{pe}\, T_e/T_i}{8n_0\, m_e\, v_{Ti}\,(1+T_e/T_i)^2}\int \frac{N_{k_1}^l\,(kk_1)^2\left(k_1^2 - k^2\right)dk_1}{(2\pi)^{5/2}\, k^2\, k_1^2\,|k-k_1|}\exp\left(-\frac{\omega_-^2}{2k_-^2\, v_{Ti}^2}\right).$$

$$\tag{4.12}$$

Two different regions of phase velocity should be distinguished. If $v_p \gg 3v_{Te}^2/v_{Ti}$, the transfer over a Δk of the order of k is substantial and the exponent in (4.12) can be dropped. Then, after integrating over angles for isotropic turbulent fluctuations, we obtain

$$\frac{dW_k^l}{dt} = W_k^l \int\limits_k^\infty Q(k, \ k_1) W_{k_1}^l \, dk_1 - W_k^l \int\limits_0^k Q(k_1, \ k) W_{k_1} \, dk_1,$$

$$W_k^l = \omega_k^l \frac{k^2}{2\pi^2} N_k^l \approx \omega_{pe} \frac{k^2}{2\pi^2} N_k^l, \quad k = |\mathbf{k}|, \qquad (4.13)$$

$$Q(k, \ k_1) = \frac{\sqrt{2\pi} \, T_e \, (k_1^2 - k^2) \, (k_1^2 + 2k^2/5)}{8n_0 \, m_e \, v_{Ti} \, T_i \, (1 + T_e/T_i)^2 \, k_1^3}. \qquad (4.14)$$

But if $v_p \ll 3v_{Te}^2/v_{Ti}$, the exponential in (4.12) allows interactions only between those oscillations for which $\frac{\Delta k}{k} \ll \frac{v_p \, v_{Ti}}{3v_{Te}^2} = \frac{\Delta k_*}{k}$.

For smooth turbulence spectra, in which there is a substantial change in the spectral density over a wave number interval which greatly exceeds Δk_*, the quantity Δk_* portrays an infinitely small physical quantity. In this approximation the spectral transfer is differential in character. It is easy to obtain an explicit expression for this transfer from (4.12), if one uses the relationship

$$\delta'(\omega_-) \to -\frac{\omega_-}{\sqrt{2\pi} \, k_-^3 \, v_{Ti}^3} \, e^{-\frac{\omega_-^2}{2k_-^2 \, v_{Ti}^2}}. \qquad (4.15)$$

As a result we have

$$\frac{\partial W_k}{\partial t} = \alpha W_k \frac{\partial W_k}{\partial k}, \quad \alpha = \frac{\omega_{pe}^3 \, \pi}{27 m_i \, n_0 \, v_{Te}^4 \left(1 + \dfrac{T_e}{T_i}\right)^2}. \qquad (4.16)$$

An estimate of the characteristic time for the spectral transfer of (4.16) is

$$\gamma \simeq \frac{1}{30} \left(\frac{v_p}{v_{Te}}\right)^2 \frac{W}{n_0 \, T_e} \cdot \frac{m_e}{m_i}. \qquad (4.17)$$

By comparing this expression with (4.10) we readily see that ion scattering dominates the electron scattering when the following condition is satisfied:

$$\frac{v_p}{v_{Te}} \gg \left(\frac{3m_i}{m_e}\right)^{1/5} \tag{4.18}$$

Since $v_p/v_{Te} > 3$ to 2, the region in which electron scattering is applicable is relatively restricted, having a significant extent such as $15v_{Te} > v_p > v_{Te}$ only for plasmas containing very heavy ions.

The vibrational energy is very efficiently transferred toward the small k. All the fluctuations will try to concentrate near k = $k_0 \rightarrow 0$. However, the phase space of the fluctuations, which is proportional to k_0^3, will also be very limited, which means that there must be strong plasmon collisions excited.[†] This is the process of (4.3), which must lead to the dispersion of the plasmons away from the small k region. The nonlinear equation which describes the plasmon interaction of (4.3) takes the form

$$\frac{\partial N_k^l}{\partial t} = \int w_{ll}^{ll}(k, \ k_1, \ k_2, \ k_3)\,(N_{k_1}^l N_{k_2}^l N_{k_3}^l + N_k^l N_{k_2}^l N_{k_3}^l -$$

$$- N_k^l N_{k_1}^l N_{k_3}^l - N_k^l N_{k_1}^l N_{k_2}^l)\,\frac{dk_1\,dk_2\,dk_3}{(2\pi)^9}. \tag{4.19}$$

The probability of a four-plasmon interaction (4.3) is found from the nonlinear current which is third order in Σ by using the methods previously described; in addition, processes which are of higher order in the field can be taken into account through iteration of the interaction [59, 80, 210] (see the graphs of Fig. 2.8). According to [210] the interaction in (4.3) drops off rapidly when $v_p \ll 3v_{Te}^2/v_{Ti}$, where, for smooth spectra, it cannot be larger than the nonlinear ion scattering in (4.16). But when $v_p \gg 3v_{Te}^2/v_{Ti}$ the probability of process (4.3) takes the form

$$w_{ll}^{ll}(k, \ k_1, \ k_2, \ k_3) = \frac{(2\pi)^6 e^4}{8m_e^4 v_{Te}^4}\left[\frac{(k_1\,k_2)\,(kk_3) + (kk_2)\,(k_1\,k_3)}{kk_1\,k_2\,k_3}\right]^2 \times$$

$$\times \delta\,(k + k_1 - k_2 - k_3)\,\delta\,(\omega_k^l + \omega_{k_1}^l - \omega_{k_2}^l - \omega_{k_3}^l). \tag{4.20}$$

An estimate of the characteristic time for the plasmon interactions

---

[†]We note that the decrease in phase volume during the scattering is accompanied by an increase in entropy due to the ion heating.

of (4.3) is

$$\frac{1}{\tau} \simeq \omega_{pe} \left(\frac{W}{n_0 T_e} \right)^2 \frac{\omega_{pe}^2}{k^2 v_{Te}^2} . \qquad (4.21)$$

The effectiveness of the interaction increases with v_p, and this does not permit the plasmons to concentrate in the region of very small k.

4.3. The Spectra of Stationary Langmuir Turbulence in Isothermal Plasma

It will be assumed that the turbulence is isotropic, and that the generation source is concentrated in a region of large k. It should be mentioned that all linear and nonlinear mechanisms for generating turbulence correspond to the excitation of fluctuations with $v_p < c$ (particle beams and electromagnetic waves, for example, do not excite fluctuations with $v_p > c$). However, the most effective excitation occurs when $v_p \ll c$ (see Chapter 3). The exciting source will be characterized by the power Q generating the turbulence (the energy generated in 1 cm^3 in 1 sec). Since the turbulent energy in an isotropic plasma is transferred toward the small k values, one of the most important dissipative mechanisms is plasmon absorption in pair collisions:

$$\left(\frac{dW_k}{dt} \right)_{bf} = -2\nu_e W_k, \qquad (4.22)$$

where W_k is the spectral density of turbulent energy,

$$W_k = \frac{k^2}{2\pi^2} \omega_k^l N_k^l, \qquad (4.23)$$

and ν_e is the average electron collision frequency in the plasma,

$$\nu_e = \frac{\omega_{pe}^4 \Lambda}{3 (2\pi)^{3/2} n_0 v_{Te}^3}, \qquad (4.24)$$

where $\Lambda = \ln 4\pi n_0 (v_{Te}^3 / \omega_{pe}^3)$ is the Coulomb logarithm.

The transformation from longitudinal oscillations into transverse oscillations in a uniform isotropic plasma takes place such that the frequencies of the transverse oscillations are smaller than those of the longitudinal oscillations. This is immediately evident from equations like (4.5) for such a transformation. These transverse oscillations have frequencies which are very close to ω_{pe},

$$\omega_k^t = \omega_{pe} + \frac{k^2 c^2}{2\omega_e}. \tag{4.25}$$

It is extremely difficult for these oscillations to escape the plasma because the index of refraction for these waves is very nearly zero. In fact, since $\omega_k^t < \omega_{k}^l$, it follows that

$$k^2 < 3\omega_{pe}^2 \frac{v_{Te}^2}{v_p^2 c^2}, \tag{4.26}$$

where v_p is the phase velocity of the transverse Langmuir fluctuations. Thus, it is convenient to call the waves of (4.25) transverse plasmons, a name which emphasizes that their frequencies are quite close to the plasmon frequencies, and that nearly half their energy is tied up in motions of the plasma, just as for the Langmuir plasmons.

The effects of interactions between longitudinal and transverse plasmons are discussed in depth in Chapter 6. Here we shall consider pure Langmuir turbulence without including the interactions between longitudinal and transverse plasmons. In a bounded or inhomogeneous plasma the interactions with transverse plasmons can lead to radiative losses of the turbulent energy at the plasmon frequency.

We begin by considering large phase velocities $v_p \gg 3v_{Te}^2/v_{Ti}$ so that only the processes of (4.22), (4.1), and (4.3) are important. We assume that the source of the turbulence is found outside this region, i.e., it is located where $v_p < 3v_{Te}^2/v_{Ti}$. This includes a large portion of the instabilities in a rather hot plasma, and it also includes the instability of a relativistic beam if

$$\frac{3v_{Te}^2}{v_{Ti}} > c, \tag{4.27}$$

i.e., for an isothermal hydrogen plasma $T_i = T_e > 15$ eV.

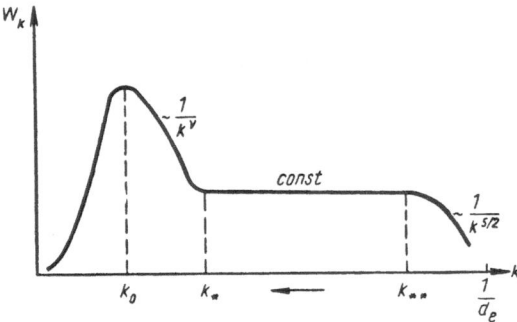

Fig. 4.1. The form of the spectrum for Langmuir fluctuations in a turbulent plasma, with $k_0 = 2\pi/L$ being
the fundamental scale of turbulence, which depends on
the size of the turbulent energy flow. The arrow
shows the direction in which the turbulent energy flows;

$$k_\bullet = \frac{\omega_{pe} v_{Ti}}{3 v_{Te}^2}; \quad k_{\bullet\bullet} = \frac{\omega_{pe}}{v_{Te}} \left(\frac{m_e}{3 m_i}\right)^{1/5}$$

From the above qualitative considerations about the interactions of turbulent fluctuations we can assume that the energy entering the region where $v_p \sim 3 v_{Te}^2 / v_{Ti}$ is transferred by ions to a very
small k; it is then dispersed by the process given in (4.3), which
means that the spectrum must have a maximum at some $k = k_0$
(Fig. 4.1). We set

$$k_0 = 2\pi/L, \qquad (4.28)$$

calling L the fundamental scale of the turbulence. In contrast to
fluids, it is determined by the generating power Q of the turbulence because the energy level of the turbulent fluctuations depends on Q (and therefore the effectiveness of the nonlinear interactions). Let us consider the asymptotic spectrum when $k \gg k_0$.
The equation describing the processes in (4.1), (4.3), and (4.22)
and the excitation of fluctuations with growth rate γ_k takes the form

$$\frac{dW_k}{dt} = -2\nu_e W_k + 2\gamma_k W_k + W_k \left[\int_k^\infty Q(k,\ k_1)\, W_{k_1}\, dk_1 - \right.$$

$$\left. -\int_0^k Q(k_1,\ k)\, W_{k_1}\, dk_1\right] + \int dk_1\, dk_2\, dk_3\, R(k,\ k_1,\ k_2,\ k_3) \times$$

$$\times \left(k^2 W_{k_1} W_{k_2} W_{k_3} + k_1^2 W_{k_2} W_k W_{k_3} - k_2^2 W_k W_{k_1} W_{k_3} - k_3^2 W_k W_{k_1} W_{k_2}\right),$$

$$(4.29)$$

where $Q(k, k_1)$ is given by 4.14 and $R(k, k_1, k_2, k_3)$ is, from (4.20),

$$R(k, k_1, k_2, k_3) = \frac{\omega_{pe}^3 \, \delta \left(k^2 + k_1^2 - k_2^2 - k_3^2\right)}{6\pi n_0^2 \, T_i^2 \, v_{Te}^2 \, 4^3 \, (1 + T_e/T_i)^2} \times$$

$$\times \int d\Omega_1 \, d\Omega_2 \, d\Omega_3 \, \delta \, (k + k_1 - k_2 - k_3) \left(\frac{(kk_3)(k_2 k_1)}{kk_3 \, k_2 \, k_1} + \frac{(k_2 k)(k_3 k_1)}{k_2 \, kk_3 \, k_1}\right)^2. \quad (4.30)$$

Here $W_k = \omega_{pe} k^2 N_k / 2\pi^3$, and $d\Omega_1$, $d\Omega_2$, $d\Omega_3$ are the solid angle elements for the vectors k_1, k_2, k_3. Equation (4.30) is valid for an isotropic turbulence. The integration over angles in (4.30) is elementary, but very tedious [211]. For $kk_1 < k_2 k_3$ and $k > k_1$ we obtain

$$R(k, k_1, k_2, k_3) = \frac{\pi \omega_{pe}^3 \, \delta \left(k^2 + k_1^2 - k_2^2 - k_3^2\right)}{48 v_{Te}^2 \, n_0^2 \, T_i^2 \left(1 + \dfrac{T_e}{T_i}\right)^2 k^3 \, k_1^3 \, k_2^3 \, k_3^3} \times$$

$$\times \left\{\frac{k_1^5}{5} \cdot \frac{(k_2^2 - k_3^2)^4}{(k^2 - k_1^2)^3} \left(k^2 - \frac{3}{7} \, k_1^2\right) + \frac{(k_2^2 - k_3^2)^2 k_1^2}{15 \, (k^2 - k_1^2)} \times \right.$$

$$\times \left(-5k^4 + 24 k^2 \, k_1^2 - 43 k_1^4\right) + \frac{k_1^3}{315} \left(622 k_1^6 + 825 k_1^4 \, k^2 + 672 k_1^2 \, k^4 + 105 k^6\right)\bigg\}.$$

$$(4.31)$$

If $k < k_1$, but $kk_1 < k_2 k_3$, one must make the substitution $k \rightleftarrows k_1$ in (4.31). If $kk_1 > k_2 k_3$ and $k_2 > k_3$ then $k \rightleftarrows k_2$, $k_3 \rightleftarrows k_1$, and if $kk_1 > k_2 k_3$, $k_2 < k_3$ then $k \rightleftarrows k_3$, $k_2 \rightleftarrows k_1$. When finding the asymptotic spectrum the form of the spectrum when $k \sim k_0$ is not very important, and we can assume that the spectrum is of the form

$$W_k = (\nu - 1) \frac{W}{k_0} \left(\frac{k_0}{k}\right)^\nu \begin{cases} 1 & k > k_0, \\ 0 & k < k_0, \end{cases} \quad (4.32)$$

or, by setting $\xi = (k_1/k)^2$ and $\eta = (k_2/k)^2$ for $kk_1 < k_2 k_3$, with $\xi = (k_3/k)^2$, $\eta = (k_2/k)^2$ for $kk_1 > k_2 k_3$, we obtain the following equation, which describes the process in (4.3):

$$\frac{1}{W_k} \cdot \frac{dW_k}{dt} = \frac{\pi}{6} \, \omega_{pe} \left[\frac{W}{8n_0 \, (T_i + T_e)}\right]^2 \frac{\omega_{pe}^2 \, (\nu - 1)^2}{k_0^2 \, v_{Te}^2} \, \xi_0^\nu \, (G_1 + G_2 + G_3) = 0, \quad (4.33)$$

$$G_1 = \int\limits_{\xi_0}^{1} d\eta \int\limits_{\xi_0}^{\eta} d\xi L\,(\xi,\ \eta)\, v\,(\xi,\ \eta), \qquad (4.34)$$

$$G_2 = \int\limits_{1}^{\infty} d\eta \int\limits_{\eta}^{\infty} d\xi M\,(\xi,\ \eta)\, v\,(\xi,\ \eta), \qquad (4.35)$$

$$G_3 = 2 \int\limits_{\xi_0}^{1} d\xi \int\limits_{1}^{\infty} d\eta N\,(\xi,\ \eta)\, u\,(\xi,\ \eta). \qquad (4.36)$$

Here

$$\xi_0 = (k_0/k)^2; \qquad (4.37)$$

$$v\,(\xi,\ \eta) = \frac{1+\xi^{\frac{\nu}{2}+1}-\eta^{\frac{\nu}{2}+1}-(1+\xi-\eta)^{\frac{\nu}{2}+1}}{\eta^{\frac{\nu}{2}+2}\,\xi^{\frac{\nu}{2}+2}\,(1+\xi-\eta)^{\frac{\nu}{2}+2}}, \qquad (4.38)$$

$$u\,(\xi,\ \eta) = \frac{1+(\xi+\eta-1)^{\frac{\nu}{2}+1}-\xi^{\frac{\nu}{2}+1}-\eta^{\frac{\nu}{2}+1}}{\eta^{\frac{\nu}{2}+2}\,\xi^{\frac{\nu}{2}+2}\,(\xi+\eta-1)^{\frac{\nu}{2}+2}}, \qquad (4.39)$$

and the coefficients L, M, and N are

$$L\,(\xi,\ \eta) = \frac{(2\eta-\xi-1)^4}{5\,(1-\xi)^3}\,\xi^{5/2}\left(1-\frac{3}{7}\,\xi\right) + \frac{\xi^{3/2}\,(2\eta-\xi-1)^2}{15\,(1-\xi)} \times$$
$$\times\,(-5+24\xi-43\xi^2) + \frac{\xi^{3/2}}{315}\,(622\xi^3+825\xi^2+672\xi+105), \qquad (4.40)$$

$$M\,(\xi,\ \eta) = \frac{(2\eta-\xi-1)^4}{5\,(\xi-1)^3}\left(\xi-\frac{3}{7}\right) + \frac{(2\eta-\xi-1)^2}{15\,(\xi-1)}\,(-5\xi^2+24\xi-43) +$$
$$+\,\frac{1}{315}\,(622+825\xi+672\xi^2+105\xi^3), \qquad (4.41)$$

$$N\,(\xi,\ \eta) = \frac{(2-\xi-\eta)^4\,\xi^{5/2}}{5\,(\eta-\xi)^3}\left(\eta-\frac{3}{7}\,\xi\right) + \frac{\xi^{3/2}}{15\,(\eta-\xi)}\,(2-\eta-\xi)^2 \times$$
$$\times\,(-5\eta^2+24\xi\eta-43\xi^2) + \frac{\xi^{3/2}}{315}\,(622\xi^3+825\xi^2\eta+672\xi\eta^2+105\eta^3) \qquad (4.42)$$

In the asymptotic region $\xi_0 \ll 1$. It follows from (4.34) and (4.38) that the primary contribution to G_1 comes from: a) the region of small ξ and small η when ξ is of order η and ξ_0; b) the region where ξ is small, but η is of order unity, with ξ of order $1-\eta$

and ξ_0. Only the second of these two regions makes a significant contribution to G_3. G_2 contains no large factors like $1/\xi \mathscr{G}$ and is negligibly small. When $\eta \ll 1$ we obtain from (4.40) that

$$L(\xi, \eta) \approx \frac{4}{3} \xi^{3/2} \left(\eta + \frac{11}{5} \xi \right) ; \tag{4.43}$$

$$v(\xi, \eta) = \frac{\left(\frac{\nu}{2} + 1 \right) (\eta - \xi)}{\xi^{\frac{\nu}{2} + 2} \eta^{\frac{\nu}{2} + 2}} . \tag{4.44}$$

When $\nu > 3/2$ the contribution of this region to G_1 is

$$\delta G_1 = \frac{16(4\nu - 3)}{15\nu \left(\frac{\nu}{2} - 1 \right) \left(\nu - \frac{3}{2} \right) \xi_0^{\nu - 3/2}} . \tag{4.45}$$

The contribution of the region $\xi \ll 1, 1 - \eta \ll 1$ to G_1 is identical with (4.45), as can be seen by changing the order of integration and making the substitution $1 + \xi - \eta = \eta$. Thus, $G_1 = 2\delta G_1$. As for G_3, it is convenient to go to an integration over $\eta' = \xi + \eta - 1$. For $\eta - 1 \sim \xi_0, \eta \ll 1$ we have u$(\xi, \eta') \approx -v(\xi, \eta')$, $N(\xi, \eta') \simeq L(\xi, \eta')$. This makes it easy to calculate G_3. We find that in this approximation G_3 and G_1 compensate for each other exactly. One must take into account the next higher terms in the expansions in ξ and η. These expansions must give $G_1 \sim 1/\xi_0^{\nu - 5/2}$ if $\nu > 5/2$, and $G_3 \sim 1/\xi_0^{\nu - 5/2}$ (there is no compensation in this case). In addition to the region of small η, one must also consider the region of finite, arbitrary η, but with ξ still small. Since L and M are proportional to $\xi^{3/2}$, while v and u are proportional to $\xi_0^{-(\nu/2 + 2)}$, the result will be $G_1 + G_3 \sim \xi_0^{-(\nu/2 - 1/2)}$. Comparing this dependence with $\xi_0^{-(\nu - 5/2)}$ we see that when $\nu > 4$ the term with $\xi_0^{-(\nu - 5/2)}$ dominates, but when $\nu < 4$ the main term contains $\xi_0^{-(\nu/2 - 1/2)}$.

Let us first consider the case with $\nu > 4$. By using the more exact expansions of L, M and u, v, we find

$$G_1 + G_3 = \frac{16 \left(\frac{\nu}{2} + 2 \right) (16\nu - 34)}{\left(\nu - \frac{5}{2} \right) \nu \left(\frac{\nu}{2} - 1 \right) \left(\frac{\nu}{2} - 2 \right) \xi_0^{\nu - 5/2}} . \tag{4.46}$$

When $\nu > 4$ Eq. (4.46) has no zero, which means that there are no solutions to the nonlinear equations. This assertion is

valid for sufficiently strong turbulence, when the effects of collisions and spectral transfer to ions are only small corrections to
the interaction in (4.3). Let us now consider the situation for
$\nu < 4$. We transform G_1 by changing the order of integration on
ξ and η and introducing the new variable $\eta' = 2\eta/(1 + \xi) - 1$. For
η' the integration runs over symmetric limits, and the integrand
is an even function of η'. By writing the resulting integral as
twice the integral from 0 to $\eta'_{\max}$ and then returning to the old
variables, we find

$$G_1 = 2 \int_{\xi_0}^{1} d\xi \int_{\xi}^{\frac{1+\xi}{2}} d\eta L(\xi, \eta) v(\xi, \eta). \qquad (4.47)$$

This integral has a singularity only for small ξ. The singularity at small η in (4.47) is compensated by a similar singularity
in G_3, so that the complete integral $G_1 + G_3$ has no singularities
for small η. This enables us to use expansions of L and M for
small ξ but with arbitrary values of η:

$$L(\xi, \eta) = \frac{4}{3} \eta(1-\eta)\xi^{3/2}. \qquad (4.48)$$

We then have

$$G_1 = \frac{8}{3} \int_{\xi_0}^{1} \frac{d\xi}{\xi^{\frac{\nu+1}{2}}} \int_{\xi}^{1/2} \frac{d\eta \left(1 - \eta^{\frac{\nu}{2}+1} - (1-\eta)^{\frac{\nu}{2}+1}\right)}{\eta^{\frac{\nu}{2}+1}(1-\eta)^{\frac{\nu}{2}+1}}. \qquad (4.49)$$

Similarly,

$$G_3 = \frac{8}{3} \int_{\xi_0}^{1} \frac{d\xi}{\xi^{\frac{\nu+1}{2}}} \int_{\xi}^{\infty} \frac{d\eta \left(1 + \eta^{\frac{\nu}{2}+1} - (1+\eta)^{\frac{\nu}{2}+1}\right)}{\eta^{\frac{\nu}{2}+1}(1+\eta)^{\frac{\nu}{2}+1}}. \qquad (4.50)$$

By using the value of the integral

$$I_{\pm} = \int_{\xi}^{\infty} \frac{d\eta}{\eta^{\frac{\nu}{2}+1}(1 \pm \eta)^{\frac{\nu}{2}+1}} = \frac{2}{\xi^{\nu/2} \nu} F\left(\frac{\nu}{2}+1, -\frac{\nu}{2}, -\frac{\nu}{2}+1, \mp\xi\right),$$

$$(4.51)$$

where F is the hypergeometric function, together with its asymptotic expansion

$$I_{\pm} \simeq \frac{2}{\xi^{\nu/2}\nu} \mp \frac{\nu/2+1}{\left(\dfrac{\nu}{2}-1\right)\xi^{\nu/2-1}},$$
(4.52)

we obtain

$$G_1 + G_3 = \frac{2^{\frac{\nu}{2}+5}}{3\nu(\nu-1)\xi_0^{\frac{\nu-1}{2}}}\left[2\left(1-\frac{1}{2^{\nu/2}}\right) - F\left(\frac{\nu}{2}+1,\ -\frac{\nu}{2},\ -\frac{\nu}{2}+1,\ \frac{1}{2}\right)\right].$$
(4.53)

Thus, the equation sought for the spectrum takes the form

$$F\left(\frac{\nu}{2}+1,\ -\frac{\nu}{2},\ -\frac{\nu}{2}+1,\ \frac{1}{2}\right) = \Gamma_0,$$
(4.54)

where

$$\Gamma_0 = 2\left(1-\frac{1}{2^{\nu/2}}\right).$$
(4.55)

Figure 4.2 shows the right (curve 2) and left (curve 1) parts of (4.54) as functions of ν. These two curves intersect when

$$\nu = 2.84.$$
(4.56)

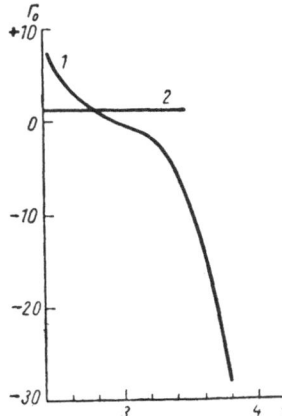

Fig. 4.2. Numerical solution to Eq. (4.54) which determines the exponent ν in the turbulence spectrum: 1) the left side of Eq. (4.54); 2) the right side of (4.54).

This spectrum is obtained in the asymptotic region when one can neglect the collisions and transfer to the ions. If these effects are taken into account, the quantity Γ_0 in (4.54) must be replaced with

$$\Gamma = \Gamma_0 - \delta\Gamma_Q - \delta\Gamma_v, \qquad (4.57)$$

where

$$\delta\Gamma_v = \frac{36 v_e \, n_0^2 \, (T_e + T_i)^2 \, v \left(\dfrac{k_0 \, v_{Te}}{\omega_{pe}} \right)^2}{\pi \omega_{pe} \, W^2 \, (v-1) \, 2^{\frac{v}{2}-1} \, \xi_0^{\frac{v+1}{2}}} \qquad (4.58$$

describes the absorption effects due to collisions, and

$$\delta\Gamma_Q = \frac{n_0 T_i \, 9v}{W \, \sqrt{\pi} \, 2^{\frac{(v+1)}{2}} \, (v-1)} \left(\frac{k_0 \, v_{Te}}{\omega_{pe}} \right)^3 \frac{v_{Te}}{v_{Ti}} \cdot \frac{1}{\xi_0^{\, v/2+1}} \qquad (4.59)$$

is the spectral transfer to the ions.

Because of the strong dependence of curve 1 on v, the spectrum remains almost a power-law spectrum, but the magnitude of the spectral index v, which characterizes the turbulence spectrum, increases until $v = 4$.

Equation (4.29) was rigorously solved in [21] by a method involving a continuous parameter. The initial form of the spectrum was found by approximately analytic methods, and in the region where $k \gg k_0$ it is a power-law spectrum with $v = 3$, while for $k \leq k_0$, $W_k = $ const $(k/k_0)^2 \exp(-k^2/k_0^2)$. These two solutions were then matched at $k \sim k_0$.

The initial solution turned out to be very close to the final solution, and the procedure converges very rapidly. The calculated results for the parameter values

$$\frac{W}{n_0 T_e} \sim 10^{-6}, \quad \omega_{pe} = 5.6 \cdot 10^3 \ \text{sec}^{-1}, \quad v_e = 2.4 \cdot 10^{-6} \ \text{sec}^{-1},$$

$$T_e = T_i, \ n_0 \sim 10^{-2}, \ v_{Te} \simeq 4.8 \cdot 10^7 \, \text{cm/sec}, \ v_{Ti} = 1.1 \cdot 10^6 \ \text{cm/sec} \qquad (4.60)$$

are shown in Fig. 4.3. In the region where $k > k_0$, which does not strictly achieve the asymptotic values of $k \sim 7k_0$, it is found by the method of least squares that the best approximation to the

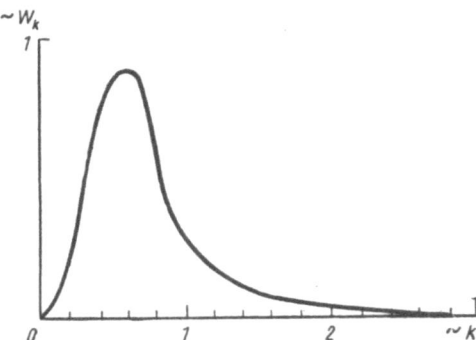

Fig. 4.3. A numerical calculation of the turbulence
spectrum.

exact spectrum is obtained with $\nu = 3.9$. We should also point out
that for the values of the parameters given in (4.60), the contribu-
tions of $\delta\Gamma_\nu$ and $\delta\Gamma_Q$ are comparable to Γ_0, and according to (4.57)
and (4.54) one ought to expect the value $\nu = 3.7$.

Thus, the numerical calculations affirm the validity of the
analytic theory for the spectrum. Furthermore, it reveals the
stability of the stable spectrum for deviations from the true distribution which correspond to the difference between the initial approximate spectrum and the final spectrum obtained through numerical experimentation. Finally, the numerical calculations let
us determine the spectrum in the region of the fundamental scale
of the turbulence. For the parameters in (4.60) the spectrum is
well approximated by the following equations:

$$W_k = \begin{cases} \sim k^2 \exp\left(1 - \dfrac{k^2}{k_0^2}\right), & 0 \leqslant k \leqslant k_0; \\[2mm] \sim \left(\dfrac{k_0}{k}\right)^\nu, & \nu = 3.9, \quad k_0 \leqslant k \leqslant 7k_0. \end{cases} \qquad (4.61)$$

We now turn to a consideration of the turbulence spectra in
the region where $v_p \ll 3v_{Te}^2/v_{Ti}$ [213]. In a sufficiently hot plasma
this can be a region outside the source generating the turbulence.
However, even in a cold plasma ($T_e < 15$ eV) the turbulence
sources can be found at phase velocities which are significantly
smaller than those of interest here. When $v_p \ll 3v_{Te}^2/v_{Ti}$ it is
characteristic to neglect the small contribution from the process

in (4.3),[†] while the differential nature of the energy transfer to the ions in (4.16) is significant. Assuming that $v_p \gg v_{Te}(3m_i/m_e)^{1/5}$, the scattering on electrons can be neglected. Then

$$\frac{dW_k}{dt} = -2\nu_e W_k + \alpha W_k \frac{\partial W_k}{\partial k} = 0 \qquad (4.62)$$

and, when ν_e can be neglected, we find W_k = const = A. If we use k• to denote the value of k at which this spectrum must be joined to $W_k = \frac{W_0(\nu-1)}{k_0}\left(\frac{k_0}{k}\right)^\nu$,

$$k_* \simeq \frac{\omega_{pe}}{3v_{Te}^2} v_{Ti}, \qquad (4.63)$$

then the value of the constant in the spectrum when k > k• is $A = \frac{\nu-1}{k_0}\left(\frac{k_0}{k_*}\right)^\nu$. If the absorption of the fluctuations is taken into account, then

$$W = \frac{2\nu_e}{\alpha} k + B, \quad B = \text{const}, \qquad (4.64)$$

so that the spectrum decreases for small k. When matching with the solution for k < k• one should be aware that the collisions increase ν. Thus, as ν increases, the dip in the spectrum increases (see curves 1-3, Fig. 4.4). For sufficiently strong absorption, when the generation region is found where k ≪ k•, this dip can go to zero, so that the maximum in the spectrum disappears and the turbulence energy cannot leak through the absorption region (curve 4, Fig. 4.4).

[†]An estimate of the growth rate in (4.3) in this case is [214, 80]

$$\gamma \sim \omega_{pe}\left(v_{Te}^2/v_p^2\right)\left(W/n_0 T_e\right)^2.$$

This growth rate is larger than either (4.17) or (4.10) only if v_p is close to $v_{Te} \cdot (m_i/m_e)^{1/5}$ and the turbulence energy W satisfies the inequalities

$$1 \gg W/nT \gg (m_e/m_i)^{1/5}.$$

Therefore, the region so obtained is so narrow that the spectrum of [214], in which only the process of (4.3) is taken into account, is not considered here.

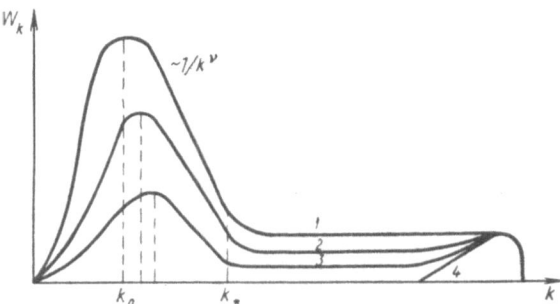

Fig. 4.4. Change in the Langmuir turbulence spectrum as the absorption (ν_e) of the turbulent fluctuations increases when the flow of turbulent energy across the spectrum remains constant.

The criterion for the disappearance of the maximum depends on the turbulence generating power Q. In order to find this criterion we consider the turbulence generation region in which $v_p > v_{Te}(3m_i/m_e)^{1/5}$. We assume that in the region

$$k_g < k < k_g - \Delta k_g \qquad (4.65)$$

the growth rate is constant, equal to γ_g. We should be mindful that γ_g is an effective growth rate, characterizing the rate at which all fluctuations grow for a given k in the interval Δk, independent of their directions. We can use this type of effective growth rate because the fluctuations rapidly become isotropic due to the ion scattering [59]. It follows from (4.12) and (4.13) that the characteristic time τ needed to establish this isotropy is of the order

$$\frac{1}{\tau} \simeq \omega_{pe} \frac{W}{n_0 T_e},$$

where W is the energy found in the interval $\Delta k = \Delta k_g$.

Assuming that $\gamma_g \gg \nu_e$ (which is essentially a supercritical excitation regime), we find that in the excitation region

$$W_k = \frac{\gamma_g}{\alpha}(k_g - k).$$

From the matching condition with (4.64) we obtain

$$B + \frac{2\nu_e}{\alpha} k_g = \frac{\gamma_g}{\alpha} \Delta k_g,$$

and the generating power is

$$Q = \int_{k_g - \Delta k_g}^{k_g} \gamma_g W_k dk = \frac{\gamma_g^2 (\Delta k_g)^2}{2\alpha}, \qquad (4.66)$$

i.e., $B + 2\nu_e k_g / \alpha = (2Q/\alpha)^{1/2}$. Thus, in place of (4.64) we can write

$$W = \frac{2\nu_e}{\alpha} (k - k_g) + \sqrt{\frac{2Q}{\alpha}}. \qquad (4.67)$$

When $k_g \gg k_*$ the condition for disappearance of the maximum in the spectrum is, from (4.67),

$$\frac{2\nu_e}{\alpha} (k_g - k_*) \approx \frac{2\nu_e}{\alpha} k_g > \sqrt{\frac{2Q}{\alpha}}$$

or

$$Q < Q_g = \frac{2\nu_e^2 k_g^2}{\alpha} = \frac{2\nu_e^2 \, 27 \, m_i \left(1 + \frac{T_e}{T_i}\right)^2}{\pi \omega_{pe}^2 m_e} \frac{\nu_{Te}^2}{\nu_{p,g}^2} \omega_{pe} n_0 T_e. \qquad (4.68)$$

In other words, this condition can be written in terms of the growth rate:

$$\gamma_g < 2\nu_e \frac{k_g}{\Delta k_g}. \qquad (4.69)$$

Let us now estimate the fundamental scale of the turbulence or the parameter k_0, which characterizes the maximum in the spectrum when the inequalities of (4.68) and (4.69) are not satisfied by a wide margin. The absorption in (4.64) can then be neglected, giving A = B or

$$\sqrt{\frac{2Q}{\alpha}} = \frac{W_0 (\nu - 1)}{k_0} \left(\frac{k_0}{k_*}\right)^\nu. \qquad (4.70)$$

We shall assume that the bulk of the turbulence energy is found where $k \ll k_*$. For simplicity we will assume that when $k < k_0$ the spectrum is approximated by $(1/k_0)(k/k_0)^s W_0(\nu - 1)$. Then the total turbulence energy will be of order

$$W = \int_0^\infty W_k \, dk = \frac{\nu + s}{s + 1} W_0. \tag{4.71}$$

On the other hand, from the balance equation we have

$$Q = 2\nu_e W = 2\nu_e \frac{s + \nu}{1 + s} W_0. \tag{4.72}$$

By eliminating W_0 from both (4.72) and (4.70) we obtain

$$k_0 = k_* \left(\frac{\nu + s}{(s + 1)(\nu - 1)} \right)^{\frac{1}{\nu - 1}} \left(\frac{Q_*}{Q} \right)^{\frac{1}{2(\nu - 1)}}, \tag{4.73}$$

where

$$Q_* = \frac{\nu_e^2 k_*^2}{2\alpha}. \tag{4.74}$$

It is impossible to assign a numerical value to the factor in front of k_* in (4.73) in such an estimate; therefore we simply have

$$k_0 \approx k_* \left(\frac{Q_*}{Q} \right)^{\frac{1}{2(\nu - 1)}} \tag{4.75}$$

We note that Q_* in $(k_*/k_g)^2$ is much smaller than Q_g [see (4.68)], which is where the maximum in the spectrum disappears. In order for (4.75) to be correct we must have $k_0 \ll k_*$. One should also keep in mind the very slow dependence of k_0 on Q.

We now turn to the region of very large values of k, such that $v_p \ll v_{Te} (3m_i/m_e)^{1/5}$, in which the main process for transporting energy across the spectrum is induced scattering by plasma electrons. This region is narrow because $v_p > (2-3) v_{Te}$. However, if the turbulence source excites oscillations having phase velocities very close to v_{Te}, and the mass of the plasma ions is large, it must be taken into account. By integrating (4.9) over angles, we obtain the following for an isotropic turbulence:

$$\frac{dW_k}{dt} = \alpha^* W_k \left\{ \int_k^\infty W_{k_1} dk_1 \frac{k^2}{k_1^3} \left(k_1^2 - k^2\right) \left(\frac{1}{3} k_1^2 + \frac{4}{7} k^2\right) - \right.$$

$$\left. - \int_0^k W_{k_1} dk_1 \frac{k_1^2}{k^3} \left(k^2 - k_1^2\right) \left(\frac{1}{3} k^2 + \frac{4}{7} k_1^2\right) \right\} = 0, \qquad (4.76)$$

where

$$\alpha^* = \frac{6 v_{Te} \sqrt{2\pi}}{5 m_e n_0 \omega_{pe}^2}. \qquad (4.77)$$

After differentiating a number of times and multiplying by $1/k$, Eq. (4.76) can be reduced to the differential equation

$$2 \frac{d^3 W_k}{dk^3} + \frac{21}{k} \cdot \frac{d^2 W_k}{dk^2} + \frac{46}{k^2} \cdot \frac{dW_k}{dk} + \frac{10 W_k}{k^3} = 0. \qquad (4.78)$$

The three linearly independent solutions to (4.78) are of the form $W_k = \text{const}/k^\nu$, where ν is one of the roots of the equation

$$2\nu^3 - 15\nu^2 + 29\nu - 10 = 0, \qquad (4.79)$$

i.e.,

$$\nu_1 = \frac{5}{2}, \qquad \nu_{2,3} = \frac{5 \pm \sqrt{17}}{2}. \qquad (4.80)$$

It is easily seen that only ν_1 satisfies the condition for localizability of the energy transfer along the spectrum, that is, the condition that the spectrum be independent of the turbulence sources and damping. This condition is expressed as $2 < \nu < 3$. If ν is outside this interval, then formally it is not a solution to the integral equation of (4.76) since either the first or the second integral will diverge. But when $2 < \nu < 3$ the largest contribution in both integrals of (4.76) comes from those k_1 which are close to k (this is the condition that the transfer interaction be local). Equation (4.76) is limited in application since the value of k is limited from above by the requirement that the Landau absorption be taken into account, and from below by the contributions of the ions to the transfer process. Thus, the solutions $\nu_{2,3}$ in fact require the use of other equations.

Consequently, the only acceptable solution which satisfies the assumption that the spectrum is determined solely by electron scattering is

$$W_k = \frac{\text{const}}{k^{5/2}} . \tag{4.81}$$

4.4. The Spectra of Low-Frequency Fluctuations Excited by Langmuir Turbulence in an Isothermal Plasma

As mentioned in §2.5, the correlation of Langmuir fluctuations can produce low-frequency fluctuations at frequencies equal to the difference between the Langmuir fluctuation frequencies, and there need not be any unique relationship between the frequency and the wave number. These correlation tails play an important role in induced scattering for interactions with the plasma particles. However, these fluctuations can be directly detected experimentally, and a number of beam-interaction experiments indicate that the maximum frequency in the spectrum of low-frequency coincides with the difference in the frequencies of the high-frequency fluctuations. An approximate expression for the nonlinear current S_{k,k_1,k_2} when $(\omega_1 + \omega_2) \ll \omega_1$ (ω_2 is of opposite sign relative to ω_1) takes the form

$$S_{k,\,k_1,\,-k_2} = \frac{(k_1 k_2)\, e \omega k}{8\pi m_e\, k_2 k_1 \omega_{pe}^2} \left(\varepsilon_k^{l\,(e)} - 1 \right). \tag{4.82}$$

Therefore, according to (2.124)

$$\frac{I_k}{8\pi} = \frac{|E_{k,\omega}|^2}{8\pi} = \frac{k^2}{2} \left| \frac{\varepsilon_k^{l\,(e)} - 1}{\varepsilon_k^l} \right|^2 \int dk_1 dk_2 \frac{(k_1\, k_2)^2}{4\, k_1^2\, k_2^2\, \omega_{pe}^2\, n_0\, m_e} \times$$

$$\times W_{k_1}^l\, W_{k_2}^l\, \delta\left(\omega - \omega_{k_1}^l + \omega_{k_2}^l \right) \delta\left(k - k_1 + k_2 \right). \tag{4.83}$$

For isotropic turbulence, integrating this expression over the angles gives

$$\frac{I_k}{8\pi} = \frac{k}{512\pi m_e\, n_0\, \omega_{pe}^2} \left|\frac{\varepsilon_k^{l\,(e)}-1}{\varepsilon_k^l}\right|^2 \int\limits_0^\infty \frac{W_{k_1}^l\, dk_1}{k_1^3} \times$$

$$\times \int\limits_{(k-k_1)^2}^{(k+k_1)^2} W_{k_2}^l\, \frac{dk_2^2}{k_2^4}\, (k_1^2 + k_2^2 - k^2)^2\, \delta\,(\omega - \omega_{k_1}^l + \omega_{k_2}^l). \qquad (4.84)$$

Here $I_{k,\omega}$ characterizes the distribution of electric fields in the low-frequency fluctuation, and $W^l = \int\limits_0^\infty W_k^l\, dk = \int dk W_k^l$. Let $\omega > 0$. Then, because of the δ -function in (4.84) it follows that

$$k_2^2 = k_1^2 - \frac{2\omega\omega_{pe}}{3v_{Te}^2} < k_1^2.$$

This value of k_2 lies within the range of integration if

$$k_1 > \frac{k}{2} + \frac{\omega\omega_{pe}}{3v_{Te}^2\, k}. \qquad (4.85)$$

Thus

$$\frac{I_k}{8\pi} = \frac{k\left|\frac{\varepsilon_k^{l\,(e)}-1}{\varepsilon_k^l}\right|^2}{192\pi\omega_{pe}n_0 T_e} \int\limits_{\frac{k}{2}+\frac{\omega\omega_{pe}}{3v_{Te}^2\, k}}^\infty \frac{W_{k_1}^l}{k_1^3}\, dk_1 W^l \sqrt{k_1^2 - \frac{2\omega\omega_{pe}}{3v_{Te}^2}} \times$$

$$\times \frac{\left(k_1^2 - \frac{k^2}{2} - \frac{\omega\omega_{pe}}{3v_{Te}^2}\right)^2}{\left(k_1^2 - \frac{2\omega\omega_{pe}}{3v_{Te}^2}\right)^2}. \qquad (4.86)$$

Let us consider $k \gg k_*$ and $\omega \ll k_*^2\,(3v_{Te}^2/\omega_{pe})$. Then $\omega/kv_{Ti} \ll 1$.

$$\frac{I_k}{8\pi} = \frac{kQ}{96\pi\left(1+\frac{T_e}{T_i}\right)^2 \omega_{pe} n_0\, T_e\, \alpha} \int\limits_{k/2}^\infty \frac{dk_1}{k_1^7}\left(k_1^2 - \frac{k^2}{2}\right)^2. \qquad (4.87)$$

Here we have used (4.67) and $W_k = \sqrt{2Q/\alpha}$. Equation (4.87) gives

$$\frac{I_k}{8\pi} = \frac{3}{16\pi^2} \cdot \frac{m_i v_{Te}^2 Q}{m_e k\omega_{pe}^4} \tag{4.88}$$

or, for $W_{k,\,\omega} = 4\pi k^2 I_k/8\pi$ $(I_k = I_{k,\,\omega}, \; W_{k,\omega} = W_{|k|,\omega})$,

$$W_{k,\,\omega} = \frac{3Qm_i}{4\pi m_e \omega_{pe}^4} kv_{Te}^2. \tag{4.89}$$

The total energy in the electric fields of the fluctuations is obtained from (4.89) using the equation

$$W = \int W_{k,\,\omega}\, dk d\omega. \tag{4.90}$$

In the frequency range we are considering here the spectrum is white, which means that it is independent of frequency. The energy increase with k is not without bounds, and in fact occurs only when $k/2 \ll k_g$, where k_g is the wave number for the Langmuir fluctuations which corresponds to the generation region. We see from (4.87) that $W_{k,\,\omega} = 0$ when $k > 2k_g$. Moreover, the spectrum in terms of ω is white only when $\omega \ll 3k^2 v_{Te}^2/2\omega_{pe}$. The energy maximum of the fluctuations is reached for $\omega \sim 3k^2 v_{Te}^2/2\omega_{pe}$. By assuming that $\omega = 3k^2 v_{Te}^2/2\omega_{pe} + \Delta\omega$, where $\Delta\omega \ll \omega$, we find

$$\frac{I_k}{8\pi} = \frac{k \left|\frac{\varepsilon_k^{l\,(e)} - 1}{\varepsilon_k^l}\right|^2}{192\pi\omega_{pe}n_0 T_e} \int\limits_{k\left(1+\frac{\Delta\omega}{2\omega}\right)}^{\infty} \frac{W_{k_1}^l}{k_1^3} W^l \sqrt{k_1^2 - k^2\left(1+\frac{\Delta\omega}{\omega}\right)}\, dk_1 \times$$

$$\times \left(k_1^2 - k^2\left(1+\frac{\Delta\omega}{2\omega}\right)\right)^2 \left(k_1^2 - k^2\left(1+\frac{\Delta\omega}{\omega}\right)\right)^{-2}. \tag{4.91}$$

It follows from this equation that sharp increase in the energy $I_k/8\pi$ is possible since $\dfrac{W^l}{\sqrt{k_1^2 - k^2\,(1+\Delta\omega/\omega)}}$ approaches a maximum in the spectrum when $k \sim k_0$. The minumum value of $k_1^2 - k^2(1 + \Delta\omega/\omega)$ in (4.91) is $k^2(\Delta\omega)^2/4\omega^2$. Therefore, the increase occurs in the frequency range $\Delta\omega/\omega \approx k_0/k$. Assuming that $\Delta\omega = 0$, we obtain the fluctuation energy at the maximum:

$$\frac{I_k}{8\pi} = \frac{k\sqrt{2Q/\alpha}}{384\pi\omega_{pe}n_0T_e} \int_0^\infty \frac{W_{k_2}\,dk_2^2}{(k_2^2+k^2)^2} \left| \frac{\varepsilon_k^{l\,(e)}-1}{\varepsilon_k^l} \right|^2. \qquad (4.92)$$

Assuming that the main contribution to the integral in (4.92) comes when $k_2 \ll k$, and that most of the energy in the Langmuir fluctuation is concentrated at $k \simeq k_0$, $W_k = W_0 \frac{(\nu-1)}{k_0}\left(\frac{k_0}{k}\right)^\nu$, we obtain the following estimate for $W_{k,\omega}$ from (4.72), (4.74), and (4.75):

$$W_{k,\omega} \approx \frac{Q}{2\pi\omega_{pe}^2 k}\left(\frac{Q_*}{Q}\right)^{\frac{1}{2\,(\nu-1)}}\sqrt{\frac{Q}{Q_*}}. \qquad (4.93)$$

Here $|(\varepsilon_k^{l(e)}-1)/\varepsilon_k^l|$ is dropped since it is of order unity in magnitude. The value of (4.93) is $(k_*/k_0)^\nu k_0^2/k^2$ times greater than that of (4.89). The integral $\int W_{k,\omega}d\omega$ taken over the frequency interval $\Delta\omega$ is

$$W_k = \int W_{k,\omega}d\omega \approx W_{k,\omega}\Delta\omega \approx W_{k,\omega}\omega\frac{k_0}{k} \simeq \frac{3}{2}\times$$

$$\times \frac{kk_0 v_{Te}^2}{\omega_{pe}}W_{k,\omega} = \frac{Qv_{Ti}}{4\pi\omega_{pe}^2}\left(\frac{Q_*}{Q}\right)^{\frac{1}{(\nu-1)}}\sqrt{\frac{Q}{Q_*}}. \qquad (4.94)$$

At the same time, the value of $\int W_{k,\omega}d\omega$ outside the range $\Delta\omega$ when $T_e = T_i$ is, according to (4.89),

$$\int W_{k,\omega}d\omega \approx \frac{9Qm_i}{8\pi m_e \omega_{pe}^5}k^3 v_{Te}^4 \approx \frac{v_{Ti}}{24\pi}\cdot\frac{Q}{\omega_{pe}^2}\left(\frac{k}{k_*}\right)^3 \qquad (4.95)$$

and increases rapidly with k. The maximum value for k is k_g; i.e., the ratio of the total energies of (4.95) to (4.94) is of order $(k_g/k_*)^3(k_*/k_0)^{-3}$, and it is always large when $\nu = 3$ ($k_g \gg k_*$). For frequencies much greater than $3k^2v_{Te}^2/2\omega_{pe}$ the fluctuation spectrum rapidly decreases:

$$W_{k,\omega} = \frac{81Qm_i}{4m_e}\left|\frac{\varepsilon_k^{l\,(e)}-1}{\varepsilon_k^l}\right|^2\left(1+\frac{T_e}{T_i}\right)^2\frac{k^5 v_{Te}^6}{\omega^2\omega_{pe}^6}.$$

4.5. Spectra of Langmuir
Fluctuations in a Nonisothermal
Plasma

In a nonisothermal plasma $(T_e \gg T_i)$ the decay of Langmuir waves into ion-acoustic oscillations is allowed [59]:

$$v_p < 3v_{Te} \sqrt{\frac{m_i}{m_e}} ; \quad k > k_*^s = \frac{\omega_{pe}}{3v_{Te}} \sqrt{\frac{m_e}{m_i}} = k_* \sqrt{\frac{T_e}{T_i}} \gg k_*.$$

In an isothermal plasma this corresponds to a region in which there is differential transfer to ions or transfer to electrons. Since the decay interaction is a resonant interaction, one need not consider the scattering effects. When $k_s \ll \omega_{pe}/v_{Te}$ the decay probability is [69]

$$w_l^{l,\,s}(\mathbf{k_1},\ \mathbf{k_2},\ \mathbf{k}) = \frac{e^2 k v_s (2\pi)^6}{8\pi m_e^2 v_{Te}^2} \left(\frac{\mathbf{k_1 k_2}}{k_1 k_2}\right)^2 \delta(\mathbf{k_1} - \mathbf{k_2} - \mathbf{k}) \, \delta \left(\omega_{\mathbf{k_1}}^l - \omega_{\mathbf{k_2}}^l - \omega_{\mathbf{k}}^s\right).$$

The equation for the ion-acoustic waves (s) takes the form

$$-\gamma_{\mathbf{k}}^s N_{\mathbf{k}}^s = \int w_l^{l,\,s}(\mathbf{k_1},\ \mathbf{k_2},\ \mathbf{k}) \, \frac{d\mathbf{k_1}\, d\mathbf{k_2}}{(2\pi)^6} \left(N_{\mathbf{k_1}}^l N_{\mathbf{k_2}}^l + N_{\mathbf{k}}^s N_{\mathbf{k_1}}^l - N_{\mathbf{k}}^s N_{\mathbf{k_2}}^l\right). \quad (4.96)$$

Here we have written the stationary balance equation in which the energy influx of the ion-acoustic plasmons takes place because of the nonlinear interactions with the Langmuir fluctuations, while the energy decrease occurs due to the absorption of ion-acoustic fluctuations by the plasma electrons; $\gamma_{\mathbf{k}}^s$ is the damping rate for ion-acoustic fluctuation damping.

We shall assume that the direct excitation of ion-acoustic waves either does not occur or that it can be neglected, and that the only source of ion-acoustic turbulence is the conversion of energy from Langmuir fluctuations (this source of turbulence is only for l-waves). We will further assume that both the Langmuir fluctuations and the ion-acoustic turbulence are isotropic. Equation (4.94) can be written as

$$-(\gamma_{\mathbf{k}}^s + \gamma_{\mathbf{k}}^N) N_{\mathbf{k}}^s = \int w_l^{ls}(\mathbf{k_1},\ \mathbf{k_2},\ \mathbf{k}) \, \frac{N_{\mathbf{k_1}}^l N_{\mathbf{k_2}}^l}{(2\pi)^6} \, d\mathbf{k_1} \, d\mathbf{k_2} = \frac{Q_{\mathbf{k}}^s (2\pi)^3}{\omega_{\mathbf{k}}^s} , \quad (4.97)$$

where γ_k^N is the nonlinear growth rate, given by

$$- \gamma_k^N = \int w_l^{ls}(\mathbf{k_1},\ \mathbf{k_2},\ \mathbf{k})\,\frac{dk_1\,dk_2}{(2\pi)^6}\,(N_k^l - N_{k_1}^l).\tag{4.98}$$

When the Langmuir turbulence is isotropic $\gamma_k^N < 0$. It is therefore clear that, in order to have direct excitation of turbulence by some external source, the linear growth rate must be larger than the nonlinear growth rate γ_k^N: that is, $\gamma_k^l > \gamma_k^N$. Thus, the direct excitation of low-frequency oscillations can be strongly suppressed by the high-frequency oscillations. This effect is much more important in the suppression of drift oscillations, which prevent the plasma from being confined. For example, it is known that in the development of a beam instability or an instability of the rf fields, the growth rate for the high-frequency oscillations can be many orders of magnitude larger than the growth rates for the low-frequency oscillations. Statements concerning stability must be understood only in the sense that the characteristic linear growth rates of the low-frequency oscillations have no effect on their spectra. Nevertheless, the excitation of low-frequency oscillations is related to their generation by the high-frequency oscillations. The source of this generation is described by the right side of (4.97). The right side of (4.97) is related to that excitation source which was considered in the previous paragraph in the absence of a decay resonance.[†]

In spite of the fact that in the aforementioned resonance case which involves decay the level of the low-frequency oscillations is much higher than in the absence of resonance, because of the strong linear absorption of the ion-acoustic oscillations ($\gamma_k^s \sim \omega_k^s \times \sqrt{m_e/m_i}$), together with the nonlinear absorption, N_k^s can be much less than N_k^l. Then, in this case, in the general equation for the decay interaction of l-waves,

$$\frac{\partial N_{k_1}^l}{\partial t} = -2\nu_e N_{k_1}^l - \int w_l^{ls}(\mathbf{k_1},\mathbf{k_2},\ \mathbf{k})\,\frac{1}{(2\pi)^6}\times$$
$$\times (N_k^s N_{k_1}^l - N_{k_1}^l N_k^l + N_k^l N_{k_1}^l)\,dk\,dk_2 +$$
$$+ \int w_l^{ls}(\mathbf{k_2},\ \mathbf{k_1},\ \mathbf{k})\,(N_{k_1}^l N_{k_2}^l - N_k^s N_{k_1}^l + N_k^s N_{k_1}^l)\,\frac{dk\,dk_2}{(2\pi)^6}\tag{4.99}$$

---

†Because $\mathrm{Re}\,\varepsilon_k^l \gg 1$ in Sec. 4.4, the imaginary parts of ε and γ_k^N can be neglected. This is clear from the fact that, for weak turbulence, $\gamma_k^N \ll \omega \ll \omega\,\mathrm{Re}\,\varepsilon_k^l$. Under resonance conditions $\mathrm{Re}\,\varepsilon_k^l = 0$, and we now must take γ_k^N into account.

one can neglect those terms containing N_k^s, and the balance equation can be written as

$$2\nu_e W_{k_1}^l = W_{k_1}^l \int \left[w_l^{ls} (\mathbf{k_2}, \mathbf{k_1}, \mathbf{k}) - w_l^{ls} (\mathbf{k_1}, \mathbf{k_2}, \mathbf{k}) \right] W_{k_2}^l \frac{dk_2 \, dk}{\omega_{pe} (2\pi)^3} \cdot \frac{d\Omega_2}{4\pi} . \qquad (4.99a)$$

We shall see in the future that the inequality $N_k^s \ll N_k^l$ is fulfilled.

By integrating over $\mathbf{k}$ and Ω_2, we obtain an equation which determines the spectrum of the Langmuir turbulence:

$$2\nu_e = \frac{\pi}{32} \cdot \frac{\omega_{pe}}{(k_*^s)^2 n_0 T_e} \left\{ \int_{k_1}^{k_1 + 2k_*^s} \frac{(k_1^2 - k_2^2)^2}{4 k_1^3 k_2^3} \left(k_1^2 + k_2^2 - \frac{(k_1^2 - k_2^2)^2}{4 (k_*^s)^2} \right)^2 W_{k_2}^l \, dk_2 - \right.$$

$$\left. - \int_{k_1 - 2k_*^s}^{k_1} \frac{1}{4 k_1^3 k_2^3} (k_1^2 - k_2^2)^2 \left(k_1^2 + k_2^2 - \frac{(k_1^2 - k_2^2)^2}{4 (k_*^s)^2} \right)^2 W_{k_2}^l \, dk_2 \right\}, \qquad (4.100)$$

where $k_*^s = \omega_{pe} \sqrt{\dfrac{m_e}{m_i}} / 3 v_{Te}$, $k_2 > 2 k_*^s - k_1$. Let us examine the limiting case, in which $k_1 \gg k_*^s$. Expanding in terms of k_*^s / k_1 we find

$$2\nu_e W_k = \alpha' W_k \frac{\partial W_k}{\partial k} , \qquad (4.101)$$

where

$$\alpha' = \alpha \left(1 + \frac{T_e}{T_i} \right)^2 . \qquad (4.102)$$

Because $T_e \gg T_i$, α' is $(T_e / T_i)^2$ times larger than α.

In the interval from $k_2 > k_*^s$ to $k_* = k_*^s (T_i / T_e)^{1/2}$ fluctuations for which $k_2 > k_*$ participate in the interaction. Since they are not resonant, they interact because of the induced scattering by ions. If collisions are neglected,

$$\frac{\partial W_k}{\partial k} = - \left(\frac{T_e}{T_i} \right)^2 \frac{\pi \omega_{pe}}{3 \alpha' n_0 T_e} k_*^s W_{k_*^s} . \qquad (4.103)$$

According to (4.101), when $k \ll k_*^s$, $W_k = \text{const} = (2Q/\alpha')^{1/2}$, if ν_e is small. This level of turbulence is a factor of 2 smaller than in an isothermal plasma for the same generation power. But when $k_* < k \ll k_*^s$, the level rises. We can obtain a rough estimate

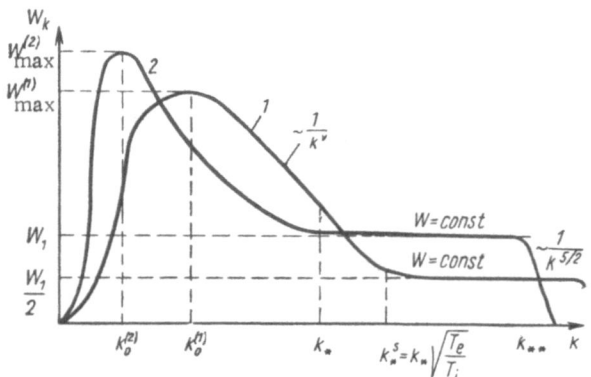

Fig. 4.5. The spectra for turbulence in both an isothermal and a nonisothermal plasma for a given turbulent energy flow: 1) $T_e \gg T_i$; 2) $T_e = T_i$. For curves 1 and 2 the quantities T_e, n_0, Q, and so on are the same, but T_i is different.

of this effect by assuming that

$$W_{k_*^s} = \sqrt{\frac{2Q}{\alpha'}} = \sqrt{\frac{2Q}{\alpha} \cdot \frac{T_i}{T_e}}.$$

According to (4.103) we obtain

$$W_{k_*} = \frac{(k_*^s)^2 \pi \omega_{pe}}{3\alpha' n_0 T_i} \sqrt{\frac{2Q}{\alpha}} = \cdot \frac{T_e}{T_i} \sqrt{\frac{2Q}{\alpha}}. \qquad (4.104)$$

Four-plasmon interactions become important at small k.† A schematic representation of the turbulent spectrum in a highly nonisothermal plasma is shown in Fig. 4.5 (curve 1), where for comparison (curve 2) we show the spectrum for turbulence in an isothermal plasma for the same generating power. When compared with the scale for an isothermal plasma, the fundamental scale of the turbulence in a nonisothermal plasma is smaller:

$$k_0 \simeq k_* \left(\frac{Q_*^s}{Q} \right)^{\frac{1}{2(\nu-1)}} \qquad Q_*^s = Q_* \left(\frac{T_e}{T_i} \right)^2. \qquad (4.105)$$

---

†Generally speaking the four-plasmon interaction decreases when $T_e \gg T_i$, but this only indicates the level of energy at which this interaction becomes important. In addition, such interactions can take place even when $k_* < k < k_*^s$.

Turning now to the expression for the nonlinear damping rate of (4.98), we obtain

$$-\gamma_k^N = \frac{\pi}{32} \cdot \frac{k_*^s \omega_{pe}}{n_0 T_e} \left\{ \int_{|k_*^s - k/2|}^{\infty} \frac{dk_2 W_{k_2}^l}{k_2^3} \cdot \frac{(2k_2^2 + 2kk_*^s - k^2)^2}{k_2^2 + 2kk_*^s} - \right.$$

$$\left. - \int_{k_*^s + k/2}^{\infty} dk_2 \frac{W_{k_2}^l}{k_2^3} \cdot \frac{(2k_2^2 - 2kk_*^s - k^2)^2}{k_2^2 - 2kk_*^s} \right\}. \qquad (4.106)$$

If $k \gg k_*^s$, then

$$-\gamma_k^N = \frac{\pi (k_*^s)^2 \omega_{pe}}{2n_0 T_e} \left(\frac{W_{k/2}^l}{k} + \frac{k^3}{4} \int_{k/2}^{\infty} \frac{(2k_2^2 - k^2)}{k_2^7} W_{k_2}^l \, dk_2 \right). \qquad (4.107)$$

Assuming that $W_k = \text{const} = (2Q/\alpha')^{1/2}$ we have

$$-\gamma_k^N = \frac{\pi}{6} \cdot \frac{(k_*^s)^2 \omega_{pe}}{n_0 T_e k} \sqrt{\frac{2Q}{\alpha'}}. \qquad (4.108)$$

Thus, γ_k^N increases as k decreases, and it reaches a maximum when $k \sim k_*^s$. For $k \ll k_*^s$,

$$-\gamma_k^N \approx \frac{\pi}{8} \cdot \frac{k\omega_{pe}}{n_0 T_e} W_{k_*}^l{}_s. \qquad (4.109)$$

An estimate of the maximum growth rate is

$$-(\gamma_k^N)_{\text{max}} \simeq \omega_{pe} \sqrt{\frac{\pi Q}{6n_0 T_e \omega_{pe}}}. \qquad (4.110)$$

Since the k dependence of the the linear growth rate and Eq. (4.109) is the same, the fact that (4.110) dominates the linear growth rate at the maximum implies that it is the larger of the two at all smaller values of k. The criterion for (4.110) dominating γ_k^s is[†]

$$Q > Q_{\text{cr}} = \frac{\omega_{pe} n_0 T_e}{12} \left(\frac{m_e}{m_i} \right)^3. \qquad (4.111)$$

[†]The number of particles N_D in a sphere having the Debye radius must satisfy the inequality $N_D < 12 \ln \Lambda (m_i/m_e)^3$.

The nonlinear growth rate in either (4.109) or (4.110) can be larger than ω_k^s. This occurs when

$$Q > \omega_{pe} n_0 T_e \left(\frac{m_e}{m_i} \right)^2. \tag{4.112}$$

In this case it generally no longer makes sense to speak of ion-acoustic oscillations; at high frequencies the turbulence radically alters the plasma properties for low frequencies.[†] Chapter 8 deals with these effects in great detail.

We are also interested in the size of the nonlinear damping rate for the maximum k, which is of order ω_{pe}/v_{Te}, when there is a maximum in both the linear damping and the linear excitation. The energy density of the Langmuir fluctuations $W_{k/2}^l$ cannot be large at such k, in view of the heavy Landau damping on the electrons. However, when $k \sim \frac{1}{4}(\omega_{pe}/v_{Te})$, $k/2 \sim \frac{1}{8}(\omega_{pe}/v_{Te})$ and the Landau absorption is small. Incidentally, when $T_e \gg T_i$ the regions in which there is transformation of the energy from the l-plasmons to the electrons do not exist, and the spectrum W_k = const of the high-frequency fluctuations is continued up to k values on the order of ω_{pe}/v_{Te}; the minimal k values are limited by the sources and Landau absorption. Thus, assuming that $k \sim \frac{1}{3}(\omega_{pe}/v_{Te})$, we obtain

$$\gamma_k^N \approx \frac{3 v_{Te} (k_*^s)^2 \pi}{8 n_0 T_e} \sqrt{\frac{2Q}{\alpha'}}. \tag{4.113}$$

The condition for dominating over the linear absorption takes the form

$$Q > \omega_{pe} n_0 T_e \frac{m_e}{m_i}. \tag{4.114}$$

Let us now consider the power Q_k^s of the source of Langmuir turbulence which excites the ion-acoustic turbulence:

$$Q_k^s = \frac{\omega_k^s}{2 n_0 T_e} \int_0^\infty dk_1 \int_{(k_1-k)^2}^{(k_1+k_2)^2} dk_2 \frac{W_{k_1}^l W_{k_2}^l k^s}{k_1^3 k_2^4 \cdot 48} (k_1^2 + k_2^2 - k^2)^2 \delta (k_1^2 - k_2^2 - 2kk_*^s). \tag{4.115}$$

[†]More to the point, one must compare both ω_k^N (the nonlinear frequency shift; see Chapter 8) and γ_k^N with ω_k^s.

Although similar to (4.84), this expression differs from the earlier one in that frequency and wave number of the low-frequency fluctuations are related by the equation $\omega_k^s = k\, v_s$. The total power for generating ion-acoustic waves is

$$Q^s = \int Q_k^s \, dk. \tag{4.116}$$

It is convenient to write Eq. (4.115) as

$$Q_k^s = \frac{\omega_k^s\, k_*^s}{96 n_0\, T_e} \int\limits_{\frac{k}{2}+k_*}^{\infty} dk_1\, W_{k_1}^l\, W_{\sqrt{k_1^2 - 2kk_*^s}}^l\, \frac{(2k_1^2 - 2kk_*^s - k^2)^2}{k_1^3\,(k_1^2 - 2kk_*^s)^2}. \tag{4.117}$$

It is easily seen that when either $k \gg k_*^s$ or $k \ll k_*^s$ in (4.117), the values of $W_{k_1}^l$ and $W_{\sqrt{k_1^2 - 2kk_*^s}}^l$ are significant only if $k_1 \gg k_*^s$. The narrow region about $k = 2k_*$ shows similar qualities, where those k values close to the fundamental scale, which is of order k_0, make a contribution to $W_{\sqrt{k_1^2 - 2kk_*^s}}^l$. Since the analysis of this region follows the same lines as that given above, there must be a single concentration of the energy in the ion-acoustic oscillations about the point $k \simeq 2k_*^s$. In the regions where $k \gg k_*^s$ and $k \ll k_*^s$ one must replace W_{k_1} and $W_{\sqrt{k_1^2 - 2kk_*^s}}$ in (4.117) by the appropriate values of $(2Q/\alpha')^{1/2}$. Then, after integrating over k_1, we find

$$Q_k^s = \frac{\omega_k^s\, k_*^s\, Q}{48 n_0\, T_e\, \alpha'} \left\{ \frac{2}{\left(\frac{k}{2} + k_*^s\right)^2} - \frac{k^2}{\left(\frac{k}{2} + k_*^s\right)^4} + \frac{k^4}{6\left(\frac{k}{2} + k_*^s\right)^6} \right\}. \tag{4.118}$$

If $k \ll k_*^s$,

$$Q_k^s = Q\, \frac{k}{k_*^s} \sqrt{\frac{m_i}{m_e}} \cdot \frac{27}{24\pi} \cdot \frac{v_{Te}^3}{\omega_{pe}^3}, \tag{4.119}$$

and if $k \gg k_*^s$

$$Q_k^s = \frac{3}{2\pi}\, Q\, \frac{k_*^s}{k} \sqrt{\frac{m_i}{m_e}} \cdot \frac{v_{Te}^3}{\omega_{pe}^3}. \tag{4.120}$$

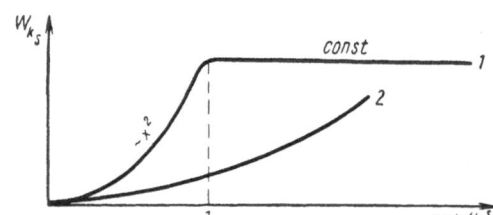

Fig. 4.6. Spectra for ion-acoustic turbulence excited by Langmuir turbulence. Curve 1, $Q \lll Q_{cr}$; curve 2, $Q \ggg Q_{cr}$.

The energy distribution of the ion-acoustic turbulence can be found from (4.97):

$$W_k^s = \frac{4\pi k^2 \omega_k^s}{(2\pi)^3} N_k^s = \frac{-4\pi k^2 Q_k^s}{\gamma_k^s + \gamma_k^N} . \tag{4.121}$$

If the inequality in (4.111) is not satisfied, and linear damping is dominant, then

$$W_k^s = \frac{1}{3} \sqrt{\frac{2}{\pi}} \cdot \frac{Q}{k_*^2 v_{Te}} \delta, \tag{4.122}$$

$$\delta = \begin{cases} \dfrac{k^2}{(k_*^s)^2}, & k \ll k_*^s, \\[2mm] \dfrac{4}{3}, & k \gg k_*^s. \end{cases} \tag{4.123}$$

But when (4.111) is fulfilled, we have

$$W_k^s = \frac{2\sqrt{2m_e}}{3m_i \sqrt{3\pi}} \cdot \frac{k^2}{(k_*^s)^3} n_0 T_e \sqrt{\frac{Q}{n_0 T_e \omega_{pe}}} \tag{4.124}$$

both when $k \ll k_*^s$ and when $k \gg k_*^s$.

Figure 4.6 shows the characteristic shapes of the spectra for $Q \ll Q_{cr}$ and $Q \gg Q_{cr}$. When $Q \ll Q_{cr}$ the greater part of the energy is found in the region where $k > k_*^s$, which then plays the role of the fundamental scale for the ion-acoustic turbulence, where $W^s \simeq Qm_i/\omega_{pe}, m_e \ll W^l \approx Q/\nu_e$ if $\nu_e/\omega_{pe} \ll m_e/m_i$, that is, if $N_D \gg (m_i/m_e) \ln \Lambda$.

When $Q \gg Q_{cr}$ the total energy of the ion-acoustic oscillations is estimated to be

$$\frac{W^s}{n_0 T_e} \approx \frac{m_e}{m_i} \sqrt{\frac{Q}{Q_{cr}}}, \quad \frac{W^s}{W^l} \approx \frac{\nu_e}{\omega_{pe}} \left(\frac{m_i}{m_e}\right)^2 \sqrt{\frac{Q_{cr}}{Q}}. \qquad (4.125)$$

We can now ask if the assumption that $N_k^s \ll N_k^l$ is valid or not. For comparable wave numbers for both the ion-acoustic and the Langmuir oscillations, the ratio $\varepsilon = N_k^s / N_k^l$ for $Q \gg Q_{cr}$ when $k \gg k_*^s$ is of the order $\varepsilon \sim k_*^s / k_l$. When the k_s and k_l are comparable and $k \gg k_*^s$, this is impossible since this type of interaction is forbidden. When $k_s \ll k_l$ the action of s-waves on l-waves is reduced by a factor of k_s^2 / k_l^2, and according to (4.123) $\varepsilon_{eff} \approx \varepsilon k_s^2 / k_l^2 \approx k_s / k_* \ll 1$. In similar fashion it can be shown that when $Q \ll Q_{cr}$,

$$\varepsilon_{max} \approx \sqrt{\frac{Q}{Q_{cr}}}. \qquad (4.126)$$

We must finally consider the narrow band of wave numbers for the ion-acoustic oscillations near $k = 2k_*^s$, having a width $\Delta k \sim k_0$; the contribution from this region can determine the fundamental scale of the Langmuir turbulence. From (4.106) we have, in order of magnitude,

$$\gamma_k^N \simeq \frac{k_*^s}{k_0} \omega_{pe} \frac{W}{n_0 T_e}, \qquad (4.127)$$

$$Q_k^s \simeq \frac{v_{Te}^2}{k_*^s \omega_{pe}} W \sqrt{\frac{2Q}{\omega_{pe} n_0 T_e} \frac{k_0}{k_*^s}}, \qquad (4.128)$$

that is, when $\gamma_k^N \gg \gamma_k^s$ and $Q \gg Q_{cr}$, there is a dip in the spectrum for the ion-acoustic oscillations at $k = 2k_*^s$, where the intensity drops by a factor of $(k_0/k_*^s)^2$.

4.6. Correlations and Nonlinear Frequency Shifts of Langmuir Fluctuations in Turbulent Plasma

In view of the fact that the decay interaction of three Langmuir oscillations is forbidden, the correlation of Langmuir oscillations is of special importance. We therefore will dwell on this problem, taking into account the effects of four-plasmon interactions since they play an important part in the formation of the spectrum for Langmuir fluctuations.

We will calculate the effects of the four-plasmon interactions on the correlation effects. Omitting the calculational details, which are similar to those presented in §2.5, we note that the terms proportional to I^2_k must be included in the nonlinear permeability ε^N_k; this then also includes the four-plasmon interactions. The result is

$$I_k = \frac{32\pi^2}{\omega^2} \int \frac{\left| \sum^{eff}_{k,k_1,k_2,k_3} \right|^2 I_{k_1} I_{k_2} I_{k_3}}{\left| \varepsilon^l_k + \varepsilon^N_k \right|^2} \delta(k-k_1-k_2-k_3)\, dk_1\, dk_2\, dk_3. \qquad (4.129)$$

When the four-plasmon interactions can be neglected, the equation for I_k reduces to

$$I_k \left(\varepsilon^l_k + \varepsilon^N_k \right) = 0. \qquad (4.130)$$

This result can be used when $k \gg k_*,\ k^s_*$. It has the form of a non-linear dispersion relation

$$\varepsilon^l_k I_k = \frac{8\pi i}{\omega} I_k \int \sum^{eff}_{k,k_1} I_{k_1}\, dk_1; \qquad (4.131)$$

$$\sum^{eff}_{k,k_1} = \sum_{k,k_1 \ k,-k_1} - \frac{8\pi i\, S_{k-k_1,\, k,-k_1} S_{k,k_1,\, k-k_1}}{(\omega - \omega_1)\, \varepsilon^l_{k-k_1}}. \qquad (4.132)$$

Equation (4.130) differs from the more common dispersion relations in that neither the frequency nor the wave vector can be complex (the ω, k which appear in the real I_k must themselves be real). By equating the real and imaginary parts of (4.131), we obtain the two equations

$$\operatorname{Re} \varepsilon^l_k = -\int \operatorname{Im} \sum^{eff}_{k,k_1} I_{k_1} \frac{8\pi}{\omega}\, dk_1, \qquad (4.133)$$

$$\operatorname{Im} \varepsilon^l_k = \int \operatorname{Re} \sum^{eff}_{k,k_1} I_{k_1} \frac{8\pi}{\omega}\, dk_1. \qquad (4.134)$$

A valid approximate expression for $\Sigma_{k,k_1}^{\text{eff}}$ is easily calculated if we assume that $\omega \gg k v_{T_e}$, $\omega_1 \gg k_1 v_{T_e}$:

$$\frac{8\pi}{\omega} \Sigma_{k,k_1}^{\text{eff}} = \frac{i(k-k_1)^2}{4\pi\omega_{pe}^2 \, n_0 \, m_e} \cdot \frac{(kk_1)^2}{k^2 \, k_1^2} \cdot \frac{\varepsilon_{k-k_1}^{l(e)} \, \varepsilon_{k-k_1}^{l(i)}}{\varepsilon_{k-k_1}^{l(e)} + \varepsilon_{k-k_1}^{l(i)}} \tag{4.135}$$

We will consider Eq. (4.133) since (4.134) gives the energy balance already discussed. The expression for $\Sigma_{k,k_1}^{\text{eff}}$ is very large for small $\omega - \omega_1$, but I_{k_1} has a maximum when $\omega_1 = \omega_{k_1}$, at which point we have $\operatorname{Re} \varepsilon_k^l \approx (\omega - \omega_k^l) \dfrac{\partial \varepsilon_k^l}{\partial \omega}\bigg|_{\omega = \omega_k^l} \approx \dfrac{2(\omega - \omega_k^l)}{\omega_{pe}}$. Putting $\omega' = \omega - \omega_k^l$, we have

$$-G_k = \operatorname{Re} \frac{\varepsilon_k^{l(e)} \varepsilon_k^{l(i)} k^2}{\omega_{pe} \varepsilon_k^l 8\pi n_0 m_e} \approx \frac{\omega_{pe}}{8n_0 T_e} \cdot \frac{\varepsilon_k^{l(i)}}{\varepsilon_k^l}. \tag{4.136}$$

Then (4.20) can be rewritten as

$$\omega' = \int W_{k_1,\omega_1'}^l \, dk_1 \, d\omega_1' \, x_1^2 G_{\omega'-\omega_1'+\omega_k^l-\omega_{k_1}^l, \, |k-k_1|} \, dx_1, \tag{4.137}$$

$$x_1 = \cos(k, k_1),$$

where

$$W_{k,\omega'}^l = I_{k,\omega'+\omega_k^l} k^2, \quad W^l = \int W_{k,\omega}^l \, dk \, d\omega. \tag{4.138}$$

Note that the left side of (4.137) is not a function of k while the right side is. Therefore ω' must be a function of k also. This is possible if $W_{k,\omega'} = W_k \delta(\omega' - \omega_k^N)$, i.e., the cancellation of I_k in (4.131) will be possible only if $\omega' = \omega_k^N$. This means that the k dependence of the frequency will be preserved, but that there will be a nonlinear shift in the frequency. An argument in favor of this situation is provided by the balance between the imaginary parts as given in (4.134), which results in $\gamma_k^l + \gamma_k^N = 0$. Assuming that $T_e = T_i$, and neglecting both ω' and ω_1 on the right side of (4.136), we obtain the following estimate for ω_k^N:

$$\omega_k^N \approx \omega_{pe} \sqrt{\frac{2Q}{\alpha}} \cdot \frac{k_*}{2n_0 T_e}, \tag{4.139}$$

which is much smaller than ω_{pe}; furthermore, ω_k^N does not depend on k. This justifies our neglect of $\omega' - \omega_1$ in (4.137).

In an isothermal plasma, decays are forbidden and the fluctuation correlations are connected only to the four-plasmon interactions. When $k \gg k_*$ these processes are strongly suppressed by the compensation between the two kinds of four-plasmon interactions, and an estimate for γ^N is [210]

$$\gamma_k^N \approx \frac{\omega_{pe}}{4} \left(\frac{W^{k>k_*}}{nT_e} \right)^2 \left(\frac{k_*}{k} \right)^4 \frac{\omega_{pe}^2}{k^2 v_{Te}^2} \simeq v_e \frac{W}{nT_e} \frac{m_i}{m_e} \cdot \frac{k_g^2}{k_*^2} \cdot \frac{k_*^6}{k^6}, \qquad (4.140)$$

and it is smaller than the difference between the Langmuir frequencies $\omega_{pe}(k^2/k_*^2) \cdot (m_e/m_i)$ when the following inequality is satisfied:

$$\frac{W}{nT_e} \ll \left(\frac{m_e}{m_i} \right)^2 \left(\frac{\omega_{pe}}{v_e} \right) \left(\frac{k}{k_*} \right)^8 \left(\frac{k_*}{k_g} \right)^2 = \left(\frac{m_e}{m_i} \right)^2 \frac{\omega_{pe}}{v_e} \begin{cases} \left(\dfrac{k_g}{k_*} \right)^6, & k = k_g, \\[2mm] \left(\dfrac{k_*}{k_g} \right)^2, & k = k_*. \end{cases}$$

We will now consider the correlation when $k \ll k_*$. We shall assume that the four-plasmon processes dominate.

In order to evaluate the γ_k^N connected with the four-plasmon interaction we must return to Eq. (4.33) and keep just the induced processes which are proportional to W_k^l. This means that one must replace the quantities v and u in (4.34) with v' and u', defined as follows:

$$\left. \begin{array}{l} v' = \dfrac{\xi^{\frac{\nu}{2}+1} - \eta^{\frac{\nu}{2}+1} - (1+\xi-\eta)^{\frac{\nu}{2}+1}}{\eta^{\frac{\nu}{2}+2} \, \xi^{\frac{\nu}{2}+2} \, (1+\xi-\eta)^{\frac{\nu}{2}+2}}, \\[5mm] u' = \dfrac{(\xi+\eta-1)^{\frac{\nu}{2}+1} - \xi^{\frac{\nu}{2}+1} - \eta^{\frac{\nu}{2}+1}}{\eta^{\frac{\nu}{2}+2} \, \xi^{\frac{\nu}{2}+2} \, (\xi-\eta-1)^{\frac{\nu}{2}+2}}, \end{array} \right\} \qquad (4.141)$$

Then, in place of (4.33) we write

$$\gamma_k^N = \frac{\pi}{6} \omega_{pe} \left(\frac{W}{8n_0(T_e+T_i)} \right)^2 \frac{\omega_{pe}^2 (\nu-1)^2}{k_0^2 v_{Te}^2} \xi_0^\nu (G_1' + G_2' + G_3'), \qquad (4.142)$$

where $G_1^{\prime}$, $G_2^{\prime}$, $G_3^{\prime}$ differ from G_1, G_2, G_3 in that v and u are replaced by v' and u'. Because the latter variables have a different structure, only the region of small $\xi \sim \xi_0$ and $\eta \sim \xi_0$ makes a contribution to $G_1^{\prime}$. Making the change of variable $\eta' = \xi + \eta - 1$ in $G_3^{\prime}$, one readily sees that it is only the region where $\xi \sim \xi_0$, $\eta' \sim \xi_0$ that is important and $G_3^{\prime} = 2G_1^{\prime}$. Finally,

$$G_1^{\prime} + G_2^{\prime} + G_3^{\prime} \simeq G_1^{\prime} + G_3^{\prime} \simeq 3G_1^{\prime} = -\frac{8(8v+5)}{5\xi_0^{v-1/2}\,v\left(\frac{v}{2}+1\right)\left(\frac{v}{2}-1\right)}.$$

(4.143)

Then, with $T_e = T_i$

$$\gamma_k^N = \omega_{pe}\frac{\omega_{pe}^2}{kk_0 v_{Te}^2}\left(\frac{W}{n_0 T_e}\right)^2\xi,$$

(4.144)

where

$$\xi = \frac{\pi(v-1)^2(8v+5)}{960v(v-1/2)\left(\frac{v}{2}+1\right)} = \begin{cases}0.06, & v=3\\0.025, & v=4.\end{cases}$$

(4.145)

We estimate γ_k^N for $k \sim k_0$.

Using (4.75) we obtain the condition that γ_k^N does not exceed the difference between the frequencies of the interacting waves, which is of order $\omega_{pe}(k_0^2 v_{Te}^2/\omega_{pe}^2)$:

$$k_0 > k_*\left(\frac{W}{n_0 T_e}\cdot\frac{m_i}{m_e}\right)^{1/2} = k_*\left(\frac{Q}{Q_*}\cdot\frac{24}{\pi}\cdot\frac{m_i}{m_e}\cdot\frac{v_e}{\omega_{pe}}\right)^{1/2}$$

(4.146)

or, if $k_0 \ll k_*$, $\frac{W}{nT} < \frac{m_e}{m_i}$, $k_0 = k_*\left(\frac{Q_*}{Q}\right)^{\frac{1}{2(v-1)}}$, then

$$Q_* < Q < Q_*\left(\frac{m_e}{m_i}\cdot\frac{\omega_{pe}}{v_e}\right)^{\frac{v-1}{v}}.$$

(4.147)

It is now clear that when there is a maximum at $k \sim k_0$ and the spectrum has the form described in §4.3, we must have

$$\ln\Lambda\frac{m_i}{m_e} < N_D.$$

(4.148)

However, the collisions are expressed in the fact that all the interactions are significantly altered when the difference in the frequencies of the interacting fluctuations becomes comparable with the collision frequency (see §4.7). Then, for a fundamental scale $k_0^2 v_{Te}^2 / \omega_{pe} < \nu_e$, we obtain the condition

$$Q_* < Q < Q_* \left(\frac{\omega_{pe}}{\nu_e} \cdot \frac{m_e}{m_i} \right)^{(\nu-1)} , \qquad (4.149)$$

which also requires that (4.148) be satisfied.

4.7. The Effect of Pair Collisions of Particles on the Correlations and Spectra of Langmuir Turbulence

In addition to the trivial influence of pair collisions considered earlier relative to the spectra of plasma turbulence, connected with the absorption of fluctuation energy, pair collisions can alter the nature of the interactions which the fluctuations have with each other. This will take place when the frequency difference between the interacting waves can be smaller than the frequency of the pair collisions. More precisely, the following inequality must be satisfied [215]:

$$\max (\omega_1 - \omega_2, |\mathbf{k}_1 - \mathbf{k}_2| v_{T\alpha}) < \nu_\alpha, \quad \alpha = e, i. \qquad (4.150)$$

Here ν_e and ν_i are the total collision frequencies for electrons and ions, respectively, for collisions with all plasma particles (electrons, ions, and neutral particles).

We should point out that, according to modern concepts [216], pair collisions are related to the fluctuation processes of the electromagnetic fields. However, the frequencies of the appropriate fluctuation fields can be much larger than the turbulent frequencies, and one will have to use the usual collision integral. Inclusion of the collective effects (such as plasmon exchange and spontaneous emission of plasmons) makes small corrections to the collisions, inducing slight changes in the Coulomb logarithm. One can use here the form of the collision integral which was obtained by Landau [217].

The problem which the theory must face is the calculation of the nonlinear plasma currents with the pair collisions included. If the nonlinear currents are known, the turbulence spectra can be obtained by using either (4.131) or (4.129). The nonlinear currents were computed in [215] using the kinetic equations.

For the sake of clarity and in order to develop the physical meaning behind the result, it is convenient to start with the hydrodynamic equations which were obtained in [218] from the kinetic equations with the help of the Landau collision integral. To justify this we note that in (4.132) only one of the frequencies of the nonlinear currents S is found in the hydrodynamic region of (4.146), whereas the other two are located in the region of infrequent collision. The hydrodynamic equations assume that all the frequencies are found in a region of frequent collision. In spite of this, if one throws away all the dissipative terms, which are negligibly small in the high-frequency regions, one can obtain the spectra of Langmuir fluctuations from the hydrodynamic equations. The effects of thermal motion in the nonlinear currents are small (just as in the collisionless case). As for the oscillation frequency, it is arbitrary when calculating the nonlinear current, but one obtains an explicit function for $\omega = \omega_k^l$ (neglecting correlations) only after I_k in (4.131) is specified through Eq. (2.51). The frequencies enter into (2.51) and the thermal motion is properly taken into account. A rigorous proof of the possibility of using the hydrodynamic interactions entails showing agreement between these calculations and the results obtained through the kinetic treatment [215].

The hydrodynamic equations which must be used to calculate the nonlinear currents are the continuity equations

$$\frac{\partial n_e}{\partial t} + \frac{\partial}{\partial \mathbf{r}} n_e \mathbf{v}_e = 0, \quad \frac{\partial n_i}{\partial t} + \frac{\partial}{\partial \mathbf{r}} n_i \mathbf{v}_i = 0 \tag{4.151}$$

and the hydrodynamic equations for both the electrons and the ions:

$$m_e n_e \left(\frac{\partial}{\partial t} + \left(\mathbf{v}^{(e)} \frac{\partial}{\partial \mathbf{r}} \right) \right) v_j^{(e)} = -\frac{\partial}{\partial x_j} n_e T_e - \frac{\partial \pi_{lj}^{(e)}}{\partial x_l} -$$
$$- e n_e \left(E_j + \frac{1}{c} [\mathbf{v}^{(e)} \, \mathbf{H}]_j \right) + R_j; \tag{4.152}$$

$$m_i n_i \left(\frac{\partial}{\partial t} + \left(\mathbf{v}^{(i)} \frac{\partial}{\partial \mathbf{r}} \right) \right) v_j^{(i)} = -\frac{\partial}{\partial x_j} n_i T_i - \frac{\partial \pi_{lj}^{(i)}}{\partial x_l} +$$
$$+ e n_i \left(E_j + \frac{1}{c} [\mathbf{v}^{(i)} \, \mathbf{H}]_j \right) - R_j, \tag{4.153}$$

in which the terms $\dfrac{\partial}{\partial x_j} nT$ describe the pressure contributions from the gas of electrons and ions, and $\pi_{ij}^{(e)}$ and $\pi_{ij}^{(i)}$ are the electron and ion viscosity tensors:

$$\left.\begin{aligned}
&\pi_{ij}^{(e)} = -0.73 \frac{n_e T_e}{\nu_e} w_{ij}^{(e)}, \quad \pi_{ij}^{(i)} = -0.96 \frac{n_i T_i}{\nu_i} w_{ij}^{(i)}, \\
&w_{ij} = \frac{\partial}{\partial x_j} v_i + \frac{\partial}{\partial x_i} v_j - \frac{2}{3} \delta_{ij} \frac{\partial}{\partial \mathbf{r}} \mathbf{v},
\end{aligned}\right\} \tag{4.154}$$

R is the frictional force between the electrons and ions,

$$\mathbf{R} = \mathbf{R_u} + \mathbf{R_t}; \tag{4.155}$$

$\mathbf{R_u}$ is the force of relative friction, which depends only on the relative velocities of the electrons and ions $\mathbf{u} = \mathbf{v}^{(e)} - \mathbf{v}^{(i)}$,

$$\mathbf{R_u} = -0.51 n_e m_e \nu_e \mathbf{u}, \quad \omega \ll \nu_e, \tag{4.156}$$

$$\mathbf{R_u} = -n_e m_e \nu_e \mathbf{u}, \quad \omega \gg \nu_e. \tag{4.157}$$

It is easy to obtain the difference between the numerical coefficients for $\omega \gg \nu_e$ and $\omega \ll \nu_e$ if one uses perturbation theory to compute the frictional force between electrons and ions from the kinetic equation. The coefficient 0.51 was found in [218] in the hydrodynamic limit. $\mathbf{R_t}$ is the thermal force, given by

$$\mathbf{R_t} = -0.71 n_e \frac{\partial}{\partial \mathbf{r}} T_e. \tag{4.158}$$

Finally, one must also use the balance equation for the temperatures of the electrons and ions:

$$\frac{3}{2} n_e \left(\frac{\partial}{\partial t} + \mathbf{v}^{(e)} \frac{\partial}{\partial \mathbf{r}} \right) T_e + n_e T_e \frac{\partial}{\partial \mathbf{r}} \mathbf{v}^{(e)} = -\frac{\partial}{\partial \mathbf{r}} q^{(e)} - \pi_{ij}^{(e)} \frac{\partial v_i^{(e)}}{\partial x_j} + Q_e, \tag{4.159}$$

$$\frac{3}{2} n_i \left(\frac{\partial}{\partial t} + \mathbf{v}^{(i)} \frac{\partial}{\partial \mathbf{r}} \right) T_i + n_i T_i \frac{\partial}{\partial \mathbf{r}} \mathbf{v}^{(i)} = -\frac{\partial}{\partial \mathbf{r}} q^{(i)} - \pi_{ij}^{(i)} \frac{\partial v_i^{(i)}}{\partial x_j} + Q_i. \tag{4.160}$$

Here $\mathbf{q} = \mathbf{q_u} + \mathbf{q_t}$ is the thermal current,

$$q_u^{(e)} = 0.71 n_e T_e \mathbf{u}, \quad q^{(e)} = -3.16 \frac{n_e T_e}{m_e \nu_e} \cdot \frac{\partial}{\partial \mathbf{r}} T_e, \tag{4.161}$$

$$q^{(i)} = -3.9 \frac{n_i T_i}{m_i \nu_i} \cdot \frac{\partial}{\partial r} T_i, \quad Q_e = -(\mathbf{Ru}) - Q_i, \qquad (4.162)$$

$$Q_i = 3 \frac{m_e}{m_i} n_e \nu_e (T_e - T_i). \qquad (4.163)$$

For now we shall consider just the case where

$$\omega_1 - \omega_2 \gg 3 \frac{m_e}{m_i} \nu_e, \qquad (4.164)$$

i.e., we assume that the difference between the fluctuation frequencies is smaller than the collision frequency but larger than characteristic frequency $3(m_e/m_i)\nu_e$, which is equal to the inverse of the time needed to bring about equality between the electron and ion temperatures. Let us consider potential fluctuations $(\mathbf{H} = 0)$. We will expand all quantities in powers of the electric field E. In S_{k-k_1,k,k_1} the frequencies k and k_1 are large; to obtain quantities of first order in the field one can neglect all the terms in (4.152) with the exception of $-enE$ on the right and the nonlinear current on the left, and write

$$v_k^{(e)\,(1)} = \frac{e}{m_e\, i\omega} \mathbf{E}_k = \frac{ek}{m_e\, i\omega k} E_k = \frac{k}{k} v_k^{(e)\,(1)}. \qquad (4.165)$$

The continuity equation gives

$$n_{e,k}^{(1)} = n_0 \frac{k v_k^{(e)(1)}}{\omega}, \qquad (4.166)$$

and the energy balance equation (4.159) is

$$\frac{3}{2} n_e \frac{\partial T_e^{(1)}}{\partial t} + n_e T_e \frac{\partial}{\partial r} v^{(e)\,(1)} \approx -0.71 n_e T_e \frac{\partial u^{(1)}}{\partial r}; \qquad (4.167)$$

$$u^{(1)} \approx v^{(e)\,(1)}, \quad T_{e,k}^{(1)} = \frac{2}{3} 1.71 \frac{T_{e,\,0}\, k v_k^{(e)\,(1)}}{\omega}. \qquad (4.168)$$

It is now easy to obtain a relationship which couples $v_k^{(2)}$ and $v_k^{(1)}$ from the above set of equations. We have used the fact that the frequency and wave number in $v_k^{(2)}$ are very small, and that they satisfy (4.150). This can be used to immediately throw away a number of terms in the equations for $v_k^{(2)}$ which are of relative order $(\omega_1 - \omega_2)/\omega$ or of higher order in $(\omega_1 - \omega_2)/\omega$, where ω is a high frequency of order ω_{pe}, $\omega_1 - \omega_2 = \omega$ satisfying (4.150). We will first consider the energy balance equation for the electron gas (4.159). For frequencies $\omega \gg \frac{3m_e}{m_i} \nu_e$ the exchange of energy between the electrons and ions can be neglected, $Q_i \ll Q_e$, while on the other hand $\nu^{i(1)} \approx \frac{m_e}{m_i} \nu^{(e)}$ (1), according to (4.165), so that

$$Q_e \approx -(\mathbf{R}\mathbf{v}^{(e)}) = 0.51 \, n_e \, m_e \, \nu_e \, (\mathbf{v}^{(e)})^2. \qquad (4.169)$$

The Fourier components of Q_e are

$$Q_{e,k} = 0.51 \, n_e \, m_e \, \nu_e \int v_{k_1}^{(e)(1)} v_{k_2}^{(e)} \, ^{(1)} \frac{(\mathbf{k_1 \, k_2})}{k_1 \, k_2} \, \delta \, (k - k_1 - k_2) \, dk_1 \, dk_2 =$$

$$= 0.51 \, \frac{e^2 \, n_e \, \nu_e}{m_e^2} \int \frac{E_{k_1} E_{k_2}}{\omega_1 \, \omega_2} \cdot \frac{(\mathbf{k_1 \, k_2})}{k_1 \, k_2} \, \delta \, (k - k_1 - k_2) \, dk_1 \, dk_2. \qquad (4.170)$$

This nonlinear term is the principal one in the equations for $v_k^{(2)}$.

When estimating the remaining nonlinear terms it must be kept in mind that Q_k is proportional to E^2/ω_{pe}^2. According to (4.166) the nonlinear term $n_e^{(1)} T_e \frac{\partial}{\partial r} \mathbf{v}^{(e)(1)}$ is $\sim (v^{(1)})^2 \, \omega_{pe}^{-1} \sim E^2/\omega_{pe}^3$, and the same holds true for $n_{e,0} T_e^{(1)} \frac{\partial}{\partial r} \mathbf{v}^{(e)}$ (1) and $\mathbf{v}^{(e)}$ (1) $\frac{\partial}{\partial r} T_e^{(1)}$. Thus, the left side of (4.159) contains only terms linear in the second-order perturbation. It is easily seen that this is also true of the term $\frac{\partial}{\partial r} \mathbf{q}_e$. Finally, because of (4.150) $\pi_{lj}^{(e)} \, ^{(1)} \, \partial v_l^{(e)} \, ^{(1)}/\partial x_j$ is of order $\frac{k^2 v_{Te}^2}{\nu_e} Q_e \ll Q_e$, and this term can be neglected. Then

$$\left(-\frac{3}{2} i\omega + 3.16 \frac{k^2 v_{Te}^2}{\nu_e}\right) \frac{T_{e,k}^{(2)}}{T_e^{(0)}} = -ikv_k^{(e) \, (2)} - 0.71 iku_k^{(2)} + \frac{1}{n_0 \, T_e^{(0)}} Q_{e,k}. \qquad (4.171)$$

In like manner, we find for the ions that

$$\left(-\frac{3}{2} i\omega + 3.9 \frac{k^2 v_{Ti}^2}{\nu_i}\right) \frac{T_{ik}^{(2)}}{T_i^{(0)}} = -ikv_k^{(i) \, (2)}. \qquad (4.172)$$

In the equation of motion for the electrons the term

$\dfrac{\partial}{\partial x_j} n_e^{(1)} T_e^{(1)} \sim \dfrac{E^2}{\omega_{pe}^4}$ can be neglected, and $\pi_{ij}^{(e)}$ is set equal to

$$\dfrac{-0.73 n_{e,0} T_e^{(0)}}{\nu_e} w_{ij}^{(e)\,(2)}. \tag{4.173}$$

By comparing this with the first term in the equation for the frictional force,

$$\mathbf{R} \approx -n_0 m_e \nu_e\, 0.51 \mathbf{u}^{(2)} - 0.71 n_0 \dfrac{\partial T_e}{\partial r}, \tag{4.174}$$

we see that it is smaller by a factor of $k^2 v_{Te}^2/\nu_e^2$ and can be dropped. The final equation for the electrons, including the nonlinear term $\mathbf{v}^{(1)}\dfrac{\partial}{\partial r}\mathbf{v}^{(1)}$, is then written as

$$\left(-i\omega + i\dfrac{k^2 v_{Te}^2}{\omega}\right)\mathbf{v}_k^{(e)\,(2)} = -0.51\nu_e \mathbf{u}_k^{(2)} - 1.71 i k v_{Te}^2 \dfrac{T_k^{(e)\,(2)}}{T_e^{(0)}} -$$

$$-i\int \dfrac{(\mathbf{k}\mathbf{k}_2)\,(\mathbf{k}_1\,\mathbf{k}_2)}{k k_1 k_2}\, v_{k_1}^{(e)\,(1)} v_{k_2}^{(e)\,(1)} \delta\,(k - k_1 - k_2)\, dk_1\, dk_2. \tag{4.175}$$

Let us compare the nonlinear term of (4.175) with the nonlinear term in (4.173). If, as an order of magnitude, we equate the frictional force in (4.175) with the nonlinear term, then

$$v_k^{(2)} \sim \dfrac{k}{\nu_e}\int v_{k_1}^{(e)\,(1)} v_{k_2}^{(e)\,(1)}\, dk_1\, dk_2\, \delta\,(k - k_1 - k_2), \tag{4.176}$$

and, by equating $k v_k^{(e)\,(2)}$ to the nonlinear term $Q_{e,k}/n_0 T_e^{(0)}$, we obtain

$$v_k^{(2)} \sim \dfrac{\nu_e}{k v_{Te}^2}\int v_{k_1}^{(e)\,(1)} v_{k_2}^{(e)\,(1)} \delta\,(k - k_1 - k_2)\, dk_1\, dk_2,$$

which is a quantity $v_e^2/k^2 v_{Te}^2 \gg 1$ times larger. Thus, the nonlinear term in (4.175) can be neglected and we have

$$\left(-i\omega + i\dfrac{k^3 v_{Te}^2}{\omega}\right)\mathbf{v}_k^{(e)\,(2)} = -0.51\nu_e \mathbf{u}_k^{(2)} - 1.71 i k v_{Te}^2\, T_{e,k}^{(2)}/T_e^{(0)} \tag{4.177}$$

Similarly, we have for the ions

$$\left(-i\omega + 1.28\frac{k^2 v_{Ti}^2}{\nu_i} + i\frac{k^2 v_{Ti}^2}{\omega}\right)v_k^{(i)\,(2)} = -ikv_{Ti}^2\frac{T_{i,k}^{(2)}}{T_i^{(0)}} + 0.71ikv_{Ti}^2\frac{T_{e,k}^{(2)}}{T_i^{(0)}} = 0.$$

(4.178)

One of the important results of this analysis is the change in the nature of the nonlinear interactions in those regions where the frequency difference or the frequency of one of the interacting fluctuations lies in the region of frequent collisions; that is, it satisfies (4.150). But even more important are the nonlinearities due to the dissipative heating of the plasma by the fluctuations; these are more important than the usual nonlinearities which arise because of the terms like $\left(v\frac{\partial}{\partial r}\right)v$ in the equations of motion for the particles (see [219]).

The system of equations involving (4.171), (4.172), (4.177), and (4.178) are easily solved, and together they determine the current:

$$j_k^{(2)} = -en_0(v_k^{(e)\,(2)} - v_k^{(i)\,(2)}) = \int S_{k,k_1,k_2} E_{k_1} E_{k_2} \delta(k-k_1-k_2)\,dk_1\,dk_2.$$

(4.179)

Note that because of (4.166) the term $e\int n_{k_1}^{(2)} v_{k_2}^{(1)} \delta(k-k_1-k_2) \sim \dfrac{E^2}{\omega_{pe}^3}$ is negligibly small. We shall use $S_{k,k_1,k_2}^{(1)}$ to denote the components of the nonlinear current which are symmetric in k_1 and k_2; k corresponds to the low frequencies of (4.150) and k_1 and k_2 are large frequencies which are comparable to the frequencies of the turbulent fluctuations. Further, let $S_{k,k_1,k_2}^{(2)}$ be the component in which one of the frequencies of the vectors k_1 and k_2 is low and the other two are high. Then

$$S_{k,k_1,-k_2}^{(1)} = i\frac{kn_0 e^3 1.71\nu_e (k_1 k_2)}{m_e^2 \omega_{pe}^2 \Omega\Omega_e k_1 k_2}.$$

(4.180)

Here

$$\left.\begin{aligned}\Omega &= -i\omega + 0.51\nu_e + i\frac{k^2 v_{Te}^2}{\omega}\left(1 - 2.96\frac{i\omega}{\Omega_e}\right); \\ \Omega_e &= -\frac{3}{2}i\omega + 3.16\frac{k^2 v_{Te}^2}{\nu_e}.\end{aligned}\right\}$$

(4.181)

It is somewhat easier to compute $S^{(2)}_{k,k_1,k_2}$ than to calculate $S^{(1)}_{k,k_1,k_2}$. Since k corresponds to a high frequency we must use the hydrodynamic equations for $v^{(2)}_k$ with all the dissipative terms and the gas pressure force omitted. Then $v^{(e)\,(2)}_k$ is proportional to E^2/ω^2_{pe}. In the equation

$$j^{(2)}_k = -en_0(v^{(e)\,(2)}_k - v^{(i)\,(2)}_k) = -e\int n^{(1)}_{k_1} v^{(1)}_{k_2} \frac{(kk_2)}{kk_2} \delta_{,}(k-k_1-k_2)\,dk_1\,dk_2$$

the second term contains one low-frequency factor and is therefore of order $E^2/\omega_{pe}\omega$.

Thus, to the accuracy desired we have

$$j^{(2)}_k = -e\int v^{(e)\,(1)}_{k_2} n^{(1)}_{e,k_1}\delta(k-k_1-k_2)\frac{(kk_2)}{kk_2}\,dk_1\,dk_2. \qquad (4.182)$$

It is easy to understand that the only case of importance in (4.182) is that one in which $n^{(1)}_{ek}$ corresponds to a low frequency and $v^{(e)\,(1)}_{k_2}$ goes with a high frequency. This follows from (4.166), which shows that $n^{(1)}_{k_1}$ is proportional to $1/\omega^2_{pe}$ for high frequencies. Then, by using (4.166) one can write

$$j^{(2)}_k = \frac{ie}{m_e}\int \frac{E_{k_2}}{\omega_{k_2}} n^{(1)}_{k_1} \frac{(kk_2)}{kk_2} \delta(k-k_1-k_2)\,dk_1\,dk_2. \qquad (4.183)$$

We must now find $n^{(1)}_k$ for low frequencies; this is easily accomplished by linearizing our hydrodynamic equations. It is more convenient to write the result for the component of the nonlinear current when the substitution $k \rightleftarrows k_1$ is made:

$$S^{(2)}_{k_1,k_2,k} = \frac{-ie^3 k\,(k_1\,k_2)}{m^2_e\,\omega_{pe}\,\varkappa\omega_e\,\omega k_1 k_2}, \qquad (4.184)$$

where

$$\begin{aligned}
\varkappa &= 1 + \left(0.51\nu_e + 1.22\,\frac{k^2 v^2_{Te}}{\Omega_e}\right)\left(\frac{1}{\omega_e} + \frac{m_e}{m_i}\cdot\frac{1}{\omega_i}\right), \\
\omega_e &= -i\omega + i\frac{k^2 v^2_{Te}}{\omega}\left(1 - 1.71\frac{i\omega}{\Omega_e}\right), \\
\omega_i &= -i\omega + i\frac{k^2 v^2_{Ti}}{\omega}\left(1 - i\frac{\omega}{\Omega_i} - 1.28\frac{i\omega}{\nu_i} + 0.71i\frac{T_e}{T_i}\cdot\frac{\omega}{\Omega_e}\right), \\
\Omega_i &= -\frac{3}{2}i\omega + 3.9\frac{k^2 v^2_{Ti}}{\nu_i}.
\end{aligned}\right\} \qquad (4.185)$$

From the linearized system of equations it is also easy to find the linear dielectric permeability for the plasma in the low-frequency region:

$$\varepsilon_k^l = 1 + i\,\frac{\omega_{pe}^2}{\varkappa\omega\omega_e} + i\,\frac{\omega_{pi}^2}{\varkappa\omega\omega_i}. \qquad (4.186)$$

It is necessary to find the components of the nonlinear current to third order for the final derivation of the nonlinear equations. The arguments which were used to obtain (4.183) give

$$j_k^{(3)} = \frac{ie^2}{m_e\,\omega} \int \frac{(kk_1)}{kk_1}\, E_{k_1}\, n_{k_2}^{(e)\,(2)}\delta(k - k_1 - k_2)dk_1\,dk_2, \qquad (4.187)$$

where $n_{k_2}^{(e)\,(2)}$ is the perturbation on the density at low frequency, and is easily obtained from Eqs. (4.170)-(4.178). We finally obtain

$$\frac{1}{2}\left(\Sigma_{k_1,\,k_2,\,k_1,\,-k_2} + \Sigma_{k_1,\,k_2,\,-k_2,\,k_1}\right) = -\frac{n_0\,e^4\,k_-^2\,(k_1,\,k_2)^2\,1{,}71\nu_e}{\omega_-\,k_1^2\,k_2^2\,m_e^3\,\omega_{pe}^3\,\Omega\Omega_e}, \qquad (4.188)$$

$$\mathbf{k}_- = \mathbf{k}_1 - \mathbf{k}_2,$$
$$\omega_- = \omega_1 - \omega_2.$$

These results can be used to study the the role of particle collisions in the interactions between turbulent fluctuations [216]. We note that primarily in the low-frequency range, where

$$\max\left(\omega,\,kv_{T\alpha}\right) \ll \nu_\alpha, \qquad (4.189)$$

there are always sound-type oscillations—these are usually acoustic oscillations (of course, these require long wavelengths or high collision frequencies: $kv_{T\alpha} \ll \nu_\alpha$). There is an important difference between those frequencies which are less than the collision frequency and those in the kinetic region, where acoustic oscillations are possible only if $T_e \gg T_i$. Under the condition of (4.189), weakly attenuated acoustic oscillations are possible when $T_e = T_i$. It is reasonable to assume that $T_e \neq T_i$ for such oscillations if their frequencies are much higher than $3(m_e/m_i)\nu_e$ — the frequency at which the electron and ion temperatures are equalized. It is easy to convince oneself of of the validity of these statements by solving the dispersion equation $\varepsilon_k^l = 0$, where ε_k^l is given in (4.186).

When $\omega_k^s v_e \ll k^2 v_{Te}^2$ we obtain (i.e., $\omega^s \gg (m_e/m_i)\nu_e$ approximately, and we neglect terms of order unity in comparison)

$$\omega = \omega_k^s = k v_s; \quad v_s = v_{Te} \sqrt{\frac{m_e}{m_i}\left(1 + \frac{5}{3}\cdot\frac{T_i}{T_e}\right)}. \qquad (4.190)$$

The damping of these oscillations is determined by

$$\operatorname{Im}\omega = \gamma_k^s = \gamma_k^{(e)} + \gamma_k^{(i)},$$

where $\gamma_k^{(e)}$ and $\gamma_k^{(i)}$ are the electron and ion damping rates, respectively:

$$\gamma_k^{(e)} = -0.41\,\frac{m_e}{m_i}\,\nu_e, \quad \gamma_k^{(i)} = -\frac{1.92 + 0.64\,\dfrac{T_e}{T_i}}{\dfrac{5}{3} + \dfrac{T_e}{T_i}}\cdot\frac{k^2 v_{Ti}^2}{\nu_i}. \qquad (4.191)$$

In the region $\omega_k^s v_e \gg k^2 v_{Te}^2$ (low-frequency acoustic oscillations with $\omega_k^s \ll \frac{m_e}{m_i}\nu_e$ so that $T_e = T_i$) we have

$$\omega_k^s = k v_s, \quad v_s = v_{Ti}\sqrt{\frac{10}{3}}, \qquad (4.192)$$

$$\gamma_k^{(e)} = -0.08\,\frac{k^2 v_{Te}^2}{\nu_e}, \quad \gamma_k^{(i)} = -0.9\,\frac{k^2 v_{Ti}^2}{\nu_i}. \qquad (4.193)$$

These oscillations are damped primarily because of the electronic thermal conductivity.

When Langmuir fluctuations interact the acoustic oscillations play an important role because the decays into acoustic oscillations can determine the spectral transfer of their energy. This will occur only when $k > k_*^s[1 + {}^5/_3(T_i/T_e)]^{1/2}$ for Eq. (4.190) and when $k > k_*^s(10/3)^{1/2}$ for (4.192); $k_*^s = (\omega_{pe}/3v_{Te})(m_e/m_i)^{1/2}$. The resonance conditions for decay are satisfied only under these conditions, when $\varepsilon_{k-k_1}^l$ of (4.132) goes through zero. With $\operatorname{Im}\dfrac{1}{\varepsilon_{k-k_1}^l} = -\dfrac{\omega - \omega_1}{|\omega - \omega_1|}\pi\delta(\varepsilon_{k-k_1}^l)$ we can use the explicit expressions for

$S_{k,\,k_1,\,k_2}^{(1)}; \; S_{k,\,k_1,\,k_2}^{(2)}$ to obtain

$$2\nu_e W_k^l = \frac{\pi}{2} W_k^l \int W_{k_1}^l dk_1 \frac{\delta(\varepsilon_{k-k_1}^l)(\omega_1 - \omega) v_e^2\, 1{,}71\,\omega_{pe}^3\,(k_1 k)^2}{|\omega - \omega_1|\,3{,}16\,|k-k_1|^4 v_{Te}^4 n_0 T_e k^2 k_1^2} \qquad (4.194)$$

when $\omega_k^s \gg \dfrac{m_e}{m_i} \nu_e$, and

$$2\nu_e W_k^l = -0.05 \mathrm{Im}\, W_k^l \int \frac{W_{k_1}^l\, dk_1\, i\nu_e\, \omega_{pe}^3\, (k_1 k)^2}{(\omega - \omega_1)\, \varepsilon_{k-k_1}^l\, n_0 T_e\, |k-k_1|^2 v_{Te}^2 k^2 k_1^2} \tag{4.195}$$

when $\omega_k^s \ll \dfrac{m_e}{m_i} \nu_e$. The derivation of (4.195) requires the more exact equations for $S_{k,\ k_1,\ k_2}^{(1)}$ and $S_{k,\ k_1,\ k_2}^{(2)}$, which take into account the equalization of the temperatures; these expressions are easily found from the hydrodynamic equations using the methods previously described.

One ought to compare the result of (4.194) with that obtained in a nonisothermal plasma ($T_e \gg T_i$) for the transfer of Langmuir waves in the decay into ion-acoustic waves (4.99a). If we use the approximate expression for the dielectric function in the low-frequency region, as obtained from (4.186),

$$\left.\begin{aligned}
\varepsilon_{k-k_1}^l &\approx \frac{\omega_{pe}^2}{|k-k_1|^4 v_{Te}^4} \cdot \frac{m_i}{m_e} \left((\omega - \omega_1)^2 - (k-k_1)^2 v_s^2 \right); \\
v_s &= v_{Te} \sqrt{\frac{m_e}{m_i} \left(1 + \frac{5}{3} \cdot \frac{T_i}{T_e} \right)},
\end{aligned}\right\} \tag{4.196}$$

it is not difficult to write (4.194) in the form

$$2\nu_e W_{\omega,k}^l = 0.13\,\pi W_{\omega,\,k}^l \int W_{\omega_1,\,k_1}^l\, dk_1\, \frac{v_e^2}{|k-k_1|^2 v_{Te}^2}\, \omega_{pe}\, \frac{(kk_1)^2}{k^2 k_1^2} \times$$

$$\times\, \frac{|k-k_1|\, v_s}{n_0 \left(T_e + \dfrac{5}{3} T_i \right)}\, [\delta(\omega - \omega_1 + |k-k_1|\, v_s) - \delta(\omega - \omega_1 - |k-k_1|\, v_s)]. \tag{4.197}$$

For easy comparison with (4.99a) we write that equation as

$$\nu_e W_{\omega,\,k}^l = \frac{\pi}{4}\, W_{\omega,\,k}^l \int W_{\omega_1,\,k_1}^l\, d\omega_1 dk_1 \omega_{pe}\, \frac{(kk_1)^2}{k^2 k_1^2} \times$$

$$\times\, \frac{|k-k_1|\, v_s}{n_0 T_e}\, [\delta(\omega - \omega_1 + |k-k_1|\, v_s) - \delta(\omega - \omega_1 - |k-k_1|\, v_s)]. \tag{4.198}$$

Here $v_s = (m_e/m_i)^{1/2} v_{Te}$ and coincides with v_s in (4.196) only when $T_e \gg T_i$.

Equation (4.198) is also somewhat more exact than (4.99a) because it does not assume δ-function forms for the correlations. The most important difference between (4.197) and (4.198) is the presence of the large factor $0.54 \dfrac{v_e^2}{|k-k_1|^2 v_{Te}^2}$ in (4.197). The integration in (4.197) runs over a region satisfying the inequality in (4.150), whereas in (4.198) the integration is over the region which fulfills just the opposite inequality. Although the first region is smaller than the second, the factor $v_e^2/|k-k_1|^2 v_{Te}^2$ can compensate for this difference, so that particle collisions can exert a substantial influence on the interactions of Langmuir fluctuations. But we must now distinguish between at least two cases.

First, acoustic fluctuations have, as a rule, rather long wavelengths, which means that in order to have the interaction of (4.197) it is necessary that the characteristic dimensions a of the volume containing the plasma, or its density, be sufficiently large. If for $k \sim 1/a$ the conditions ω, $kv_{T\alpha} \ll \nu_\alpha$ are not satisfied, acoustic oscillations do not exist and the interaction in (4.197) does not take place. Second, the pair collisions appear in the quasistationary spectra if they exist for a characteristic time which is much longer than the pair collision frequency. If the source of the plasma turbulence acts over a period of time which is smaller than $1/\nu_e$, then the collisions will not be important to the spectra. Formally this is clear from the fact that the nonstationary nature of the fluctuations leads to the dependence of W on two ω [or $W_k(t, t')$], and the dependence on $(\omega + \omega')/2$ produces a "spreading" which is larger than the frequencies ν_α if the turbulence is unstable for a time smaller than $1/\nu_\alpha$. Finally, the correlation of turbulent fluctuations smears out the fluctuation spectra with respect to frequencies and decreases the relative number of waves satisfying (4.150) even when the turbulence is completely stationary.

We will write the interaction of (4.195) using the specific expression for $\varepsilon_{k-k_1}^l$:

$$2\nu_e W_{\omega, k}^l = 0.05\, v_e\, W_{\omega, k}^l\, \mathscr{P} \int \frac{W_{\omega_1, k_1}}{\omega_1 - \omega} \cdot \frac{2.14\, \omega_{pe}}{\left[(\omega - \omega_1)^2 - |k - k_1|^2 v_s^2\right]} \times$$

$$\times \left[(\omega - \omega_1)^2 - 1.19\,|k - k_1|^2 v_{Ti}^2\right] \frac{(kk_1)^2}{k^2 k_1^2}\, d\omega_1 dk_1. \qquad (4.199)$$

By comparing with (4.198) we see that even in this case the effectiveness of the interaction has increased by a factor of $\dfrac{v_e}{|k-k_1|v_s} \gg 1$. The relative number of waves whose frequency differences satisfy the condition $|\omega-\omega_1| \ll 3\,\dfrac{m_e}{m_i}\,v_e$ [which is necessary in order that (4.199) be valid] is significantly smaller than the number of waves which satisfy (4.150).

Let us limit our attention to the interaction given in (4.197). After integrating (4.198) by parts, we can set up a balance equation

$$2v_e\,W_{\mathbf{k}}^l = 2v_e \int W_{\omega,\,\mathbf{k}}^l\,d\omega = 0.13\,\pi \int W_{\omega,\,\mathbf{k}}^l\,W_{\omega_1,\,\mathbf{k}_1}^l\,d\omega\,d\omega_1\,dk_1 \times$$

$$\times \frac{v_e^2}{|k-k_1|^2\,v_{Te}^2}\,\omega_{pe}\,\frac{(kk_1)^2}{k^2\,k_1^2}\cdot\frac{|k-k_1|\,v_s}{n_0\left(T_e+\dfrac{5}{3}\,T_i\right)} \times$$

$$\times\,[\delta\,(\omega-\omega_1+|k-k_1|\,v_s)-\delta\,(\omega-\omega_1-|k-k_1|\,v_s)]. \qquad (4.200)$$

We will assume that $|k-k_1|v_s$ fulfills the condition in (4.150). The integration over ω and ω_1 in (4.200) runs over the interval where $\omega-\omega_1$ is larger than $|k-k_1|v_s$. Under these conditions we can use the following expansion of the δ-function:

$$\delta\,(\omega-\omega_1\pm|k-k_1|\,v_s) \approx \delta\,(\omega-\omega_1)\pm|k-k_1|\,v_s\,\delta'\,(\omega-\omega_1). \qquad (4.201)$$

We obtain

$$2v_e\,W_{\mathbf{k}}^l = \omega_{pe}\,\frac{0.54\,\pi v_e^2}{2}\int W_{\omega,\,\mathbf{k}}^l\,\frac{\partial W_{\omega,\,\mathbf{k}_1}^l}{\partial\omega}\,\frac{d\omega m_e}{n_0\,T_e m_i}\,\frac{(kk_1)^3}{k^2\,k_1^2}\,dk_1. \qquad (4.202)$$

From here on the course of the calculation depends on the relationship between the correlation width γ_k^N and the frequency $|k-k_1|v_s$. We shall assume that, in accord with our above development, the frequency dependence of the correlation functions is

$$W_{\omega,\,\mathbf{k}}^l = \frac{1}{\pi}\,W_{\mathbf{k}}^l\,\frac{\gamma_k^N}{(\omega-\omega_k)^2+(\gamma_k^N)^2}, \qquad (4.203)$$

so that $\int W_{\omega, \mathbf{k}} d\omega = W_{\mathbf{k}}$. By using the relationship

$$\int_{-\infty}^{\infty} \frac{1}{\pi^2} \cdot \frac{\gamma_{\mathbf{k}}^N \gamma_{\mathbf{k}_1}^N d\omega}{\left[(\omega - \omega_{\mathbf{k}_1})^2 + (\gamma_{\mathbf{k}_1}^N)^2\right]\left[(\omega - \omega_{\mathbf{k}})^2 + (\gamma_{\mathbf{k}}^N)^2\right]} =$$

$$= \frac{1}{\pi} \cdot \frac{\gamma_{\mathbf{k}_1}^N + \gamma_{\mathbf{k}}^N}{\left[(\omega_{\mathbf{k}_1} - \omega_{\mathbf{k}})^2 + (\gamma_{\mathbf{k}}^N + \gamma_{\mathbf{k}_1}^N)^2\right]}, \qquad (4.204)$$

we find

$$2\nu_e = 0.54 \omega_{pe} \, v_e^2 \int \frac{m_e \, W_{\mathbf{k}_1}^l \, d\mathbf{k}_1 \, (\mathbf{k}\mathbf{k}_1)^2 \, (\omega_{\mathbf{k}_1} - \omega_{\mathbf{k}}) \, (\gamma_{\mathbf{k}}^N + \gamma_{\mathbf{k}_1}^N)}{m_l \, k^2 \, k_1^2 \, n_0 \, T_e \left[(\omega_{\mathbf{k}_1} - \omega_{\mathbf{k}})^2 + (\gamma_{\mathbf{k}}^N + \gamma_{\mathbf{k}_1}^N)^2\right]^2}. \qquad (4.205)$$

The $\mathbf{k}_1$ integration in (4.205) must be over just those values of k_1 for which $|\mathbf{k} - \mathbf{k}_1| v_s < \nu_e, \nu_i$ and for which (4.150) is satisfied.

We now assume that $T_e = T_i$. Then $v_s \sim v_{Ti}$ and $\gamma_i \sim (m_e/m_i)^{1/2} \nu_e$, i.e., when

$$|\mathbf{k} - \mathbf{k}_1| v_s < \nu_i \qquad (4.206)$$

all the conditions of (4.150) are fulfilled.

If $kv_s \gg \nu_i$, (4.206) is satisfied only by those k_1 close to k, and when $k = k_1$ if the angle between $\mathbf{k}$ and $\mathbf{k}_1$ is such that $\theta < \sqrt{2} \, \nu_i/kv_s = \theta_{max}$. The difference between the absolute values of the wave numbers, $\Delta k = k - k_1$, also cannot exceed ν_i/v_s; as a consequence $\Delta\omega = \omega_{\mathbf{k}_1} - \omega_{\mathbf{k}} \approx \frac{3v_{Te}^2}{\omega_{qe}} k \frac{\nu_i}{v_s}$. When $\Delta\omega$ is much larger than the correlation width $\gamma_{\mathbf{k}}^N$, the right side of (4.205) has the value

$$m_e \omega_{pe} \, v_e^2 \frac{\partial}{\partial \omega_{\mathbf{k}}} \int \frac{W_{\mathbf{k}_1} \, \delta(\omega_{\mathbf{k}} - \omega_{\mathbf{k}_1}) \, d\mathbf{k}_1}{m_i n_0 T_e} \theta_{max}^2 \simeq \frac{\omega_{pe}^3}{18} \cdot \frac{v_e^4}{k^3 v_{Te}^6} \cdot \frac{\partial}{\partial k} \cdot \frac{W_h}{k} \cdot \frac{m_e}{m_i n_0 T_e}. \qquad (4.207)$$

By comparing the coefficient in (4.207) with the coefficient α in (4.62) we see that the collisions will determine the interactions only if

$$\frac{v_e^4}{k^4 v_{Te}^4} > 1. \qquad (4.208)$$

This condition can be met only in a very dense plasma. The

dense-plasma condition is found in laser flash experiments [190, 194] or in experiments in which the plasma is produced through the interaction of a laser beam with a solid surface.

Note that if $\Delta\omega \sim \dfrac{3v_{Te}}{\omega_{pe}} k v_e \ll \gamma_k^N$ the interaction of (4.205) is considerably reduced, by about a factor of $(\Delta\omega/\gamma_k^N)^3$. Assume that $k v_s \ll \nu_i$. For $k = k_*$ this means that $N_D < m_i/m_e$, so that in (4.205) we must carry out the integration over angles. If at the same time we have

$$\frac{3v_{Te} k v_e}{\omega_{pe}} > \gamma_k^N, \tag{4.209}$$

then Eq. (4.205) takes the form

$$2\nu_e = 0.54 \, \frac{v_e^2 \, \omega_{pe}^3}{72 \, k} \cdot \frac{\partial}{\partial k} \cdot \frac{W_k}{k} \frac{1}{n_0 T_e} \cdot \frac{m_e}{m_i \, v_{Te}^4}. \tag{4.210}$$

The solution to (4.209) is easily found. In particular, if the turbulence is sufficiently strong and the left side of (4.209) can be equated to zero, then

$$W_k = k \sqrt{\frac{288 m_i \, Q n_0 \, T_e \, v_{Te}^4}{m_e 0.54 \, \omega_{pe}^3 \, v_e^2}} \quad \text{when } k \ll k_g. \tag{4.211}$$

Pair collisions can be especially important for large phase velocities since the prospects for satisfying (4.150) increase as the phase velocities increase. It is known that as the generating power Q increases, the maximum in the turbulence spectrum shifts toward the higher phase velocities. One might guess that the role of the collisions becomes more important as Q increases, too. However, when $k > k_*$ the interactions can be resonant and then correspond to a decay into acoustic oscillations. The nonresonant character of the interaction is indicated by the fact that (4.188) begins to make a contribution which is comparable with $S^{(1)} \dfrac{1}{\mathrm{g}} S^{(2)}$, and there is a significant reduction in the size of Σ^{eff} [see (4.132)] because of the compensation of two terms:

$$\Sigma_{k_1, \, k_2}^{\mathrm{eff}} \simeq - \, \mathrm{i} \, \frac{(k_1 k_2)^2 \, 1.71 \, \nu_e \, e^2 \, (k_1 - k_2)^2 \, \varepsilon_{k_1 - k_2}^{l \, (i)}}{k_1^2 \, k_2^2 \, \omega_{pe} \, (\omega_1 - \omega_2) \, \Omega\Omega_e \, m_e^2 \, \varepsilon_{k_1 - k_2}^{l}}. \tag{4.212}$$

The imaginary part of $\Sigma^{\text{eff}}_{k_1, k_2}$ plays a role similar to the scattering on electrons and ions. However, if the scattering is connected with collisionless Landau damping of a virtual wave of frequency $\omega_1 - \omega_2$, the interaction described by (4.212) corresponds to the damping of a virtual wave by pair collisions. We must point out that in many instances the characteristic transformation time turns out to be less than ν_e^{-1} when $W/n_0 T_e \ll 1$, so that the relative contribution of pair collisions to the interaction is not large [215]. The most important effect turns out to be the nonlinear frequency shift, as described by the real part of (4.212). It depends heavily on k. In fact, when $|\omega_1 - \omega| \ll \dfrac{|k_1 - k|^2 \, v_{Te}^2}{\nu_e}$, (4.212)

gives

$$\omega_k^N = -\omega_{pe} \int \frac{1.71 \, v_e^2 \, W^l_{k_1} \, dk_1 \, (kk_1)^2}{v_{Te}^2 \, |k - k_1|^2 \, k^2 \, k_1^2 \, n_0 \, T_e \left(1 + \dfrac{T_i}{2.14 \, T_e} \right)} , \qquad (4.213)$$

while when $|\omega_1 - \omega| \gg |k_1 - k|^2 \, v_{Te}^2/\nu_e$,

$$\omega_k^N = \omega_{pe} \int \frac{W^l_{k_1} \, dk_1 \, 3.84 \, |k - k_1|^2 \, v_{Te}^2}{(\omega_1 - \omega)^2 \, n_0 \, T_e \left(2.14 + \dfrac{T_i}{T_e} \right)} . \qquad (4.214)$$

It follows from (4.213) and (4.214) that the frequency shift increases by the large factor $\dfrac{v_e^2}{|k - k_1|^2 \, v_{Te}^2} \gg 1$ when $|\omega_1 - \omega| \ll \dfrac{|k_1 - k|^2 \, v_{Te}^2}{\nu_e}$ and then decreases as the frequency difference increases. Therefore, for an intense turbulence in the region $kv_{Te} \ll \nu_e$, when the interacting waves are found in the region of frequent collisions $(|\omega^l_{k_1} - \omega^l_k| < \nu_e)$ for any propagation angles and a difference in waves numbers Δk which is of order k, the order of magnitude of the frequency difference for the interacting fluctuations is

$$\Delta \omega \approx \frac{k^2 \, v_{Te}^2}{\nu_e} . \qquad (4.215)$$

This is much larger than the collisionless frequency difference $\Delta \omega \sim 3k^2 v_{Te}^2/\omega_{pe}$, and is always smaller than ν_e since $k^2 v_{Te}^2/\nu_e^2 \ll 1$.

This situation is encountered if

$$\frac{W}{n\overline{l}} \gtrsim \frac{k^4 v_{Te}^4}{v_e^4} \cdot \frac{v_e}{\omega_{pe}} \qquad (4.216)$$

in the neighborhood of the spectrum maximum.

An increase in the frequency difference of the interacting fluctuations is quite important for many reasons. First, this frequency difference can become comparable to the correlation width [see (4.209)] and an increase in $\Delta\omega$ eases the corresponding conditions. When the dispersion effects are determined by the nonlinear effect the interaction in (4.210), which is written for linear dispersion $\left(\omega_k^l = \omega_{pe} + 3 \, \frac{k^2 v_{Te}^2}{2\omega_{pe}} \right)$, is also changed. Second, an increase in the frequency difference has an impact on our estimate of the role of correlation effects in the four-plasmon interaction (see §4.6). However, the collisions also change the intensity of the four-plasmon interaction, the probabilities of which are determined by Σ^{eff} as computed above [see (4.212)]. Using the expression for Σ^{eff}, the resulting balance equation can be written in terms of the probability for the four-plasmon interaction in standard form

$$2v_e N_{\mathbf{k}}^l = \int w_{ll}^{ll} (\mathbf{k}, \, \mathbf{k}_1, \, \mathbf{k}_2, \, \mathbf{k}_3) \, (N_{\mathbf{k}_1}^l N_{\mathbf{k}_2}^l N_{\mathbf{k}_3}^l +$$

$$+ N_{\mathbf{k}}^l N_{\mathbf{k}_2}^l N_{\mathbf{k}_3}^l - N_{\mathbf{k}}^l N_{\mathbf{k}_1}^l N_{\mathbf{k}_2}^l - N_{\mathbf{k}}^l N_{\mathbf{k}_1}^l N_{\mathbf{k}_3}^l) \, \frac{dk_1 dk_2 dk_3}{(2\pi)^9} , \qquad (4.217)$$

where

$$w_{ll}^{ll} (\mathbf{k}, \, \mathbf{k}_1, \, \mathbf{k}_2, \, \mathbf{k}_3) = \frac{(2\pi)^9 \, e^4}{(4\pi)^3 \, m_e^4} \left| \frac{1.71 \, v_e \, k_-^2}{\omega_- \, \Omega\Omega_e} \cdot \frac{\varepsilon_{k_-}^{l \, (l)}}{\varepsilon_{k_-}^l} \cdot \frac{(\mathbf{k}_1 \, \mathbf{k}_2)}{k_1 \, k_2} \cdot \frac{(\mathbf{k}_3 \, \mathbf{k})}{k_3 \, k} + \mathbf{k} \rightleftarrows \mathbf{k}_1 \right|^2 \times$$

$$\times \, \delta \, (\mathbf{k} + \mathbf{k}_1 - \mathbf{k}_2 - \mathbf{k}_3) \, \delta \, (\omega_{\mathbf{k}}^N + \omega_{\mathbf{k}_1}^N - \omega_{\mathbf{k}_2}^N - \omega_{\mathbf{k}_3}^N), \qquad (4.218)$$

$$k_- = \{\omega_{\mathbf{k}_1}^N - \omega_{\mathbf{k}_2}^N, \quad \mathbf{k}_1 - \mathbf{k}_2\}. \qquad (4.219)$$

In the region where $k \gg k_*$; $\varepsilon_{k.}^{l \, (l)} \approx \varepsilon_{k_-}^l$, the difference between the probability described by (4.218) and the collisionless probability is just the factor

$$v_{Te}^4 \left| \frac{1.71 \, v_e \, k_-^2}{\omega_- \, \Omega\Omega_e} \right|^2 , \qquad (4.220)$$

which, when $\omega_- \gg (k^2 - v_{Te}^2)/\nu_e$, takes the form

$$0.29 \frac{v_e^4}{k_-^4 \, v_{Te}^4} . \qquad (4.221)$$

This factor is very large since

$$\frac{\nu_e}{k v_{Te}} \gg 1; \qquad (4.222)$$

physically this indicates a strong increase in plasmon repulsion.

We assume that the turbulence generating power is sufficient-ly large that the maximum in the spectrum, from estimates cor-responding to the collisionless limit, approaches or is actually found in the region where the conditions of (4.150) are satisfied. Here we are neglecting the differential energy transformation. Then, roughly speaking, when $k v_{Te} \sim \nu_e$, i.e., when

$$\frac{v_p}{v_{Te}} \simeq \frac{\omega_{pe}}{\nu_e} \sim N_D, \qquad (4.223)$$

the collision effects come into play. In fact, they can be important much earlier, and the relay transfer smooths the transition. When v_p is smaller than the limit given in (4.223), the four-plasmon interaction, although it distributes the plasmons over various wave vectors, is not in a state in which it can set up a systematic en-ergy flow connected with scattering on ions. If v_p exceeds the limit in (4.223), the systematic current is sharply decreased and becomes negligible when compared with the trivial absorption of fluctuations by pair collisions. A collisionless flow can be set up before reaching the limit in (4.223) with a small generating power. If this does not occur, it must be set up at the v_p which corresponds to (4.223).

We should mention that the interaction of (4.217) describes the subdividing of the scales of the turbulent fluctuations.[†] The

---

[†]For the collisionless region enlargement of the scales is accompanied by an in-crease in entropy because of the collisionless heating of the plasma ions.

energy flow appropriate to the scales which determine (4.223) acts as if it encounters some effective wall and is stopped. The relay processes can significantly reduce the critical phase velocities at which this situation arises [216].

4.8. The Spectra of Ion-Acoustic Turbulence

In contrast to the above analysis, here we will assume that the source of the turbulence excites ion-acoustic oscillations. The occurrence of other types of fluctuations results from inter-actions of these fluctuations with each other. To begin, we shall consider the case in which the excitations are just ion-acoustic oscillations (this takes place in an isotropic plasma). As shown in §2.8, the interactions of these fluctuations are determined by the ions, for which the nonlinear current is

$$\Sigma^{(i)}_{k,\,k_1,\,k,\,-k_1} + \Sigma^{(i)}_{k,\,k_1,\,-k,\,k} = \frac{\pi e^4 \, \omega \, (kk_1)^2}{m_i^2 \, k^2 \, k_1^2} \times$$

$$\times \int \frac{dp}{(2\pi)^3} \cdot \frac{\delta(\omega - \omega_1 - (k - k_1)v)}{(\omega - kv)^4} \left((k - k_1) \frac{\partial f^{R\,(i)}}{\partial p} \right). \qquad (4.224)$$

Because of our earlier discussion the entire effect here reduces to the induced scattering of ion-acoustic waves on plasma ions, and can be found using the scattering probability. The quantity Σ^{eff} which appears in the nonlinear interaction differs from (4.224) in that the term $1/(\omega - kv)^4$ must be replaced with

$$\left[\frac{1}{(\omega - kv)^2} - \frac{1}{\omega^2} \right]^2. \qquad (4.225)$$

The first term in this expression accounts for the Compton scattering as described in (4.224); the second term accounts for the nonlinear scattering. In the limit $T_e \gg T_i$, when just the ion-acoustic waves exist,

$$\Sigma^{\text{eff}}_{k,\,k_1} = \frac{2e^4 \, \pi \, (k,\,k_1)^2}{m_i^2 \, k_1^2 \, \omega^2 \, \omega_1^3 k^2} \int (kv)^2 \, \delta \left(\omega - \omega_1 - (k - k_1) \, v \right) \times$$

$$\times \left((k - k_1) \frac{\partial f^{R\,(i)}}{\partial p} \right) \frac{dp}{(2\pi)^3} \approx \frac{\omega_{pi}^4 \, T_i \, (kk_1)^2 \, [kk_1]^2}{8\pi \, \omega^2 n_0 \, m_i^2 \, k^2 \, k_1^2 \, \omega_1^3} \, \delta'(\omega - \omega_1). \qquad (4.226)$$

It follows that

$$-2\gamma_k^s I_k = I_k \int^\circ \frac{8\pi \operatorname{Re} \Sigma_{k,\,k_1}^{\mathrm{eff}} I_{k_1}\, dk_1}{\left| \omega \dfrac{\partial \varepsilon_k^l}{\partial \omega} \right|_{\omega=\omega_k^s}}. \tag{4.227}$$

Here γ_k^s is the linear Landau damping of the ion-acoustic waves plus the growth rate for the instability which leads to the excitation of ion-acoustic waves. If the correlation of ion-acoustic waves can be neglected, then

$$I_{k_1} = \frac{2\pi (\omega_{k_1}^s)^2}{\omega_{pi}^2} \{ W_{k_1}^s \, \delta(\omega_1 - \omega_{k_1}^s) + W_{-k_1}^s \, \delta(\omega_1 + \omega_{k_1}^s) \} \tag{4.228}$$

and Eq. (4.227) takes the form

$$-2\gamma_k^s W_k^s = W_k^s \int dk_1\, W_{k_1}^s\, 2\pi\, \frac{(kk_1)^2}{k^2 k_1^2} \,[kk_1]^2\, \frac{T_i\, \delta'(\omega_k^s - \omega_{k_1}^s)}{m_i^2\, n_0\, \omega_{k_1}^s}. \tag{4.229}$$

For isotropic turbulence we easily find that

$$-2\gamma_k^s = \frac{4\pi T_i}{15\, m_i^2\, n_0} \cdot \frac{k^2}{\dfrac{d\omega_k^s}{dk}} \cdot \frac{d}{dk}\left(\frac{W_k^s}{\omega_k^s} \cdot \frac{k^2}{\dfrac{d\omega_k^s}{dk}} \right). \tag{4.230}$$

The interaction of (4.229) was first presented in [220], and an analysis of the oscillation spectra for a weakly-ionized plasma was presented in [6].

Let us examine the region in which there is only Landau damping and the nonlinear transfer of the oscillations; this means that there is no generation (which occurs for large wave numbers). It then follows from (4.230) that

$$W_k^s = \frac{15}{4\sqrt{2\pi}} \cdot \frac{T_e}{T_i} \sqrt{\frac{m_e}{m_i}} \cdot \frac{1}{k}\, n_0 T_e \ln \frac{k}{k_0}. \tag{4.231}$$

The wave vector $k = k_0$ corresponds to the point at which the turbulence energy goes to zero. k_0 is close to the maximum in the spectrum, which is attained for $k = k_0 e$, where e is the base of the

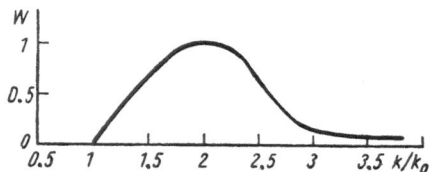

Fig. 4.7. The spectrum for the ion-acoustic oscillations which arise because of the energy balance of the nonlinear transformation of energy across the spectrum and the Landau

damping $\left(W = \dfrac{4 W_k^s \sqrt{2\pi}\, T_i}{15 T_e^2\, n_0 k_0} \sqrt{\dfrac{m_i}{m_e}} \right).$

natural logarithms (Fig. 4.7). The total energy contained in the ion-acoustic oscillations is

$$\frac{W^s}{n_0 T_e} = \int_{k_0}^{k_g} \frac{W_k^s\, dk}{n_0 T_e} = \frac{15}{8\sqrt{2\pi}}\, \ln^2 \frac{k_g}{k_0} \cdot \frac{T_e}{T_i} \sqrt{\frac{m_e}{m_i}} . \qquad (4.232)$$

k_g is determined by the wave numbers for which the generation exceeds the Landau absorption. If the generation region is very narrow, running from k_g to $k_g + \Delta k_g$, then in this region the growth rate can be approximately replaced with its average value γ_g. By assuming that $\gamma_g \gg \gamma_k^s$, it is readily found that in this region

$$W_k^s = \frac{15\, m_i\, n_0\, T_e\, \gamma_g\, (k_g + \Delta k_g - k)}{T_i\, 4\pi k_g^3} \sqrt{\frac{T_e}{m_i}} , \qquad (4.233)$$

while the turbulence generating power is

$$Q = \int_{k_g}^{k_g + \Delta k_g} \gamma_g W_k^s\, dk = \frac{15\, m_i n_0 T_e}{T_i\, (4\pi)\, k_g^3} \cdot \frac{\gamma_g^2\, \Delta k_p^2}{2} \sqrt{\frac{T_e}{m_i}} \qquad (4.234)$$

and

$$W_{k=k_g} = \sqrt{\frac{Q\, 15\, m_i\, n_0\, T_e\, v_s}{2 T_i\, \pi k_g^3}} . \qquad (4.235)$$

When this is matched with (4.231) we obtain the dependence of k_0 on Q:

$$\ln \frac{k_g}{k_0} = \sqrt{\frac{Q}{Q_*^s}} \; ; \tag{4.236}$$

$$Q_*^s = \frac{15}{16} \, k_g \, v_s \, n_0 \, T_e \, \frac{m_e}{m_i} \cdot \frac{T_e}{T_i} \, . \tag{4.237}$$

Therefore, the total turbulence energy can be expressed in terms of the generating power in the following manner:

$$\frac{W^s}{n_0 \, T_e} = \frac{15}{8\sqrt{2\pi}} \cdot \frac{Q}{Q_*^s} \cdot \frac{T_e}{T_i} \, \sqrt{\frac{m_e}{m_i}} \, , \tag{4.238}$$

$$k_0 = k_g \exp \left(-\sqrt{\frac{Q}{Q_*^s}} \, \right) . \tag{4.239}$$

These results are valid only when the generation of the turbulence takes place in a limited range of wave numbers; this is possible when the turbulence generation is due to high-frequency fields and other nonlinear processes. It is important that the growth rate should not depend on k, like the damping rate, and that there be a region where γ_k^s and γ_g intersect.

The damping rate, which is linearly dependent on k, does not belong to the above case when there is no intersection with the collisionless Landau damping. If for some k the growth rate is greater than the damping rate, it will be larger everywhere in the collisionless region.

When the plasma contains electrons which have a directed velocity relative to the ions, it contains a current. It is easy to see that in a constant external electric field the electrons' directed velocity is m_i/m_e times as large as the ion velocities; in such a case it is convenient to consider just the electron motion. Thus, the nonlinearity which is determined by the ions remains just as before; the electrons' directed velocity affects just γ_k^s , in which one must take into account the Doppler shift of the frequencies, where ω_k^s is replaced with $\omega_k^s - ku$, where u is the drift velocity of the electrons (the directional motion):

$$\gamma_k^s = -\sqrt{\frac{\pi}{8} \cdot \frac{m_e}{m_i}} \, k \left(v_s - \frac{(ku)}{k} \right).$$ (4.240)

If u is significantly larger than v_s then instability arises for all k, but not for all angles. To create a stationary spectrum, the nonlinear processes, in conjunction with the frequency changes, must alter the angular distribution of the fluctuations, transforming them from a generation region into an absorption region.

If the spectrum is evaluated from (4.229), then (see [220])

$$W_k = \frac{\text{const}}{k} \ln \frac{k_0}{k}.$$ (4.241)

But a region for absorption of fluctuations appears only for small k, when the absorption effects due to particle collisions come into play. According to [130] the damping of ion-acoustic oscillations due to ion—ion collisions takes the form

$$\nu_s = \frac{2\sqrt{2} \, \Lambda \omega_{pe}^4}{15 \, (2\pi)^{3/2} \, n_0 \, v_{Te}^3} \left(\frac{m_e}{m_i} \cdot \frac{T_e}{T_i} \right)^{1/2}$$ (4.242)

This is the damping found in a completely ionized plasma. In a weakly ionized plasma the damping is determined by collisions between ions and neutral particles [220]. The magnitude of those k for which collisions are important can be estimated from

$$k_\nu \simeq \frac{\nu_s}{v_{Te}} \cdot \frac{m_i}{m_e} \sqrt{\frac{8}{\pi}}.$$ (4.243)

The spectrum presented above in (4.231) is possible when $k_0 \gg k_\nu$. In the region where collision damping is dominant, we have

$$W_k^s = \frac{15 \, \nu_s \, m_i \, n_0 \, v_s \, T_e}{2\pi T_i \, k} \left(\frac{1}{k_*} - \frac{1}{k} \right).$$ (4.244)

Assume that $k_0 \ll k_\nu$. Matching (4.244) with (4.231) we find

$$\frac{1}{k_*} - \frac{1}{k_\nu} \simeq \frac{1}{k_\nu} \ln \frac{k_\nu}{k_0},$$ (4.245)

i.e., k_* can be smaller than k_y only by a factor of $\ln(k_y/k_0)$. In other words, when $k < k_y$ the spectrum goes rapidly to zero. Thus, the fundamental scale $1/k_*$ of the turbulent fluctuations in this case is of order $1/k_y$. The same estimate is valid in the presence of directional drift of the electrons [220]. However, according to [137] an external electric field can create an anisotropic distribution of the ion-acoustic oscillations with an intensity maximum along the field direction. The maximum fluctuation amplitudes are determined primarily by the nonlinear effects in this direction.

Correlations of ion-acoustic oscillations lead to a three-plasmon decay interaction [142]:

$$s \to s' + s''. \tag{4.246}$$

This same process (4.246) can cause a broadening of the correlation resonance curve. The decay processes are determined by the values of the real parts of S_{k, k_1, k_2}, quantities which are hardly affected by the saturation of resonances as treated in Chapter 2. Thus,

$$S^{(e)}_{k, k_1, k_2} \simeq -\frac{e^3 \omega n_0}{2k_1 k_2 kT_e^2}. \tag{4.247}$$

We shall show that the correlation broadening due to the process in (4.248) is greater, for an isotropic turbulence, than the broadening resulting from induced scattering on ions. The nonlinear decay growth rate γ_k^N is

$$\gamma_k^N = \mathrm{Im}\, \frac{S^2 \delta}{\left(\dfrac{\partial \varepsilon_k^i}{\partial \omega}\right)^2 \omega^3 \dfrac{\partial \varepsilon_k^i}{\partial \omega} (\omega - \omega_1 - \omega_{k-k_1})}, \tag{4.248}$$

where δ is the relative phase volume in which the resonance decay condition can be fulfilled. The prohibition on decay in the region where $k^2 d_e^2 \ll 1$ is connected with the fact that the oscillation frequencies differ from kv_s by a factor of order $k^2 d_e^2$. Therefore, in order to have decay it is necessary that

$$\gamma_k^N > \omega k^2 d_e^2. \tag{4.249}$$

Under these conditions the resonance denominator is of order γ_k^N;

δ is also of order γ_k^N. In fact, for waves propagating in the same direction this identity is satisfied for any k_1 and k:

$$kv_s \equiv k_1 v_s + v_s (k - k_1) \qquad (4.250)$$

resonance is destroyed only because of a difference in the angles for k_1 and k. When $k_1 \sim k$ the resonance denominator is of the order $\theta^2 \omega_1$, where θ is the angle between k_1 and k. Resonance is lost when

$$\theta^2 \omega \sim \gamma_k^N. \qquad (4.251)$$

The relative phase volume in which the decay is allowed is of order γ_k^N / ω, i.e., γ_k^N separates out of the right side of (4.248). We then have

$$\gamma_k^N \simeq \omega \, \frac{W}{n_0 T_e}. \qquad (4.252)$$

This result is valid only for small k which satisfy the condition

$$k \ll \frac{\omega_{pe}}{v_{Te}} \sqrt{\frac{W}{n_0 T_e}}. \qquad (4.253)$$

We note also that (4.252) determines the order of magnitude of the intensity of the nonlinear interaction, which, because the decay is forbidden, is T_e / T_i times greater than the scattering on ions.

When the inequality in (4.253) is satisfied the decay resonance acts like a δ-function; therefore one can write the usual decay equations while neglecting the difference between the frequencies and kv_s. For clarity we shall write them as the balance equations for the number of ion-acoustic quanta [142]:

$$\frac{\partial N_k^s}{\partial t} = \frac{\pi \omega_k^s}{2n_0 \, m_i} \int \frac{dk_1 \, | \, k - k_1 \, |}{(2\pi)^3} \, \{ \delta \, (\omega_k^s + \omega_{k_1}^s - \omega_{k-k_1}^s) \times$$

$$\times 2 \, (N_{-k_1}^s N_{k-k_1}^s + N_k^s N_{k-k_1}^s - N_k^s N_{-k_1}^s) + (N_{k_1}^s N_{k-k_1}^s -$$

$$- N_k^s N_{k-k_1}^s - N_k^s N_{k_1}^s) \, \delta \, (\omega_k^s - \omega_{k_1}^s - \omega_{k-k_1}^s) \}. \qquad (4.254)$$

An expression for the probability of a decay process [the coefficient in (4.254)] is easily obtained from (4.251); it can also be directly obtained from the hydrodynamic equations. If the turbulent fluctuations are isotropic, Eq. (4.254) gives

$$\frac{\partial N_k^s}{\partial t} = \int_0^k \frac{k_1^2 (k-k_1)^2 \, dk_1}{8\pi^2 n_0 m_i} \left(N_{k_1}^s N_{k-k_1}^s - N_k^s N_{k-k_1}^s - N_k^s N_{k_1}^s \right) +$$

$$+ 2 \int_0^{k_{\max}} \frac{k_1^2 (k+k_1)^2 \, dk_1}{8\pi^2 n_0 m_i} \left(N_{k_1}^s N_{k+k_1}^s + N_k^s N_{k+k_1}^s - N_k^s N_{k_1}^s \right). \tag{4.255}$$

Here $k_{\max}$ corresponds to (4.253). A detailed analysis of the solution to (4.255), including damping and the buildup of (4.240), is given in [142]. Here we shall consider in a qualitative fashion the effects described by the interaction in (4.255), under the assumption that the largest part of the vibrational energy is concentrated in the region where $k \sim k_0$. When $k \gg k_0$, i.e., in the asymptotic region, the interaction of fluctuations with fluctuations of the energy-containing region is easily found by expanding (4.255) in terms of the small parameter k_0/k, assuming that the terms containing N_k^s are small:

$$\frac{\partial N_k^s}{\partial t} = \frac{1}{k^2} \cdot \frac{\partial}{\partial k} D k^4 \frac{\partial N_k^s}{\partial k}, \quad D = \int_{k_{\min}}^k \frac{k_0^4 \, dk_0 \, N_{k_0}^s}{4\pi^2 n_0 m_i}. \tag{4.256}$$

Strictly speaking, (4.256) is valid when $k_0/k \ll 1$, but for qualitative estimates one can extend the integration to $k_0 \simeq k$. By balancing the Landau damping with the nonlinear interaction of (4.256), we find that when $\ln (k_* /k) \gg 1$,

$$\frac{W_k^s}{n T_e} = \frac{1}{4} \sqrt{\frac{\pi}{2} \frac{m_e}{m_i}} \frac{1}{k} \ln \frac{k_*}{k}. \tag{4.257}$$

This spectrum has a structure which is similar to that of (4.231); however (if changes in the logarithmic factor are not included) the large factor $(15/\pi)(T_e/T_i)$ is missing. The decrease in the energy density of the ion-acoustic waves is related to the amplification of the nonlinear interaction and is found to be in accord with the measured width of the collisionless shock waves. In the presence of electron drift in (4.257) in the generation region, there appears a factor of order u/v_s, the spectrum becomes aniso-tropic, and the logarithmic factor is altered [142]. The diffusion in (4.256) is bounded from above by a "wall" in wave-number space, as is clear from (4.253). The diffusion causes the spec-trum to be transformed toward smaller wave numbers, with the simultaneous diffusion with respect to angles, finally transforming the fluctuations into a region which is outside the Cerenkov cone $\cos \theta = v_s/u$, where they can then be absorbed. The diffusion pace for the angles is determined by (4.251). The spectrum in (4.257) corresponds to the same type which has been measured in a strong electric field [151] and at the front of shock waves [150]. The pre-diction of anisotropy also corresponds to the observations: the main distribution of the oscillations is in the direction of the elec-tric field. Using the angular distribution of the fluctuations [142], one can find the magnitude of the anomalous plasma conductivity

$$\sigma = \frac{\omega_{pe}^2}{4\pi \nu_{\text{eff}}}, \quad \nu_{\text{eff}} \simeq \frac{u}{v_s} \frac{\omega_{pi}}{100}; \qquad (4.258)$$

this is in good agreement with observation [150, 151].

In a relatively weak electric field, such that $E^2/8\pi n T_e < (m_e/m_i)^2 \cdot 10^{-3}$, it was shown in [137] that in the quasilinear approx-imation the oscillations are concentrated around the modulus at which $k \sim k_0$; however the anisotropies will remain and the energy of the oscillations in the direction of the external electric field grows in time until it is limited by the nonlinear effects.

4.9. The Effect of a Magnetic Field on the Interactions and Spectra of Longitudinal Plasma Oscillations

A magnetic field results both in new oscillation branches and in alterations in the interactions of oscillations within a given branch. When dealing with high-frequency fluctuations, one may use Eqs. (2.242)-(2.244) for the nonlinear currents. For longitudinal waves we find

$$S_{k,\,k_1,\,k-k_1} = \frac{e\,|k-k_1|\,\left(\varepsilon^{l\,(e)}_{k}\frac{}{}_{k_1}-1\right)}{8\pi m_e\,kk_1\left(\omega^2-\omega^2_{He}\right)} \times$$

$$\times\left\{\,(kk_1)\omega - \frac{\omega^2_{He}}{\omega}(k_1 h)\,(kh) + i\omega_{He}\,([kk_1]\,h)\right\}. \tag{4.259}$$

Here $h = H_0/H_0$. Equation (2.238) gives the probability of scattering on ions by means of a virtual longitudinal wave.

Using these expressions, it is not difficult to obtain the following equation for the nonlinear interactions between longitudinal waves:

$$\frac{\partial N^\sigma_k}{\partial t} = -\int \frac{\left(\omega^\sigma_k-\omega^\sigma_{k_1}\right)\omega^4_{pe}\,N^\sigma_k\,N^\sigma_{k_1}\,dk_1\left((\omega^\sigma_k)^2-\omega^2_{He}\right)^{-1}}{|k_z-k_{1z}|\,v_{Ti}\left(1+\dfrac{T_e}{T_i}\right)^2 n_0 T_i\,(2\pi)^{5/2}\,\omega^\sigma_k\,\omega^\sigma_{k_1}\left.\dfrac{\partial\varepsilon^l_k}{\partial\omega}\right|_{\omega=\omega^\sigma_k}} \times$$

$$\times\frac{\left((\omega^\sigma_{k_1})^2-\omega^2_{He}\right)^{-1}}{\left.\dfrac{\partial\varepsilon^l_{k_1}}{\partial\omega_1}\right|_{\omega_1=\omega^\sigma_{k_1}}}\left\{\omega^2_{He}\frac{([kk_1]\,h)^2}{k^2 k^2_1}+\left(\omega^\sigma_k\frac{(kk_1)}{kk_1}-\frac{\omega^2_{He}}{\omega^\sigma_k}\cdot\frac{(k_1 h)\,(kh)}{kk_1}\right)^2\right\}. \tag{4.260}$$

Here $\left|\omega^\sigma_k-\omega^\sigma_{k_1}-v\omega_{Hi}\right|\ll\left|k_{1z}-k_z\right|v_{Ti}$. For unmagnetized ions ($kv_{Ti}\gg\omega_{Hi}$) this last inequality is satisfied for all v if $|\omega^\sigma_k-\omega^\sigma_{k_1}|$ is sufficiently small. For magnetized ions the inequality is fulfilled only for $v = 0$. However, it is easily seen that for magnetized ions the cyclotron scattering with $v \neq 0$ is small, which means that (4.260) is valid in this case.

It is not hard to obtain expressions for the spectral transfer. In a weak magnetic field in which $\omega_{He}\ll\omega_{pe}$, the transfer in the branch $\omega^\sigma_k\approx\omega_{pe}$ agrees approximately with (4.12), while for the branch $\omega^\sigma_k=\omega_{He}\cos\theta$ we find

$$\frac{\partial N^\sigma_k}{\partial t} = -N^\sigma_k\int\frac{N^\sigma_{k_1}\,dk_1\left(\omega^\sigma_k-\omega^\sigma_{k_1}\right)\omega^2_{He}}{4(2\pi)^{5/2}|k_z-k_{1z}|\,v_{Ti}\,n_0 T_i\left(1+\dfrac{T_e}{T_i}\right)^2} \times$$

$$\times\left\{\left(\cos\theta\,\frac{(kk_1)}{kk_1}-\cos\theta\cos\theta_1\right)^2+\frac{1}{k^2 k^2_1}\,([kk_1]\,h)^2\right\}. \tag{4.261}$$

In a strong field $(\omega_{He} \gg \omega_{pe})$ for the branch where $\omega_k^\sigma = \omega_{pe} \cos\theta$, we have the following expression for magnetized ions $(kv_{Ti} \ll \omega_{Hi})$:

$$\frac{\partial N_k^\sigma}{\partial t} = -N_k^\sigma \int \frac{(\omega_k^\sigma - \omega_{k_1}^\sigma)\omega_{pe}^2 \cos^2\theta \, |\cos^2\theta_1| \, N_{k_1}^\sigma \, dk_1}{|k_z - k_{1z}| \, v_{Ti} \left(1 + \dfrac{T_e}{T_i}\right)^2 n_0 \, T_i \, 4 \, (2\pi)^{5/2}} \times$$

$$\times \exp\left\{-\frac{(\omega_k^\sigma - \omega_{k_1}^\sigma)^2}{2 \, (k_z - k_{1z})^2 \, v_{Ti}^2}\right\}. \tag{4.262}$$

When treating unmagnetized ions $|k_z - k_{1z}|$ must be replaced with $|k - k_1|$. Similarly, for the branch where $\omega_k^\sigma \approx \omega_{He}$, we have $(kv_{Ti} \ll \omega_{Hi})$

$$\frac{\partial N_k^\sigma}{\partial t} = -N_k^\sigma \int \frac{(\omega_k^\sigma - \omega_{k_1}^\sigma) \, N_{k_1}^\sigma \, dk_1 \, \omega_{pe}^4 \sin^2\theta_1 \sin^2\theta}{4 \, (2\pi)^{5/2} \, |k_z - k_{1z}| \, v_{Ti} \left(1 + \dfrac{T_e}{T_i}\right)^2 \omega_{He}^2} \times$$

$$\times \exp\left\{-(\omega_k^\sigma - \omega_{k_1}^\sigma)^2/2 \, |k_z - k_{1z}|^2 \, v_{Ti}^2\right\}. \tag{4.263}$$

We note that the spectral transfer of (4.262) decreases as the angle between the Langmuir waves and the magnetic field increases, i.e., the magnetic field inhibits the isotropization of the flucutations. It is significant that the ion scattering in a magnetic field can take place not only when the frequency difference is small but even when it is sufficiently close to $v\omega_{Hi}$.

But the Langmuir fluctuations can be made isotropic even in an infinitely strong magnetic field. In fact, the limitation on the spread of angles for those waves which effectively interact with one another is given by an exponential factor. Consider the interaction of waves for which $k = k_1$. The exponential factor then has the form

$$\exp\left\{-\frac{\omega_{pe}^2}{2k^2 \, v_{Ti}^2} \left(\frac{|\cos\theta| - |\cos\theta_1|}{\cos\theta - \cos\theta_1}\right)^2\right\}. \tag{4.264}$$

Thus, only those waves moving in opposite directions effectively interact with each other. As a result of the two-stage process in such an interaction the change in the angle of the waves is $(\Delta\theta) \sim$

$(kv_{Ti}/\omega_{pe})(\cos\theta/\sin\theta)$. If we do not consider angles close to $\pi/2$, we can say that the isotropization process is slowed down by a factor of v_p/v_{Ti}.

An increase in the angle between the magnetic field and the fluctuations can produce a significant reduction in the frequency and a slipping of the fluctuations into a hybrid resonance with $\omega = \sqrt{\omega_{Hi}\omega_{He}}$, which takes place when $|\theta - \pi/2| < (m_e/m_i)^{1/2}$. Although this effect can produce marked turbulent heating of the plasma, there is a parallel increase in the phase velocities of the fluctuations which tends to inhibit the heating. The one-dimensional transfer of the Langmuir waves energy along H for ionic scattering is small when compared with the same transfer in the case of isotropic turbulence without a magnetic field. When the interaction is accomplished by means of magnetized particles $(\nu = 0)$, a new effect called saturation of the interaction appears [221]. It appears quite clearly for electrons during the transfer of energy between waves propagating along H. Then only those electrons for which $v_z \approx v_{Te}^2/v_p \ll v_{Te}$ participate in the interactions. The number of these electrons is small when compared with the total number of electrons, and the energy transferred by them in the interaction $(\sim W)$ can be comparable to their energy. This means that for low velocities there can be a plateau with respect to v_z in the electron distribution function, which implies that the energy transfer, being proportional to $\partial f^R/\partial v_z$, is sharply reduced.

Such effects begin when $v_p \gg 3v_{Te}^2/v_{Ti}$ for ions. They are not possible in the region of differential spectral transfer, for then all the ions participate in the interaction. Similarly, when $v_p \gg 3v_{Te}^2/v_{Ti}$ the transfer with respect to angles can be accomplished by all the ions, and this means that the transfer of the oscillations into the region of hybrid resonance is not very sensitive to these saturation effects. The magnetic field, by creating a new region for effective absorption in the hybrid resonance, has provided a new channel for dissipating turbulent energy; it has opened up the possibility of turbulent plasma heating. Another possibility for heating is related to a change in the direction of the spectral transfer in the case of anisotropically distributed electrons and ions. The appearance of such distributions is much more probable in the presence of magnetic fields. Because of the loss cone, they arise naturally in magnetic traps. An anisotropic instability can develop during a period which is much longer than the characteristic times for transferring the energy along the spectrum; therefore, the fluctuations can be absorbed more rapidly than the anisotropic instability can develop. A change in the transfer direction means that the oscillations can be dissipated by Landau absorption. In order to get a change in the transfer direction for scattering on electrons, our estimates show that it requires a degree of anisotropy, defined such that $\alpha \sim \Delta\theta/\theta$, which satisfies the inequality $\alpha_e > v_{Te}/v_p$; for ions the requirement translates into the condition $\alpha_i > v_{Te}^2/v_p v_{Ti}$.

The most advantageous situation is found for traps with hot ions when the loss-cone instability corresponds to the Cerenkov excitation by ions of Langmuir oscillations of frequency $\omega_{pe}\cos\theta$. Scattering on ions is ineffective. The argument is the same as that which shows that scattering on electrons is ineffective for an ion-acoustic instability. Thus, the nonlinear transformation of the energy is accomplished by the electrons. For them the conditions for changing the transfer direction are somewhat eased. In traps with hot ions one can satisfy the conditions necessary for the turbulent heating of the electrons. But the ions are found in a . :-gime of statistical acceleration, so only a small fraction of them can obtain energy from the oscillations.

We postpone to Chapter 6 the discussion of the effects due to interactions of oscillations with frequency $\omega \sim \omega_{He}$, because the strongest nonlinear effect for them is the transformation of the oscillations into electromagnetic radiation.

The interaction of unmagnetized ion-acoustic oscillations remains about the same as it is in the absence of a magnetic field, so that all the computations of §4.8 remain valid. However, we can point out two new and important aspects. First, there is a strong absorption of the fluctuations by the ions near the ion-cyclotron resonance at $k \sim \omega_{Hi}/v_s$. This absorption is the mechanism for the effective turbulent heating of the ions.

The practical questions concerning the effectiveness of the turbulent heating of the ions in an ion-acoustic instability, and the relationship between heating of electrons and heating of ions depend on the magnitude of the magnetic field. In order to heat the ions it is essential that during the course of transferring the energy to wave numbers of order ω_{Hi}/v_s, the energy of the fluctuations is not lost by electron absorption. The energy transfer is accomplished by scattering by the ions; therefore, in order to conserve the number of quanta the change $\Delta\omega$ in their frequencies of order ω leads to ion heating in the amount W. The time required for this increase is $\frac{1}{\gamma_k^s} \sim \frac{T_e}{T_i} \cdot \frac{1}{\omega_k^s} \cdot \frac{n_0 T_e}{W^s}$, i.e., $\frac{dT_i}{dt} \sim \omega_k^s T_i \cdot \left(\frac{W^s}{n_0 T_e}\right)^2$. But the heating of electrons because of nonlinear damping is estimated to be $\frac{dT_e}{dt} \approx \gamma_k^s \frac{W^s}{n_0} \sim \omega_k^s \sqrt{\frac{m_e}{m_i}} T_e \left(\frac{W^s}{n_0 T_e}\right)$. We may therefore conclude that ion heating can be more important when $W^s/n_0 T_e > (m_e/m_i)^{1/2}$. A magnetic field produces an additional energy drain in the neighborhood of the ion-cyclotron resonance. Here the plasmons are absorbed by the ions, which thereby increase their energies. The time needed for this transformation is $(1/\omega_{Hi})(n_0 T_e^2/T_i W)$ and decreases as the magnetic field increases.

The second new aspect is that the magnetic field now has a significant influence on the ion-acoustic turbulence because of the appearance of a channel by which the oscillational energy can be turned into whistlers. As pointed out in [222, 223] this can serve as an effective method for diagnosing ion-acoustic turbulence. Finally, magnetized acoustic oscillations for which $k \ll \omega_{Hi}/v_s$ interact among themselves in much the same way as do unmagnetized oscillations. In particular, the ion scattering is of the same order of magnitude in this instance as in the absence of a magnetic field.

4.10. Spectra of Turbulence

for Whistlers

Whistlers, which have the frequency

$$\omega_k^w = \frac{k^2 c^2 |\omega_{He}| |\cos\theta|}{\omega_{pe}^2 + k^2 c^2},\qquad(4.265)$$

can be excited because of the anisotropic instability due to the conversion from ion-acoustic and Langmuir oscillations, and so on. They are weakly damped when propagating primarily along the magnetic field and when $\omega_k^w > k v_{Te}$. We shall consider the long-wavelength part of the spectrum given in (4.265), in which $k \gg \omega_{pe}/c$ so that $\omega_k^w \ll \omega_{He}$. When $k \gg \omega_{pe}/c$ the spectrum of

(4.265) coincides with the spectrum for a longitudinal gyrofrequency plasma wave, and the interactions of such oscillations are described by (4.261). Thus,

$$\omega_k^w = \frac{k^2 c^2 \, |\omega_{He}| \, |\cos\theta|}{\omega_{pe}^2} .$$ (4.266)

The decay process

$$w \to w' + w''$$ (4.267)

is forbidden for this spectrum only if all the waves are propagating along H_0. In an isothermal plasma, where $T_e = T_i$, the processes of decay into ion-acoustic oscillations are not possible. If $kv_{Te}/\omega \gg 1$ the linear damping by electrons is not small (when $\theta \neq 0$) and the turbulence spectra can be formed as the result of excitation of whistlers along H following a nonlinear transfer to $\theta \neq 0$ with subsequent damping by electrons. Estimates show (see [224]) that scattering by electrons is important when

$$v_{Te} \left(\frac{m_e}{m_i} \right)^{1/4} \gg v_A > v_{Te} \sqrt{\frac{m_e}{m_i}} ,$$ (4.268)

and, just as with the ion-acoustic oscillations, scattering by electrons can be neglected in those regions where there is simultaneous linear damping and electron scattering possible. The magnetic fields determined by (4.268) lie within a very narrow range, so we shall concern ourselves here with decay interactions, ion scattering, and linear damping by electrons. Note that if the decays are allowed by the conservation laws, they will dominate the scattering by ions.

In the general case the solution of the decay equations is a rather difficult problem. However, if the decays do in fact determine the spectra, and the oscillational energy is concentrated in some region of wave numbers $k \approx k_0$, which will be called the energy-containing region, then it can be assumed that the spectra in the asymptotic region $k \gg k_0$ are determined by their interactions with the most intense oscillations found in the energy-containing region. Then the equations for the spectra describe the processes of the diffusion scattering of the fluctuations:

$$\frac{\partial N_k^w}{\partial t} = \frac{\partial}{\partial k_i} D_{ij} \frac{\partial N_k^w}{\partial k_j} ;$$ (4.269)

$$D_{ij} = \int k_{1i} k_{1j} w (kk_1) N_k^\omega \frac{dk_1}{(2\pi)^3}, \qquad (4.270)$$

where $w(kk_1)$ is the decay probability:

$$w(kk_1) = \frac{\pi}{2} \cdot \frac{\omega_{pe}^4 \omega_1 \omega^2}{|k_z| n_0 m_e \omega_{He}^3 c^2 \cos^2\theta} \delta\left(\frac{k_1 k}{k} + \frac{k_{1z} k}{k_z}\right)\frac{[k_1 k]^2}{k^2 k_1^2}. \qquad (4.271)$$

We shall not discuss the solution to Eqs. (4.270) (see [224]). Instead, we merely point out that according to (4.265), for those oscillations which make an angle θ with the magnetic field, only the energy-containing oscillations for which

$$\cos^2\theta_1 < \frac{\cos^2\theta \sin^2\theta}{1 + 3\cos^2\theta} < \frac{1}{9} \qquad (4.272)$$

interact. Therefore, if the whistlers in the energy-containing region are directed along H_0, their effect on the asymptotic vibrations turns out to be zero. In those cases where this is not the situation, the whistler spectra in the asymptotic region can take the form $\sim W(x)/\omega^\nu$, in which the exponent ν depends on the details of the oscillation distribution in the energy-containing region [224].

Let us consider in more detail the ion scattering processes. According to (2.242) and the table of unit vectors given in Chapter 2, the nonlinear current which determines the ion scattering is set up by the electrons:

$$S_{kk_1 k_2} = - \frac{\omega_{pe}^2 k_2 (\varepsilon_{k_2}^{l(e)} - 1)(1 - k_{1z} v_z/\omega_1)}{2(4\pi)^2 e n_0 (1 + \cos^2\theta)^{1/2}(1 + \cos^2\theta_1)^{1/2} \omega_{He}} \times$$

$$\times (i\cos(\varphi - \varphi_1)(|\cos\theta| + |\cos\theta_1|) - \sin(\varphi - \varphi_1)(1 + |\cos\theta||\cos\theta|)).$$

$$(4.273)$$

Here, through perturbation theory, some additional small corrections† of order $k_{1z} v_z/\omega_1$ have been added to refine (2.273). The largest matrix element of S is obtained with $\theta = \theta_1 = 0$; $\theta = 0$, $\theta_1 = \pi$; and $\theta = \pi$, $\theta_1 = 0$.

It is easily seen that the most probable process will be whistler scattering in which there is approximate equality of the fre-

---

†These terms must be included because in the first approximation with respect to this parameter, the matrix element turns out to be equal to $(\omega - \omega_1)\delta(\omega - \omega_1)$, which is zero.

quencies of the interacting waves:

$$k_1^2 |\cos \theta_1| = k^2 |\cos \theta|. \qquad (4.274)$$

Thus the turbulence is basically non-one-dimensional, and the balance between generation and absorption can be achieved by transfer with respect to angles.

For magnetized ions it is easily found that

$$\frac{\partial N_k^w}{\partial t} = N_k^w \int \frac{dk_1 N_{k_1}^w (\omega_{k_1}^w - \omega_k^w)(\omega_{k_1}^w)^2 \left(1 - 2k_{1z} \frac{\omega - \omega_1}{\omega_1(k_z - k_{1z})}\right)}{4(2\pi)^{5/2} n_0 T_i |k_z - k_{1z}| v_{Ti} \left(1 + \frac{T_e}{T_i}\right)^2} \times$$

$$\times \frac{1}{2|\cos \theta \cos \theta_1|} [(|\cos \theta| + |\cos \theta_1|)^2 + (1 + |\cos 0 \cos \theta_1|)^2] \times$$

$$\times \exp \left\{-\frac{(\omega_k^w - \omega_{k_1}^w)^2}{2v_{Ti}^2 |k_z - k_{1z}|^2}\right\}. \qquad (4.275)$$

For unmagnetized ions we must replace $|k_z - k_{1z}|$ with $|k - k_1|$ and k_{1z} by the quantity $\frac{k_1(k - k_1)}{|k - k_1|}$. Equation (4.275) does not have singularities when $\theta, \theta_1 \to \pi/2$ because $\omega_{k_1}^w$ and ω_k^w are proportional to $|\cos \theta_1||\cos \theta|$. If we introduce the distribution function $W_{\omega\Omega}$ for whistlers in terms of angles and frequencies,

$$\int W_{\omega\Omega}\, d\omega d\Omega = \int \frac{N_k^w \omega_k^w dk}{(2\pi)^3} \qquad (4.276)$$

[where because of the condition that $k \ll \omega_{pe}/c$ in (4.266), the integration over angles is restricted by the constraint $\cos \theta > \omega/\omega_{He}$], Eq. (4.275) can be written as

$$\frac{\partial W_{\omega, \Omega}}{\partial t} = a^w W_{\omega, \Omega} \int d\Omega_1 \frac{1}{|\cos \theta||\cos \theta_1|} [(|\cos \theta| + |\cos \theta_1|)^2 +$$

$$+ (1 + |\cos \theta \cos \theta_1|)^2] \left[\omega^2 \left(\frac{\partial}{\partial \omega} W_{\omega\Omega_1} \left(\frac{\cos \theta}{\sqrt{|\cos \theta|}} - \frac{\cos \theta_1}{\sqrt{|\cos \theta_1|}}\right)\right)^2 +$$

$$+ \omega W_{\omega\Omega_1} \frac{\cos \theta}{\sqrt{|\cos \theta|}} \left(\frac{\cos \theta}{\sqrt{|\cos \theta|}} - \frac{\cos \theta_1}{\sqrt{|\cos \theta_1|}}\right)\right], \qquad (4.277)$$

where

$$\alpha^w = \frac{\pi}{4} \cdot \frac{\omega_{pe}^2}{\omega_{He} n_0 m_i c^2 \left(1 + \dfrac{T_e}{T_i}\right)^2}. \qquad (4.278)$$

Note that the ions can be assumed magnetized, $kv_{Ti}/\omega_{Hi} < 1$, up to the maximum k in the spectrum of (4.265) if $v_A \gg v_{Te}$. If $v_{Ti} \ll v_A \ll v_{Te}$ the magnetization condition is fulfilled if $\omega < \omega_{Hi} \cdot v_A^2/v_{Ti}^2$. The equation similar to (4.277), which describes the scattering on unmagnetized ions, has the form

$$\frac{\partial W_{\omega\Omega}^w}{\partial t} = \alpha^w W_{\omega\Omega}^w \int d\Omega_1 \frac{[(|\cos\theta| + |\cos\theta_1|)^2 + (1 + |\cos\theta\cos\theta_1|)^2]}{|\cos\theta||\cos\theta_1|} \times$$

$$\times \left[\omega^2 \frac{\partial}{\partial\omega} W_{\omega\Omega_1}^w \left(\frac{1}{|\cos\theta_1|} + \frac{1}{|\cos\theta|} - \frac{2\cos\theta\cos\theta_1}{\sqrt{|\cos\theta\cos\theta_1|}} \right) + \right.$$

$$\left. + \omega W_{\omega\Omega_1}^w \left(\frac{1}{|\cos\theta|} - \frac{\cos\theta\cos\theta_1}{\sqrt{|\cos\theta\cos\theta_1|}} \right) \right]. \qquad (4.279)$$

It is easy to see that the stationary solutions to (4.279) must be of the form

$$W_{\omega\Omega}^w = \frac{1}{\omega^\nu} W(\cos\theta). \qquad (4.280)$$

From this we find

$$\nu \int_{-1}^{1} W(x_1) dx_1 \frac{1}{|x_1|} [(|x| + |x_1|)^2 + (1 + |xx_1|)^2] \left(\frac{x}{\sqrt{|x|}} - \frac{x_1}{\sqrt{|x_1|}} \right)^2 =$$

$$= \int_{-1}^{1} W(x_1) dx_1 \frac{1}{|x_1|} \cdot \frac{x}{\sqrt{|x|}} \left(\frac{x}{\sqrt{|x|}} - \frac{x_1}{\sqrt{|x_1|}} \right) [(|x| + |x_1|)^2 + (1 + |xx_1|)^2]. \qquad (4.281)$$

With the same assumptions as for (4.277), the equation which replaces (4.281) for scattering by unmagnetized ions takes the form

$$\nu \int_{-1}^{1} W(x_1) dx_1 [(|x| + |x_1|)^2 + (1 + |xx_1|)^2] \frac{1}{|x_1|} \left(\frac{1}{|x|} + \frac{1}{|x_1|} - \frac{2xx_1}{\sqrt{|x||x_1|}} \right) =$$

$$= \int_{-1}^{1} W(x_1)\, dx_1\, [(|x|+|x_1|)^2 + (1+|xx_1|)^2] \frac{1}{|x_1|} \left(\frac{1}{|x|} - \frac{xx_1}{\sqrt{|xx_1|}} \right)$$

$$(4.282)$$

For a given ν, Eqs. (4.281) and (4.282) provide integral conditions on $W(x_1)$ which limit the possible kinds of angular dependences the turbulence spectra can have. It is clear from (4.281) and (4.282) that the angular dependence becomes a δ-function; that is, all the waves are concentrated in a specific direction (say, form a "stream").

When $\nu = 1/2$ the wave packets along the field and against the field are stable, but packets propagating at an angle $x = x_0$ are unstable relative to exciting oscillations with $|x| > |x_0|$. This means that packets with $|x_0| \neq 1$ are concentrated in two beams, one propagating along the field and one against the field. For $\nu = 1$ packets with $|x| = |x_0|$ are unstable relative only to exciting waves for which $x > -|x_0|$, while the packets with $x = -|x_0|$ are unstable relative to waves with $x < -|x_0|$.

Thus, spectra with $\nu = 1/2$ must transform into stable spectra propagating along the field. This effect is the self-channeling of whistlers along the magnetic field, where there is no linear damping. It is of interest for interpreting geophysical phenomena in the earth's magnetosphere, where the excitation of whistlers plays an important role; it is also of interest with respect to laboratory experimentation. We must also mention that for propagation along the field the decay interactions are unimportant.

Let us now consider the role of damping due to electrons. This damping is significant when $\omega < k v_{Te}$, i.e.,

$$\omega < \omega_{Hi} \frac{v_{Te}^2}{v_A^2},$$

$$(4.283)$$

and takes the form

$$\gamma \simeq -\frac{\sqrt{2\pi}}{2} \cdot \frac{\omega^{3/2}}{|\omega_{He}|^{3/2}} \cdot \frac{v_{Te}}{c} \omega_{pe} \frac{1-x^2}{\sqrt{|x|}}.$$

The maximum damping in the region of (4.283) is estimated as

$$\gamma_{k\,max}^{\omega} \simeq \omega_{Hi} \frac{m_i}{m_e} \left(\frac{v_s}{v_A} \right)^4,$$

while the characteristic time for spectral transfer by an amount $\Delta\omega \sim \omega$ is of the order

$$\gamma^N \sim \omega_{Hi} \frac{W}{n_0 m_i v_A^2} \sim \omega_{Hi} \frac{4\pi W}{H_0^2} \tag{4.284}$$

If $W < H_0^2/4\pi$ and $v_A < v_s(m_i/m_e)^{1/4}$ the nonlinear interaction will not be able to balance the damping, and this means that the oscillations will be absorbed. Thus, with $v_A < v_s(m_i/m_e)^{1/4}$ under the conditions where the generation of oscillations takes place for $\omega > \omega_{Hi} v_{Te}^2/v_A^2$, the whistlers are transferred to $\omega \sim \omega_{Hi} v_{Te}^2/v_A^2$ where they will be absorbed. The energy of the oscillations is concentrated near $\omega \simeq \omega_{Hi} v_{Te}^2/v_A^2$, which corresponds to the wave vector

$$k = k_0 = \frac{\omega_{pi}}{c} \cdot \frac{v_{Te}}{v_A} . \tag{4.285}$$

k_0 acts as the fundamental scale for the turbulence. However, we must also be aware of the fact that the oscillations are damped by the electrons, causing the electrons to heat. This heating then leads to the condition $T_e \gg T_i$ and this brings into play more powerful interactions between the whistlers and the ion-acoustic oscillations (see below).[†]

It is now easy to see that the damping rate will be of order ω_{Hi}, if the frequency of the whistlers is of order

$$\omega \sim \omega_* \sim \omega_{Hi} \left(\frac{v_A}{v_s}\right)^{2/3} \left(\frac{m_i}{m_e}\right)^{1/3} \tag{4.286}$$

This quantity is larger than $\omega_{Hi} v_A^2/v_{Ti}^2$ when $v_A < v_s(m_i/m_e)^{1/4}$. When $\gamma^N \gtrsim \omega_{Hi}$, the linear damping rate can be compensated by the nonlinear effects in (4.284) if they are sufficiently strong.

Let us now consider a nonisothermal plasma with $T_e \gg T_i$ in which there exist ion-acoustic waves. Under these conditions a new channel for energy transfer appears even without ion-acoustic oscillations present. The question naturally arises, will ion-acoustic waves be excited in this case? If there is a source for exciting ion-acoustic oscillations, those waves will then be able to

---

[†]The interaction $w \to s$, as it turns out, does not lead to the generation of s-waves for the spectra of (4.280); instead, it generates w-waves from the s-waves.

transfer their energy to the whistlers, so that the spectra of both
the whistlers and the ion-acoustic oscillations will depend in some
measure on their interaction. This situation arises because the
frequencies of the whistlers and the ion-acoustic oscillations can
be identical. There will always be regions in which their frequencies overlap if $\omega_{pi} > \omega_{Hi}$ or

$$\frac{v_A}{c} < 1. \tag{4.287}$$

When the frequencies coincide, induced scattering by electrons and ions now becomes effective. Decay processes are also
possible. However, they are allowed only in a relatively narrow
range of frequencies and angles (see below). If the ion-acoustic
oscillations are isotropic, scattering by ions can be obtained by
using (2.242) (see also [225]):

$$\frac{\partial W^w_{\omega,\,\Omega}}{\partial t} = \frac{W^w_{\omega,\,\Omega}\,\pi\,(1-\cos^2\theta)\,\omega}{2n_0\,m_i\,v^2_A\left(1+\dfrac{T_e}{T_i}\right)^2}\,\frac{\partial}{\partial\omega}\,\omega^2\,W^s_\omega, \tag{4.288}$$

$$\frac{\partial W^s_\omega}{\partial t} = W^s_\omega\,\frac{\pi}{2}\int\frac{d\Omega\,(1-\cos^2\theta)\,\omega^3}{n_0\,m_i\,v^2_A\left(1+\dfrac{T_e}{T_i}\right)^2}\cdot\frac{\partial}{\partial\omega}\,W^w_{\omega,\,\Omega}. \tag{4.289}$$

If ion-acoustic turbulence is present, but there are no whistlers, we may conclude that there is the possibility of a buildup
of the whistlers due to the ion-acoustic oscillations. In fact, if
we use the spectrum for ion-acoustic fluctuations given in (4.238),

$$W^s_\omega = \frac{2W^s}{\ln^2\dfrac{\omega_*}{\omega_0}}\cdot\frac{1}{\omega}\,\ln\frac{\omega}{\omega_0}, \tag{4.290}$$

we will then, according to (4.288), obtain a buildup of the whistlers.
This means that ion-acoustic turbulence is accompanied by whistler turbulence, and the complete stationary turbulent state in a
magnetic field must reflect this fact. The characteristic times
for transfering whistlers are determined by the relationship for
the growth rate of their excitation:

$$\gamma^N_k \simeq \omega^w_k\left(\frac{T_i}{T_e}\right)^2\frac{W^s}{n_0\,m_i\,v^2_A}. \tag{4.291}$$

We should also point out that the maximum effects appear when $\omega_{He} = \omega_{pi}$ or when

$$\frac{v_A}{c} \sim \frac{m_e}{m_i}. \tag{4.292}$$

If the magnetic field is reduced, all s-waves, whose frequencies are greater than ω_{He}, adjust their spectra because of the s → s interaction, and the picture remains just as if there were no magnetic field. The buildup of whistlers continues until the output of energy from w → w scattering no longer balances the energy uptake from the s-waves, or until the intensity of the s-waves does not decrease because of the generation of w-waves. If the interaction in (4.289) becomes stronger than the linear damping of s-waves and the s → s interaction, and (4.288) exceeds the interaction given in (4.275), then the stationary condition corresponds to setting the right sides of (4.288) and (4.289) equal to zero, i.e.,

$$W_\omega^s = \frac{\text{const}}{\omega^2}, \qquad W_{\omega,\,\Omega}^w = \text{const}. \tag{4.293}$$

Now consider the decay processes. In principle the following interactions between whistlers and ion-acoustic oscillations are possible: $w \rightleftarrows w' + s$; $w + w' \rightleftarrows s$; $s \rightleftarrows s' + w'$; $w \rightleftarrows s + s'$. In practice the first and second processes are forbidden by the conservation laws. From the equation

$$|\mathbf{k}_1 \pm \mathbf{k}_s| |k_{1z} \pm k_z^s| \mp k_1 |k_{1z}| = \frac{\omega_{pi}}{c} \cdot \frac{v_s}{v_A} k_s \tag{4.294}$$

and the fact that the maximum value of k_s cannot be larger than $2k_1$, the left side of (4.294) is less than, or of the order of, $2k_1k_s$. That is, $k_1 < \frac{1}{2}(\omega_{pi}/c)(v_s/v_A)$, which is not possible when $v_A > v_s$. For the process $s \rightleftarrows s' + w$ we find

$$|\mathbf{k}_1 + \mathbf{k}| - k_1 = \frac{k^2 v_A c |\cos \theta|}{\omega_{pi} v_s}. \tag{4.295}$$

Since the k for whistlers is much smaller than for ion-acoustic oscillations, and $v_A \gg v_s$, (4.295) can be written as

$$\cos \theta \cos \theta_1 + \sin \theta \sin \theta_1 \cos(\varphi - \varphi_1) = \frac{kc}{\omega_{pi}} \cdot \frac{v_A}{v_s} |\cos \theta|. \tag{4.296}$$

Since $kc/\omega_{pi} \gg 1$, $v_A/v_s \gg 1$ the right side of (4.296) is large, but the left side is some quantity less than unity. Therefore, $|\cos \theta| \ll 1$. On the other hand, $\cos(kk_1)$ cannot be small, i.e., the ion-acoustic waves will propagate at an angle of nearly $\pi/2$ relative to the direction of the magnetic field. The total phase volume for such oscillations is small, and they do not play any significant role.

The only remaining process is $s + s' \rightleftarrows w$, and the probability for this one is easily found from (2.242). The expression for the probability is a maximum when k_1, $k_2 \approx \omega_{pe}/v_{Te}$ and then takes the form

$$w_w^{s,\,s}\left(\mathbf{k},\,\mathbf{k}_1,\,\mathbf{k}_2\right) = (2\pi)^4\,\delta\left(\mathbf{k}_1 + \mathbf{k}_2 - \mathbf{k}\right)\,\delta\left(\omega_{\mathbf{k}_1}^s + \omega_{\mathbf{k}_2}^s - \omega_{\mathbf{k}}^w\right) \times$$

$$\times\,\frac{\sin^2\theta\,\omega_{\mathbf{k}}^w\,\omega_{\mathbf{k}_1}^s\,\omega_{\mathbf{k}_2}^s}{n_0 m_i v_A^2\left(1 + k_1^2\,d_e^2\right)\left(1 + k_2^2\,d_e^2\right)}. \tag{4.297}$$

From this we obtain an estimate of the generating power for whistlers when their reabsorption by the inverse process $w \rightarrow s + s'$ can be neglected for isotropic turbulence and $k_1 \gg k$:

$$\frac{\partial W_{\omega,\,\Omega}^w}{\partial t} = \frac{v_s\,\omega_{pi}^3\,\sqrt{\omega}\,\sin^2\theta}{c^3\,|\cos\theta|\,\sqrt{|\cos\theta|}\,n_0 m_i v_A^2\,\omega_{Hi}^{3/2}}\,\left(W_{\frac{\omega}{2v_s}}^s\right)^2. \tag{4.298}$$

The total energy generated in 1 cm^3 in 1 sec is

$$\frac{\partial W^w}{\partial t} = \left(\frac{v_{Te}}{c}\right)^3\,\omega_{pi}\,\frac{\omega_{pi}^3}{\omega_{He}^3}\,\frac{(W^s)^2}{n_0 m_i v_A^2}. \tag{4.299}$$

The observation of the frequency $2\omega_{pi}$ in the whistler spectrum can serve as a good method of diagnosing ion-acoustic turbulence because measurements of the amplitudes of whistler oscillations are much simpler than measuring the amplitudes of ion-acoustic waves.

4.11. The Spectra of Magnetohydrodynamic Plasma Turbulence

We now turn to the spectra of the lowest-frequency vibrations in a homogeneous plasma — the magnetohydrodynamic oscillations [226].

Assuming that $v_A \gg v_s$ in an isothermal plasma, one need consider just the interactions of Alfven waves and fast magnetoacoustic waves:

$$\omega_k^M = kv_A, \quad \omega_k^A = |k_z| v_A. \tag{4.300}$$

If we use the equations from magnetohydrodynamics, as was done in [227, 228], we can obtain the decay interactions of Alfven waves and magnetoacoustic waves. The conservation laws allow only $M \rightleftarrows M + A$, $A \rightleftarrows A + M$ decays. Here we will treat just the special case of a one-dimensional magnetohydrodynamic turbulence, where decays are forbidden when correlations are neglected. Ion scattering is then the decisive factor. The characteristic time for the interaction can be evaluated from the non-linear growth rate

$$\gamma_k^N \approx \omega \frac{W}{n_0 m_i v_A^2} = \omega \frac{4\pi W}{H_0^2} \tag{4.301}$$

and is of about the same order of magnitude as for the decay interaction of waves propagating at an angle of about unity relative to H_0. In addition to ion scattering one must also include the Landau damping of magnetohydrodynamic waves, which can affect spectra for the case $v_A < v_{Te}$.

For the processes

$$A \rightleftarrows A, \quad M \rightleftarrows M, \quad A \rightleftarrows M, \tag{4.302}$$

the nonlinear equations which describe this scattering by ions take the form

$$\frac{\partial W_\omega^+}{\partial \tau} = W_\omega^+ \omega^2 \left(\omega \frac{\partial W_\omega^-}{\partial \omega} + W_\omega^- \right), \tag{4.303}$$

$$\frac{\partial W_\omega^-}{\partial \tau} = W_\omega^- \omega^2 \left(\omega \frac{\partial W_\omega^+}{\partial \omega} + W_\omega^+ \right), \tag{4.304}$$

where

$$\tau = t\pi \left(1 + \frac{v_A^2}{c^2} \right)^{-1} \frac{1}{4(1 + T_e/T_i)^2} \cdot \frac{v_{Ti}^2}{v_A^2 n_0 T_i}. \tag{4.305}$$

Here $W_\omega = W_\omega^A + W_\omega^M$ is the sum of the spectral functions for the Alfven and magnetohydrodynamic oscillations; the plus and minus superscripts correspond to waves propagating either along the magnetic field or opposite to it. The total energy of the magnetic disturbances is

$$W = \int_0^\infty (W_\omega^+ + W_\omega^-)\, d\omega. \qquad (4.306)$$

It follows from (4.304) and (4.305) that the stationary solutions are of the form $W_\omega^\pm = \dfrac{\text{const}}{\omega}$, i.e.,

$$\nu = 1. \qquad (4.307)$$

Solutions are then possible when 1) $W_\omega^- = 0$, $W_\omega^+ = \dfrac{\text{const}}{\omega}$, 2) $W_\omega^+ = 0$, $W_\omega^- = \dfrac{\text{const}}{\omega}$, and 3) $W_\omega^+ = \dfrac{\text{const}}{\omega}$, $W_\omega^- = \dfrac{\text{const}}{\omega}$.

We will now briefly consider the case of a nonisothermal plasma when interactions with magnetized ion-acoustic oscillations are possible. Since these interactions depend strongly on the ratio of the ion-acoustic energy to the magnetic fluctuation energy, when a sufficiently high level of magnetoacoustic oscillation is reached, the decay of magnetic fluctuations into magnetic fluctuations can be neglected as compared with ion scattering into ion-acoustic waves. Decay interactions of the type $Ms \rightleftarrows Ms' + A$ are either forbidden or the relative phase volume in which they are allowed is small. The indicated interactions can be written in the following form:

$$\frac{\partial W_{\omega,\,x}^{Ms}}{\partial t} = \frac{\pi}{4} \cdot \frac{W_{\omega,\,x}^{Ms}}{m_i\, n_0\, v_A^2} \cdot \frac{(1-x^2)}{x^2} \left(\frac{T_i}{T_e}\right)^2 \frac{\omega^4}{\omega_{Hi}^2} \int x_1^2\, dx_1 \left(\omega\, \frac{\partial W_{\omega,\,x_1}^M}{\partial \omega} + 2 W_{\omega,\,x_1}^M\right),$$

$$(4.308)$$

$$\frac{\partial W_{\omega,\,x}^{M}}{\partial t} = \frac{\pi}{4} \cdot \frac{W_{\omega,\,x}^{M}}{m_i\, n_0\, v_A^2} \cdot \frac{\omega^4}{\omega_{Hi}^2}\, x^2 \left(\frac{T_i}{T_e}\right)^2 \int \frac{1-x_1^2}{x_1^2}\, dx_1 \left(\omega\, \frac{\partial W_{\omega,\,x_1}^{Ms}}{\partial \omega} + 2 W_{\omega,\,x_1}^{Ms}\right),$$

$$(4.309)$$

$$\frac{\partial W_{\omega,\,x}^{A}}{\partial t} = \frac{\pi}{4} \cdot \frac{W_{\omega,\,x}^{A}}{n_0\, m_i\, v_A^2} \cdot \frac{\omega^4}{\omega_{Hi}^2} \cdot \frac{1}{x^2} \left(\frac{T_i}{T_e}\right)^2 \int \frac{(2x_1^2 + x^2 - 3x^2 x_1^2)}{x_1^2}\, dx_1 \times$$

$$\times \left(\omega\, \frac{\partial W_{\omega,\,x_1}^{Ms}}{\partial \omega} - 2 W_{\omega,\,x_1}^{Ms}\right), \qquad (4.310)$$

$$\frac{\partial W_{\omega,x}^{Ms}}{\partial t} = \frac{\pi}{4} \cdot \frac{W_{\omega,x}^{Ms}}{n_0 m_i v_A^2} \cdot \frac{1}{x^2} \left(\frac{T_i}{T_e}\right)^2 \frac{\omega^4}{\omega_{Hi}^2} \int dx_1 \left(\frac{1-x_1^2}{x_1^2} x^2 - \frac{1-x^2}{2}\right) \times$$

$$\times \left(\omega \frac{\partial W_{\omega,x_1}^{A}}{\partial \omega} + 4 W_{\omega,x_1}^{A}\right). \tag{4.311}$$

First, it is interesting to note that the spectrum for the magnetized ion-acoustic oscillations,

$$W_{\omega}^{Ms} = \frac{\text{const}}{\omega} \ln \frac{\omega}{\omega_0}, \tag{4.312}$$

is set up because of the interactions of Ms-waves among themselves, according to Eqs. (4.310) and (4.309); it is stable relative to the excitation of Alfven waves, but unstable relative to the excitation of magnetoacoustic waves. The accumulation of the latter oscillations can lead to the opposite action on the ion-acoustic oscillations. The spectrum of M-waves must be reconstructed so that this action is reduced to zero; we then obtain from (4.308)

$$W_{\omega,x_1} = \frac{W(x_1)}{\omega^2} \tag{4.313}$$

or $\nu = 2$.

As ω decreases the role of interactions between Ms-waves and M-waves is reduced, and the spectrum goes to (4.307).

To conclude, we must discuss some of the directions in which the theory of stationary turbulence is developing. One such direction is the study of the role played by the turbulence anisotropy, especially near the generation region. Another approach is the study of the uniqueness of the stationary states. Frequently it is important to know if there can be a number of stationary states for a given turbulence generation power, and if this is the case, one would like to know which initial conditions on the system lead to a given stationary state. Further, one would like to know if there can be a situation in which the system does not go to any of these stationary states.

These questions are related to a more general problem, that of the dynamics in the development of a turbulent plasma state. The study of second sound (Chapter 8) can give some information about these dynamics near a state of stationary turbulence.

Chapter 5

Stochastic Particle Acceleration in Turbulent Plasma

5.1. General Problems

The interaction of charged particles with turbulent plasma is characterized by an effective exchange of energy between the particles and the fluctuations which can lead to the acceleration of the particles [229]. It is a characteristic feature that the energy exchange can be quite large for fast particles since their energies are much higher than the mean thermal energy of the majority of particles. Since there is no charge separation during collective motions in fluids, in order to produce interactions with fast particles the turbulence fluctuations must carry magnetic fields. A moving fluid element with a magnetic field frozen in creates electric fields which are proportional to its velocity. These fields can accelerate particles.

Historically the study of effects due to acceleration of fast particles began with these rather ineffective mechanisms, which are some sort of expression of the Fermi mechanism [230], in which particles are accelerated during collisions with randomly moving magnetic clouds. It should be clear that under these conditions the acceleration arises because of gradients in the magnetic field. In magnetohydrodynamic fluctuations these gradients are very small, being of the order of a mean-free path length (see below); therefore the effectiveness of acceleration by such fluctuations is very small. However, even the earliest works on the theory of high-frequency turbulence showed [56] that there are effective, direct interactions between particles and turbulent fluctuations [229]. The effectiveness of statistical acceleration increases rapidly as the oscillation frequency increases. This is because the electric field associated with turbulent fluctuations captures an ever increasing fraction of the total oscillation energy as the frequencies increase. On the other hand the characteristic wavelengths of the high-frequency fluctuations are, as a rule, significantly shorter than those of the low-frequency waves. The stochastic nature of the interactions with the particles in a turbulent plasma is ensured by the random nature of the fields associated with turbulent fluctuations.

A simple calculation of the "number of turbulent collisions" shows that the acceleration is more efficient if the time between such collisions is small. For example, if $\Delta\varepsilon_+$ is the energy gained by a particle in a single interaction, while $\Delta\varepsilon_-$ is the energy lost in a single interaction, then as the result of a large number of interactions N in which the particle can randomly either gain or lose energy, it obtains an energy $\nu\Delta\varepsilon$, where $0 < \nu < N$.

The average energy of the accelerated particles can increase with time either because $\Delta\varepsilon_+$ is somewhat larger than $\Delta\varepsilon_-$ or because the number of interactions ν_+ which increase the particle energy is slightly larger than the number ν_- which cause the energy to drop. When dealing with Fermi acceleration $\Delta\varepsilon_+ = \Delta\varepsilon_-$ and $\nu_+ > \nu_-$. An individual interaction act, in this case, is the collision between a charge and a moving magnetic wall. Elementary collision kinetics show that in a head-on collision the charge picks up energy in the amount $\Delta\varepsilon_+ = 2\varepsilon\,(u/c^2)\,v$, where u is the wall velocity, ε and v are the charge's energy and velocity. In overtaking collisions the charge loses the same amount of energy. If L is the average distance between magnetic walls, the number of head-on collisions is $\nu_+ = \nu + u/L$, the number of overtaking collisions is $\nu_- = \nu - u/L$, and we have

$$\frac{d}{dt}\langle\varepsilon\rangle = \Delta\varepsilon\,(\nu_+ - \nu_-) = \frac{4\varepsilon v u}{Lc^2}. \tag{5.1}$$

Equation (5.1) is the formula for Fermi acceleration. Under certain specific assumptions (see the monograph [46]), L is taken to be equal to the mean distance between two turbulent elements which carry the magnetic field. We must point out that the statistical acceleration of the particles can be accomplished in a laboratory experiment by using a specially constructed field which varies according to the laws of change. This acceleration mechanism was suggested by Burshtein, Veksler, and Kolomenskii [232]. Particle accelerators operating as stochastic accelerators employ this principle† [233]. In a plasma random fields are created in a natural way when turbulence is excited.

Characteristically, the acceleration [see Eq. (5.1)] is more effective the smaller the turbulence scale is. The factors indicated render the high-frequency fluctuations most effective for accelerating the fast particles. Among these fluctuations are the Langmuir oscillations. We note that Langmuir oscillations are excited very fast and quickly result in stationary turbulence. Moreover, all the high-frequency waves have exponentially small damping by thermal particles; this is reflected in the fact that they are not in resonance with the thermal particles. These fluctuations are resonant with the fast particles and are therefore fitted for accelerating the fast particles. Low-frequency oscillations are usually resonant with the thermal particles, and are therefore strongly damped; this leads to the turbulent heating of the great bulk of particles. A fast particle differs from the other particles because it can have resonant interactions which are not possible for the thermal particles. For example Langmuir fluctuations whose phase velocities are higher than v_{Te} can accelerate only fast particles with velocities greater than v_{Te}.

We should also point out that from a more general point of view, it is difficult to isolate the transverse electromagnetic waves from the other turbulent oscillations. First, in the limit of low frequencies such that $\omega \to \omega_{pe}$, they differ very little from the plasmons. Second, when accelerated in a turbulent plasma, the fast particles become a source of electromagnetic radiation at frequencies $\omega \gg \omega_{pe}$ because of their scattering by turbulent fluctuations. This radiation is in quasiequilibrium with the fast particles and the turbulence. Therefore the final quasiequilibrium distribution of fast particles and turbulence is set up in conjunction with the quasiequilibrium distribution of the high-frequency radiation. Since the effectiveness of the acceleration increases as the frequencies of the stochastic field grow, special account must be taken of the effects of acceleration on the high-frequency radiation.

†And are called "stochatrons" in the Russian literature.

In this instance it is important to know if this radiation is in equilibrium with the fast particles and the turbulence; this depends on the characteristic length L over which the equilibrium can be established. In astrophysical situations the characteristic plasma dimensions are much larger than L; in laboratory cases they are smaller. There are significant differences between the kinds of turbulence found in the laboratory and in astrophysical plasmas.

The statistical acceleration of particles in a turbulent plasma is a very widespread phenomenon. Suffice it to say that the acceleration of cosmic rays is connected with the presence of plasma turbulence. This has been the starting point of many investigations of the acceleration of cosmic rays [46, 229]. As laboratory experimentation progressed, turbulence in the plasmas appeared for whatever reason (either in response to the experimenter's wishes, or in spite of them), and fast accelerated particles were immediately observed [42-45]. Practical possibilities also appeared for using this effect to fill magnetic traps, obtain powerful radiation sources, and so on. But the energy problem is the most important: how much of the turbulent energy can be given over to the fast particles under a given set of experimental conditions, and how much goes into heating the plasma? The answer to this question depends on the type of turbulent fluctuations, the specific conditions of the experiment, and its goal. Heating is most efficient when the path of the turbulent energy currents runs through the "traps," regions of intensive cyclotron and Cerenkov absorption. Acceleration, on the other hand, requires a weak absorption of the fluctuations by thermal particles. Both conditions can be realized experimentally.

In general, particle acceleration is not harmless to the turbulence; a large number of accelerated particles can have a significant impact on the turbulence spectra. With a small number of accelerated particles, one may assume that the turbulence spectra are regulated by internal generation, absorption, and spectral transfer processes. The number of accelerated particles depends on the capacity for injecting particles into the accelerating mode. If there is no specific injection mechanism, injection will be accomplished in a natural manner through the tails of the Maxwell distributions of the turbulence itself (because the effects of interactions between particles and fluctuations is proportional to W^2, W^3, and so on), or through pair collisions between the particles. We must also point out that the energy density of the accelerated particles can be significantly higher than the energy density of the turbulent oscillations. In fact, if the turbulence is stationary, there will be a constant flow of turbulent energy across the spectrum. The accelerated particles will have little effect on W and the turbulence spectrum if the energy which they extract during the time needed for the turbulent oscillations to move from the generation region to the absorption region is smaller than W.

As shown in Chapter 2, the interaction between particles and turbulent fluctuations leads to effects described by diffusion equations for the particles in either the velocity space or the momentum space of those particles. These same equations can be used to study the exchange of energy between the fast particles and the turbulent fluctuations:

$$\frac{df_\mathbf{p}}{dt} = \frac{\partial}{\partial p_i} D_{ij} \frac{\partial f_\mathbf{p}}{\partial p_j}.$$ (5.2)

The diffusion coefficient can be expanded in powers of the turbulent energy W. The term linear in W describes the induced emission and absorption of waves by particles; the quadratic term gives the induced scattering, and so on down the series. The probabilities for these processes were found earlier (see Chapter 2). Diffusion [Eq. (5.2)] can be considered as those processes which produce changes in the modulus of the momentum $|\mathbf{p}| = p$ and in the direction of particle motion. If the turbulence is isotropic, then

$$D_{ij} = D_p^l \frac{p_i p_j}{p^2} + D_p^t \left(\delta_{ij} - \frac{p_i p_j}{p^2} \right).$$ (5.3)

The coefficient D^l gives the diffusion relative to the modulus p, while D^t describes diffusion for angles. Diffusion relative to angles leads to isotropization of the accelerated particles. In an isotropic plasma $D^t \gg D^l$ if $v \gg v_p$, that is, before picking up energy the particles undergo rapid isotropization. If the particles are isotropically distributed, the term containing D^t falls out of (5.2), and the equation is †

$$\frac{df_p}{dt} = \frac{1}{p^2} \cdot \frac{\partial}{\partial p} p^2 D_p^l \frac{\partial f_p}{\partial p}.$$ (5.4)

---

†For Fermi acceleration, $D_p^l = \frac{2u^2}{3L} \varepsilon p$ [234].

Assume that $\overline{L}_p$ is the mean value of L_p relative to the accelerated particles[†]:

$$\overline{L}_p = \int f_p L_p \frac{d\mathbf{p}}{(2\pi)^3} \Big/ \int f_p \frac{d\mathbf{p}}{(2\pi)^3}. \tag{5.5}$$

The rate at which the average particle energy increases is easily found from (5.4):

$$\frac{\partial}{\partial t}\,\overline{\varepsilon_p} = \overline{\dot{\varepsilon}_p} \tag{5.6}$$

and the rate of increase of the average quadratic energy scattering is

$$\frac{\partial}{\partial t}\,\{\,\overline{\varepsilon_p^2} - (\overline{\varepsilon_p})^2\,\} = \overline{(\Delta\dot{\varepsilon}_p)^2}. \tag{5.7}$$

Then

$$\dot{\varepsilon}_p = \frac{1}{p^2}\cdot\frac{\partial}{\partial p}\,p^2 D_p^l\,\frac{d\varepsilon_p}{dp}\;; \tag{5.8}$$

$$(\Delta\dot{\varepsilon}_p)^2 = 2D_p^l\left(\frac{d\varepsilon_p}{dp}\right)^2 + 2\left(\varepsilon_p - \overline{\varepsilon}_p\right)\dot{\varepsilon}_p. \tag{5.9}$$

For nonrelativistic velocities it is convenient to write (5.8) as

$$\dot{\varepsilon}_p = \frac{2}{m\sqrt{\varepsilon_p}}\cdot\frac{\partial}{\partial\varepsilon_p}\,\varepsilon_p^{3/2} D^l\,(\varepsilon_p),\qquad \varepsilon_p = \frac{p^2}{2m}. \tag{5.8a}$$

We shall now assume that when t = 0, particles of energy $\varepsilon = \varepsilon_0$ appear in the turbulent plasma, i.e., $f(\varepsilon, 0) = n_1\delta(\varepsilon - \varepsilon_0)$. Thus, when t = 0, $\overline{\varepsilon}_p = \varepsilon_0$ and the rate at which energy is acquired is $\dot{\varepsilon}_p|_{\varepsilon=\varepsilon_0}$, while the energy spread is $(\Delta\dot{\varepsilon}_p)^2|_{t=0} = 2D_p^l\left(\frac{\partial\varepsilon_p}{\partial p}\right)^2\Big|_{\varepsilon=\varepsilon_0} = G(\varepsilon_0)$. Since the values of the averages do not depend on the number of accelerated particles n_1, we can say that in the limit $n_1 \to 0$ $\dot{\varepsilon}_p(\varepsilon_0)$ gives the systematic acceleration of a particular particle while $G(\varepsilon_0)$ gives the fluctuation acceleration of that particle.

[†]Averaging over a distribution of fast particles is denoted by a line above the symbol; this is contrasted with averaging over a statistical ensemble, which is indicated by $\langle\ \rangle$. The mass of an accelerated particle is given by m.

A distribution function normalized to energy, f_ε, is frequently used to describe acceleration effects:

$$f_p = \frac{1}{4\pi p^2} \cdot \frac{d\varepsilon_p}{dp} f_\varepsilon (2\pi)^3; \tag{5.10}$$

$$\int f_\varepsilon \, d\varepsilon = n_1. \tag{5.11}$$

For this distribution function Eq. (5.4) is written as

$$\frac{df_\varepsilon}{dt} = \frac{\partial^2}{\partial \varepsilon^2} D^l \left(\frac{d\varepsilon}{dp} \right)^2 f_\varepsilon - \frac{\partial}{\partial \varepsilon} \left[\frac{1}{p^2} \cdot \frac{\partial}{\partial p} p^2 D^l \frac{d\varepsilon}{dp} \right] f_\varepsilon. \tag{5.12}$$

The physical meaning of the separate terms in (5.12) is quite clear. The last term gives the systematic acceleration, and the first term is the fluctuation acceleration. Equation (5.12) can also be written in terms of $\dot{\varepsilon}_p$ and $G(\varepsilon)$, which characterize change in the average parameters of the accelerated particles with time [9]:

$$\frac{df_\varepsilon}{dt} = \frac{\partial^2}{\partial \varepsilon^2} \cdot \frac{1}{2} G(\varepsilon) f_\varepsilon - \frac{\partial}{\partial \varepsilon} (\dot{\varepsilon}_p f_\varepsilon). \tag{5.13}$$

We should mention that (5.4) can be obtained for nonisotropic turbulence if one assumes that the distribution of accelerated particles is isotropic. Then, averaging (5.2) over the directions of p, we obtain Eq. (5.4), in which the diffusion coefficient D is averaged over the angles of the accelerated particles. Thus, Eq. (2.195) gives the following result when averaged in the manner described for induced scattering and absorption of fluctuations by particles in a magnetoactive plasma:

$$D_p^l = \sum_\sigma \sum_{\nu = -\infty}^{\infty} \int \frac{N_k^\sigma dk}{(2\pi)^3} \cdot \frac{(\omega_k^\sigma)^2}{v^2} w_p^\sigma (k, \nu) \frac{d\Omega_p}{4\pi}; \tag{5.14}$$

$d\Omega_p$ is the solid angle element for the vector p. For isotropic turbulence we can write

$$D_p^l = \sum_\sigma \sum_{\nu = -\infty}^{\infty} \frac{1}{8\pi v^2} \int_0^\infty W_k^\sigma dk \int_{-1}^1 \omega_k^\sigma dx \int_{-1}^1 dy \int_0^{2\pi} d\varphi w_{p,\,y}^\sigma (k, x, \nu), \tag{5.14'}$$

where $x = \cos (k, H_0)$, $y = \cos (p, H_0)$. When considering magnetized particles it is convenient to use the probabilities given in

(2.220) whereas for unmagnetized particles we should use (2.223). It should be mentioned that fast particles can be considered as unmagnetized more often than can thermal particles.

Equation (5.4) can also be obtained for the case of a nonisotropic particle distribution. However, now the coefficient D^l is given by an expression which is averaged over the angular distribution of the accelerated particles if their distribution function is a product of an angular part and an energy distribution function. When dealing with isotropic particles and anisotropic fluctuations, D^l is replaced with a quantity which is averaged with functions describing the angular distribution of the oscillations. The rapid isotropization of the fluctuations because of the nonlinear effects and the accelerated particles in the field of the isotropic oscillations makes the analysis of acceleration effects on the isotropic particles and fluctuations much more interesting. As explained earlier, there are no isotropic distributions for some of the fluctuation spectra in a magnetic field. Anisotropy effects must be taken into account in this instance.[†] The accelerated particles observed in experiment can be explained by the effects of acceleration on the different turbulent modes, but as a rule the distribution of accelerated particles with respect to direction turns out to be more or less isotropic. Therefore we will discuss the acceleration for different types of turbulent oscillations under the assumption that the fast particles and the turbulence are both isotropic (this will be the assumption whenever possible, that is). Accelerated fast particles were observed in the very first experiments using powerful pulsed self-constricting discharges [236-238] in experiments in which the plasma was excited by a fast beam of charged particles [43, 239-241], in turbulent heating experiments [26], and in other experiments wherein the plasma is either in an external electric field [31, 148] or is acted on by powerful high-frequency fields [44].

As a rule the spectra of accelerated particles have a maximum with a relatively slow decay at higher energies (Fig. 5.1). It is curious that such spectra are a sort of miniature version of the spectra for cosmic electrons and ions; this seems to indicate that we are dealing with one specific phenomenon which is charac-

_____
[†]See Ref. 235 for the effects of turbulent anisotropy on acceleration when $H_0 = 0$.

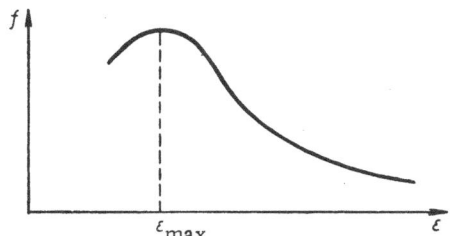

Fig. 5.1. Spectrum of particles accelerated in a turbulent plasma.

teristic of turbulent plasma, and differs only in scale. It has frequently been found in laboratory experiments that the acceleration effects are correlated with the excitation of a particular kind of turbulent fluctuation (either high or low frequency) [43, 148, 98]. There remains no doubt that the acceleration arises because of the energy exchange between fast particles and turbulent fluctuations.

The theory is faced with the problem of studying the relative effectiveness of fast particle acceleration by different kinds of turbulent fluctuations, the effects of acceleration on the turbulent spectra, and explaining the observed spectra for the accelerated particles. Some of these questions are entangled with problems of turbulent plasma radiation (see Chapter 6).

5.2. Stochastic Acceleration of Charged Particles by Langmuir Oscillations

Oscillations which are due to only plasma electrons will be called high-frequency oscillations; the ions are treated as an immobile background. When $H_0 = 0$, Eq. (5.14) gives us the following expression for the Langmuir oscillations:

$$D^l = \frac{2\pi^2 e^2 \omega_{pe}^2}{v^3} \int_{\frac{\omega_{pe}}{v}}^{\infty} \frac{W_k^l}{k^3} \, dk. \tag{5.15}$$

The significant dependence of the diffusion coefficient on the particle energy appears for nonrelativistic particles. It follows from

(5.8) that [242, 229]

$$\dot{\varepsilon}_p = \frac{2\pi^2 e^2}{p} \left\{ \frac{m^2 c^4}{\varepsilon_p^2} W^l_{\omega_{pe}} + 2 \frac{\omega^2_{pe}}{c^2} \int\limits_{\frac{\omega_{pe}}{v}}^{\infty} \frac{W^l_k}{k^3} dk \right\}. \tag{5.16}$$

At nonrelativistic velocities the second term in (5.16) is of first order in $v^2/c^2 \ll 1$, so that the acceleration rate is completely determined by the turbulence spectrum

$$\dot{\varepsilon}_p = \frac{\pi \omega^2_{pe} m_e}{2n_0 \sqrt{2m\varepsilon_p}} W^l_{\omega_{pe}} \sqrt{\frac{m}{2\varepsilon_p}}. \tag{5.17}$$

Here $\varepsilon_p = p^2/2m$ is the nonrelativistic kinetic energy of the particle. On the other hand, at the ultrarelativistic limit the acceleration rate decreases as the energy increases:

$$\dot{\varepsilon}_p = \frac{\omega^4_{pe} \pi}{\varepsilon_p n_0 c} \int\limits_{\frac{\omega_{pe}}{c}}^{} \frac{W^l_k dk}{k^3}. \tag{5.18}$$

This effect is simply explained. It takes place because a particle with a given velocity interacts with all oscillations whose phase velocities are smaller than the particle velocity. As the particle velocity grows the number of turbulent oscillations participating in acceleration grows, and the effect can increase rapidly as the particle energy increases. The velocity of the ultrarelativistic particles is close to c, and they interact with the fluctuations irrespective of the dependence on particle energy. The coefficient for energy diffusion is a constant, independent of energy. The simple equation $\varepsilon^2 \simeq 4Dt$ gives us $\varepsilon/t \sim D/\varepsilon$, which is in fact contained in (5.18). For ultrarelativistic particles (5.12) takes the form

$$\frac{df_e}{dt} = D \frac{\partial^2 f_e}{\partial \varepsilon^2} - 2D \frac{\partial}{\partial \varepsilon} \cdot \frac{f_e}{\varepsilon}, \quad D = D^l c^2. \tag{5.19}$$

For nonrelativistic energies (5.12) can be written as

$$\frac{df_e}{dt} = \frac{2}{m} \cdot \frac{\partial}{\partial \varepsilon} \varepsilon^{3/2} D^l(\varepsilon) \frac{\partial}{\partial \varepsilon} \cdot \frac{f_e}{\sqrt{\varepsilon}}, \quad \varepsilon = \frac{p^2}{2m}. \tag{5.20}$$

We will now consider the acceleration effects, by taking explicit account of the turbulence spectra (Chapter 4) and neglecting the effects of acceleration on the spectra at first. The ratio

$$\eta = \frac{3 v_{Te}^2}{c v_{Ti}} \tag{5.21}$$

has particular importance for the acceleration of fast particles. If $\eta \ll 1$, relativistic particles exist which are accelerated by the spectrum const/k^ν, where $2.84 < \nu < 4$ as found in [211]. Then $D \sim \varepsilon^{\frac{\nu-1}{2}}$ and $\dot{\varepsilon}_p \sim \varepsilon^{\frac{\nu-1}{2}}$.

When $\nu = 2.84$ we have $\dot{\varepsilon}_p \sim \varepsilon^{0.92}$, while if $\nu = 4$, $\dot{\varepsilon} \sim \varepsilon^{3/2}$. These acceleration mechanisms are considerably more effective than the Fermi acceleration, for which $\dot{\varepsilon} \sim \sqrt{\varepsilon}$ at nonrelativistic velocities (even without considering the fact that the dimensionless constant for Fermi acceleration is much smaller than its counterpart for acceleration by Langmuir oscillations). When $\varepsilon \ll mc^2 \eta^2$ the spectrum W_k = const participates in the acceleration, i.e., $\nu = 0$ and $\dot{\varepsilon} \sim 1/\sqrt{\varepsilon}$. When $\eta > 1$, that is, a rather hot plasma, part of the effective acceleration disappears because the spectrum with $\nu = 0$ is stretched to include phase velocities equal to the velocity of light $(k \sim \omega_{pe}/c)$. Finally, at low energies $\varepsilon \ll T_e \times (m_i/m_e)^{2/5}$, the acceleration is determined by the spectrum with $\nu = 5/2$ and $\varepsilon \sim \varepsilon^{3/4}$. For ultrarelativistic particles with $\dot{\varepsilon} \sim 1/\varepsilon$ the rate of acceleration for electrons and ions is identical when $\varepsilon \gg m_i c^2$. Figures 5.2 and 5.3 give schematic representations of the acceleration rates for electrons and ions for the cases $\eta \ll 1$ and $\eta \gg 1$. The acceleration curve has two maxima, the first in the region where the particle velocities are of order $(3m_i/m_e)^{1/5} \times v_{Te}$, which is just a few times larger than the plasma electrons' thermal velocity, and the second in the region where $\varepsilon = mc^2$. For the ions this maximum corresponds to an energy which is a factor of m_i/m_e greater, but the rate of acceleration at the maximum $\varepsilon = m_i c^2$ for electrons is m_i/m_e times greater than the rate at the ion maximum energy $\varepsilon = m_i c^2$. In the region where $v < v_{Te}(m_i/m_e)^{1/5}$ the electron acceleration is $(m_i/m_e)^{7/4}$ times more effective, but from the region where $v > v_{Te}(m_i/m_e)^{1/5}$ until $v = \eta c$ it is $(m_i/m_e)^{1/2}$ times more effective. The proportionality coefficients in the acceleration laws $\varepsilon_p \sim \varepsilon^{\frac{\nu-1}{2}}$ are easily found from the turbulence generating power Q if one uses the specific spectra found in Chapter 4. For example, with W = const = $(2Q/\alpha)^{1/2}$ Eq. (5.17) gives us

$$\frac{\dot{\varepsilon}}{mc^2} = \Gamma \sqrt{\frac{mc^2}{\varepsilon}}; \quad \Gamma = \frac{3\sqrt{3\pi}}{2} \omega_{pe} \frac{v_{Te}^3}{c^3} \cdot \frac{m_e^{3/2} m_i^{1/2}}{m^2} \left(1 + \frac{T_e}{T_i}\right) \sqrt{\frac{Q}{Q_0}},$$

$$Q_0 = \omega_{pe} n_0 T_e.$$

(5.22)

The spectra of accelerated particles is determined when there is a balance between acceleration and deceleration. Deceleration is connected with many effects, and these can be divided into those which produce deceleration in turbulent plasma and those more common mechanisms which occur in both turbulent and nonturbulent plasmas. An example of the first type of deceleration effect is the reverse Compton effect for accelerated particles and turbulent fluctuations, while the second, common type of mechanism is typified by ionization losses. The reverse Compton effect sets in at high energies and is related to radiation processes, which will be treated in another chapter. Ionization losses can be supplemented by other processes, such as those related to the adiabatic expansion of the plasma after its turbulent heating or after it has been created by a laser; in astrophysical conditions these losses can result from the dispersion of supernova clouds [243-245]. Stable equilibrium between acceleration and deceleration is possible if the deceleration curve $\varepsilon_{diss} = f(\varepsilon)$ is steeper than the acceleration curve[†] (see Figs. 5.2 and 5.3). If the deceleration is due to ionization losses, then in a highly rarefied plasma $(\varepsilon \ll mc^2)$ [246], we have

$$\varepsilon_{p, \text{ion}} \simeq \frac{\sqrt{m} e^2 \omega_{pe}^2}{\sqrt{2} \varepsilon^{1/2}} \ln \frac{\varepsilon}{\hbar \omega_{pe}}.$$

(5.23)

We see from Figs 5.2 and 5.3 that the sections in front of the maximum in the acceleration curve cannot give stable distributions — the particles will "slip through" these regions. This does not mean that there cannot be equilibrium distributions of accelerated particles in the indicated regions. However, such stationary distributions exist only when the acceleration is much greater

---

[†]If this takes place, a particle with energy greater than the equilibrium energy ε_* (which corresponds to the point of interaction of the acceleration and deceleration curves) will be decelerated, while a particle with energy less than ε_* is accelerated.

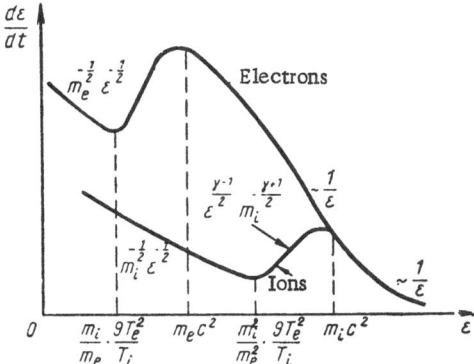

Fig. 5.2. Rate of acceleration of both electrons and ions for Langmuir turbulence with $\eta \ll 1$: $\eta = 3v_{Te}^2/cv_{Ti}$.

than the deceleration. In fact, Eq. (5.4) has a stationary solution

$$p^2 D_p^i \frac{\partial f_p}{\partial p} = \text{const.} \tag{5.24}$$

This solution indicates a constant particle flow with respect to energy, and holds if there is continuous injection. However, it is clear that this flow is set up because of a balance between acceleration and deceleration. In other words, the deceleration curve in the range of Eq. (5.24) is everywhere much lower than the acceleration curve in Figs. 5.2 and 5.3. This means that the ionization deceleration and acceleration curves will not intersect

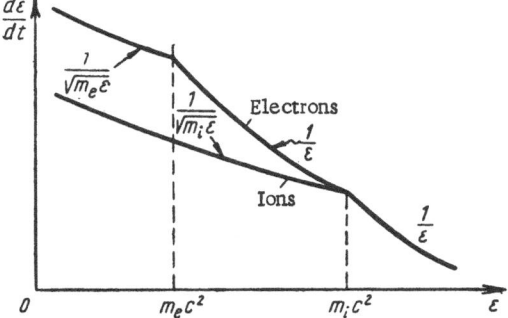

Fig. 5.3. Rate of acceleration of electrons and ions for Langmuir turbulence with $\eta \gg 1$; $\eta = 3v_{Te}^2/cv_{Ti}$.

in the region where the curves have their maximum energies, where $(d/d\varepsilon)\dot{\varepsilon} > 0$.

It is easy to find the spectra of the accelerated particles from (5.24) since with $D \sim \varepsilon^{\frac{\nu-1}{2}}$, $f_p \sim \varepsilon^{-\nu/2}$ and $f_e \sim f_p\varepsilon^{1/2} \sim \varepsilon^{\frac{-\nu+1}{2}}$; that is, when $\nu = 2.84$, $f_e \sim 1/\varepsilon^{0.92}$; when $\nu = 0$, $f_e \sim \sqrt{\varepsilon}$; when $\nu = 4$, $f_e \sim 1/\varepsilon^{3/2}$; and when $\nu = 5/2$, $f_e \sim 1/\varepsilon^{3/4}$. We see that the spectra of the accelerated particles are power functions, and they gradually decrease in the high-energy region; however, in a number of cases they increase instead ($\nu = 0$). One observes a rapid decrease in the number of accelerated particles as the energy increases in the region of effective acceleration whose rate rapidly increases with energy. This is because the acquisition of energy by the particles in such regions makes the subsequent acceleration more efficient. Thus, the stable flow described by Eq. (5.24), like the balance in the number of particles entering this region of energies $d\varepsilon$ from the low-energy side of ε and leaving from $d\varepsilon$ to larger ε, appears when the number of high-energy particles is smaller than the number of low-energy particles. Equilibrium with ionization losses for nonrelativistic energies is possible when $\dot{\varepsilon} \sim 1/\varepsilon^{1/2}$, when according to (5.15)

$$D^l(\varepsilon) = \frac{3\sqrt{3\pi}\,\omega_{pe}\,v_{Te}^3}{4\sqrt{\varepsilon}}\sqrt{mm_i}\,m_e^{3/2}\left(1 + \frac{T_e}{T_i}\right)\sqrt{\frac{Q}{Q_0}}. \tag{5.25}$$

Here $W_k = \text{const} = (2Q/\alpha)^{1/2}$, $\varepsilon = mv^2/2$, and Q_0 is given in (5.22).

When ionization losses are included, Eq. (5.20) takes the form[†]

$$\frac{\partial f_e}{\partial t} = \frac{3\sqrt{3\pi}\,m_e^{3/2}\,m_i^{1/2}}{2m^{1/2}}\left(1 + \frac{T_e}{T_i}\right)\omega_{pe}v_{Te}^3\sqrt{\frac{Q}{Q_0}}\cdot\frac{\partial}{\partial\varepsilon}\,\varepsilon\,\frac{\partial}{\partial\varepsilon}\,\frac{f_e}{\sqrt{\varepsilon}} +$$

$$+ \frac{\nu_e}{\sqrt{2}}\sqrt{m}\,m_e v_{Te}^3\,\frac{\partial}{\partial\varepsilon}\left(\frac{f_e}{\sqrt{\varepsilon}} + T_e\frac{\partial}{\partial\varepsilon}\cdot\frac{f_e}{\sqrt{\varepsilon}}\right) = 0. \tag{5.26}$$

If there were no turbulence, $Q = 0$, the ionization losses would then lead the fast particles into equilibrium with the electrons with $f_e = \text{const}\,\sqrt{\varepsilon}\,e^{-\varepsilon/T_e}$. When there is turbulence, the condition

---

†The second term in (5.26) in the ionization collisions gives a fluctuation spread in the energies, i.e., $\dot{\varepsilon} = 0$. The first term gives a systematic deceleration with $\dot{\varepsilon} \neq 0$.

that the total flow be zero gives

$$\frac{\sqrt{m_e\, m_i}}{m} \cdot \frac{3\sqrt{3\pi}}{\sqrt{2}} \left(1 + \frac{T_e}{T_i}\right) \frac{\omega_{pe}}{v_e} \sqrt{\frac{Q}{Q_0}}\; \varepsilon\, \frac{\partial}{\partial\varepsilon} \cdot \frac{f_\varepsilon}{\sqrt{\varepsilon}} + \frac{f_\varepsilon}{\sqrt{\varepsilon}} + T_e\, \frac{\partial}{\partial\varepsilon}\, \frac{f_\varepsilon}{\sqrt{\varepsilon}} = 0.$$

(5.27)

We then have

$$f_\varepsilon = \text{const}\, \sqrt{\varepsilon}\, \frac{1}{(\varepsilon + \gamma T_e)^\gamma},$$

(5.28)

where

$$\gamma = \frac{m}{\sqrt{m_e\, m_i}} \sqrt{\frac{2}{3\pi}} \cdot \frac{T_i}{3\,(T_e + T_i)} \cdot \frac{v_e}{\omega_{pe}}\, \sqrt{\frac{Q_0}{Q}}.$$

(5.29)

If $\varepsilon \gg T_e\gamma$ the spectrum is very nearly a power function. We also note here that (5.29) is strictly speaking correct when $T_e \lesssim T_i$. When $T_e \gg T_i$, $\alpha \to \alpha' = \alpha\,(1 + T_e/T_i)^2$ according to Chapter 4, so that the factor $T_i/(T_e + T_i)$ in (5.29) must be dropped. Equation (5.28) shows that in a turbulent plasma the distribution function for high-velocity particles is nowhere Maxwellian; rather, it grows long tails for the accelerated particles, and smoothly decreases at high energies according to some power law. The power spectrum index γ depends on the generating power Q and is markedly different for electrons and ions. For electrons $\gamma \sim (1 - 5)$ when $Q \sim Q_0 \frac{1}{N_D^2} \frac{m_e}{m_i} = Q_e^{\text{cr}}$, while for ions it has this value when $Q \sim Q_0 \frac{1}{N_D^2} \frac{m_i}{m_e} = Q_i^{\text{cr}}$. Since N_D is a very large number in real cases, even Q^{cr} is not large. If $Q \gg Q_{e,i}^{\text{cr}}$, $\gamma \to 0$ which means that $f_\varepsilon \sim \sqrt{\varepsilon}$. But this is just the solution which appears if collisions are omitted. This in turn means that there is a constant flow of accelerated particles with respect to energy at nonrelativistic energies, and that equilibrium between acceleration and deceleration is achieved only when the accelerated particles have ultra-relativistic energies. Equation (5.19) must be used in this case to find the spectrum for the accelerated particles:

$$\frac{\partial f_\varepsilon}{\partial t} = D\, \frac{\partial}{\partial\varepsilon}\, \varepsilon^2\, \frac{\partial}{\partial\varepsilon} \cdot \frac{f_\varepsilon}{\varepsilon^2} + v_e^{\text{eff}}\, \frac{m_e\, v_{Te}^3}{c} \cdot \frac{\partial f_\varepsilon}{\partial\varepsilon} = 0.$$

(5.30)

Here we have included just the effect of the systematic decrease in particle energy brought about by ionization losses †; ν_e^{eff} is the effective collision frequency, which differs from ν_e only by a numerical factor of order unity. ν_e^{eff} describes the effective frictional force of the relativistic particles on the plasma. From (5.30) we obtain

$$f_\varepsilon = \text{const } \varepsilon^2 e^{-\frac{\varepsilon}{mc^2 T_*}}, \tag{5.31}$$

where

$$T_* = \frac{Dc}{mc^2 \nu_e^{\text{eff}} m_e v_{Te}^3} \tag{5.32}$$

plays the role of the effective temperature of the accelerated particles in units of mc^2. If $\eta > 1$ [see (5.21)], we have according to (5.15)

$$D = \frac{3\sqrt{3\pi} \, m_e^{3/2} \omega_{pe}}{2\sqrt{2}} \, v_{Te}^3 \, m_i^{1/2} \left(1 + \frac{T_e}{T_i}\right) \sqrt{\frac{Q}{Q_0}} \, c \tag{5.33}$$

so that $T_* = 1/2\gamma$, where γ is given by Eq. (5.29).

This last equation also shows that when $\gamma \ll 1$, and when the collisions do not produce a stationary flow of accelerated particles when $f_\varepsilon \sim \sqrt{\varepsilon}$ for nonrelativistic particles, the effective temperature $T_* mc^2$ turns out to be ultrarelativistic. If $\eta \ll 1$, T_* is $\frac{2}{2+\nu} \cdot \frac{1}{\eta^\nu}$ times greater than that given by Eq. (5.33), where ν is the exponent for the turbulence spectrum. The value of Q^{cr} for which $\gamma < 1$ is very small; as a consequence the accelerating processes will automatically produce relativistic particles in a turbulent plasma. Thus, if $N_D \sim 10^8$ to 10^{10}, which is characteristic of astrophysical conditions, $Q_e^{\text{cr}} \sim (10^{-20}$ to $10^{-22})Q_0$. If we now return to Eq. (4.66), we can easily estimate the generating power for a beam with $\Delta v_p \sim v \sim v_{Te}\left(\frac{m_i}{m_e}\right)^{1/5}$; $Q \sim Q_0 \left(\frac{m_i}{m_e}\right)^{3/5} \frac{108}{\pi}\left(\frac{n_1}{n_0}\right)^2$; $\gamma \sim \frac{n_1}{n_0} \omega_{pe}$. Comparing this with Q_e^{cr} and Q_i^{cr} at the same time shows that even beams with very small concentrations give $T_* \gg 1$; i.e., we have ultrarelativistic accelerated electrons and ions. In fact, the minimal value of n_1 is determined by the condi-

†The fluctuation term is negligible if $T_e \ll m_e c^2$.

tion that there be an instability. The stability boundary is esti-
mated as $\gamma > v_e$; i.e., $n_1 > n_0 \dfrac{\omega_{pe}}{v_e} \sim \dfrac{n_0}{N_D}$ or $Q_{\min} \sim Q_e^{\mathrm{cr}} \dfrac{108}{\pi} \left(\dfrac{m_i}{m_e} \right)^{8/5} \gg$
Q_e^{cr} . Of course, when $Q \sim Q_{\min}$ we cannot use the spectrum $W_k =$
$(2Q/\alpha)^{1/2}$. However, the results obtained from this estimate are
very clear, showing that for any developed subcritical turbulence
the appearance of relativistic accelerated particles is absolutely
necessary if only the turbulence persists for a time long enough
to establish stationarity in the turbulence spectrum and the spectrum
of the accelerated particles. This result is very important to the
problems arising from the origin of cosmic rays, and in fact explains why a cosmic plasma, which is usually turbulent, is the universal source which generates relativistic electrons and ions.

It must also be mentioned that for high particle energies, deceleration due to the inverse Compton effect with turbulent fluctuations comes into play, and we have (see Chapter 6)

$$\varepsilon = -\left(\frac{m_e}{m} \right)^2 \frac{\omega_{pe}^4}{\pi n_0^2 c^3} \left(\frac{\varepsilon}{mc^2} \right)^2 W^l, \quad W^l = \frac{Q}{v_e}. \qquad (5.34)$$

When $\varepsilon/mc^2 = T_*$ these losses are greater than the ionization
losses if

$$Q > Q^{\mathrm{cr}} \sqrt{\frac{m_i}{m_e}} \cdot \frac{c}{v_{Te}} \sqrt{\frac{\omega_{pe}}{v_e}} = Q_{\mathrm{cr}}^* . \qquad (5.35)$$

If $Q \gg Q_{\mathrm{cr}}^*$ the losses given in (5.34) exceed the ionization
losses when $\varepsilon/mc^2 \ll T_*$. In a dense high-temperature plasma
Q_{cr}^* is only a few orders of magnitude different from Q_{cr}. However,
the absolute value of Q_{cr}^* (which is rather small) is important. For,
the requirement that $W \ll n_0 T_e$ means that $Q < Q_{\max} \approx v_e n_0 T_e \approx$
$Q_0 \dfrac{v_e}{\omega_{pe}}$ and $Q_{\mathrm{cr}}^*/Q_{\max} \simeq \sqrt{\dfrac{v_e}{\omega_{pe}} \cdot \dfrac{c}{v_{Te}}} \sqrt{\dfrac{m_e}{m_i}}$. Thus, at high ultrarelativistic energies the spectrum of the accelerated particles
is already determined by the deceleration due to the reverse
Compton effect, and as we shall soon see, even more powerful radiation acceleration mechanisms will appear (see Chapter 6).†
Thus, acceleration by Langmuir oscillations leads to the generation of ultrarelativistic electrons and ions, or on a cosmic scale,

---

†The accelerated particle spectra then are of the form $1/\varepsilon\gamma$.

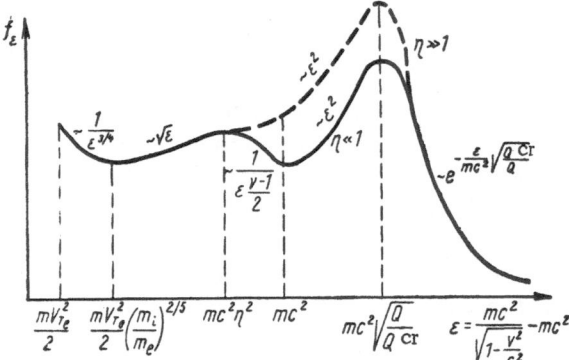

Fig. 5.4. Form of the spectra for electrons and ions accelerated
by Langmuir turbulence when $Q^{cr} < Q < Q^*_{cr}$.

to cosmic and subcosmic rays. It can be considered the injection
mechanism for the radiation acceleration mechanisms which
operate when $Q > Q^*_{cr}$ (Chapter 6). The fast accelerated particles
then become a source of radiation in the turbulent plasma (Chapter
6). Figure 5.4 shows the spectra for accelerated electrons and
ions when $Q^{cr} < Q < Q^*_{cr}$, and Fig. 5.5 shows these same spectra
when $Q \ll Q^{cr}$ and $Q > Q^*_{cr}$. Under laboratory conditions a very
high energy cannot be reached either because the plasma does not
last long enough, or because the accelerated particles are in a
state which is not contained by the magnetic field and therefore

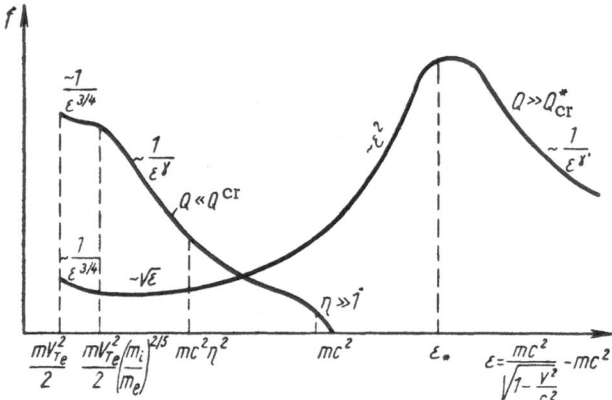

Fig. 5.5. Form of the spectra of electrons and ions accelerated
by Langmuir turbulence for $Q \ll Q_{cr}$ and $Q \gg Q^*_{cr}$.

they can leave the plasma volume. The latter effect can be included in (5.26) in a rough fashion by the term f_ε / τ, which accounts for the "catastrophic losses." The maximum particle energy will then be on the order of the energy at which the Larmor radius equals the dimension of the system.

5.3. Stochastic Acceleration by High-Frequency Fluctuations in a Magnetoactive Plasma

In the presence of magnetic fields a new parameter appears, one which characterizes the magnetization of the accelerated particles:

$$\beta = \frac{kv}{\omega_H}. \tag{5.36}$$

If $\beta \ll 1$ the particles are magnetized; if $\beta \gg 1$, they are unmagnetized. In a weak magnetic field such that $\omega_{pe} \ll \omega_{He}$, the Langmuir oscillations and their spectra are changed very little. On the other hand, $v > \omega_{pe}/k \gg \omega_{He}/k \geq \omega_H/k$ (the last inequality holds only for nonrelativistic electrons), so that $\beta \gg 1$. It therefore appears that we can use the probability of Eq. (2.223) for unmagnetized particles, and the results will coincide with those obtained in § 5.2. This is confirmed by direct calculation using the diffusion coefficient (5.14). When applied to longitudinal oscillations Eq. (5.14) can be written as

$$D^l = \sum_{\nu = -\infty}^{\infty} \frac{2\pi^2 e^2}{v^2} \int_0^\infty dk \int_{-1}^1 dx \int_{-1}^1 dy \ \frac{W_k^l \, \omega_k^l}{k^2 \left. \dfrac{\partial \varepsilon_k^l}{\partial \omega} \right|_{\omega = \omega_k^l}} J_\nu^2(\xi) \, \delta\left(\omega_k^l - kvxy - \nu\omega_H\right),$$
$$\tag{5.37}$$

$$\xi = \frac{kv}{\omega_H} \sqrt{1 - x^2} \sqrt{1 - y^2} = \beta \sqrt{1 - x^2} \sqrt{1 - y^2}. \tag{5.38}$$

When $\beta \gg 1$ the conservation laws allow processes for very large values of ν $(\beta \sim \nu)$; thus, the sum over ν can be replaced by an integration because $\Delta\nu/\nu \sim 1/\nu \ll 1$. We obtain

$$D^l = \frac{2\pi^2 e^2}{v^2} \int_0^\infty dk \ \frac{W_k}{k^2} \int_{-1}^1 dx \int_{-1}^1 dy \ \frac{\omega_k^l}{\left. \dfrac{\partial \varepsilon_k^l}{\partial \omega} \right|_{\omega = \omega_k^l}} \times$$

$$\times \frac{1}{|\omega_H|} J^2_{\alpha-\beta xy} \left(\beta \sqrt{1-x^2} \sqrt{1-y^2}\right), \quad \alpha = \omega^l_k/\omega_H . \qquad (5.39)$$

The largest contribution comes from those points at which the argument of the Bessel function is equal to the index $\omega^l_k = kvxy + kv\sqrt{1-x^2}\sqrt{1-y^2} = k_z v_z + k_\perp v_\perp$. But this also corresponds to the Cerenkov resonance condition without magnetic fields. In fact, when the Bessel function in (5.39) is replaced by a delta function, the resonance at $\omega = kv$ has a finite width. From the asymptotic form of the Bessel function [247]

$$J_{\frac{\omega^\sigma - kvxy}{\omega_H}} \left(\frac{kv}{\omega_H} \sqrt{1-x^2} \sqrt{1-y^2}\right) = \frac{1}{\pi} \sqrt{\frac{(\omega^\sigma_k - kvxy)^2 - k^2 v^2 (1-x^2)(1-y^2)}{3(\omega^\sigma_k - kvxy)}} \times$$

$$\times K_{1/3} \left(\frac{[(\omega^\sigma_k - kvxy)^2 - k^2 v^2 (1-x^2)(1-y^2)]^{3/2}}{3\omega_H (\omega^\sigma_k - kvxy)^2}\right), \qquad (5.40)$$

written for $(\omega^\sigma_k - kvxy)^2 > k^2 v^2 (1-x^2)(1-y^2)$, we can estimate the resonance width to be

$$\frac{\Delta\omega}{kv} \sim \left(\frac{\omega_H}{kv}\right)^{2/3} = \frac{1}{\beta^{2/3}} . \qquad (5.41)$$

The asymptotic form (5.40) gives an exponential decrease if the displacement from resonance is greater than (5.41). This is because the integrand in (5.39) falls in a region forbidden by $\delta(\omega - \mathbf{kv})$ $(\omega > \mathbf{kv})$. When the reverse inequality $(\omega^\sigma_k - kvxy)^2 < k^2 v^2 (1 - x^2)(1 - y^2)$ is satisfied, one must use the asymptotic form of the Bessel function which employs $J_{1/3}$ and $J_{-1/3}$ (see [248]). This gives damped oscillations. This corresponds to the region allowed by $\delta(\omega - \mathbf{kv})$. The diffusion coefficient can be written in a form close to that of (5.15) [249]:

$$D^l = \frac{4\pi e^2 \omega^2_{pe}}{v^3} \int_0^\infty dk \frac{W_k}{k^3} \Phi_1(\alpha, \beta); \quad \alpha = \frac{\omega_{pe}}{\omega_H}, \quad \beta = \frac{kv}{\omega_H}, \qquad (5.42)$$

where when $\alpha = \omega_{pe}/\omega_H \to \infty$ we have $\Phi_{1,\infty}(\alpha, \beta) \to \begin{cases} \pi/2, & \beta > \alpha, \\ 0, & \beta < \alpha. \end{cases}$

The function $\Phi_1(\alpha, \beta)$ is shown in Fig. 5.6 as obtained by numerical calculation with respect to β for $\alpha = 5; 10; 20$. This figure shows that the curves extend only slightly into the regions forbidden by the resonance. But in the allowed region the oscilla-

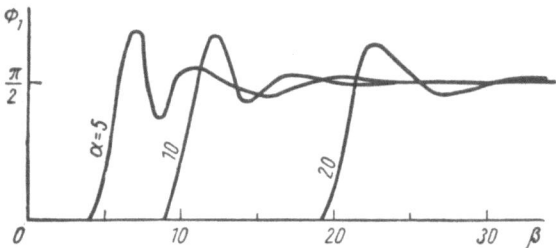

Fig. 5.6. Graphs of $\Phi_1(\alpha, \beta)$.

tions do not damp out very fast for smaller values of α. For the other branch of longitudinal oscillations, where $\omega_k^\sigma = \omega_{He} |\cos\theta|$, we obtain

$$D^l = \frac{4\pi e^2 \omega_{He}^4}{\omega_{pe}^2 v^3} \int_0^\infty dk \, \frac{W_k}{k^3} \, \Phi_2(\alpha, \beta); \qquad \alpha = \frac{\omega_{He}}{\omega_H}, \quad \beta = \frac{kv}{\omega_H}. \qquad (5.43)$$

In this instance the condition that $\omega = kv$ means in fact that $k > \dfrac{\omega_{He} |x|}{v(xy + \cos\varphi \sqrt{1-x^2} \sqrt{1-y^2})}$, and very small values of k are possible in principle when $x \to 0$. However, the minimum values of k are determined either by the magnetization condition $k > \omega_H/v$ or by the stipulation that there exist longitudinal gyrofrequency plasma waves with $k > \omega_{pe}/c$. The magnetization condition is most easily found from (5.37). When $\omega_k^\sigma \to 0$ only the term with $\nu = 1$ operates and we have $kv > \omega_H$. Thus $\Phi_2(\alpha, \beta)$ does not go rapidly to zero when $\beta < \alpha$, unlike the situation in Fig. 5.6. Figure 5.7 shows graphs of Φ_2, making it clear that Φ_2 goes to zero when $\beta < 1$. The asymptotic value of Φ_2 when $\beta \to \infty$ is $\pi/15$. It is attained when $\beta \gg \alpha$. In the region where $k \ll \omega_{pe}/c$ the branch $\omega_k^\sigma = \omega_{He} |\cos\theta|$ goes continuously into the whistlers. For electrons the largest value of this parameter is $\beta \approx (\omega_{pe}/\omega_{He})(v/c)$, and they can be considered unmagnetized when $v/c \gg \omega_{He}/\omega_{pe}$. But for ions we have $\beta_{max} \simeq (m_i/m_e)(\omega_{pe}/\omega_{He})(v/c)$. The minimum value of β is $(m_e/m_i)^{1/2}$ times smaller, so that in moderate magnetic fields the electrons can be magnetized. For the unmagnetized condition we have

$$D^l = \frac{2\pi^2 \omega_{He}^2 e^2 c^2}{3v\omega_{pe}^4} \int_{\frac{\omega_H}{v}} W_k^\omega \, k \, dk. \qquad (5.44)$$

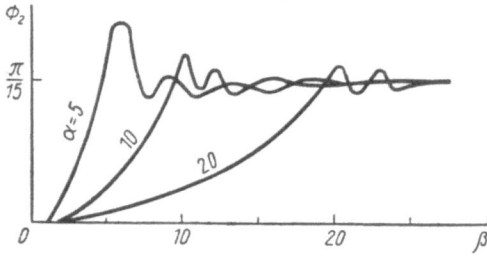

Fig. 5.7. Graphs of $\Phi_2(\alpha, \beta)$.

Here the lower limit is approximate and would correspond to changing Φ_2 to its asymptotic limit of $\pi/15$ in (5.43) down to $\beta \sim 1$ (see the graphs of Φ_2 in Fig. 5.7). However, to estimate the effect, (5.44) is a convenient equation. Figure 5.8 shows the value of $\Phi_3(\alpha, \beta)$ for whistlers in the formula

$$D^l = \frac{4\pi \omega_{He}^2 e^2}{v \omega_{pe}^4} \int \Phi_3(\alpha, \beta) W_k^w \, k \, dk; \qquad \alpha = \frac{\omega_k^w}{\omega_H}, \qquad \beta = \frac{kv}{\omega_H}. \qquad (5.45)$$

One should also keep in mind that the lower limit in (5.44) can be dropped if $v/c \gg \omega_H/\omega_{pi}$ because the smallest k for whistlers is ω_{pi}/c, and this case the diffusion coefficient given by (5.44) is a good approximation to the exact result. The condition

$$\frac{v}{c} \gg \frac{\omega_H}{\omega_{pi}} \qquad (5.46)$$

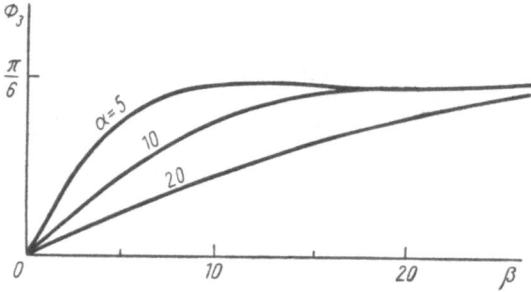

Fig. 5.8. Graphs of $\Phi_3(\alpha, \beta)$.

can be easily satisfied by ions, especially heavy ones. The effect
of acceleration on whistlers in the limit of (5.46) does not depend
on the whistlers' turbulence spectrum. Here D is proportional to
$1/\sqrt{\varepsilon}$, so that according to (5.8') $\dot{\varepsilon} \sim 1/\sqrt{\varepsilon}$ and, generally speaking,
acceleration by whistlers is much less effective than acceleration
by Langmuir oscillations. But they do have the advantage that
very low-energy particles can be accelerated, heavy ions in particular, whereas Langmuir oscillations can efficiently accelerate
only particles with velocities higher than v_{Te}. For magnetized
particles the fundamental contribution to the diffusion coefficient
comes from the resonances at $\nu = \pm 1$, and

$$D^l = \frac{\pi^2 e^2}{3\omega_{pe}^2} \omega_{He} \int\limits_{k^2 > \frac{\omega^2_{pe}}{c^2} \left| \frac{\omega_H}{\omega_{He}} \right|} W_k^w \, dk \left(1 - \frac{|\omega_H| \omega_{pe}^2}{|\omega_{He}| k^2 c^2} \right)^2. \tag{5.47}$$

Because $k \ll \omega_{pe}/c$, the diffusion coefficient of (5.47) does not go
to zero only for ions or ultrarelativistic electrons.

In a strong magnetic field $\omega_{He} \gg \omega_{pe}$ the acceleration in the
branch $\omega = \omega_{He}$ usually corresponds to unmagnetized particles,
and

$$D^l = \frac{4\pi e^2 \omega^2_{pe}}{v^3} \int\limits_{\frac{\omega_{He}}{v}}^{\infty} \frac{W_k}{k^3} \, dk \, \Phi_4 (\alpha, \beta). \tag{5.48}$$

For nonrelativistic electrons the lower limit is approximate,
because in this limit the electrons are unmagnetized. The exact
value of the form factor $\Phi_4(\alpha, \beta)$ when $\beta \to 1$ is $\frac{1}{6}(1 - \beta)^3$, but
when $\beta \gg 1$ it is $\pi/3$. For relativistic electrons and ions $\alpha \gg 1$,
so that Eq. (5.48) can approximate this result quite well with
$\Phi_4 = \pi/3$. Figure 5.9 shows the function $\Phi_4(\alpha, \beta)$. For branches
of magnetized Langmuir oscillations with $\omega = \omega_{pe}|\cos \theta|$ in a strong
magnetic field,

$$D^l = \frac{4\pi e^2 \omega_{pe}^2}{v^3} \int \frac{W_k \, dk}{k^3} \Phi_5 (\alpha, \beta). \tag{5.49}$$

The magnitude of Φ_5 when $\alpha > \beta$, $\beta \gg 1$ is $\pi/6$. However, for the
oscillations with frequency $\omega_{He} |\cos \theta|$ (in a weak field) the lower

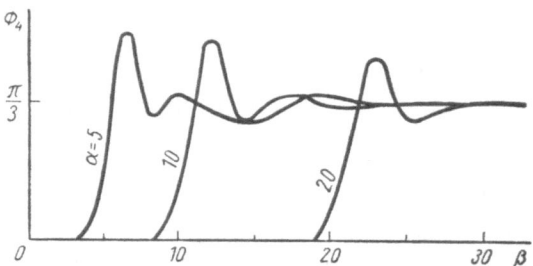

Fig. 5.9. Graphs of $\Phi_4(\alpha, \beta)$.

limit of (5.49) is, roughly speaking, determined by the particle magnetization condition $k > \omega_H/v$. Figure 5.10 shows the function $\Phi_5(\alpha, \beta)$. It also shows that the magnetization of the oscillations reduces the prohibition on acceleration of low-energy particles.

Finally, we shall present the result for magnetized, potential oscillations which corresponds to the limit $\beta \ll 1$, which is in the region where the particles are magnetized. The curves in Figs. 5.7 and 5.10 enter this region very slightly, but the results which are thereby obtained are very important to the injection problem. In this region one can find explicit analytic expressions for the diffusion coefficient.

It is easily found from Eq. (5.14') that in a weak magnetic field, for $\omega_k^\sigma = \omega_{He}|\cos\theta|$ only $\nu = 1$ gives any contribution, and

$$D^l = \frac{\pi^2}{3} \cdot \frac{\omega_{He}\, e^2}{\omega_{pe}^2} \left(1 - \frac{\omega_H^2}{\omega_{He}^2}\right)^2 \int W_h\, dk. \qquad (5.50)$$

For ultrarelativistic electrons and ions with $\omega_H \ll \omega_{He}$, the diffusion coefficient turns out to be identical. In a strong magnetic

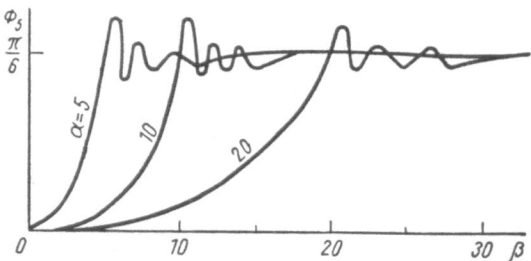

Fig. 5.10. Graphs of $\Phi_5(\alpha, \beta)$.

field we find that for the branch $v < \omega_{pe}/k$

$$D^l = \frac{2\pi^2 e^2}{3\omega_{pe}} \left(1 - \frac{\omega_H^2}{\omega_{pe}^2}\right) \int W_k \, dk \qquad (5.51)$$

under conditions of strong magnetization and when $\omega_H < \omega_{pe}$. The injection rate of (5.51) is difficult to compare with (5.50) because they belong to different values of the magnetic field. The conditions $\omega_H < \omega_{pe}$ and $\omega_{He} \gg \omega_{pe}$ indicate that in very strong magnetic fields, such that $\omega_{Hi} \gg \omega_{pe}$, only heavy ions with $m > m_i$ can be injected by the mechanism given in (5.51). Neither (5.50) nor (5.51) depends on the detailed form of the turbulence spectrum.

5.4. Stochastic Acceleration of Particles by High-Frequency Electromagnetic Radiation in a Magnetoactive Plasma

As has already been demonstrated, the appearance of relativistic electrons in turbulent plasma is a phenomenon which is completely natural for such a plasma. By necessity, intense synchrotron radiation must appear in the presence of magnetic fields [46]. However, particles located in the field of this radiation can not only spontaneously lose energy to this radiation, but they can also acquire energy from it because of induced absorption and emission [250]. In fact, the fact that there must be acceleration follows directly from (5.14). In the limit of very high frequencies $\omega \gg \omega_{pe}$, one can neglect the presence of the plasma. By projecting Γ_ν onto two orthogonal transverse unit vectors [see (2.216)] we find

$$\Gamma_{1,\nu} = \frac{\omega k_z - k^2 v_z}{k_\perp k} J_\nu(\xi), \qquad \Gamma_{2,\nu} = -i v_\perp J_\nu'(\xi). \qquad (5.52)$$

The emission probability averaged over the polarizations takes the well-known form [247]

$$w_p^t(\mathbf{k}, \nu) = (2\pi)^2 \frac{e^2}{\omega} \left\{ \frac{(\cos\theta - v_z)^2}{\sin^2\theta} J_\nu^2 \left(\frac{kv_\perp \sin\theta}{\omega_H}\right) + \right.$$

$$\left. + v_\perp^2 J_\nu'^2 \left(\frac{kv_\perp \sin\theta}{\omega_H}\right) \right\} \delta(\omega - \nu\omega_H - kv_z \cos\theta). \qquad (5.53)$$

For high frequencies the radiation spectrum is practically continuous, and the sum over ν in (5.14) can be replaced by an integral

$$
\dot{\varepsilon}_p = \frac{e^2}{2} \int\limits_{-1}^{1} dx \int\limits_{-1}^{1} dy \int\limits_{0}^{\infty} W_\omega^t \, d\omega \left\{ \omega x \frac{\partial}{\partial p_z} + \frac{\omega \omega_H}{v_\perp} \cdot \frac{\partial}{\partial p_\perp} \cdot \frac{(1 - v_z x)}{\omega_H} \right\} \times
$$

$$
\times \frac{2\pi^2}{\omega \, \omega_H} \left\{ \frac{(x - vy)^2}{1 - x^2} J^2_{\frac{\omega}{\omega_H}(1 - vxy)} \left(\frac{\omega}{\omega_H} v \sqrt{1 - x^2} \sqrt{1 - y^2} \right) + \right.
$$

$$
\left. + v^2 (1 - y^2) J'^2_{\frac{\omega}{\omega_H}(1 - vxy)} \left(\frac{\omega}{\omega_H} v \sqrt{1 - x^2} \sqrt{1 - y^2} \right) \right\}. \qquad (5.54)
$$

Here $W^t = \int W_\omega^t \, d\omega$ is the energy density of the electromagnetic radiation.

We shall consider two cases: 1) relativistic particles moving nearly perpendicular to the magnetic field, and 2) relativistic particles with an isotropic distribution. In the first case one can drop the integration over y in (5.54), setting y = 0. If one uses the asymptotic expression

$$
J_{\frac{\omega}{\omega_H}} \left(\frac{\omega}{\omega_H} v \sqrt{1 - x^2} \right) = \frac{\sqrt{2\rho}}{\pi \sqrt{3}} K_{1/3} \left(\frac{\omega}{3\omega_H} (2\rho)^{3/2} \right), \quad 2\rho = \frac{m^2 c^4}{e^2} + x^2 \qquad (5.55)
$$

($x^2 \ll 1$ because the relativistic particle radiates primarily along the direction of its motion), Eq. (5.54) takes the form [250]

$$
\dot{\varepsilon}_p = \frac{9 \sqrt{3} \, \pi}{2e} e^2 \omega_H^3 \left(\frac{\varepsilon}{mc^2} \right)^7 \int\limits_{0}^{\infty} \frac{W_\omega^t}{\omega^3} c^2 \, \xi^3 K_{5/3} (\xi) \, d\xi, \qquad (5.56)
$$

where

$$
\xi = \frac{2}{3} \cdot \frac{\omega}{\omega_H} \left(\frac{mc^2}{\varepsilon} \right)^3. \qquad (5.57)
$$

The frequency distribution of the electromagnetic radiation can be different, but the greatest interest centers on the spectrum which generates in a natural way the most relativistic electrons. As shown in [46] (see Chapter 6 also), in the low-frequency region, where reabsorption plays an important part,

$$
W_\omega^t \simeq \text{const } \omega^{5/2}. \qquad (5.58)
$$

It is then easy to see that $\dot{\varepsilon}_p \simeq$ const ε^2 from Eq. (5.56); i.e., the response of the acceleration to energy changes is considerably faster than or Fermi acceleration. The spectrum in (5.58) cannot be extended to $\omega \to \infty$; it is terminated usually at some ω_* which depends on both the dimensions and the radiation reabsorption length. When the radiation spectrum W_ω^t terminates at ω_*, or decreases according to the function $W_\omega = $ const$/\omega^\nu$, $\nu > 1$, the effective acceleration $\sim \varepsilon^2$ ceases when $(\varepsilon_*/mc^2)^2 \simeq \omega_* mc/eH$, the diffusion coefficient becomes a constant with respect to energy, and $\dot{\varepsilon} \simeq$ const$/\varepsilon$. Similar results are obtained for an isotropic distribution of relativistic particles. In this case, according to (5.53) and (5.14), the diffusion coefficient can be written as

$$D^l = \sum_{\nu=-\infty}^{\infty} \frac{e^2 \pi^2}{v^2} \int_0^\infty d\omega \, W_\omega^t \left(\frac{v\omega_H}{\omega}\right)^2 \int_{-1}^1 dx' \int_{-1}^1 dy \times$$

$$\times \, \delta \left(v\omega_H - \omega \frac{1-v^2 y^2}{1-vx' y}\right) \left[v^2 \frac{1-y^2}{1-v^2 y^2} J_\nu'^2 (\xi') + \frac{x'^2}{1-x'^2} J_\nu^2 (\xi')\right], \qquad (5.59)$$

$$\xi' = v v \frac{\sqrt{1-x'^2}\,\sqrt{1-y^2}}{\sqrt{1-v^2 y^2}}, \qquad (5.60)$$

where x' is the cosine of the angle between the directions k and $\mathbf{H}_0$ in a coordinate system where v = 0 for the particles. The integration over the angle x' is performed just as for an individual particle in a vacuum (see [247]) if it is assumed that x' $\ll$ 1 because the particle radiates primarily along its direction of motion,

$$\dot{D}^l = \sum_{\nu=-\infty}^{\infty} 2 \frac{e^2 \pi^2}{v^2} \int_0^\infty d\omega W_\omega^t \left(\frac{v\omega_H}{\omega}\right)^2 \int_{-1}^1 dy \, \delta (v\omega_H - \omega + \omega v^2 y^2) \times$$

$$\times \left[\frac{1}{\nu} u J_{2\nu}' (2\nu u) - \frac{1-u^2}{u} \int_0^u J_{2\nu} (2\nu \xi) \, d\xi\right]; \qquad (5.61)$$

$$u = \sqrt{1 - \frac{m^2 c^4 \omega}{\varepsilon^2 v \omega_H}}. \qquad (5.62)$$

Using the asymptotic form of the Bessel functions, we find [251]

$$D^l = \frac{e^2 \pi}{\sqrt{3}} \omega_H^{1/2} \frac{m^2 c^4}{\varepsilon^2} \int_0^\infty d\omega \frac{W_\omega^t}{\omega^{3/2}} \int_\xi^\infty d\xi' \, K_{5/3} (\xi') \sqrt{\frac{\omega}{\omega_H}} \sqrt{1 - \frac{\xi^2}{\xi'^2}}. \qquad (5.63)$$

If we write

$$W_\omega^t = \frac{1}{\omega_*} I \left(\frac{\omega}{\omega_*} \right)^{\nu'},$$ (5.64)

in place of (5.58), then when $\varepsilon/mc^2 \ll (mc\omega_*/eH)^{1/2}$ we obtain [251]

$$D^l = \frac{\pi^{3/2} 3^{\nu'-1/2} Ie^2}{8eH} mc \left(\frac{eH}{mc\omega_*} \right)^{\nu'-1} \times$$

$$\times \left(\frac{\varepsilon}{mc^2} \right)^{2\nu'-2} \frac{\Gamma \left(\frac{\nu'}{2} \right) \Gamma \left(\frac{\nu'}{2} + \frac{4}{3} \right) \Gamma \left(\frac{\nu'}{2} - \frac{1}{3} \right)}{\Gamma \left(\frac{\nu'}{2} + \frac{3}{2} \right)}.$$ (5.65)

Thus, when $\nu' = {}^5/_2$, $D \sim \varepsilon^3$ and $\dot{\varepsilon} \sim \varepsilon^2$, i.e., the isotropic case is qualitatively similar to that considered above (they differ in the numerical coefficients in D). However, the anisotropic distributions of relativistic particles can be rapidly liquidated, even if because of the synchrotron instability [252], and isotropic distributions correspond more closely to the real distributions in cosmic objects.

Figure 5.11 shows a schematic representation of the behavior of $\dot{\varepsilon} = f(\varepsilon)$ for acceleration by high-frequency radiation. This curve is similar in shape to the curve obtained for acceleration by Langmuir fluctuations but the maximum is located in the region of ultrarelativistic energies and depends heavily on the size of the object, the radiation density, and other characteristics. The increase in the acceleration rate as the energy increases is analogous to the case of Langmuir fluctuations; this comes from an increase in intensity with the frequency, and is therefore related

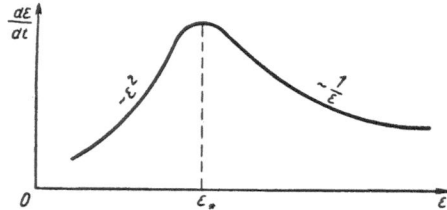

Fig. 5.11. Acceleration rate of relativistic electrons by synchrotron radiation.

to the fact that a greater total radiation density acts on those particles with the larger energies. The decrease in the acceleration effect is explained by the saturation of D.

Let us now discuss the role of the plasma in the acceleration mechanisms which we have been examining. We shall assume that $\omega \gg \omega_{He}$, as before. The ω_{He} terms can be neglected in the dielectric susceptibility tensor:

$$\varepsilon(\omega) = 1 - \frac{\omega_{pe}^2}{\omega^2}. \tag{5.66}$$

In spite of the fact that the index of refraction is only slightly different from unity, the plasma can have a significant effect on the acceleration of ultrarelativistic particles [253]. In this instance one is interested in how close $v^2 = c^2 \left(1 - \frac{m^2 c^4}{\varepsilon^2}\right)$ is to $\frac{c^2}{\varepsilon(\omega)} = c^2 \left(1 + \frac{\omega_{pe}^2}{\omega^2}\right)$. By comparing these two expressions we see that the plasma has a significant impact when $\omega < \omega_{pe}(\varepsilon/mc^2)$. The emission probability when $\varepsilon(\omega) \neq 1$ is easily found from the general equations if we use Eq. (5.52) and $k^2 c^2 = \omega^2 \varepsilon(\omega)$; the rather simple, but general result was first found in [248] (see [254] also):

$$w_{p_z, p_\perp}(k,v) = \frac{(2\pi e)^2}{\varepsilon(\omega)\,\omega}\,\delta\left(\omega - \omega_H v - \omega v_z x \sqrt{\varepsilon(\omega)}\right) \times$$

$$\times \left\{\frac{(x - \sqrt{\varepsilon(\omega)}\,v_z)^2}{1 - x^2}\,J_v^2(\xi') + v_\perp^2\,\varepsilon(\omega)\,J_v'^{2}(\xi')\right\},$$

$$\xi' = \left(1 - \sqrt{\varepsilon}\,v_z x\right)^{-1} v \sqrt{\varepsilon}\,v_\perp \sqrt{1 - x^2}, \quad x = \cos\theta. \tag{5.67}$$

Because qualitatively the results obtained for the isotropic case and the case of particle motion perpendicular to the magnetic field are the same, we will consider the simpler case where $v_z = 0$. We then obtain relationships which are similar to those of (5.56):

$$\dot{\varepsilon}_p = \frac{e^2 \omega_H^3 2\pi^2}{\varepsilon \sqrt{3}} \int_0^\infty v^2\,dv \left\{2(\eta_1 + \eta)\,\xi_1\,K_{5/3}(\xi_1) - \right.$$

$$\left. - 3\eta\xi_1\,K_{1/3}(\xi_1) - 2\eta \int_{\xi_1}^\infty K_{1/3}(\xi_1)\,d\xi_1\right\} \frac{w_\omega^t}{\omega^3}, \tag{5.68}$$

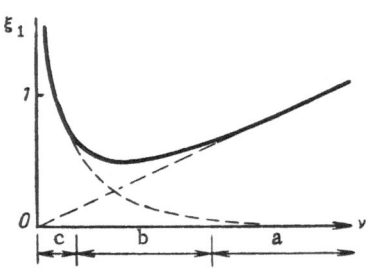

Fig. 5.12. An example of the function $\xi_1(\nu)$ (the dashed lines are the asymptotic limits of the curves). Three regions are marked for different variations in ξ_1 (a, b, and c).

$$\omega = \omega_H \nu, \quad \eta_1 = \frac{m^2 c^4}{\varepsilon^2}, \quad \eta = 1 - \varepsilon^2(\omega) = \frac{\omega_{pe}^2 m^2 c^2}{\nu^2 e^2 H^2 \eta_1}, \quad \xi_1 = \frac{2\nu}{3}(\eta_1 + \eta)^{3/2}.$$

Figure 5.12 shows the schematic dependence of ξ_1 on ν. Three different regions of ν can be distinguished:

a) $\nu \gg \dfrac{\omega_{pe} \, mc}{eH}\left(\dfrac{\varepsilon}{mc^2}\right)^2$; $\eta_1 > \eta$; $\xi_1 = \xi = \dfrac{2\nu}{3}\eta_1^{3/2}$. In this region the plasma has no effect on the acceleration and the result agrees with (5.56):

b) $\nu \ll \dfrac{\omega_{pe} \, mc}{eH}\left(\dfrac{\varepsilon}{mc^2}\right)^2$; $\eta_1 \ll \eta$; $\xi_1 = \dfrac{2}{3\nu^2}\left(\dfrac{\omega_{pe} \, \varepsilon}{eHc}\right)^3$. Here

$$\dot{\varepsilon} = \frac{2\pi^2 e^2 \omega_{pe}^{7/2}}{3\pi \sqrt{2}}\left(\frac{mc}{eH}\right)^{1/2}\left(\frac{\varepsilon}{mc^2}\right)^{7/2}\int_0^\infty \frac{d\xi_1 \, W_\omega^t}{\omega^3 \, \xi_1^{3/2}}\, \Phi(\xi_1), \qquad (5.69)$$

$$\Phi(\xi_1) = 2\int_{\xi_1}^\infty K_{5/3}(\xi_1')\, d\xi_1' - \xi_1 K_{5/3}(\xi_1).$$

A graph of $\Phi(\xi_1)$ is shown in Fig. 5.13. When $\xi_1 \gg 1$, $\Phi(\xi_1)$ is negative and the acceleration is replaced with deceleration. This means that the particles give energy to the radiation, and the

Fig. 5.13. Graph of $\Phi(\xi_1)$.

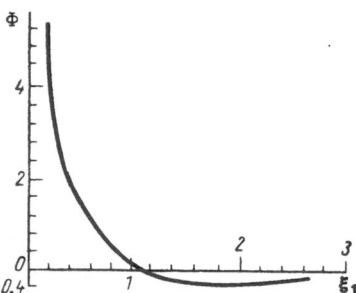

possibility of generating transverse waves arises [255]. However, the relative contribution of the region $\xi_1 \gg 1$ is not large for many distributions.

c) $v \sim \dfrac{\omega_{pe}\, mc}{eH} \left(\dfrac{\varepsilon}{mc^2}\right)^2$ or $\eta_1 \sim \eta$. This region is the minimum

of the curve in Fig. 5.12. The smallest value of ξ_1 is $\xi_{1\min}=$

$= \dfrac{\omega_{pe}\, \sqrt{3}\, m^2\, c^3}{eH}$, where $v = v_m = \sqrt{2}\, \omega_{pe} \dfrac{mc}{eH} \left(\dfrac{\varepsilon}{mc^2}\right)^2$. If $\dfrac{\varepsilon}{mc} \gg \dfrac{\omega_{pe}\, mc}{eH}$, then

$\xi_{1\min} \ll 1$ and one can use this asymptotic form in (5.68):

$$\int_{\xi}^{\infty} K_{5/3}(\xi)\, d\xi = \frac{3}{2} \cdot \frac{2^{2/3}\, \Gamma(5/3)}{\xi^{2/3}},$$

where

$$\dot{\varepsilon} = \frac{4\pi e^2\, \omega_H^3}{\varepsilon}\, 3^{1/6}\, \Gamma(5/3) \int \frac{\omega^{4/3}\, d\omega}{\omega^3}\, \omega_H^{7/3}\, W_\omega^t.$$

The region in which this formula can be used overlaps with (5.56) and (5.68). It follows that (5.56) is applicable when $v \gg (\omega_{pe}\varepsilon/eHc)^{1/2}$, and (5.68) can be used for $v \ll (\varepsilon/mc^2)^3$.

In addition to the synchronous acceleration effects in turbulent plasma, there are other effects related to the turbulent fields present. In many respects they are similar to those effects already described, and they all must be discussed together. We shall therefore postpone our discussion of the spectra for particles accelerated by radiation until a later chapter, at which time we will also treat the spectra for the radiation from a turbulent plasma.

5.5. Stochastic Particle Acceleration by Low-Frequency Ion-Acoustic Oscillations

We shall now examine the fluctuations connected with those collective motions involving the plasma ions. For ion-acoustic oscillations, $\omega_k^s = \dfrac{k v_s}{\sqrt{1 + k^2 d_e^2}}$, where $v > v_s$ and $kv \gg \omega_H$, we obtain

$$D^s = \frac{2\pi^2 e^2}{v^3} \int \frac{kW_k^s\, dk\omega_{pi}^2\, v_{Te}^4}{(\omega_{pe}^2 + k^2 v_{Te}^2)^2} \equiv D_0\, \frac{e^2\, m^{3/2}}{4\varepsilon_p^{3/2}}. \tag{5.70}$$

The limits of the integration in (5.70) are not dependent on the velocities of the accelerated particles if those velocities are larger than the sound velocity; therefore the acceleration by ion-acoustic oscillations is independent of the spectrum for the ion-acoustic turbulence. In particular, it follows from (5.70) that in the first approximation the systematic acceleration is zero. But the fluctuating acceleration is very large. Since $\dot{\varepsilon}_p = 0$, $(\Delta\dot{\varepsilon})^2 = \dot{\varepsilon}^2$:

$$\dot{\varepsilon}^2 = 2D^s v^2 = D_0\, \frac{e^2\, \sqrt{m}}{\sqrt{\varepsilon_p}}, \quad \varepsilon_p = \frac{p^2}{2m}. \tag{5.71}$$

This last equation shows that the acceleration rate (or the heating) of fast particles is larger the greater their mass and charge, so that heavy, multiply-charged ions will be accelerated the easiest. This result is important to the production of cosmic rays [46]. The essential difference between acceleration by ion-acoustic waves and by Langmuir oscillations is that acceleration by ion-acoustic waves is possible at very low velocities ($v \sim v_s \sim v_{Te} \times (m_e/m_i)^{1/2}$, and this is important for ion acceleration and injection into the acceleration regime of the high-frequency oscillations. As shown in Chapter 4, the development of high-frequency turbulence leads to the heating of the plasma electrons and the generation of ion-acoustic turbulence even if there is no direct source of excitation for it. It then follows that the effects of (5.71), which lead to injection, arise with the necessity for a long turbulence existence. It is useful to compare the diffusion coefficients D^l and D^s for conditions in which ion-acoustic and Langmuir oscillations are produced by a source of Langmuir vibrations with power Q.

Therefore, when $\omega_{pe}/k_*^s > c$, i.e., for rather hot electrons, $W_k^l = \sqrt{\dfrac{2Q}{\alpha'}}$ and

$$D^l = \frac{3\sqrt{3\pi}}{2\sqrt{2}}\, m_e^{3/2}\, m_i^{1/2}\, Z^2 \sqrt{\frac{m}{2\varepsilon_p}}\, v_{Te}^3\, \sqrt{\frac{Q}{Q_0}}, \quad e_m = Ze. \tag{5.72}$$

On the other hand, by using (4.123), Eq. (5.70) gives us

$$D^s = \frac{m_e^{5/2}}{m_i^{1/2}} \, \omega_{pe} \sqrt{\frac{Q}{Q_0}} \, v_{Te}^5 \left(\frac{m}{2\varepsilon_p}\right)^{3/2} Z^2, \tag{5.73}$$

$$Q \gg Q_0 \frac{1}{12}\left(\frac{m_e}{m_i}\right)^3. \tag{5.74}$$

Equations (5.73) and (5.74) have been written to within an accuracy of a numerical coefficient of order unity. When $v \sim v_{Te}$, D^s is m_e/m_i times smaller than D^l and decreases much more rapidly as the particle energy increases. Thus, the mechanism of (5.73) can, in fact, act as a natural injection mechanism.

If the ion-acoustic instability is excited directly, one can use Eqs. (4.231) and (4.236) to give

$$D^s = \frac{15\sqrt{\pi}\,Z^2}{8\sqrt{2}} \, m_e^2 v_{Te}^5 \left(\frac{m_e}{m_i}\right)^{3/2} \frac{T_e}{T_i} k_g v_{Te} \left(\frac{m}{2\varepsilon_p}\right)^{3/2} \left(\sqrt{\frac{Q}{Q_*^s}} + e^{-\sqrt{\frac{Q}{Q_*^s}}} - 1\right). \tag{5.75}$$

When $Q \ll Q_*^s$ the diffusion coefficient D^s is linear in Q, while if $Q \gg Q_*^s$ it is proportional to $(Q/Q_*)^{1/2}$. If we use the spectrum given in (4.257), then

$$D^s = \left(\frac{\pi}{2}\right)^{3/2} \frac{Z^2}{4} k_* v_{Te} m_e^2 v_{Te}^5 \left(\frac{m}{2\varepsilon_p}\right)^{3/2} \left(\frac{m_e}{m_i}\right)^{3/2}, \tag{5.76}$$

which is smaller than (5.75) in general, i.e., most of the acceleration is accomplished by fluctuations with frequencies close to ω_{pi}. If an equilibrium particle distribution arises as a balance between the acceleration of (5.70) and the deceleration due to collisions, then

$$\frac{\partial f_e}{\partial t} = \frac{e^2 \sqrt{m}\,Z^2}{2} D_0 \frac{\partial^2}{\partial \varepsilon^2} \cdot \frac{f_e}{\sqrt{\varepsilon}} + \frac{\nu_{e,i} \sqrt{m}\,m_{e,i}}{\sqrt{2}} v_{Te,i}^3 \frac{\partial}{\partial \varepsilon} \cdot \frac{Z^2 f_e}{\sqrt{\varepsilon}} = 0, \tag{5.77}$$

i.e.,

$$f_\varepsilon = \text{const} \, \sqrt{\varepsilon} \, e^{-\varepsilon/T_{\text{eff}}}. \tag{5.78}$$

For fast electrons and ions ($v \gg v_{Te}$) one must take into account the collisions with electrons, and we obtain for (5.73)

$$T_{\text{eff}} = T_e \frac{\omega_{pe}}{\nu_e} \sqrt{\frac{m_e}{m_i}} \sqrt{\frac{Q}{Q_0}}. \tag{5.79}$$

This gives us some idea of what the turbulent energy flux must be in order to have efficient injection; i.e., $T_{\text{eff}} \gg T_e(m/m_e)$, and

$$Q \gg Q_0 \left(\frac{\nu_e}{\omega_{pe}}\right)^2 \frac{m_i}{m_e} \cdot \frac{m^2}{m_e^2} \simeq \nu_e n_0 T_e \frac{m_i m^2 \nu_e}{m_e^3 \omega_{pe}}. \tag{5.80}$$

Thus it is clear that for high power Q the accelerative mechanism of the dissipation becomes a necessary element in the development of turbulence. When conditions are such that (5.74) is valid, the effective injection temperature is

$$T_{\text{eff}} \approx \frac{\omega_{pe}}{\nu_e} T_e \frac{T_e}{T_i} \left(\frac{m_e}{m_i}\right)^{3/2} \left(\sqrt{\frac{Q}{Q_*^s}} + e^{-\sqrt{\frac{Q}{Q_*^s}}} - 1\right). \tag{5.81}$$

Then because of (4.236), $Q \gg Q_*^s$ and T_{eff} is always larger than T_e if

$$\frac{\omega_{pe}}{\nu_e} > \frac{T_i}{T_e} \left(\frac{m_i}{m_e}\right)^{3/2}. \tag{5.82}$$

This last condition is easily met, and ion-acoustic turbulence is a good injector of fast particles.

We will now give qualitative consideration to the effects which originate when turbulence is excited by a constant external electric field. If the velocity of the electron streaming motion is close to v_s (see [137]), then

$$Q \simeq eEn_0 v_s, \tag{5.83}$$

so that if $Q \gg Q_*^s$, or according to (4.237), if

$$\frac{E^2}{8\pi n_0 T_e} \gg \frac{m_e \varepsilon}{m_i}, \qquad \varepsilon = \frac{m_e}{m_i} \cdot \frac{T_e^2}{T_i^2} \ll 1, \tag{5.84}$$

we obtain

$$T_{\text{eff}} = T_e \frac{\omega_{pe}}{\nu_e} \cdot \frac{T_e}{T_i} \left(\frac{m_e}{m_i}\right)^{1/2} \left(\frac{E m_i T_i}{m_e T_e \sqrt{4\pi n_0 T_e}}\right)^{1/2}. \tag{5.85}$$

We note that because of (4.243) $Q < Q_*^s/\sqrt{\varepsilon}$. If $Q < Q_*^s$ the non-linearity is small and the equilibrium can arise by taking into account the quasilinear decrease in the growth rate. It is then evident that $2W\gamma_k^s = Q$, i.e.,

$$W \approx \frac{m_i e E n_0 v_s}{m_e \omega_{pe}} \sqrt{\frac{2}{\pi}}, \quad D_0 \simeq \frac{v_{Te}^3}{\omega_{pe}^2} e E n_0 v_s. \tag{5.86}$$

Therefore

$$T_{\text{eff}}^{(i)} = \frac{E}{E_k} \sqrt{\frac{m_i}{m_e}} T_e, \quad E_k = \nu_e v_{Te} \frac{m_e}{e}. \tag{5.87}$$

Equation (5.87) was derived in [256].

5.6. Acceleration by Alfven Waves and Magnetoacoustic Oscillations

The accelerated particles will not be magnetized if $\omega = k v_A > \omega_H v_A /v$. Thus, for nonrelativistic electrons $\omega > \omega_{He} v_A /v$, but $\omega < \omega_{Hi}$, i.e., $v > v_A(m_i/m_e) = v_{inj}$ and the injection threshold is very high. Even if $v_A = v_s$ we will have $v_{inj} \approx (m_i/m_e)^{1/2} v_{Te}$, and the electron energy will be at the very least 10^3 times greater than the thermal energy. However, usually $v_A \gg v_s$ and the electron injection energy is still large [when $v_A \sim v_{Te}$ it is $(m_i/m_e)^2 T_e$]. This tells us that nonrelativistic electrons are not, in practice, accelerated efficiently by such oscillations. Ultrarelativistic electrons will be accelerated. Conversely, the condition for demagnetization is almost always satisfied for nonrelativistic ions if $v > v_A$, and it is very well satisfied for heavy ions such that $m \gg m_i$. For Alfven waves and magnetoacoustic oscillations the diffusion coefficient takes the form [257, 249]

$$D_A = \sum_{\nu=-\infty}^{\infty} \frac{2\pi^2 e^2 v_A^4 Z^2 \omega_H^2}{v^2 \left(1+\frac{v_A^2}{c^2}\right) c^2} \int_0^\infty d\omega \, \frac{W_\omega^A}{\omega^2} \int_{-1}^1 dx \, x^2 \, W^A(x) \times$$

$$\times \int_{-1}^1 dy \, \frac{v^2 J_\nu^2(\xi)}{1-x^2} \, \delta(\omega - kv\,xy - \nu\omega_H), \qquad (5.88)$$

$$\xi = \frac{\kappa v}{\omega_H} \sqrt{(1-x^2)(1-y^2)},$$

$$2\pi W_{\omega,\Omega}^A = W_\omega^A \, W^A(x), \quad \int_{-1}^1 W^A(x)\,dx = 1, \quad x = \cos\theta,$$

$$D_M = \sum_{\nu=-\infty}^{\infty} \frac{2\pi^2 e^2 v_A^2 Z^2}{1+\frac{v_A^2}{c^2}} \int_0^\infty d\omega \, W_\omega^M \int_{-1}^1 dy \, (1-y^2) \, W^M(x) \times$$

$$\times (J_\nu'(\xi))^2 \, \delta(\omega - kvxy - \nu\omega_H),$$

$$2\pi W_{\omega,\Omega}^M = W_\omega^M \, W^M(x), \quad \int W^M(x)\,dx = 1, \quad x = \cos\theta. \qquad (5.89)$$

The probability given in 2.223 can be used for unmagnetized particles:

$$D_A = \frac{2\pi^2 e^2 v_A^3 Z^2}{c^2 v \left(1+\frac{v_A^2}{c^2}\right)} \int_{\omega_H \frac{v_A|x|}{v}} \frac{W_\omega^A}{\omega} d\omega \int_{-1}^1 |x|^3 \, dx \, W^A(x) \left[\frac{v_A^2}{v^2} \cdot \frac{(2-3x^2)}{2} + \frac{1}{2}\right],$$

$$|x| < \frac{v}{v_A}. \qquad (5.90)$$

The upper limit in (5.90) is unimportant since the energy density for Alfven waves goes to zero when $\omega \to \omega_{Hi}$ (from the definition of the Alfven branch). The lower limit $\omega_H v_A / v$ is written approximately from the demagnetization condition. We note that the angular dependence of the Alfven oscillations is not very important when $v \gg v_A$. In this instance only the integral $\int_{-1}^1 |x|^3 W^A(x)\,dx$

enters into D_A, and it merely determines the numerical constant.†
Because of normalization this constant is of order unity. If $v \ll v_A$,
(5.90) takes the approximate form

$$D_A = \frac{\pi^2 e^2 v_A Z^2 W^A (0) v}{2 \left(1 + \frac{v_A^2}{c^2}\right) c^2} \int_{\omega_H}^{\infty} \frac{W_\omega^A}{\omega} d\omega. \tag{5.91}$$

It follows from (5.88) that when there is magnetization $kv \ll \omega_H$
we have $\omega = \omega_H$, i.e., $v \ll v_A x < v_A$ and the integration extends into
the region which is not covered by (5.90). From (5.88) it follows
that

$$D_A = \frac{2\pi^2 e^2 v_A^2 Z^2}{3c^2 \left(1 + \frac{v_A^2}{c^2}\right)} W_{\omega_H}^A \int_{x \gg \frac{v}{v_A}} W^A (x) \, dx. \tag{5.92}$$

By comparing (5.92) with (5.91) we find that the region of unmagnetized particles makes a very small contribution, and we can
use (5.92) (of course, under conditions when the angular distribution is more or less isotropic; more precisely, if there is no
situation in which all the waves move along the field). Equations
(5.92) and (5.90) are approximations (limiting cases). Even if
$v \gg v_A$ the lower limit in (5.90) is approximate, and the exact
solution contains a form factor which depends on $\alpha = \omega/\omega_H$, and

$$\beta = \frac{kv}{\omega_H} = \frac{v}{v_A} \cdot \frac{\omega}{\omega_H |x|}, \qquad D_A = \frac{3\pi e^2 Z^2 v_A^3}{c^2 v \left(1 + \frac{v_A^2}{c^2}\right)} \int \Phi_A (\alpha, \beta) \frac{W_\omega^A}{\omega} d\omega. \tag{5.93}$$

Figure 5.14 shows the function Φ_A for the case of isotropic turbulence.

For magnetohydrodynamic oscillations and unmagnetized
particles we find

$$D_M = \frac{3\pi^2 v_A^3 e^2 Z^2}{c^2 v \left(1 + \frac{v_A^2}{c^2}\right)} \int_{\omega_H \frac{v_A}{v}} \frac{W_\omega^M}{\omega} d\omega \left(1 - \frac{v_A^2}{v^2}\right). \tag{5.94}$$

---

†The lower limit can be approximated by $\omega_H v_A/v$ since $x \sim 1$.

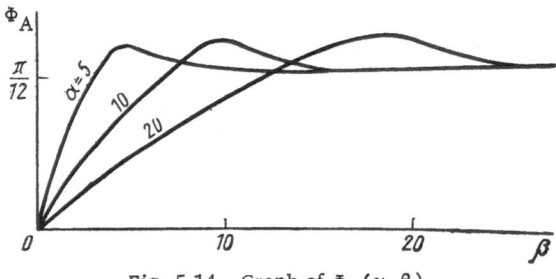

Fig. 5.14. Graph of $\Phi_A(\alpha, \beta)$.

Here the integral is nonzero only if $v > v_A$. In general the angular dependence in (5.94) drops out because $\int W^M(x)dx = 1$, and the acceleration effects do not depend on the angular distribution of turbulent fluctuations. In the general case D_M can be written in a form analogous to (5.93). The function $\Phi_M(\alpha, \beta)$ is shown in Fig. 5.15. Under magnetization conditions $kv \ll \omega_H$ we have $\omega_H = \omega(\nu = 1)$ and $v \ll v_A$.

We then find from (5.89) that

$$D_M = \frac{2\pi^2 e^2 Z^2 v_A^2}{3c^2 \left(1 + \dfrac{v_A^2}{c^2}\right)} W_{\omega_H}^M \int\limits_{-1}^{1} dx\, W^M(x). \qquad (5.95)$$

Equations (5.92) and (5.95) tell us that injection into the acceleration regime of the Alfven waves and magnetoacoustic oscillations is determined by the spectral density of the Alfven and magnetoacoustic waves when $\omega = \omega_H$. According to Chapter 4, for a

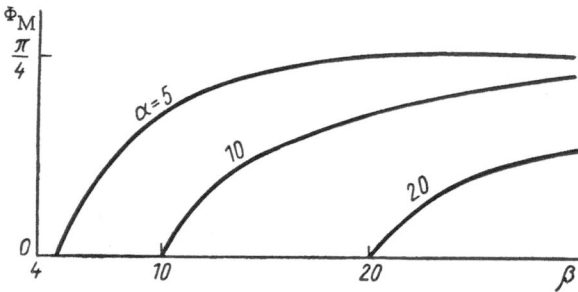

Fig. 5.15. Graph of $\Phi_M(\alpha, \beta)$.

one-dimensional spectrum $W_\omega^A = \frac{const}{\omega^\nu}$, $\nu = 1$, so that

$$D = const\, \frac{e^2\, v_A^2\, mcZ^2}{\left(1 + \frac{v_A^2}{c^2}\right)\, |eH|\, Z}, \tag{5.96}$$

and therefore $\dot{\varepsilon} \sim (3/2m)D \sim Z$ does not depend on the particle mass. For magnetoacoustic waves with $T_e \gg T_i$ it is possible to have $\nu = 2$, giving $D \sim m\varepsilon$ and $\dot{\varepsilon} = const\,\varepsilon$. This mechanism is more effective than the Fermi acceleration when $\dot{\varepsilon} = const\sqrt{\varepsilon}$. Evidently, since $\omega \ll \omega_{Hi}$ and $\omega = \omega_H$, neither protons nor electrons can be injected by this mechanism.

The general form of the acceleration which follows from (5.92) and (5.96) is

$$\dot{\varepsilon} = 2\pi^2 \frac{e^2\, Z^2\, v_A^2}{\left(1 + \frac{v_A^2}{c^2}\right) m}\left(W_{\omega_H}^A + W_{\omega_H}^M\right)\frac{1}{c^2}; \tag{5.97}$$

it was first obtained in [258] (for $v_A \ll c$), where it was also shown that predominant acceleration of heavy ions takes place when $\nu > 1$. If it is assumed in the one-dimensional case that

$$W_\omega = \frac{W}{\ln\frac{\omega_{max}}{\omega_{min}}\,\omega}, \tag{5.98}$$

the equilibrium distribution of ions will then be

$$f_\varepsilon = const\,\sqrt{\varepsilon}\,e^{\left(\frac{T_{eff}}{\varepsilon}\right)^{1/2}},$$

$$T_{eff} = 18\,\frac{v_i^2 mc^2}{\omega_{pi}^2\,\pi^2}\left(1 + \frac{v_A^2}{c^2}\right)^2 \frac{v_{Ti}^2}{v_A^2}\left(\frac{n_0 T_i}{W}\right)^2\left(\ln\frac{\omega_{max}}{\omega_{min}}\right) Z^2. \tag{5.99}$$

The value of T_{eff} denotes that energy for which the distribution goes to $f = const\sqrt{\varepsilon}$. If the injection is due to magnetoacoustic oscillations with $\nu = 2$,

$$f_\varepsilon = const\,\sqrt{\varepsilon}\,e^{\left(\frac{T_{eff}}{\varepsilon}\right)^{1/2}},$$

$$T_{eff}^{1/2} = \frac{H^2}{W^M}\cdot\frac{3v_i m_i\sqrt{m}\,v_{Ti}^3\,L}{v_A^3\,8m\pi^3}\,Z^2\left(1 + \frac{v_A^2}{c^2}\right). \tag{5.100}$$

The distributions of (5.100) and (5.99) hold only if $mv_A^2/2 > \varepsilon > mv_{Ti}^2/2$. In (5.100) it is assumed that the spectrum $W^M = const/\omega^2$ holds until $k > k_0 = 2\pi/L$, where L is the fundamental scale of the turbulence, i.e.,

$$W_\omega^M = \frac{2\pi}{L} v_A \frac{W^M}{\omega^2}. \qquad (5.101)$$

The values of T_{eff} play the role of those energies for which collisions cannot counteract the acceleration, and the particles, being continuously accelerated, reach velocities such that $v = v_A$.

When $v \ll v_{Ti}$ we find instead of (5.99) that

$$f_\varepsilon = const \sqrt{\varepsilon} \exp(-\varepsilon/T_{eff}), \qquad (5.102)$$

$$T_{eff} = T_i \frac{4\pi W v_A^2 m}{Zv_i m_i n_0 cv_A v^2 T_i m_i \ln\frac{\omega \max}{\omega \min}} \qquad (5.103)$$

This indicates that the effective temperatures of the heavy ions are high, so that there is a preferential acceleration of heavy ions. Similar results have also been obtained for Fermi acceleration [46]. We will now consider the acceleration of high-energy particles with $v \gg v_A$. Assuming that the one-dimensional spectrum is $\nu = 1$, we find, to within a coefficient of order unity,

$$D = \frac{We^2 v_A^2 Z^2}{\left(1 + \frac{v_A^2}{c^2}\right)\omega_H} \begin{cases} 1, & \varepsilon \ll \varepsilon_*, \\ \frac{\varepsilon_*}{\varepsilon}, & \varepsilon \gg \varepsilon_*, \end{cases} \qquad (5.104)$$

where $\varepsilon_* = \left(\frac{Z}{2\pi}\right)^2 \frac{m_i}{m} \frac{H^2 \omega_{pi}^2 L^2}{8c^2 n_0 \pi}$ for nonrelativistic particles, and $\varepsilon_* = LeHZ$ for ultrarelativistic energies. ε_* is the energy at which the fundamental scale of the turbulence is comparable to the Larmor radius of the accelerated particle.

If $v_A > c\frac{m}{Zm_i}\left(\frac{c}{\omega_{pi} L}\right)$, then $\varepsilon_* \gg mc^2$. One usually finds that $\varepsilon_* \gg mc^2$ in astrophysical situations. In the laboratory L can be of the order of the system dimensions, and the corresponding ε_* is nonrelativistic. Assuming that $\varepsilon_* \gg mc^2$, we obtain for (5.104) that when $\varepsilon \ll \varepsilon_*$, $\dot{\varepsilon} = const$, and when $\varepsilon \gg \varepsilon_*$, $\dot{\varepsilon} \sim \frac{1}{\varepsilon}$.

For magnetoacoustic oscillations with $T_e \gg T_i$ the picture is much different if $\nu = 2$. For then, according to (5.94) we have, up to a constant of order unity,

$$D_M = \frac{v \varepsilon^2}{L} v_A^2 \frac{W^M}{H^2} \begin{cases} 1, & \varepsilon \ll \varepsilon_*, & (5.105) \\ \dfrac{\varepsilon_*^2}{\varepsilon^2}, & \varepsilon \gg \varepsilon_*. & (5.106) \end{cases}$$

Equation (5.105) was obtained in [258, 45]. A similar result was obtained [259] for the one-dimensional case. Equation (5.105) is analogous to the equation for Fermi acceleration, where v_A replaces the cluster velocity, and the distance between clusters is replaced by the fundamental scale of the turbulence. Because the spectral transfer in the collisionless region decreases the vibrational frequencies, the fundamental scale must be determined by the mean free path length if it is smaller than the system dimensions; that is, it is determined by the region in which the collisionless theory is applicable, with $\omega \sim (2\pi/L)v_A \sim \nu_i$ or $L \sim v_A/\nu_i$. This is true, of course, if the source of the turbulence acts for a time which is long enough that the turbulence energy is transferred into this region. In the collision region the transfer direction is just the opposite. Therefore, the original estimate, first made in [260], which gave $L \sim v_A/\nu_i$, turns out to be correct. Figure 5.16 shows a characteristic curve which describes the effectiveness of the acceleration as a function of the particle energy. The explanation of the drop in the acceleration for $\varepsilon > \varepsilon_*$ is similar to the explanation given above. It has to do with the fact that all the fluctuations accelerate the particles and the diffusion co-

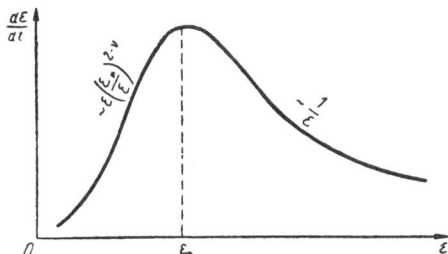

Fig. 5.16. The acceleration rate for ions due to magnetohydrodynamic and Alfven turbulent fluctuations ($\varepsilon_* = ZeHL$).

efficient becomes constant [see (5.106)]. According to (5.105) when $\varepsilon \ll \varepsilon_*$ the accelerated particle distribution function is a power law $\sim 1/\varepsilon^\gamma$, where γ depends heavily on the type of particle, and so on.

Although ε_* is usually relativistic, the mechanism of (5.105) cannot explain the orgin of cosmic rays for two reasons: 1) the value of γ in $1/\varepsilon^\gamma$ is not universal; 2) the effectiveness of the acceleration is very small because L is so large, $L \sim v_A/\nu_i$ (see [46]). In a subsequent chapter we will examine the specific mechanisms for acceleration which are related to the radiation from a turbulent plasma, and which also can be effective for ultrarelativistic energies.

5.7. Charged-Particle Acceleration
Due to Induced Scattering on
Turbulent Fluctuations

The acceleration effects in induced scattering are proportional to the second power of the turbulent energy, and are therefore weaker than the effects treated above, which are due to induced emission and absorption. These new effects are of interest because in some cases the acceleration due to the emission and absorption of fluctuations is either strictly forbidden by the conservation laws, or is very ineffective because the conservation laws permit these processes only for a small range of angles between the particles and waves, or in a very narrow range of velocities, and so on.

For example, the conservation rules forbid (in the absence of magnetic fields) the interaction between Langmuir oscillations and particles with velocities lower than $v < v_{Te}$. These processes are of interest as mechanisms for injecting ions into the accelerating mechanism, which is related to the action of Langmuir fluctuations. Langmuir oscillations themselves can also be cited as an example. As shown in Chapter 4, Langmuir oscillations in an isotropic plasma are transferred into a region of high phase velocities. For nonlinear interactions the value $v_p = c$ is not excluded or eliminated, and the oscillations are transferred into a region where $v_p > c$, where the Cerenkov interaction and particle acceleration is impossible.

If $k \gg \omega_{pe} v / v_{Te}^2$, $v \ll v_{Te}$, then for ions of mass m we find $D = \text{const}$ and

$$\dot{\varepsilon} = \frac{4\pi^2 e^2 Z^4}{9 m n_0 T_e \omega_{pe}} \int_{\frac{\omega_{pe} v}{3 v_{Te}^2}} k \, (W_k^l)^2 \, dk \left(1 + \frac{T_e}{T_i}\right)^{-2}. \qquad (5.107)$$

If $v > v_{Ti}$ the lower limit is greater than $\omega_{pe}(v_{Ti}/3v_{Te}^2)$ and we can use $W_k = \sqrt{\frac{2Q}{\alpha}}$, i.e., assuming that $k_{max} \approx \frac{\omega_{pe}}{v_{Te}} \left(\frac{m_e}{3m_i}\right)^{1/5}$,

$$\dot{\varepsilon} = \frac{8\pi e^2 Z^4}{\pi^2 m \omega_{pe}^2} \left(\frac{3m_i}{m_e}\right)^{3/5} Q. \qquad (5.108)$$

The injection mechanism of (5.108) is very effective for protons. If $k \ll \omega_{pe} v/v_{Te}^2$, the acceleration drops sharply.

The effects of induced scattering for $v \gg v_{Te}$ and $v_p < c$ are considered in [60]. Here we will dwell on the case where $v_p \gg c$ because it is in this region of phase velocities where the fundamental energy of the Langmuir oscillations condenses, and the Cerenkov acceleration is forbidden. We obtain

$$(\dot{\varepsilon})^2 \simeq Z^4 \frac{W^2}{n_0^2} \left(\frac{k_0 v_{Te}}{\omega_{pe}}\right)^3 \omega_{pe} \sqrt{\frac{m v_{Te}^2}{2\varepsilon}} =$$

$$= \frac{Z^4 Q^2}{n_0^2 v_e^2} \omega_{pe} \left(\frac{m_e}{m_i}\right)^{3/2 \, (\nu - 1)} \left(\frac{Q_*}{Q}\right)^{3/2 \, (\nu - 1)} \sqrt{\frac{m v_{Te}^2}{2\varepsilon}}. \qquad (5.109)$$

Here we have used $k_0 = k_*(Q_*/Q)^{1/2(\nu - 1)}$. This equation also describes the predominant acceleration of nonrelativistic heavy, multiply-charged ions.

5.8.   Effectiveness of Various Acceleration Mechanisms and Their Influence on Turbulence Spectra

All the mechanisms which have been described have characteristic curves containing maxima. Therefore, it is useful to compare the acceleration effectiveness in terms of $\dot{\varepsilon}$ at the maximum. By comparing (5.109) with acceleration due to resonance plasmons, we find that (5.109) is smaller. When comparing acceleration effects for high-energy particles one must keep in mind that acceleration due to ion-acoustic waves has a maximum at $v \sim v_s$ and is ineffective for particles of very high velocities. But when the acceleration due to magnetohydrodynamic oscillations is compared with that due to Langmuir waves we obtain

$$\frac{(\dot{\varepsilon}^l)_{max}}{(\dot{\varepsilon}^M)_{max}} = \frac{Zv_{Te}}{v_A} \cdot \frac{m_i}{m} \cdot \frac{n_0 T_e}{W^M} \sqrt{\frac{W^l}{n_0 T_e}} \sqrt{\frac{v_e}{\omega_{pe}}}. \qquad (5.110)$$

Here $Q^l = W^l v_e$. The time scale of the acceleration is now important; roughly speaking, it is the time t_{max} needed for the particle to reach its maximum energy, with $(\dot{\varepsilon}^l)_{max} = \frac{\varepsilon_{max}}{t_{max}} = \varepsilon^l_{max}\gamma^l_{max}$.

It is clear that

$$\frac{\gamma^l_{max}}{\gamma^M_{max}} \simeq \frac{(\dot{\varepsilon}^l)_{max}\,\varepsilon^M_{max}}{(\dot{\varepsilon}^M)_{max}\,\varepsilon^l_{max}}, \qquad (5.111)$$

and

$$\frac{\gamma^l_{max}}{\gamma^M_{max}} \simeq \frac{eHL}{mc^2} \cdot \frac{v_{Te}}{v_A} Z^2 \frac{m_i}{m} \cdot \frac{nT_e}{W^M} \sqrt{\frac{W^l}{n_0 T_e}} \sqrt{\frac{v_e}{\omega_{pe}}}. \qquad (5.112)$$

If

$$L = \frac{v_A}{v_i} \simeq \frac{v_A}{v_e} \sqrt{\frac{m_i}{m_e}},$$

then

$$\frac{eHL}{mc^2} = \frac{m_i}{m} v_A^2 \frac{\omega_{pe}}{v_e} c^{-2}$$

and

$$\frac{\gamma^l_{max}}{\gamma^M_{max}} = \sqrt{\frac{\omega_{pe}}{v_e}} Z^2 \frac{n_0 T_e}{W^M} \sqrt{\frac{W^l}{n_0 T_e}} \, v_A v_{Te} \, c^{-2}. \qquad (5.113)$$

As a rule, the large factor $(\omega_{pe}/v_e)^{1/2}$ makes the acceleration due to high-frequency Langmuir oscillations most effective in astrophysical situations. Acceleration due to magnetohydrodynamic oscillations acts as the injector most frequently.

On the other hand, acceleration due to Langmuir oscillations serves as the injector for acceleration due to electromagnetic radiation (see Chapter 6 also). A cascade of interactions leads to the appearance of particles of very high energy in the plasma (cosmic rays), and to the subsequent change of state of the high-frequency oscillations. The number of accelerated particles increases because the turbulence itself accomplishes the injection. The high-frequency oscillations have no other mode of absorption except absorption by fast particles, which are in turn accelerated. The absorption increases and continues until the acceleration effects become essential for the absorption of the oscillations. The number of accelerated particles increases slowly because the injection mechanisms are not very efficient. This is in fact a necessary state, for otherwise a large number of particles would be heated rapidly and the turbulence would decrease. Thus, strictly speaking, the high-frequency turbulence is quasi-stationary and gradually transfers into a regime of accelerative turbulence, where all the power Q is expended on particle acceleration. The successive stages of this deformation are shown in Fig. 5.17 along with the disappearance of the spectrum maximum. The absorption of Langmuir oscillations by relativistic particles with $f_\varepsilon = \text{const } \varepsilon^2 d^{-\varepsilon/T}$, and

$$\gamma = -\frac{\omega_{pe}}{\pi - 1} Z^2 \frac{n_1}{n_0} \cdot \frac{m_e c^2}{T_{\text{eff}}} \cdot \frac{\omega_{pe}^3}{k^3 c^3} \begin{cases} 1, & \omega_{pe} < kc, \\ 0, & \omega_{pe} > kc, \end{cases} \qquad (5.114)$$

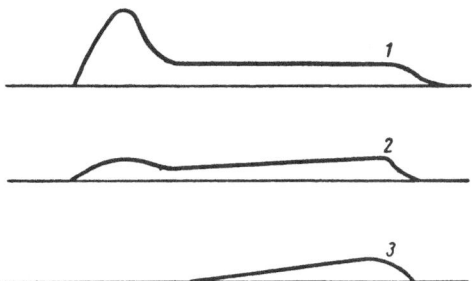

Fig. 5.17. Changes in the spectrum of Langmuir turbulence due to particle acceleration. The successive states 1, 2, and 3 correspond to various increases in the number of accelerated particles.

reaches its maximum when $k \sim \omega_{pe}/c$ and grows when T_{eff} falls and the number of fast particles n_1 increases. Since the accumulation of accelerated particles causes just the maximum in the spectrum to disappear, the acceleration is sharply curtailed and the decrease in energy density W (at constant Q) slows down considerably. The effects of the radiation processes on the fast particles continue to act right up to the time the maximum disappears, and the energy given to the fast particles goes into radiation. The equilibrium thus achieved in the radiation leads to particle spectra of the form $1/\varepsilon^{\gamma}$, and in general the particles are produced until the maximum in the turbulence spectrum disappears. Absorption of the turbulence by cosmic rays plays an important role in equalizing the energy distribution between cosmic rays and turbulence.

Chapter 6

Radiation from a Turbulent Plasma

6.1. The Problem

The radiation emitted by a turbulent plasma is one of its most important manifestations and has many applications. Radiation from a turbulent plasma was encountered in the very first laboratory experiments which employed beams of charged particles, electric fields, and other means to excite plasma turbulence (see [24, 31, 32]). One of the most important uses of the radiation from a turbulent plasma revolves around attempts to draw conclusions about the plasma parameters from the radiation (quantities such as the temperature and density, to name two); it is also useful in diagnosing the spectra of the plasma turbulences. A radiating turbulent plasma can be used as a source of electromagnetic radiation. The radiation spectrum is usually quite broad. The effects of the radiation from a turbulent plasma can lead to the radiative dissipation of the plasma turbulence and the possibility of establishing stationary turbulence which is dissipated as radiation. When taken with the mechanism of spectral transfer, radiation from a turbulent plasma is an aid to damping (stabilizing) the oscillations.

Among the mechanisms for radiation from a turbulent plasma, we can define two categories: those connected with the nonlinear conversion of oscillations into electromagnetic radiation by the majority of particles, and those related to the conversion due to fast accelerated particles (see Chapter 5). In actuality the distribution function for particles in a turbulent plasma is not Maxwellian; it has long tails in the high-energy region. Thus any division of the particles

331

into groups composing the bulk of the plasma and a group of fast
particles is certainly conditional. Moreover, the radiation coming from the fast particles has some specific properties worthy
of note. One such feature for the emission by fast particles is
the substantial increase in the frequency of the emitted waves,
which can be much higher than the frequency of the turbulent fluctuations in the case of ultrarelativistic particles. On the other
hand, when the bulk of the thermal plasma particles are involved
in converting the electromagnetic radiation, the frequency is usually altered very little; it can be doubled, but it does not change by
an order of magnitude. For example, the process of nonlinear
mixing and the decay of turbulent fluctuations can produce frequencies of the order

$$\omega^t = \omega^\sigma \pm \omega^{\sigma'}. \tag{6.1}$$

Thus, when two Langmuir waves mix one obtains frequencies
$\sim 2\omega_{pe}$; when two waves with frequencies $\omega \sim \omega_{He}$ in a strong magnetic field merge, waves with frequency $2\omega_{He}$ appear. The mixing
of a plasma wave and any low-frequency wave produces radiation
with a frequency close to the plasma frequency. When waves are
scattered on thermal particles the frequency changes can only be of
the order of the Doppler frequency shift due to the particles' thermal motion kv_{Te}, kv_{Ti}. For Langmuir oscillations this shift is
much smaller than the original frequency: $\Delta\omega/\omega_{pe} \sim v_{T\alpha}/v_p \ll 1$.
This indicates that there is practically no change in frequency.
By equating the frequencies of the transverse and longitudinal
waves, we find

$$\omega_{pe} + \frac{k^2 c^2}{2\omega_{pe}} = \omega_{pe} + \frac{3k_1^2 v_{Te}^2}{2\omega_{pe}}, \tag{6.2}$$

where $k = \sqrt{3}(v_{Te}/c)k_1$, i.e., the wavelength of the transverse waves
is c/v_{Te} times longer than the longitudinal wavelength. These
transverse waves, with frequencies close to ω_{pe}, will be called
transverse plasmons. The excitation of transverse plasmons does
not mean that there will be radiation from the plasma at frequencies close to ω_{pe} because the emission of such waves from the
plasma is complicated by the fact that their index of refraction n
is nearly zero. Furthermore, the nonlinear effects in the spectral

transfers only serve to move the index even closer to zero. But there are a number of circumstances which do make possible the efficient radiation of waves at frequencies close to ω_{pe}. The first is the plasma inhomogeneity, which results in a flow tending to increase the wave numbers. However, the density gradient must be very large in order to "resist" the nonlinear energy flux due to the spectral transfers in the opposite direction. The second is the excitation of turbulence with large wave numbers, where $\omega_{pe} + 3k_{T}^{2}v_{Te}^{2}/2\omega_{pe}$ is much different from ω_{pe} ($\omega - \omega_{pe}$ can be of the order of $^{1}/_{6}\omega_{pe}$). Of course, when dealing with intense turbulence the fundamental energy is rapidly transferred to the region where $v_p \gg c$, and the amount of energy entering the region where $v_p \sim v_{Te}$ is relatively small. Third, anisotropy effects in the distribution of thermal particles can change the transfer direction. Finally, for transverse plasmons the spectral transfer can be suppressed by their emission from the system. According to (6.2) the group velocity of the transverse plasmons is increased by a factor of c/v_{Te}, and if γ^N is the nonlinear growth rate for the transfer, then the length over which the process described by γ^N takes place will be c/v_{Te} times longer. If the system dimension L is smaller than this conversion length, the transfer of the transverse plasmons is not possible. But this does not prevent the longitudinal plasmons from concentrating in the region where their conversion into transverse plasmons is inhibited.

Because of the small changes in the frequencies for the thermal particles, fluctuations other than the Langmuir oscillations cannot be converted into electromagnetic radiation unless $\omega_{He} \ll \omega_{pe}$. In a strong magnetic field $\omega_{He} \gg \omega_{pe}$, the fluctuations with frequencies such that $\omega^\sigma \approx \omega_{He}$ can be converted rather easily into radiation because $n \approx 1$ for the waves which are produced in these conversions; as a consequence the difficulties encountered in getting the radiation out of the plasma do not arise and conversion on thermal particles can produce intense radiation from the turbulent plasma at frequencies close to ω_{He}.

The presence of fast accelerated particles in a turbulent plasma considerably simplifies the whole problem of emitting radiation when $\omega_{He} \ll \omega_{pe}$. The Doppler shift in the frequencies for nonrelativistic particles is $kv \sim (v/v_p)\omega_{pe}$, and when $v \gg v_p$ it can be much greater than ω_{pe}. Interestingly, the experimental data on

emissions from the turbulent plasma in the sun [261] show that the
radiation frequencies are, as a rule, close to ω_{pe} but still system-
atically larger than ω_{pe}.

It was shown in the previous chapter that a turbulent plasma
must, of necessity, generate relativistic particles. The frequency
change during conversion by relativistic particles turns out to be
even more substantial.

If the particles are not magnetized, then the energy conservation law

$$\omega - \omega^\sigma = (\mathbf{k} - \mathbf{k}^\sigma)\,\mathbf{v} \tag{6.3}$$

for the scattering gives us

$$\omega = \frac{\omega^\sigma - \mathbf{k}^\sigma\,\mathbf{v}}{1 - v\cos\theta}, \tag{6.4}$$

where θ is the angle between the direction of the emitted wave and
the particle velocity. When $\theta \leq mc^2/\varepsilon$, we have

$$\omega_{max} \simeq |\,\omega^\sigma - \mathbf{k}^\sigma\,\mathbf{v}\,|\,2\,\frac{\varepsilon^2}{(mc^2)^2}. \tag{6.5}$$

If $\omega^\sigma/k^\sigma = v_p$, the phase velocity of the fluctuation, is much less
than c,

$$\omega_{max} \simeq \frac{2c}{v_p}\,\omega^\sigma\left(\frac{\varepsilon}{mc^2}\right)^2. \tag{6.6}$$

But if $v_p \gg c$,

$$\omega_{max} \simeq 2\omega^\sigma\left(\frac{\varepsilon}{mc^2}\right)^2. \tag{6.7}$$

The big difference between the maximum radiation frequency
and the frequency of the turbulent fluctuations raises the possibility of exciting radiation from the low-frequency oscillations. As
ε/mc^2 increases even the lowest-frequency magnetohydrodynamic
and Alfven waves can radiate. Such radiation is similar to the
synchrotron radiation in random magnetic fields. However, since
(6.6) is possible for unmagnetized particles, the effects which

arise can be qualitatively different from the effects due to synchrotron radiation.

Thus, we now see that a turbulent plasma can be the source
of electromagnetic radiation over a very broad band of frequencies. In order to describe the effects of the excitation and the
propagation of the electromagnetic waves in the turbulent plasma,
we can use the equations for the number of quanta, and describe
their interactions with the fluctuations and particles by using decay and scattering probabilities. The general form of such equations, written for I_k, is

$$\langle E_k^t E_{k'}^t \rangle = I_k \, \delta \, (k+k'), \qquad \int_0^\infty \frac{I_k}{4\pi} \left(\frac{1}{\omega} \cdot \frac{\partial}{\partial \omega} \, \omega^2 \varepsilon \right) d\omega \;=\; \omega_k \, N_k^t \, (2\pi)^{-3},$$

$$\tag{6.8}$$

$$I_k = \int_0^\infty I_k d\omega, \qquad \frac{\partial I_k}{\partial t} + v_g \frac{\partial I_k}{\partial r} = \gamma_k \, I_k + Q_k \left. \frac{4\pi\omega}{\frac{\partial}{\partial \omega} \, \omega^2 \varepsilon} \right|_{\omega=\omega_k}.$$

$$\tag{6.9}$$

Here we have isolated the term proportional to the radiation intensity I_k. It describes the effects due to reabsorption of the emitted waves (if $\gamma_k < 0$), while the last term of (6.9) gives the effects
of spontaneous emission. In addition to reabsorption the buildup
of electromagnetic waves $(\gamma_k > 0)$ is also possible when the radiation waves are amplified by the turbulent plasma. It is frequently
convenient to introduce the radiation intensity

$$I_{\omega, \, \Omega} = I_k \, k^3 \left(\frac{dk}{d\omega} \right)^2 \frac{1}{2\pi}.$$

$$\tag{6.10}$$

The spectral density of electromagnetic radiation is then written as

$$I_\omega = \int I_{\omega, \, \Omega} \, d\Omega,$$

$$\tag{6.10}$$

while the total intensity is

$$I = \int_0^\infty I_\omega \, d\omega.$$

$$\tag{6.12}$$

For an isotropic and uniform distribution of radiation, Eq. (6.9) can be written as

$$\frac{\partial I_\omega}{\partial t} = \gamma_\omega I_\omega + Q_\omega. \tag{6.13}$$

Here $Q_\omega = \int Q_k k^2 \frac{dk}{d\omega} d\Omega$, $\gamma_\omega = \gamma_{k\,(\omega)}$. In the general case Q_k and γ_k can be written as

$$\gamma_k = \sum_\sigma \gamma_k^\sigma + \sum_\sigma \int \gamma_{k,\,k_1}^\sigma W_{k_1}^\sigma dk_1, \tag{6.14}$$

$$Q_k = \sum_\sigma Q_k^\sigma + \sum_\sigma \int Q_{k,\,k_1}^\sigma W_{k_1}^\sigma dk_1 + \sum_{\sigma,\,\sigma'} \int Q_{k,\,k_1,\,k_2}^{\sigma\sigma'} W_{k_1}^\sigma W_{k_2}^{\sigma'} dk_1 dk_2. \tag{6.15}$$

In the expression for γ_k it is sufficient to keep just the terms linear in the turbulent energy, but in Q_k one must keep the quadratic terms in addition because the transport equations take into account all terms which are quadratic in I_k and W_k^σ. The first term in (6.15) appears because of the spontaneous scattering of oscillations on the thermal and accelerated particles. If a layer of the plasma is sufficiently thin (optically thin) and its thickness L is much smaller than the reabsorption length v_g/γ_k, then the induced processes in (6.9) are unimportant and the radiation power $\partial I_k/\partial t$ is determined by the spontaneous radiation. Q_k gives the energy radiated spontaneously in one second from one cubic centimeter of the plasma attributed to dk. Q_ω is the spectral power density of the radiation. If the plasma layer is thick, such that $L \gg \frac{v_g}{\gamma_k}$, then when $\gamma_k < 0$ the radiation intensity reaches saturation

$$\left[\frac{4\pi\omega}{\frac{\partial}{\partial\omega} \omega^2 \varepsilon} \bigg|_{\omega=\omega_k} \right] I_k = \frac{Q_k}{|\gamma_k|}. \tag{6.16}$$

If $\gamma_k > 0$ (i.e., if there is a maser effect), then under stationary conditions

$$\left[\frac{4\pi\omega}{\frac{\partial}{\partial\omega} \omega^2 \varepsilon} \bigg|_{\omega=\omega_k} \right] I_k = \frac{Q_k}{|\gamma_k|} \left(e^{|\gamma_k|\frac{L}{v_g}} - 1 \right), \tag{6.17}$$

i.e., $Q_k/|\gamma_k|$ determines the initial level of turbulence from which one can assume the exponential growth of the intensity with distance.

6.2. Radiation Processes and the Propagation of Electromagnetic Waves in Turbulent Plasma in Terms of the Stokes Parameters

It will be assumed that the radiated waves are transverse waves. In a weak magnetic field, such that $\omega_{He} \ll \omega_{pe}$, those waves with frequencies close to ω_{pe}, and all waves with $\omega \gg \omega_{pe}$, will be approximately transverse. In a strong magnetic field waves with $\omega \gg \omega_{pe}$ will be approximately transverse. We will study the polarization properties of such waves, which are emitted by a turbulent plasma. It will be useful to introduce the Stokes parameters [46, 262]. We will consider a random field of transverse electromagnetic waves E_k. Since $E_k k = 0$, E_k has only two orthogonal components which are perpendicular to k; they will be denoted as $E_{k,1}$ and $E_{k,2}$. Products of these components averaged over a statistical ensemble will be denoted as[†]

$$\langle E^t_{k, i} E^t_{k', i} \rangle = I_{k, ij} \delta (k + k'), \qquad (6.18)$$

$$i = 1, 2, \qquad k = \{k, \omega\}.$$

Further, we introduce the quantity

$$I_{k, ij} = \int\limits_0^\infty d\omega I_{k, \omega, ij}. \qquad (6.19)$$

If the correlations between the electromagnetic fields can be neglected, then

$$I_{k\omega, ij} = I_{k, ij} \delta (\omega - \omega_k) + I_{-k, ij} \delta (\omega + \omega_k). \qquad (6.20)$$

The four quantities $I_{k, ij}$ can be combined into four new elements

---

[†] The equation here has been written for a uniform and stationary radiation field; it can be generalized to the case of a slightly nonstationary inhomogeneous electromagnetic field in the geometric-optics approximation (see [262]).

labeled I_k, G_k, U_k, and V_k, the Stokes parameters:

$$
\left.
\begin{aligned}
I_k &= I_{k,11} + I_{k,22}; & G_k &= I_{k,11} - I_{k,22}; \\
U_k &= I_{k,12} + I_{k,21}; & V_k &= \frac{1}{i}(I_{k,12} - I_{k,21}); \\
I_{k,12} &= \frac{1}{2}(U_k + iV_k); & I_{k,21} &= \frac{1}{2}(U_k - iV_k).
\end{aligned}
\right\} \quad (6.21)
$$

The physical sense of the components $I_{k,11}$ and $I_{k,22}$ is clear — these quantities are proportional to the radiation intensity; the components $I_{k,12}$ and $I_{k,21} = I_{k,12}^*$ characterize the correlation of the projections of the radiation field strength on two mutually perpendicular directions, i.e., the polarizations. All four of the new quantities are real (not complex) and have the property of being additive for a combination of independent electromagnetic radiation fields. In order to clarify still further the physical meanings of each of the Stokes parameters, we will examine a number of specific examples.

It is convenient to introduce an E_k if the ω, k correlations are negligible:

$$E_{k,\omega}^t = E_k^t \delta(\omega - \omega_k) + E_{-k}^t \delta(\omega + \omega_k). \quad (6.22)$$

Then

$$I_{k,ij} \sim \langle E_{ki}^t E_{kj}^{t*} \rangle.$$

We will consider a radiation field in which the difference in phase of the projections of E_k on two specified directions is unchanged, but the phase itself undergoes random changes. It is known that [263] for a given phase difference the end of the vector E_k in a plane perpendicular to k always describes an ellipse. If the phase of such an oscillation undergoes a rapid change, the vector E_k discontinuously (i.e., fast when compared with the period) changes its value, while remaining at all times on the same ellipse. When averaged over a sufficiently long time period the end of the vector E_k can be found at any point of the ellipse with equal probability. The average value of the field is zero. But the averages of the quadratic combinations are not zero. The time average is equal to the ensemble average.

Assume that axis 1 is directed along the ellipse's major axis and that axis 2 is along the minor axis. In this case the phase shifts of the two oscillations are $\pm\pi/2$, i.e.,

$$I_{k,12} \sim \langle |E_{k,1}^{(0)}| e^{-i\varphi_1} |E_{k,2}^{(0)}| e^{-i\varphi_2}\rangle = \langle |E_{k,1}^{(0)}| |E_{k,2}^{(0)}|\rangle \frac{1\pm i}{\sqrt{2}}$$

or

$$U_{k,0}^2 = V_{k,0}^2 \sim \frac{1}{2} |E_{k,1}^{(0)}|^2 |E_{k,2}^{(0)}|^2,$$

where $E_{k,1}^{(0)}$, $E_{k,2}^{(0)}$ are the semiaxes of the ellipse. If axes 1 and 2 are not directed along the major axes of the ellipse, the projections $E_{k,1}$, $E_{k,2}$ can be written in terms of the projections along the principal axes:

$$E_{k,1} = E_{k,1}^{(0)} \cos\varphi + E_{k,2}^{(0)} \sin\varphi,$$
$$E_{k,2} = -E_{k,1}^{(0)} \sin\varphi + E_{k,2}^{(0)} \cos\varphi. \tag{6.23}$$

We then have

$$I_{k,12} \sim \langle E_{k,1} E_{k,2}^*\rangle = \frac{1}{2}\sin 2\varphi \left(\langle E_{k,2}^{(0)} E_{k,2}^{(0)*}\rangle - \langle E_{k,1}^{(0)} E_{k,1}^{(0)*}\rangle\right) +$$

$$+ \cos^2\varphi \langle E_{k,1}^{(0)} E_{k,2}^{(0)*}\rangle - \sin^2\varphi \langle E_{k,2}^{(0)} E_{k,1}^{(0)*}\rangle \sim -\frac{1}{2}\sin 2\varphi Q_{k,0} +$$

$$+ \frac{\cos^2\varphi}{2}(U_{k,0}+iV_{k,0}) - \frac{\sin^2\varphi}{2}(U_{k,0}-iV_{k,0}) =$$

$$= -\frac{1}{2}\sin 2\varphi Q_{k,0} + \frac{1}{2}\cos 2\varphi U_{k,0} + \frac{1}{2}iV_{k,0} = \frac{1}{2}(U_k + iV_k).$$

Thus,

$$U_k = -\sin 2\varphi Q_{k,0} + \cos 2\varphi U_{k,0},$$
$$V_k = V_{k,0}. \tag{6.24}$$

In like fashion it is easy to show that

$$Q_k = Q_{k,0}\cos 2\varphi + U_{k,0}\sin 2\varphi. \tag{6.25}$$

These equations are also valid for transformations between any pair of axes 1 and 2, and they show that a number of invariants do exist. For instance, V is an invariant quantity. Only the invariants have a definite physical interpretation. The meaning of V can be revealed in this manner. Let one of the ellipse's axes go to zero. Then $E_{k,1}^{(0)} = 0$ and $V_k = V_{k,0} = 0$, i.e., the radiation will be linearly polarized. Until now it has been assumed that all elements of the ensemble describe one and the same ellipse. A more complicated case arises when the elements of different ellipses are mixed; this can be interpreted as mixing ensembles of the type discussed above. If V = 0 for each of them, then V = 0 for all the ensembles. This case corresponds to mixing linear polarizations. From these considerations one must assume that the invariant V characterizes the ellipticity of the electromagnetic field. The so-called degree of ellipticity q is given by the following ratio of two invariants:

$$q = \frac{V}{I}. \tag{6.26}$$

It also follows from (6.24) and (6.25) that the quantity $U^2 + G^2$ is invariant. It characterizes the degree of polarization for the radiation. The degree of polarization p is given by

$$p^2 = \frac{U^2 + G^2}{I^2}. \tag{6.27}$$

It is easy to understand the idea behind this definition. For, consider the "intensity" of the radiation in a given direction of polarization $I_{k,11} \sim \langle E_{k,1} E_{k,1}^* \rangle$. It follows from (6.23) that

$$I_{k,11} = I_{k,11,0} + (I_{k,22,0} - I_{k,11,0}) \sin^2\varphi + (I_{k,12,0} + I_{k,21,0}) \times$$

$$\times \sin\varphi \cos\varphi = I_{k,11,0} - G_{k,0} \sin^2\varphi + U_{k,0} \sin\varphi \cos\varphi.$$

Let us trace the value of $I_{k,11}$ as φ changes, which corresponds to a change in the position of the analyzer. The extreme values of $I_{k,11}$ are easily found:

$$I_{k,11,\max} = I_{k,11,0} - \frac{G_{k,0}}{2} + \frac{1}{2} \sqrt{U_{k,0}^2 + G_{k,0}^2},$$

$$I_{k,11,\min} = I_{k,11,0} - \frac{G_{k,c}}{2} - \frac{1}{2} \sqrt{U_{k,0}^2 + G_{k,0}^2}.$$

Thus,

$$I_{k,11,max} - I_{k,11,min} = \sqrt{U_{k,0}^2 + G_{k,0}^2},$$
$$I_{k,11,max} + I_{k,11,min} = 2I_{k,11,0} - G_{k,0} = I_{k,0}.$$

Therefore,

$$p^2 = \left(\frac{I_{k,11,max} - I_{k,11,min}}{I_{k,11,max} + I_{k,11,min}}\right)^2, \tag{6.28}$$

which corresponds to the frequently used definition of the degree of polarization in terms of the intensities I_{max} and I_{min} (which are here denoted by $I_{k,11,max}$ and $I_{k,11,min}$) recorded by the analyzers. Note that when the degree of polarization is zero (p = 0), the ellipticity can be other than zero (q ≠ 0). Sometimes the polarization properties of the radiation are characterized by the quantity π:

$$\pi^2 = p^2 + q^2 = \frac{V^2 + U^2 + G^2}{I^2}. \tag{6.29}$$

Radiation for which π = 0 is termed unpolarized; this is equivalent to having p = 0 and q = 0. In this case the radiation is characterized by a single quantity, the intensity I, and

$$I_{k,ij} = \frac{1}{2} I \delta_{ij}; \quad i, j = 1, 2. \tag{6.30}$$

Here i and j are the projections on directions perpendicular to k. In general, in a coordinate system in which no axis is directed along k we will have

$$I_{k,ij} = \frac{1}{2}\left(\delta_{ij} - \frac{k_i k_j}{k^2}\right) I_k; \quad i, j = 1, 2, 3. \tag{6.31}$$

When analyzing cosmic radiation special consideration is given to the polarization properties of the radiation from turbulent plasma. This is because the cosmic radiation received is frequently polarized.

The equations describing the changes in the Stokes parameters as a result of processes occurring in a turbulent plasma can be found in the following manner. Let $j_{k,i}$ be the current in the plasma which gives rise to the radiation. The equations for the fields of the transverse waves excited by the current $j_{k,i}$ take the form

$$(k^2c^2 - \omega^2 \varepsilon_k^t) E_{k,i}^t = 4\pi i \omega j_{k,i}^t, \qquad (6.32)$$

where $j_{k,i}^t = j_{k,i} - k_i(k j_k)/k^2$. Using (6.32) it is easy to find

$$(k^2c^2 - \omega^2 \varepsilon_k^t - k_1^2 c^2 + \omega_1^2 \varepsilon_{k_1}^{t*}) \langle E_{k,i}^t E_{k_1,j}^{t*} \rangle =$$
$$= 4\pi i \omega \langle j_{k,i}^t E_{k_1,j}^{t*} \rangle + 4\pi i \omega_1 \langle j_{k_1,j}^{t*} E_{k,i}^t \rangle. \qquad (6.33)$$

For simplicity we will assume that the plasma is homogeneous, that the time needed for the Stokes parameters to change is much greater than the frequency of the transverse waves, and that the characteristic length over which the Stokes parameters change is much larger than the wavelength of the transverse waves. Just as in Chapter 2, in place of $\langle E_{k,i}^t E_{k_1,j}^{t*} \rangle = I_{ij}(\varkappa, \Delta k)$; $2\varkappa = k + k_1$; $\Delta k = k - k_1$, it is convenient to introduce the Stokes parameters, which are time-dependent,

$$I_{ij}(x, k) = \int I_{ij}(\varkappa, \Delta k) e^{i \Delta k x} d\Delta k; \quad x = \{r, t\}; \quad \Delta k = \{\Delta k, \Delta \omega\}. \qquad (6.34)$$

Assuming that $\Delta k \ll \varkappa \approx k$ we find that the expression in brackets on the left side of (6.33) has the following approximate form:

$$\Delta k \frac{\partial}{\partial k} (k^2 c^2 - \omega^2 \varepsilon_k^t) - \Delta \omega \frac{\partial}{\partial \omega} \omega^2 \varepsilon_k^t = \frac{\partial}{\partial \omega} \omega^2 \varepsilon_k^t (-\Delta \omega + \Delta k v_g^t).$$

Here we have assumed that one can neglect the weak nonstationarity and nonuniformity of the fields for the transverse waves in the coefficients of the small terms $\Delta \omega$ and Δk; it is also assumed that the dispersion equation $k^2 c^2 - \omega^2 \varepsilon_k^t = 0$ is satisfied. Multiplying (6.33) by $e^{i \Delta k x}$ and integrating over Δk, we obtain

$$\frac{\partial}{\partial \omega} \omega^2 \varepsilon_k^t \left(\frac{\partial}{\partial t} I_{k,ij} + v_g^t \frac{\partial}{\partial r} I_{k,ij} \right) = -4\pi \int e^{i \Delta k x} d\Delta k \times$$
$$\times [\omega \langle j_{k,i}^t E_{k_1,j}^{t*} \rangle + \omega_1 \langle j_{k_1,j}^{t*} E_{k,i}^t \rangle]. \qquad (6.35)$$

It is now convenient to use (6.32) to express E in terms of j, giving us

$$\frac{\partial}{\partial t} I_{k,ij} + v_g^t \frac{\partial}{\partial r} I_{k,ij} = \frac{(4\pi)^2}{\frac{\partial}{\partial \omega} \omega^2 \varepsilon_k^t} \int e^{i \Delta k x} i \omega \omega_1 d\Delta k \times$$
$$\times \langle j_{k,i}^t j_{k_1,j}^{t*} \rangle [(k_1^2 c^2 - \omega_1^2 \varepsilon_{k_1}^{t*})^{-1} - (k^2 c^2 - \omega^2 \varepsilon_k^t)^{-1}]. \qquad (6.36)$$

Since E^t is a random field, (6.32) tells us that j^t is also random;

$$\langle j^t_{k, i} j^{t*}_{k_1, i} \rangle = R_{k, ij} \, \delta \, (k - k_1). \qquad (6.37)$$

By substituting (6.37) into (6.36) and assuming that

$$\frac{1}{k^2 c^2 - \omega^2 \varepsilon^t_k {}^*} - \frac{1}{k^2 c^2 - \omega^2 \varepsilon^t_k} = -2\pi i \delta \left(k^2 c^2 - \omega^2 \varepsilon^t_k\right),$$

we find that

$$\left(\frac{\partial}{\partial t} I_{k, ij} + v^t_g \frac{\partial}{\partial r} I_{k, ij}\right)\frac{\partial}{\partial \omega} \omega^2 \varepsilon^t_k = 4 (2\pi)^3 R_{k, ij} \omega^2 \delta \left(k^2 c^2 - \omega^2 \varepsilon^t_k\right). \qquad (6.38)$$

Since $I_{k,ij}$ satisfies the equation $k^2 c^2 - \omega^2 \varepsilon^t = 0$ under this approximation, it must be proportional to $\delta(k^2 c^2 - \omega^2 \varepsilon^t)$.

Setting $I_{k, ij} = \int_0^\infty I_{k, ij} \, d\omega$ and integrating (6.38) over ω, we find that

$$\frac{\partial}{\partial t} (I_{k, ij}) + v_g \frac{\partial}{\partial r} I_{k, ij} = \left. \frac{4 (2\pi)^3 \omega^2 R_{k, ij}}{\left(\frac{\partial}{\partial \omega} \omega^2 \varepsilon^t_k\right)^2} \right|_{\omega = \omega_k} . \qquad (6.39)$$

Note that the energy density of the radiation field is determined from the equation

$$W = \int \left. \frac{1}{\omega_k} \cdot \frac{\partial}{\partial \omega} \omega^2 \varepsilon^t_k \right|_{\omega = \omega_k} \frac{I_{k, ii} \, dk}{4\pi} = \int \omega_k \frac{N^t_k \, dk}{(2\pi)^3} . \qquad (6.40)$$

It is clear from (6.40) that the equation for the diagonal matrix elements of the matrix $I_{k,ij}$ (6.39), which determine the Stokes coefficients, coincides with the general equation for the number N^t_k of quanta, i.e., it describes the processes of spontaneous and induced scattering of turbulent oscillations by the electromagnetic radiation; it also describes the nonlinear mixing of turbulent oscillations in the electromagnetic waves. Equation (6.38) is more general because it takes into account correlation effects. Let us consider the Stokes parameters for electromagnetic radiation due to the spontaneous scattering of turbulent waves by plasma particles. The general expression for the current linear in the field of the scattered turbulent oscillations takes the form ($H_0 = 0$)

$$j_{k,\,i} = \int \Lambda_{ij}(k,\,k_1)\, e^{i\,(k-k_1)\,r_0}\, dk_1\, \delta\,(\omega - \omega_1 - (\mathbf{k}-\mathbf{k}_1)\,\mathbf{v})\, E_{k_1,\,j}^\sigma. \qquad (6.41)$$

Here we have presented the current created by an individual scattering particle which, at the initial moment, had the coordinate r_0. When computing the average value $\langle j_{k,\,i}^t\, j_{k',\,j}^{t*}\rangle$ we average the result over the initial values of $\mathbf{r}_0$ (since the scattering particle can have any r_0), $\dfrac{1}{V}\int dr_0 \langle j_{k,\,i}^t\, j_{k',\,j}^{t*}\rangle$, and over velocities, and then multiply the result by the number of particles with velocities $\mathbf{v}$ in the volume V: $f_\mathbf{p}\, dp\,(2\pi)^{-3}V$ $\left[(2\pi)^{-3}\int f_\mathbf{p}\, dp\, V = nV = N\right.$, where n is the number of particles per unit volume and N is the number of particles in the volume V]. Thus,

$$\langle j_{k,\,i}^t\, j_{k',\,j}^{t*}\rangle = \int f_\mathbf{p}\, dp\, dr_0\, \Lambda_i^t\,(k,\,k_1)\, \Lambda_j^{t*}(k',\,k_1')\, dk_1\, dk_1' \times$$

$$\times (2\pi)^{-3} \exp\left[i\,(\mathbf{k} - \mathbf{k}_1 - \mathbf{k}' + \mathbf{k}_1')\, r_0\right]\, \delta\,(\omega - \omega_1 - (\mathbf{k}-\mathbf{k}_1)\,\mathbf{v}) \times$$

$$\times \delta\,(\omega' - \omega_1' - (\mathbf{k}-\mathbf{k}_1')\,\mathbf{v})\, \langle E_{k_1}^\sigma\, E_{k_1'}^{\sigma'}\rangle, \qquad (6.42)$$

where $\Lambda_i^t(k,\,k_1) = e_{\mathbf{k}_1,\,j}^\sigma \left(\Lambda_{ij}(k,\,k_1) - \dfrac{k_i\, k_s\, \Lambda_{sj}(k,\,k_1)}{k^2}\right)$; $e_{\mathbf{k}_1\,j}^\sigma$ is a normal unit vector for the turbulent oscillations under consideration. Assuming that $\langle E_k^\sigma E_{k'}^{\sigma*}\rangle = I_k^\sigma\, \delta\,(k-k')$ we find

$$R_{k,\,ij} = \int \Lambda_i^t\,(k,\,k_1)\, \Lambda_j^{t*}(k,\,k_1)\, f_\mathbf{p} dp\, I_{k_1}^\sigma\, \delta\,(\omega-\omega_1 - (\mathbf{k}-\mathbf{k}_1)\,\mathbf{v})\, dk_1. \qquad (6.43)$$

When the scattering takes place on those fluctuations whose level corresponds to thermal equilibrium, Eq. (6.43) leads to the usual equation, which describes the scattering resulting from fluctuations in the plasma density [264, 265].

Now let us examine the Stokes parameters for radiation due to the spontaneous nonlinear coalescence of the waves σ and σ'. The nonlinear current which describes this process is

$$j_{k,\,i} = 2 \int S_{ijl}(k,\,k_1,\,k_2)\, e_{\mathbf{k}_1,\,j}^\sigma\, e_{\mathbf{k}_2,\,l}^{\sigma'}\, E_{k_1}^\sigma\, E_{k_2}^{\sigma'}\, dk_1\, dk_2\, \delta\,(k-k_1-k_2). \qquad (6.44)$$

Thus

$$R_{k,\,ij} = \int \Lambda_i(k,\,k_1,\,k_2)\, \Lambda_j^*(k,\,k_1,\,k_2)\, I_{k_1}^\sigma\, I_{k_2}^{\sigma'}\, \delta\,(k-k_1-k_2), \qquad (6.45)$$

where

$$\Lambda_i(k, k_1, k_2) = 2S_{mjl}(k, k_1, k_2) e^{\sigma}_{k_1, j} e^{\sigma'}_{k_2, l} \left(\delta_{im} - \frac{k_i k_m}{k^2}\right). \tag{6.46}$$

Now consider the Stokes parameters for spontaneous emission of radiation by the plasma particles. The current set up by charges moving along spirals in a plasma located in an external magnetic field is

$$j_{k, i} = \frac{e}{(2\pi)^3} \sum_{\nu = -\infty}^{\infty} \Gamma_{\nu i} e^{i \cdot kr_0 + i\nu\varphi_0} \delta(\omega - k_z v_z - \nu\omega_H), \tag{6.47}$$

where φ_0 is the initial phase of the charge's rotation about the circle. To find $\langle j^t_{k, i} j^{t*}_{k', j}\rangle$ we average the result over r_0, $v_\perp v_z$, and φ_0. Assuming that $\frac{1}{2\pi}\int d\varphi_0 e^{i(\nu - \nu')\varphi_0} = \delta_{\nu, \nu'}$, we obtain

$$\langle j_{k, i} j^*_{k', j}\rangle = \delta(k - k') e^2 \sum_{\nu = -\infty}^{\infty} \int \frac{f_p \, dp}{(2\pi)^6} \Gamma^t_{\nu, i} \Gamma^{*t}_{\nu, j} \times$$
$$\times \delta(\omega - k_z v_z - \nu\omega_H) \delta(\omega' - k_z v_z - \nu\omega_H), \tag{6.48}$$

i.e.,

$$\Gamma^t_{\nu, i} = \Gamma_{\nu, i}\left(\delta_{ij} - \frac{k_i k_j}{k^2}\right),$$
$$R_{k, ij} = \frac{e^2}{(2\pi)^6} \int f_p \, dp \sum_{\nu = -\infty}^{\infty} \Gamma^t_{\nu, i} \Gamma^{*t}_{\nu, j} \delta(\omega - k_z v_z - \nu\omega_H). \tag{6.49}$$

This exhausts the spontaneous processes possible for terms quadratic in the turbulent energy.

The power of the spontaneous emission which appears in (6.9) can be expressed in terms of the $R_{k,ij}$ found here by the equation

$$Q_k = \frac{2(2\pi)^2 \omega}{\left(\frac{\partial}{\partial\omega} \omega^2 \varepsilon^t_k\right)} \text{Tr } R_{k, ij}\bigg|_{\omega = \omega_k}. \tag{6.50}$$

Let us now turn to finding the equations for the Stokes parameters for induced scattering of the particles by turbulent fluctuations. The effects of induced scattering are related to a number of nonlinear processes; if will therefore be convenient to use Eq. (6.35) after the current j_k^t on the right has been replaced with the nonlinear current, which is both quadratic and cubic in the fields E. By using the well-known procedure (see Chapter 2, §3) for averaging three and four random fields, we find

$$\left(\frac{\partial}{\partial t} I_{k,\,ij} + \mathbf{v}_g \frac{\partial}{\partial \mathbf{r}} I_{k,\,ij}\right) = -\frac{4i\,\omega\,(4\pi)^2}{\dfrac{\partial}{\partial\omega}\omega^2\,\varepsilon_k^t}\left\{\int S_{ilm}^{ti}(k,\,k-k_1,\,k_1)\,(\omega-\omega_1)dk_1\times\right.$$

$$\times\,\Pi_{ln}(k-k_1)\,S_{nqp}^{t\,p}(k-k_1,\,-k_1,\,k)\,e_{k_1,\,m}^{\sigma}\,e_{-k_1\,q}^{\sigma}\,I_{k_1}^{\sigma}\,I_{k,\,pj}-$$

$$-\int S_{ilm}^{t\,j\,*}(k,\,k-k_1,\,k_1)\,dk_1\,(\omega-\omega_1)\,\Pi_{ln}^{*}(k-k_1)\times$$

$$\times\,S_{nqp}^{t\,p\,*}(k-k_1,\,-k_1,\,k)\,e_{k_1,\,m}^{\sigma}\,e_{-k_1\,q}^{\sigma*}\,I_{k_1}^{\sigma}\,I_{kpj}^{*}\Big\}-\frac{8\pi\omega}{\dfrac{\partial}{\partial\omega}\omega^2\,\varepsilon_k^t}\times$$

$$\times\left\{\int dk_1\,\Sigma_{jn\,ls}^{t_i\,t_l}(k,\,k_1,\,k,\,-k_1)\,e_{k_1\,n}^{\sigma}\,e_{-k_1,\,s}^{\sigma}\,I_{k_1}^{\sigma}\,I_{k,\,ij}+\right.$$

$$+\int dk_1\,\Sigma_{jnls}^{t_j\,t_l\,*}(k,\,k_1,\,k,\,-k_1)\,e_{k_1,\,n}^{\sigma*}\,e_{-k_1,\,s}^{\sigma*}\,I_{k_1}^{\sigma}\,I_{k_jl}^{*}\Big\}. \tag{6.51}$$

Here

$$S_{ijl}(k,\,k_1,\,k_2) = \frac{1}{2}\left(S_{ijl}(k,\,k_1,\,k_2) + S_{ilj}(k,\,k_2,\,k_1)\right)$$

are the components symmetrized in axes 1 and 2 of the nonlinear current which is second order in the field; $\Sigma_{ijls}(k,\,k_1,\,k_2,\,k_3) = \frac{1}{2}\left(\Sigma_{ijls}(k,\,k_1,\,k_2,\,k_3) + \Sigma_{ijsl}(k,\,k_1,\,k_3,k_2)\right)$ are the components symmetrized in the axes 3 and 2 of the nonlinear current which is third order in the field; e_k^{σ} is the unit vector for the fields of the turbulent fluctuations; the superscripts t_i and t_l mean that for the corresponding indices (i, l) one must take into account only the components perpendicular to k [for example, $S_i^{ti} = S_j\left(\delta_{ij} - \frac{k_ik_j}{k^2}\right)$]. Equation (6.51) takes into account only the induced emission processes which are proportional to I_{lj} (the effects of induced scattering and induced decay).

   Finally, let us examine the equations for the Stokes parameters for induced emission by plasma particles. The effects of induced scattering and absorption of electromagnetic waves by plasma particles are described by the imaginary (more precisely, the antihermitian) part of the plasma dielectric function. It makes sense to speak of transverse waves, and therefore polarization effects, only for high frequencies, for which the real part of ε_{ij} corresponds to an isotropic plasma with the necessary accuracy.

The antihermitian portion of ε_{ij} can be different for the various polarizations. Different polarizations increase or damp out with different values of γ. In order to consider the polarization effects during emission, it is sufficient to use the linear equation

$$\left(k^2 c^2 - \omega^2 \varepsilon^t\right) E^t_{k,\,i} - \omega^2 \varepsilon^A_{il} E^t_{k,\,l} = 0, \tag{6.52}$$

where ε^A_{ij} is the antihermitian part and ε^t is the hermitian part. We also use the equation for $E^{t*}_{k,\,i}$, multiply the corresponding equations by $E^{t*}_{k',\,i}$ and $E^t_{k,\,j}$, and subtract one from the other:

$$\left(k^2 c^2 - \omega^2 \varepsilon^t_k - k'^2 c^2 + \omega'^2 \varepsilon^t_{k'}\right) \langle E^t_{k,\,i}\, E^{t*}_{k',\,i}\rangle =$$

$$= \omega^2 \varepsilon^A_{i,\,l}(k) \langle E^t_{k,\,l} E^{t*}_{k',\,i}\rangle - \omega'^2 \varepsilon^{A*}_{i,\,l}(k') \langle E^{t*}_{k',\,l} E^t_{k,\,i}\rangle. \tag{6.53}$$

The left side of (6.53) coincides with the left side of (6.33) and transforms in the same manner. But on the right side of (6.53), one can approximately assume that the transverse field is stationary because ε^A is so small, and use $\langle E^t_{k,\,i} E^{t*}_{k',\,j}\rangle = I_{k,\,ij}\delta(k-k')$. As a result we have

$$\frac{\partial}{\partial\omega}\omega\varepsilon^t_k\left(\frac{\partial}{\partial t}I_{k,\,ij} + \mathbf{v}_g\frac{\partial}{\partial\mathbf{r}}I_{k,\,ij}\right) = i\,\omega^2\left(\varepsilon^A_{il}I_{k,\,lj} - \varepsilon^{A*}_{j,\,l}I^*_{k,\,li}\right). \tag{6.54}$$

Equations (6.51) and (6.54) describe all the induced processes, and, together with the $G_{k,ij}$ found above for the spontaneous processes, give a general equation for the Stokes parameters [266]. Note that the right side of (6.51) has the same form as the right side of (6.54), that is, one can introduce an effective nonlinear ε^N which depends on the energy of the turbulent fluctuations. The complete equation then has the form of (6.54), in which ε_{ij} is the sum of the linear ε^A_{ij} and nonlinear ε^N_{ij}. Furthermore, the complete equation can be transformed into a simpler form if one makes use of the equation $I^*_{li} = I_{il}$:

$$\frac{\partial}{\partial t}I_{k,\,ij} + \mathbf{v}_g\frac{\partial}{\partial\mathbf{r}}I_{k,\,ij} = \gamma_{il}I_{k,\,lj} + \gamma^*_{jl}I_{k,\,il} + Q_{kij},$$

$$\gamma_{ij} = \frac{i\,\omega^2\,\tilde{\varepsilon}_{ij}}{\dfrac{\partial}{\partial\omega}\omega^2\,\varepsilon^t_k}, \qquad \tilde{\varepsilon}_{ij} = \varepsilon^A_{ij} + \varepsilon^N_{ij}. \tag{6.55}$$

$\varepsilon^A_{i,\,l}$ will be found now for a plasma in a constant magnetic field. Note that $\operatorname{Im}\varepsilon^\sigma = e^{\sigma*}_{k,\,i}\varepsilon^A_{ij}e^\sigma_{k,\,j}$; the damping of the wave σ from

$k^2c^2 - \omega^2 \varepsilon^\sigma = 0$ for small Im ε^σ takes the form $\gamma = -\dfrac{\omega^2 \operatorname{Im} \varepsilon^\sigma}{\dfrac{\partial}{\partial \omega} \omega^2 \varepsilon^\sigma}$. On the other hand, since $\dfrac{\partial N_k^\sigma}{\partial t} = 2\gamma N_k^\sigma$,

$$\gamma = \frac{1}{2} \sum_{\nu=-\infty}^{\infty} w_{p_z, \, p_\perp}^\sigma (k\nu) \left(k_z \frac{\partial f_p}{\partial p_z} + \frac{\omega_H \nu}{v_\perp} \cdot \frac{\partial f_p}{\partial p_\perp} \right) \frac{dp}{(2\pi)^3} =$$

$$= \frac{e^2}{2\pi \frac{\partial}{\partial \omega} \omega^2 \varepsilon_k^t} \sum_{\nu=-\infty}^{\infty} \int dp e_{k\,i}^{\sigma*} \Gamma_{\nu,\,i} \Gamma_{\nu,\,j}^* e_{kj}^\sigma \, \delta \left(\omega^\sigma - k_z v_z - \nu \omega_H \right) \times$$

$$\times \left(k_z \frac{\partial f_p}{\partial p_z} + \frac{\omega_H \nu}{v_\perp} \cdot \frac{\partial f_p}{\partial p_\perp} \right).$$

Here we have used tne probability equation of (2.220). Since the $e_{k,\,i}^\sigma$ are arbitrary, we find from our comparison that

$$\omega^2 \varepsilon_{ij}^A = \frac{e^2}{2\pi} \sum_{\nu=-\infty}^{\infty} \int dp \, \Gamma_{\nu,\,i} \Gamma_{\nu,\,j}^* \, \delta \left(\omega - k_z v_z - \nu \omega_H \right) \left(k_z \frac{\partial f_p}{\partial p_z} + \frac{\omega_H \nu}{v_\perp} \cdot \frac{\partial f_p}{\partial p_\perp} \right).$$

$$(6.56)$$

Of course, this equation can be obtained by solving the kinetic equation directly [267, 83].

In the absence of turbulence, the combination of (6.55) and (6.56) describes the transport and generation of polarized synchrotron radiation [268].

Equation (6.55) is simply generalized to include the weak effects due to the rotation of the plane of polarization of the high-frequency radiation ($\omega \gg \omega_{pe}$) if the small hermitian parts of ε_{ij} are taken into account [269]. The equations thus obtained are similar to (6.9). Equation (6.9) can be derived from (6.55) if one assumes that the radiation is unpolarized, that is, it is assumed that (6.30) is satisfied for $I_{k,\,il}$. To obtain some feeling for the new effects contained in (6.55), let us consider the simplest case of homogeneous turbulence $\left(\frac{\partial}{\partial \mathbf{r}} = 0 \right)$:

$$\frac{\partial}{\partial t} I_{k,\,ij} = \gamma_{il} I_{klj} + \gamma_{jl}^* I_{kil} + Q_{kij}. \qquad (6.57)$$

If there is a buildup in the electromagnetic waves, the spontaneous

emission term Q_{ij} can be neglected. In this instance one can use
a homogeneous system of equations, whose solutions are sought in
the form $e^{i\lambda t}$, as usual, where λ is given by a fourth-order dispersion relationship in the general case. If the solution to this
equation is known, one can use the initial values of the Stokes
coefficients to determine the Stokes coefficients at time t; subsequently, the degree of polarization p and degree of ellipticity q
can then be determined. Consider a simple example, in which the
coefficients γ_{ij} are real and $\gamma_{11} = \gamma_{22}$. Assuming reality for I_{11}
and I_{22} we obtain equations for $V = \text{Im } I_{12}$ and G:

$$\frac{dV}{dt} = 2\gamma_{11} V, \quad \frac{dG}{dt} = 2\gamma_{11} G.$$

If G = V = 0 when t = 0, then V = G = 0. In this case Eq. (6.57)
gives a second-order equation for λ. We have

$$\lambda_{\pm} = 2 (\gamma_{11} \pm \gamma_{12}),$$

$$I = \frac{I_0 + U_0}{2} \exp 2t (\gamma_{11} + \gamma_{12}) + \frac{I_0 - U_0}{2} \exp 2t (\gamma_{11} - \gamma_{12}),$$

$$U = \frac{I_0 + U_0}{2} \exp 2t (\gamma_{11} + \gamma_{12}) - \frac{I_0 - U_0}{2} \exp 2t (\gamma_{11} - \gamma_{12}), \quad (6.58)$$

where I_0 and U_0 are the values of I and U when t = 0. Assume that
$U_0 > 0$; then the initial degree of polarization is $p_0 = |U_0|/I_0$, and
at time t it will be

$$p^2 = \left(1 - \frac{1 + p_0}{1 - p_0} e^{4\gamma_{12} t} \right) \left(1 + \frac{1 + p_0}{1 - p_0} e^{4\gamma_{12} t} \right)^{-1}. \quad (6.59)$$

Note that for any sign on γ_{12} the polarization goes to unity;
the radiation becomes completely polarized regardless of the initial polarization (in particular, even totally unpolarized radiation
with $p_0 = 0$ turns into polarized radiation). This example illustrates the situation which arises in a turbulent plasma. In order
that polarization effects can appear, there must be some preferred
directions. For example, when $H_0 \neq 0$ turbulent fluctuations can
be anisotropically distributed relative to the magnetic field. As
in the case of polarized radiation, spontaneous processes will determine the initial intensities I_0 and U_0.

If there is no buildup of electromagnetic waves, Eq. (6.55) determines the stationary values of the intensity and other Stokes parameters in an optically thin plasma. As a rule the polarization of this radiation is not very great.

6.3. Emission of Radiation at Frequencies Close to ω_{pe} by Langmuir Turbulence

Radiation with frequencies from ω_{pe} to $2\omega_{pe}$ can be produced without accelerated fast particles. The possibility of emission at ω_{pe} is connected with scattering by electrons and ions,

$$l + e \rightarrow t + e', \quad l + i \rightarrow t + i', \qquad (6.60)$$

in the presence of ion-acoustic waves s with the following coalescence and decay processes:

$$l + s \rightarrow t, \quad l \rightarrow t + s. \qquad (6.61)$$

Other coalescence processes are possible with other low-frequency oscillations in a magnetic field.

Radiation at $2\omega_{pe}$ is given off by the process

$$l + l' \rightarrow t. \qquad (6.62)$$

The emission processes can have relatively little effect on the turbulence spectra, but may serve as the primary dissipative mechanism. In order to assess the role of these and other mechanisms of radiation, one must first know the plasma's optical thickness for a given process, because if the plasma dimensions are smaller than the optical thickness the emission is determined by spontaneous processes, and if it is greater, the induced processes dominate. We will begin with estimates and comparisons of the processes indicated in (6.60). From the point of view of emission, the region of greatest interest is that of small v_p for the Langmuir waves generating the electromagnetic waves which more or less freely leave the plasma. Consider the scattering by ions in (6.60) when $v_p / v_{Ti} \ll 3(m_i / m_e)^{1/2}$, where this process can only take place differentially with $\Delta\omega \le k^l v_{Ti}$. By using the expression for the scattering probability (Chapter 2) we find

$$\frac{dN_k^t}{dt} = -N_k^t \int \frac{dk_1\, N_{k_1}^l\, \omega_{pe}^2\, (\omega - \omega_1)\, [k_1\, k]^2 \exp\left\{-\dfrac{(\omega - \omega_1)^2}{2(k - k_1)^2\, v_{Ti}^2}\right\}}{4 n_0\, m_i\, v_{Ti}^3\, (2\pi)^{5/2} \left(1 + \dfrac{T_e}{T_i}\right)^2 |k - k_1|\ k^2 k_1^2}. \tag{6.63}$$

Since the frequency is essentially unchanged during the transformation, it is convenient to relate the transverse waves to those k values of the longitudinal waves whose frequencies are the same as the frequencies of the transverse waves. To this end we introduce

$$W_{kl}^t = \frac{4\pi k^2\, \omega_{pe}\, N_k^t\, \sqrt{3}\, v_{Te}}{(2\pi)^3\, c} \Bigg|_{k = \sqrt{3}\,\frac{v_{Te}}{c}\, kl}, \tag{6.64}$$

so that the energy density of the transverse waves is

$$W^t = \int W_{kl}^t\, dk^l. \tag{6.65}$$

We will drop the index l on k in the future. Using

$$\delta'(\omega_-) = \frac{-\omega_-}{\sqrt{2\pi}\, k_-^3\, v_{Ti}^3}\, \exp\left\{-\frac{\omega_-^2}{2k_-^2\, v_{Ti}^2}\right\}, \tag{6.66}$$

we find

$$\frac{dW_k^t}{dt} = \alpha W_k^t\, \frac{1}{k} \cdot \frac{\partial}{\partial k}\, kW_k^l. \tag{6.67}$$

Here $W_k^l = (2\pi)^{-3}\, \omega_{pe}\, 4\pi k^2 N_k^l$ and α is a constant equal to that given in (4.16). The same equations for the changes in the l-waves are

$$\frac{dW_k^l}{dt} = \alpha W_k^l k\, \frac{\partial}{\partial k} \cdot \frac{1}{k}\, W_k^t + \alpha W_k^l\, \frac{\partial W_k^l}{\partial k}. \tag{6.68}$$

This expression also includes the term describing the l-interactions. The tt-scattering term in (6.67) is smaller by a factor of v_{Te}^2/c^2 and is therefore dropped. According to the definition of W_k^l, W_k^t given in (6.64), they differ from the number of quanta only by a factor of ω_{pe}. Using (6.67) and (6.68) we easily find that

$$\frac{d}{dt} \int (W_k^l + W_k^t)\, dk = 0,$$

which expresses conservation of the number of quanta during the scattering. One can compare the scattering on electrons with the scattering on ions. In the region where $\omega_- \ll kv_{Ti}$ times less effective. It is therefore reasonable to compare the transfer to v_{Ti}/v_{Te}. According to [59] we have the following estimate for scattering by electrons:

$$\gamma_e^N \approx \omega_{pe} \left(\frac{m_e}{m_i}\right)^2 \left(\frac{v_p}{v_{Te}}\right)^3 \frac{W}{n_0 T_e} , \qquad (6.69)$$

whereas (6.67) gives

$$\gamma_i^N \approx \frac{\omega_{pe}}{10} \cdot \frac{m_e}{m_i} \left(\frac{v_p}{v_{Te}}\right)^2 \frac{W}{n_0 T_e} . \qquad (6.70)$$

Since $v_p \ll 3v_{Te}(m_i/m_e)^{1/2}$, (6.69) is always smaller than (6.70) and electron scattering can be neglected for all practical purposes.

Equation (6.68) has the steady-state solution $W_k^l = 0$, $W_k^l = = \text{const} = \sqrt{\frac{2Q}{\alpha}}$, which was found earlier. However, according to (6.67) this solution is unstable relative to the excitation of transverse quanta in the region $v_p \gg v_{Te}(m_i/m_e)^{1/5}$. But when $v_p < v_{Te}(m_i/m_e)^{1/5}$, if $W_k^l \simeq \text{const } k^{-5/2}$ Eq. (6.67) says that transverse plasmons are damped, which means that they are not excited. It is evident that the induced excitation is important if the thickness L of the plasma (its characteristic dimension) is greater than the optical thickness for this process. By setting $W_k^l = \sqrt{\frac{2Q}{\alpha}}$ we find the optical thickness L_* to be

$$L_* = \frac{v_g k}{\alpha} \sqrt{\frac{\alpha}{2Q}} . \qquad (6.71)$$

It is characteristic that $v_g = \sqrt{3}kv_{Te}c/\omega_{pe}$ is quite large, and L_* takes the form

$$L_* = \frac{9}{\sqrt{2\pi}} \sqrt{\frac{m_i}{m_e}} \left(1 + \frac{T_e}{T_i}\right) \frac{k^2 v_{Te}^2}{\omega_{pe}^2} \cdot \frac{c}{\omega_{pe}} \sqrt{\frac{Q_0}{Q}} . \qquad (6.72)$$

The length over which equilibrium is established for the ll-scattering is a factor of v_{Te}/c smaller. If $L \ll L_*$ one need take into account just the processes of spontaneous emission of t-waves.

According to (6.50) and (6.43), if one assumes that

$$\Lambda_i = \frac{i\,e^2}{m_e\,(2\pi)^3} \cdot \frac{k_{1i}}{k_1\,\omega_{pe}} \cdot \frac{1}{\left(1+\dfrac{T_e}{T_i}\right)}, \tag{6.73}$$

the emission power is

$$Q_k^t = 4\pi k_1^2\,Q_{k_1}\,\frac{\sqrt{3}\,v_{Te}}{c}\bigg|_{k_1 = \frac{\sqrt{3}\,kv_{Te}}{c}} = \frac{(4\pi)^{-1}v_{Te}\,k\omega_{pe}^3}{\sqrt{3}\,nc^3\left(1+\dfrac{T_e}{T_i}\right)^2}\,W_k^l. \tag{6.74}$$

This is the radiative power for longitudinal waves whose frequencies equal those of the transverse waves. The result given in (6.74) agrees with that first obtained in [264, 265] for $T_e = T_i$. Setting $W_k^l = \sqrt{\dfrac{2Q}{\alpha}}$ and assuming that k of order $(\omega_{pe}/v_{Te}) \times (m_e/m_i)^{1/5}$ make the maximum possible contribution, we obtain an estimate of the total radiative power:

$$Q^t = \int\limits_0^\infty Q_k^t\,dk \;\approx\; T_e\,\frac{\omega_{pe}^3}{c^3}\,\omega_{pe}\,\sqrt{\frac{Q}{Q_0}}\,. \tag{6.75}$$

If $L \ll L_*\big|_{k \sim \frac{\omega_{pe}}{v_{Te}}\left(\frac{m_e}{m_i}\right)^{1/5}}$, intensity of the order of (6.75) is lost by the plasma to radiation. Without taking account of the radiation losses, all the power is consumed in heating the plasma because of absorption of the fluctuations in pair collisions of particles. The radiation power can be comparable to the generation power Q if Q is so small that the level of turbulence is lower than the level of the thermal fluctuations. According to (6.72), L_* is rather large because $Q \ll Q_0$, so that condition $L \ll L_*$ is frequently satisfied under laboratory conditions. But in astrophysical situations the opposite inequality is satisfied. Assuming that $L \gg L_*$ and neglecting the spontaneous processes, we will examine in greater detail the solution to (6.67) and (6.68). In the regions where there are no sources to excite turbulence, and neglecting pair collisions, we obtain the following from (6.67) when $\frac{d}{dt} = 0$:

$$W_k^l = \frac{\text{const}}{k} = \frac{A}{k}\,; \tag{6.76}$$

Eq. (6.68) gives

$$W_k^t = Bk - \frac{A}{2k} = \frac{A}{k} \left(\frac{k^2 - k_*^2}{2k_*^2} \right), \qquad B = \frac{A}{2k_*^2}, \qquad (6.77)$$

where k_* is the value of k at which the intensity of the transverse waves goes to zero.

We will now take into account the fact that the appearance of transverse plasmons is a second-order effect, and (6.67) does not give any excitation of t-waves in the generation region where W_k^l drops sharply. Therefore k_* must be equal to k_g. But when $k < k_*$, $W_k^t < 0$, which is impossible; therefore, either there is no stable solution or $W_k^t = 0$. The numerical calculations carried out in [270] showed that there is a continuous transfer from the longitudinal waves to transverse waves and back, and that these oscillations damp out very slowly. In this case the spectrum $W_k = (2Q/\alpha)^{1/2}$ can be only considered as an approximate expression, averaged over those oscillations.

The four-plasmon interactions $t + t' \rightleftarrows t_1 + t_1', t + l \to t_1 + l_1$ are of the same order of magnitude in the variables k^l [see (6.64)] as is the interaction $l + l' \rightleftarrows l_1 + l_1'$. Thus, the repulsion between transverse plasmons at small k has the same qualitative nature, and is of the same order of magnitude, as the repulsion between longitudinal plasmons described in Chapter 4.

Let us now examine the process of (6.62). The probability for this event is [39]

$$w_t^{ll}(\mathbf{k}, \mathbf{k}_1, \mathbf{k}_2) = \frac{e^2 (2\pi)^6 (k_1^2 - k_2^2)^2}{32\pi m_e^2 k^2 \omega_{pe}} \cdot \frac{[\mathbf{k}_1 \mathbf{k}_2]^2}{k_1^2 k_2^2} \delta(\mathbf{k} - \mathbf{k}_1 - \mathbf{k}_2) \times$$

$$\times \delta\left(\omega - 2\omega_{pe} - \frac{3v_{Te}^2}{2\omega_{pe}} (k_1^2 + k_2^2) \right). \qquad (6.78)$$

The spontaneous emission power will be given by

$$Q_k = \int Q_{k, k_1, k_2} W_{k_1}^l W_{k_2}^l \, dk_1 \, dk_2, \qquad (6.79)$$

$$Q_{k, k_1, k_2} = \frac{\pi\omega (k_1^2 - k_2^2)^2 [\mathbf{k}_1 \mathbf{k}_2]^2}{8m_e n_0 k^2 \omega_{pe} k_1^2 k_2^2} \delta(\mathbf{k} - \mathbf{k}_1 - \mathbf{k}_2) \times$$

$$\times \delta\left(\sqrt{k^2 c^2 + \omega_{,e}} - 2\omega_{pe} - \frac{3v_{Te}^2 (k_1^2 + k_2^2)}{2\omega_{pe}} \right). \qquad (6.80)$$

Since the thermal corrections to the spectrum of the longitudinal waves are small, $(k_1 + k_2)^2 \approx 3\omega_{pe}^2/c^2$. If $v_p \ll c$ for both of the coalescence waves, merging is possible when the waves are propagating in nearly opposite directions. If one of the waves slated for mixing has $v_p \approx c$, the second must have $v_p \approx c$. If $v_p \gg c$ for one, the second must have its v_p in a narrow region about $c/\sqrt{3}$. The relative fraction of the waves moving opposite to one another is not very large, so that the emission decreases as the phase velocities of the waves decrease. Assuming that k_1, $k_2 \gg \omega_{pe}/c$, we obtain

$$Q_{\mathbf{k}} = \int Q_{\mathbf{k},\,\mathbf{k}_1}\, W_{\mathbf{k}_1}^l\, W_{-\mathbf{k}_1}^l\, dk_1, \qquad (6.81)$$

$$Q_{\mathbf{k},\,\mathbf{k}_1} = \frac{\pi\omega\,[\mathbf{k}_1\,\mathbf{k}]^2\,(\mathbf{k}\mathbf{k}_1)^2}{2k^2\,m_e\,n_0\,\omega_{pe}\,k_1^4}\, \delta\left(\sqrt{k^2\,c^2 + \omega_{pe}^2} - 2\omega_{pe} - \frac{3k_1^2\,v_{Te}^2}{\omega_{pe}}\right). \qquad (6.82)$$

Upon introducing $W_{k_1}^l = 4\pi k_1^2\, W_{\mathbf{k}_1}^l$, $\int_0^\infty W_{k_1}^l\, dk_1 = W^l$, we have for isotropic turbulence

$$Q_{\mathbf{k}}^l = \int \frac{\omega\, dk_1\, k^2\, \delta\left(\sqrt{k^2\,c^2 + \omega_{pe}^2} - 2\omega_{pe} - \dfrac{3k_1^2\,v_{Te}^2}{\omega_{pe}}\right)}{60\,k_1^2\,m_e\,n_0\,\omega_{pe}}\,(W_{k_1}^l)^2. \qquad (6.83)$$

This equation enables us to describe the wings of the emission line at a frequency $\sim 2\omega_{pe}$. It is evident that emission frequencies smaller than $2\omega_{pe}$ are not possible, that is, the intensity for $\omega < 2\omega_{pe}$ is zero. If $\omega - 2\omega_{pe} \gg (v_{Te}^2/c^2)\omega_{pe}$, contributions come only from waves with $v_p \gg c$, and the result is described by (6.83). Taking $W_k = \text{const} = (2Q/\alpha)^{1/2}$, we find the radiation intensity in the wings to be

$$4\pi Q_{\mathbf{k}}^l\, k^2\, \frac{dk}{d\omega} = Q_\omega^l = Q\,\frac{324\,m_i\,v_{Te}^5}{5m_e\,c^5}\left(1 + \frac{T_e}{T_i}\right)^2 \frac{\sqrt{\omega_{pe}}}{(\omega - 2\omega_{pe})^{3/2}}. \qquad (6.84)$$

Here we have neglected the difference between ω and $2\omega_{pe}$ in the coefficient in front of the resonance factor $1/(\sqrt{\omega - 2\omega_{pe}})^3$. The factor $(1 + T_e/T_i)$ can be replaced by unity when $T_e \gg T_i$. Equation (6.84) is valid for v_p as large as c if $3v_{Te}^2/v_{Ti} > c$. The total power emitted at frequencies larger than ω_* takes the form

$$Q_*^l = \int_{\omega_*}^\infty Q_\omega\, d\omega = \frac{648\,Q m_i\,v_{Te}^5}{5m_e\,c^5}\left(1 + \frac{T_e}{T_i}\right)^2 \sqrt{\frac{\omega_{pe}}{\omega_* - 2\omega_{pe}}}. \qquad (6.85)$$

This equation can be used for order of magnitude estimates when $\omega_* - 2\omega_{pe} \sim 2\omega_{pe}(v_{Te}^2/c^2)$:

$$Q^t \approx 10^2 \frac{m_i}{m_e} \, Q \, \frac{v_{Te}^4}{c^4} \left(1 + \frac{T_e}{T_i}\right)^2. \tag{6.86}$$

Evidently, if $Q^t \sim Q$, all the power is dispersed as radiation when one goes through that interval of k values corresponding to $v_p < c$. In this case the turbulence begins to radiate. The large numerical coefficient in (6.86) shows that this situation is possible when $v_{Te}/c \sim 1/30$. Of course, it is also required that the plasma dimension be smaller than the optical thickness. Before estimating it, we will consider in greater detail the contribution to the radiation from waves with $v_p \sim c$ and $v_p \gg c$. Therefore we will present a general result for an isotropic turbulence obtained from (6.79) after integrating over angles:

$$Q_\omega^t = \frac{\pi}{16 \, m_e \, n_0 \, \omega_{pe}} \int_0^\infty \frac{dk_1}{k_1} \, W_{k_1}^l \int_{\left|k_1 - \frac{1}{c}\sqrt{\omega^2 - \omega_{pe}^2}\right|}^{k_1 + \frac{1}{c}\sqrt{\omega^2 - \omega_{pe}^2}} dk_2 \, W_{k_2}^l \, \frac{\omega^2}{k_2 c^2} \times$$

$$\times \left(k_1^2 - k_2^2\right)^2 \frac{1}{k^2} \left[1 - \frac{\left(k_1^2 + k_2^2 - k^2\right)^2}{4 k_1^2 k_2^2}\right] \delta\left(\omega - 2\omega_{pe} - \frac{3 v_{Te}^2}{2\omega_{pe}}\left(k_1^2 + k_2^2\right)\right). \tag{6.87}$$

It is not hard to obtain the approximate equation (6.83) from (6.87) for the region where $v_p \ll c$.

When one of the waves k_1 has $v_p \gg c$, we obtain the following approximate equation from (6.87):

$$Q_\omega^t = \frac{\pi\sqrt{3} \, \omega_{pe}^2}{3c^2 \, m_e \, n_0} \int W_{k_1} dk_1 \, W_{\sqrt{3}\,\omega_{pe}/c} \, \delta\left(\omega - 2\omega_{pe}\left(1 + \frac{9}{4}\frac{v_{Te}^2}{c^2}\right) - \frac{3 k_1^2 v_{Te}^2}{2\omega_{pe}}\right). \tag{6.88}$$

When

$$k_* < \sqrt{3} \, \omega_{pe}/c \tag{6.89}$$

we have

$$Q_\omega^t = \frac{2\sqrt{\pi} \, \omega_{pe}}{\sqrt{2}} \left(1 + \frac{T_e}{T_i}\right) \sqrt{\frac{Q}{Q_0}} \, \frac{v_{Te}}{c} \sqrt{\frac{m_i}{m_e}} \, \frac{(\nu - 1)\,\omega_{pe}}{k_0^2 c^2} \times$$

$$\times W \left(k_0 \sqrt{3} \, v_{Te}\right)^\nu \left(2\omega_{pe}\left(\omega - 2\omega_{pe}\left(1 + \frac{9}{4} \cdot \frac{v_{Te}^2}{c^2}\right)\right)\right)^{-\nu/2}. \tag{6.90}$$

Thus, the intensity increases so long as $\omega - 2\omega_p \left(1 + \dfrac{9}{4} \cdot \dfrac{v_{Te}^2}{c^2} \right)$ is not approximately equal to $k_0^2 v_{Te}^2 / \omega_{pe}^2 \left(k_0 \ll \dfrac{\omega_{pe}}{c} \right)$. The total radiation intensity can be found from (6.88):

$$Q^t = W^l \frac{\pi \sqrt{3} \, \omega_{pe}^2}{3c^3 m_e \, n_0} \sqrt{\frac{2Q}{\alpha}} . \qquad (6.91)$$

If it is assumed that the turbulence is dispersed because of collisions, then $W^l = Q/\nu_e$ and

$$Q^t = Q \frac{6\sqrt{\pi}}{\sqrt{2}} \sqrt{\frac{m_i}{m_e}} \sqrt{\frac{Q}{Q_0}} \cdot \frac{\omega_{pe}}{\nu_e} \cdot \frac{v_{Te}^3}{c^3} \left(1 + \frac{T_e}{T_i} \right) . \qquad (6.92)$$

The flux of turbulent energy Q for which the radiation given in (6.92) is greater than that of (6.86) is

$$Q_1^R = Q_0 \frac{\nu_e^2}{\omega_{pe}^2} \, 10^4 \, \frac{m_i}{m_e} \cdot \frac{v_{Te}^2}{c^2} , \qquad (6.93)$$

while the value of Q for which the emission power of (6.92) is comparable with Q is

$$Q_2^R = Q_0 \frac{\nu_e^2}{\omega_{pe}^2} \cdot \frac{m_e}{m_i} \cdot \frac{c^6}{v_{Te}^6} . \qquad (6.94)$$

Naturally, if the turbulent energy can be dissipated as radiation for $v_p < c$, i.e., $10^2 \dfrac{m_i}{m_e} \left(1 + \dfrac{T_e}{T_i} \right)^2 v_{Te}^4 / c^4 > 1$, then $Q_1^R > Q_2^R$; but if the reverse inequality holds, then $Q_1^R < Q_2^R$. In this case all the turbulent energy is concentrated where $v_p \gg c$ and when $Q > Q_2^R$ the turbulence becomes radiative.

In describing a radiative turbulence we note that, by definition, radiated waves leave the system so that their energy density in the plasma is small. This means that in the equations which describe the changes in the turbulence spectrum because of the radiation, one can neglect the terms proportional to the radiation intensity. If we are interested in $v_p \ll c$, we can use (6.78) to obtain

$$\frac{\partial W_k^l}{\partial t} = -\frac{\pi \sqrt{3} \, \omega_{pe}^4 \, (W_k^l)^2}{5 n_0 \, m_e \, c^5 \, k^2} + \alpha W_k \frac{\partial W_k}{\partial k} . \qquad (6.95)$$

Here we have also taken into account the interaction related to scattering by plasma ions. The equilibrium spectrum obtained from (6.95) takes the form

$$W_k^l = W_0^l\, e^{-k_r/k}; \quad k_r = \frac{27\sqrt{3}\,\bar{v}_{Te}^4}{5c^4}\cdot\frac{\omega_{pe}}{c}\cdot\frac{m_i}{m_e}\left(1+\frac{T_e}{T_i}\right)^2. \tag{6.96}$$

The condition $k_r > \omega_{pe}/c$ differs only by a numerical factor from the condition for radiative dissipation as obtained from (6.86). If the turbulence is generated with $k \gg k_r$, then, as is clear from (6.96), the radiation can be neglected in the generation region and, with $k_g < (\omega_{pe}/v_{Te})(m_e/m_i)^{1/5}$, we have

$$W_0^l = \sqrt{\frac{2Q}{\alpha}}. \tag{6.97}$$

Consequently, the radiation has practically no effect on the spectrum of turbulence up to $k \sim k_r$, and then the intensity drops off sharply.

Let us now consider the case where $v_p \gg c$, assuming that when $v_p \ll c$ the radiation is weak. We find that the intensity of the radiative absorption is

$$\frac{\partial W_k^l}{\partial t} = -W_k^l\,\frac{\pi}{4\sqrt{3}}\cdot\frac{\omega_{pe}^2}{c^3 n_0 m_e}\,W_{\sqrt{3}\,\frac{\omega_{pe}}{c}}^l. \tag{6.98}$$

If $\sqrt{3}\,\omega_{pe}/c > k_*$, (6.98) can be written as an effective absorption:

$$\frac{\partial W_k^l}{\partial t} = -\nu_{\text{eff}}\,W_k^l, \quad \nu_{\text{eff}} = \frac{3\sqrt{\pi}}{2\sqrt{2}}\,\omega_{pe}\,\frac{v_{Te}^3}{c^3}\left(1+\frac{T_e}{T_i}\right)\sqrt{\frac{Qm_i}{Q_0\,m_e}}. \tag{6.99}$$

ν_{eff} will dominate the pair collisions when $Q > Q_2^R$ [see (6.94)]. Thus, in the dissipative emission regime we obtain the same qualitative turbulence spectra as those studied in Chapter 4, in which ν_e must be replaced with ν_{eff} as given by (6.99). In particular, the fundamental scale of the turbulence will be determined by the equation

$$k_0 = k_*\left(\frac{27\left(1+\frac{T_e}{T_i}\right)^4 m_i\,v_{Te}^6}{16m_e\,c^6}\right)^{\frac{1}{2(\nu-1)}}. \tag{6.100}$$

This result shows that the fundamental turbulence scale stops being dependent on the generation power Q. As mentioned in Chapter 4, there is a maximum in the spectrum if $k_0 \ll k_*$. Thus, if the plasma temperature is sufficiently high (for an isothermal hydrogen plasma $v_{Te}/c \sim 10^{-1}$), and $Q > Q_2^R$, the maximum in the spectrum disappears. All these results are for an optically thin plasma layer, through which the radiation freely passes. It is now necessary to find an expression for the optical thickness, i.e., the length over which the absorption of the radiation becomes important.

If $\omega - 2\omega_{pe} \gg \omega_{pe} v_{Te}^2/c^2$, the main contribution comes from $k_1 \gg k$, and

$$\gamma_k = - \frac{\pi \omega_{pe}^2}{5 n_0 m_e c^2} \int W_{k_1}^l \, dk_1 \, \delta \left(\omega - 2\omega_{pe} - \frac{3 v_{Te}^2 k_1^2}{\omega_{pe}} \right). \qquad (6.101)$$

Taking $W_{k_1} = \sqrt{2Q/\alpha}$, we find

$$\gamma_\omega = \gamma_{k\,(\omega)} = \frac{3}{5} \sqrt{\frac{\pi}{2}} \omega_{pe} \left(1 + \frac{T_e}{T_i} \right) \sqrt{\frac{m_i}{m_e} \cdot \frac{v_{Te}^2 \sqrt{\omega_{pe}}}{c^2 \sqrt{\omega - 2\omega_{pe}}}} \sqrt{\frac{Q}{Q_0}}. \qquad (6.102)$$

This result and (6.84) means that the radiation spectrum in an optically thick plasma takes the form

$$I_\omega = \frac{Q_\omega^t}{\gamma_\omega} = 108 \sqrt{\frac{2}{\pi}} \sqrt{\frac{Qm_i}{Q_0 m_e}} \cdot \frac{v_{Te}^3}{c^3} \left(1 + \frac{T_e}{T_i} \right) \frac{n_0 T_e}{(\omega - 2\omega_{pe})}. \qquad (6.103)$$

The total radiation intensity for frequencies greater than ω_* differs from (6.103) in that $1/(\omega - 2\omega_{pe})$ is replaced by

$$\ln \frac{\omega_{max} - 2\omega_{pe}}{\omega_* - 2\omega_{pe}} \sim \ln \left[3 \left(\frac{m_e}{m_i} \right)^{2/5} \frac{c^2}{v_{Te}^2} \right].$$

Thus, as an order of magnitude the intensity of electromagnetic waves in the wings of the lines is

$$\int_{\omega_*}^{\infty} \frac{I_\omega \, d\omega}{n_0 T_e} \simeq 10^3 \sqrt{\frac{Qm_i}{Q_0 m_e}} \cdot \frac{v_{Te}^3}{c^3}, \qquad (6.104)$$

which is approximately $10^2 \, v_{Te}^3/c^3$ times smaller than the energy of the Langmuir oscillations. From (6.102) the optical thickness L_* is given by

$$L_* = \frac{\sqrt{3} \, c}{2\gamma_\omega}. \tag{6.105}$$

If $\omega - 2\omega_{pe} \approx \omega_{pe} \, v_{Te}^2/c^2$, the main contribution comes from $k_1 \ll k$ (if the spectrum has a maximum at $k_0 \gg \omega_{pe}/c$), and

$$\gamma_k = \frac{\pi \omega_{pe}^2}{4c^2 \, n_0 \, m_e} \int W_{k_1}^l \, dk_1 \, \delta \left(\omega - 2\omega_{pe}\left(1 + \frac{9}{4} \cdot \frac{v_{Te}^2}{c^2}\right) - \frac{3v_{Te}^2 \, k_1^2}{2\omega_{pe}} \right). \tag{6.106}$$

In conjunction with (6.89) we find that

$$I_\omega = \text{const} = \frac{24}{\sqrt{2\pi}} \cdot \frac{n_0 T_e}{\omega_{pe}} \cdot \frac{v_{Te}}{c} \left(1 + \frac{T_e}{T_i}\right) \sqrt{\frac{Qm_i}{\gamma_0 m_e}}. \tag{6.107}$$

This corresponds to (6.103) in order of magnitude when $\omega - 2\omega_{pe} \sim \omega_{pe} v_{Te}^2/c^2$. This indicates that the estimate given in (6.104) still holds for the total radiation intensity. Since the radiation energy in an optically thick medium is a relatively small portion of the total turbulence energy, its effect on the turbulence spectrum is insignificant.

Finally, in a nonisothermal plasma with $T_e \gg T_i$ we must consider the role of the processes $l \rightleftarrows t + s$, $l + s \rightleftarrows t$. The probability of the first of these processes is

$$w(\mathbf{k}_l, \mathbf{k}_t) = \frac{e^2 \, (2\pi)^6 \, \omega_{pe}^3 \, \omega_s^3 \, [\mathbf{k}_t, \mathbf{k}_l]^2}{8\pi m_e^2 \, v_{Te}^4 \, \omega_{pi}^2 \, k_s^2 \, \omega k_t^2 \, k_l^2},$$

$$w_l^{ts}(\mathbf{k}_l, \mathbf{k}_t, \mathbf{k}_s) = w(\mathbf{k}_l, \mathbf{k}_t) \, \delta(\mathbf{k}_l - \mathbf{k}_t - \mathbf{k}_s) \, \delta(\omega_{\mathbf{k}_l}^l - \omega_{\mathbf{k}_t}^t - \omega_{\mathbf{k}_s}^s), \tag{6.108}$$

and the probability for the second process differs from this only in the signs of $\mathbf{k}_s$ and $\omega_{\mathbf{k}_s}^s$ in the δ-functions. When computing the power of spontaneous emission both these processes must be taken into account. Neglecting ω_s as compared with $\omega_{\mathbf{k}_l}^l$ and $\omega_{\mathbf{k}}$, we find for isotropic turbulence that

$$Q_{\mathbf{k}_t}^t = \int Q_{\mathbf{k}_t, \, \mathbf{k}_l} \, W_{\mathbf{k}_l}^l \, W_{\mathbf{k}_l - \mathbf{k}_t}^s \, \delta(\omega_{\mathbf{k}_l}^l - \omega_{\mathbf{k}_t}) \, dk_l. \tag{6.109}$$

This shows that the frequencies of the radiated waves are equal to the Langmuir frequencies. Setting $k_t = (\sqrt{3}v_{Te}/c)k$ we find that the emission power can, according to (6.74), be written as $(k^t \ll k_l)$

$$Q_k^t = \frac{\sqrt{3}\, v_{Te}\, \omega_{pe}}{kc^3} W_k^l\, W_k^s\, \eta_k, \quad Q^t = \int Q_k^t\, dk, \qquad (6.110)$$

where η_k is related to the coefficient $w(k_l, k)$ of the δ-functions in the probability of (6.108) by the equation

$$\eta_k = \frac{\omega^t}{(2\pi)^3 \omega^l\, \omega^s} \int w(k_1, k)\frac{d\Omega_1}{4\pi}, \qquad (6.111)$$

where Ω_1 is the solid angle for k_1.

These results are quite general and can be used for the coalescence of high-frequency waves with any low-frequency waves in a magnetoactive plasma. For the probability given in (6.108) we have

$$\eta_k = \frac{\pi\omega_{pe}^2}{6n_0\, T_e}. \qquad (6.112)$$

Thus,

$$Q_k^t = \frac{\pi\sqrt{3}\, \omega_{pe}^3\, v_{Te}}{6n_0\, T_e\, kc^3} W_k^l\, W_k^s. \qquad (6.113)$$

Assuming that $k_s \gg k_l^s$ and using (4.123) when $Q \ll Q_0 \cdot \frac{1}{12}(m_e/m_i)^3$, we find

$$Q_k^t = Q\frac{12}{k} \sqrt{\frac{Q}{Q_0}}\left(\frac{m_i}{m_e}\right)^{3/2} \frac{v_{Te}^3}{c^3}. \qquad (6.114)$$

But if $Q \gg Q_0 \frac{1}{12}(m_e/m_i)^3$ we must then use (4.124):

$$Q_k^t = \frac{9\sqrt{3}}{2} k\, \frac{v_{Te}^2}{\omega_{pe}^2} Q\, \frac{4m_i}{m_e} \cdot \frac{v_{Te}^3}{c^3}. \qquad (6.115)$$

The last equation dictates that the fundamental emission takes place at large k, on the order of $(\omega_{pe}/v_{Te})(m_e/m_i)^{1/5}$. But in this case the emission power can be comparable to Q only in a very hot plasma, where $v_{Te}/c > (m_e/m_i)^{1/5}$. If only ion-acoustic turbu-

lence is excited in the plasma, the emission given in (6.113) arises
from thermal Langmuir plasmons. If we use Eq. (4.231), we can
write the radiation intensity as

$$Q_k = \frac{5\sqrt{3}}{8\,(2\pi)^{3/2}} \cdot \frac{\omega_{pe}^3}{c^3} \, v_{Te} \frac{T_e^2}{T_i} \sqrt{\frac{m_e}{m_i}} \, \ln \frac{k}{k_0} \,. \tag{6.116}$$

We shall now consider the effects of reabsorption of the radiation.
When Langmuir fluctuations are excited, we have

$$\gamma_{kt} = - \int N_{\mathbf{k}_1}^l \, w_t^{ls}(\mathbf{k}^t,\ \mathbf{k}_1,\ \mathbf{k}_s) \frac{d\mathbf{k}_1 \, d\mathbf{k}_s}{(2\pi)^6} + \int N_{\mathbf{k}_1}^l \times$$
$$\times w_l^{ts}(\mathbf{k}_1,\ \mathbf{k}^t,\ \mathbf{k}_s) \, d\mathbf{k}_1 \, d\mathbf{k}_s \, (2\pi)^{-3}. \tag{6.117}$$

Putting $k^t = k\sqrt{3}\,\dfrac{v_{Te}}{c}$ when $k \gg k_*^s = \dfrac{2\omega_{pe}}{3v_{Te}}\sqrt{\dfrac{m_e}{m_i}}$, we find

$$\gamma_{\mathbf{k}} = \alpha' \frac{1}{k} \cdot \frac{\partial}{\partial k} kW_k^l, \quad \alpha' = \alpha\left(1 + \frac{T_e}{T_i}\right)^2. \tag{6.118}$$

α is defined by Eq. (4.16). Just as with the dynamics for l-wave
interactions, the difference between scattering on ions and decays
results in the replacement of α by α' [see (6.67)]. The instability of
Langmuir turbulence relative to the excitation of transverse waves
follows from (6.118) [as well as (6.67)]. This is immediately clear
from (6.118) if one substitutes $W_k^l = (2Q/\alpha')^{1/2}$ into that equation.

If ion-acoustic waves with $k_s \gg \omega_{pe}/c$ are excited, then

$$\gamma_k = - \frac{\pi\omega_{pe}^3}{18n_0 \, T_e \, v_{Te}^2} \cdot \frac{W_k^s}{k} \,. \tag{6.119}$$

By using (6.116) and (6.119) we obtain the Rayleigh−Jeans equation
for the intensity of the transverse oscillations; i.e., the excitation
of transverse waves by purely ion-acoustic turbulence is impossible (see below concerning other nondecay excitations). In a
nonisothermal plasma the equation for the changes in l-waves due
to $l$ts-decays takes the form of (6.68) with α replaced by α'. The
narrow lines of transverse waves with frequencies $\sim\omega_{pe}$ must
always excite Langmuir turbulence, as has been observed in a
number of experiments. According to (6.119), external electromagnetic waves are efficiently absorbed in a plasma containing
ion-acoustic turbulence.

6.4. Emission of Electromagnetic
Waves by a Turbulent Plasma
in an External Magnetic Field

In a weak magnetic field, where $\omega_{\text{He}} \ll \omega_{\text{pe}}$, processes leading to emission at $2\omega_{\text{pe}}$, $l \rightleftarrows t$ scattering processes, and $l$ts-decays are only slightly altered. The mixing of Langmuir waves with whistlers and with plasma gyrofrequency waves $l + w \rightleftarrows t$, $l \rightleftarrows t + w$ are new processes, as is their mixing with Alfven waves and magnetohydrodynamic waves:

$$l \rightleftarrows t + A; \quad l \rightleftarrows t + M; \quad l + A \rightleftarrows t, \quad l + M \rightleftarrows t.$$

The probability for decaying into whistlers is

$$w_t^{l,\,w} = \frac{(2\pi)^4 \, k_w^2 \, \omega_{pe}^2 \, [k_l \, k_l]^2}{2n_0 \, m_e \, \omega_{He} \, k_t^2 \, k_l^2} \cdot \frac{\sin^2 \theta_w}{|\cos \theta_w|} \, \delta \, (k_t - k_l - k_w) \times$$

$$\times \, \delta \, (\omega_t - \omega_l - \omega_w) \, [\sin^2 \varphi_w \cos^2 \varphi_w + \tfrac{1}{4} \cos^2 \theta_w \, (\cos^2 \varphi_w - \sin^2 \varphi_w)^2]. \quad (6.120)$$

If $\omega_{\text{He}}/\omega_{\text{pe}} \gg v_{\text{Te}}^2/c^2$, then $|k_t| \ll |k_l|, |k_w|$ from the conservation laws, and the corrections to the thermal motion are much smaller than ω_w, i.e., $k_t = \sqrt{\dfrac{2\omega_{He}}{\omega_{pe}}} \, k_w |\cos \theta_w|$, $k_w = -k_l$, $|k_w| = |k_l|$, and the greatest difference between ω_t and ω_{pe} is ω_{He}. Under these conditions the process $l \to t + w$ does not take place. It is convenient to introduce the radiation intensity because it has been normalized to dk_l :

$$Q_{k,\Omega}^t = k_t^2 \, Q_{k_t,\Omega} \sqrt{\frac{2\omega_{He}}{\omega_{pe}} |\cos \theta_w|} \, \Bigg|_{k_t = k_l} \sqrt{\frac{2\omega_{He}}{\omega_{pe}} |\cos \theta_w|}. \quad (6.121)$$

Assuming that the fluctuations are symmetrically distributed relative to the direction of the magnetic field, we find from (6.120) that

$$Q_{k,\Omega}^t = \int d\Omega_w \sqrt{\frac{2\omega_{pe}}{\omega_{He}} \cdot \frac{\omega_{pe}^4 \sin^2\theta_w \, \rho(\theta,\theta_w) \, W_k^l \, W_{k,-\Omega_w}^w}{16\omega_{He} \, kn_0 \, m_e \, |\cos\theta_w|^{3/2} \, c^4}} \; ;$$

$$\rho(\theta,\theta_w) = \frac{1}{2}\left(1 - \cos^2\theta \cos^2\theta_w\right)\left(\cos^2\theta_w + 1\right) -$$

$$- \sin^2\theta \sin^2\theta_w \frac{1}{4}\left(\cos^2\theta_w + 1\right).$$

(6.122)

Equation (6.122) can be compared with (6.113). For comparable energies in the w- and s-waves, the power emitted by the whistlers is comparable to the power emitted by the ion-acoustic turbulence when $\omega_{He}/\omega_{pe} \sim (v_{Te}/c)^{2/3}$. It should be kept in mind that the radiated frequencies are always very close to ω_{pe} ($\omega_{max} = \omega_{pe} + \omega_{He}$) and their exit from the plasma is inhibited.

We will again write down the absorption of transverse waves by a plasma with the excitation of whistlers:

$$\gamma_k = -\frac{\pi}{8c^2}\int W_{k,-\Omega_w}^w \, d\Omega_w \, \frac{\omega_{pe}^6 \sin^2\theta_w}{k\omega_{He}^3 \, |\cos\theta_w|^3} \, \rho(\theta,\theta_w).$$

(6.123)

When the energies of the whistlers and ion-acoustic waves are comparable this absorption can be greater than the absorption given in (6.119) for ion-acoustic waves [with $\omega_{He}/\omega_{pe} \sim (v_{Te}/c)^{4/3}$], but the frequency of the transverse wave must then be quite close to ω_{pe}. In an optically thick medium the radiation intensity is estimated as

$$\int I_{k,\Omega} \, d\Omega = \int \frac{Q_{k,\Omega}^t \, d\Omega}{\gamma_{k,\theta}} \approx W_k^l \left(\frac{\omega_{He}}{\omega_{pe}}\right)^{3/2},$$

(6.124)

i.e., the total energy transferred into the transverse waves due to this process is much smaller than the energy of the Langmuir waves.

Frequencies even closer to ω_{pe} ($\omega_{max}^t \lesssim \omega_{pe} + \omega_{Hi}$) participate in the interactions $t \rightleftarrows l + A$, $t \rightleftarrows l + M$. The probability for this interaction takes the form

$$w_t^{l,M} = \frac{(2\pi)^4 \, \omega_A \omega_{pi}^2 \sin^2\theta_M}{4n_0 \, m_e \, v_A^2} \, \delta(\mathbf{k}_t - \mathbf{k}_l - \mathbf{k}_M) \, \delta(\omega_t - \omega_l - \omega_M) \, \frac{[\mathbf{k}_l \, \mathbf{k}_t]^2}{4k_l^2 \, k_t^2},$$

$$w_t^{l,A} \simeq w_t^{l,M} \frac{\omega_{Hi}^2}{\omega_A^2} \ll w_t^{l,M}.$$

(6.125)

If $k_l \ll 3\omega_{pe} v_A / 2v_{Te}^2$ (because $k_t \ll k_l$, k_M we have $k_l \approx k_M \ll \omega_{pi}/c$, and our condition is always satisfied if $v_A > v_{Ti} v_{Te}/c$), only the process $t \rightleftarrows l + M$ is important, where $k_t = \sqrt{k_l} \sqrt{\dfrac{2\omega_{pe} v_A}{c^2}}$. The radiation power, normalized to dk_l, is given by

$$Q_{k,\Omega}^t = \frac{\omega_{pe}^2}{8n_0 m_e v_A} W_k^l \int d\Omega_M W_{k,\Omega_M}^M \, \Omega_M \sin^2 \theta_M \, \frac{\omega_{pi}^2 \, \rho_{\theta,\theta_M}}{c^3 (2\omega_{pe} v_A k^3)^{1/2}},$$

$$\rho_{\theta,\theta_M} = \left[1 - \cos^2 \theta \cos^2 \theta_M - \frac{1}{2} \sin^2 \theta \sin^2 \theta_M \right], \tag{6.126}$$

and the damping rate of transverse waves in a magnetoacoustic turbulence is

$$\gamma_{k_t} = -\frac{\pi \omega_{pi}^2}{8n_0 m_e v_A^2} \int d\Omega_A W_{\omega,\,\Omega_M}^M \left. \frac{\sin^2 \theta_M}{\cos^2 \theta_M} \right|_{\omega = \frac{k_t^2 c^2}{2\omega_{pe}}} \rho_{s_\theta,\,\theta_M}. \tag{6.127}$$

Now let us turn to consideration of the case with strong magnetic fields, where $\omega_{He} \gg \omega_{pe}$. A new feature will be the appearance of plasma oscillations with frequencies close to ω_{He}:

$$\omega_k^h = \omega_{He} + \frac{\omega_{pe}^2 \sin^2 \theta}{2\omega_{He}} + \frac{2\omega_{He}}{\omega_{He}^2} k^2 v_{Te}^2 \cot^2 \theta. \tag{6.128}$$

It is characteristic of the conversion of such waves into radiation that the electromagnetic waves with frequencies close to ω_{He} will have indices of refraction which are nearly unity (with the exception of those propagating in directions very close to H_0). Therefore, the converted waves can freely leave the plasma. One of the effective methods for converting the waves given in (6.128) into electromagnetic waves is ion scattering. Another characteristic feature of conversion of the waves in (6.128) is that the nonlinear interaction of those waves is $(\omega_{pe}/\omega_{He})^2$ times smaller than their conversion into electromagnetic radiation [270].

Thus, this conversion is the only nonlinear process available, and turbulence with frequencies given by (6.128) is always of radiative type. The ions can be considered as unmagnetized if $v_p = \omega_{He}/k \ll (m_i/m_e)v_{Ti}$; h-waves will be weakly attenuated if $v_p \gg (\omega_{He}/\omega_{pe})v_{Te}$. Thus, the region in which there is scattering on unmagnetized ions exists when

$$1 \ll \frac{\omega_{He}}{\omega_{pe}} \ll \sqrt{\frac{m_i T_i}{m_e T_e}}. \tag{6.129}$$

Under the condition of (6.129) the scattering probability is

$$w_p(k_t, k_h) = \frac{\pi \omega_{pe}^2}{2n_0^2} \cdot \frac{[k_t k_h]^2}{k_t^2 k_h^2} (1 + \cos^2 \theta_t)\left(1 + \frac{T_e}{T_i}\right)^{-2} \delta\left(\omega_t - \omega_h - (k_t - k_h)v\right).$$

(6.130)

We therefore obtain the following expression for the interaction of t- and h-waves:

$$\frac{\partial W_{\omega, \Omega}^t}{\partial t} = \alpha_* W_{\omega, \Omega}^t \int d\Omega_h \, \Psi(\theta, \theta_h) \frac{d}{d\tilde{\omega}} \tilde{\omega} W_{\tilde{\omega}, \Omega_h}^h;$$

(6.131)

$$\tilde{\omega} = \omega - \omega_{He} - \frac{\omega_{pe}^2}{2\omega_{He}} \sin^2 \theta_h;$$

$$\Psi(\theta, \theta_h) = \frac{\sin^2 \theta_h}{\cos^2 \theta_h}\left(1 - \cos^2 \theta_h \cos^2 \theta_t - \frac{1}{2}\sin^2 \theta_h \sin^2 \theta_t\right);$$

$$\alpha_* = \frac{\pi \omega_{pe}^4}{4v_{Te}^2 n_0 m_i \omega_{He}^2}\left(1 + \frac{T_e}{T_i}\right)^{-2};$$

(6.132)

$$W^h = \int W_{\tilde{\omega}, \Omega_h}^h \, d\tilde{\omega} \, d\Omega_h;$$

$$\frac{\partial W_{\tilde{\omega}, \Omega_h}^h}{\partial t} = W_{\tilde{\omega}, \Omega_h}^h \int d\Omega \alpha_* \Psi(\theta, \theta_h) \tilde{\omega} \frac{d}{d\omega} W_{\omega, \Omega}^t.$$

(6.133)

It follows that

$$W_{\tilde{\omega}, \Omega_h}^h = \frac{W(\theta_h)}{\tilde{\omega}} \quad \text{or} \quad W_{k, \Omega_h}^h \sim \frac{W(\theta_h)}{k}$$

(6.134)

corresponds to a stable spectrum, and $W_{\omega, \Omega}^t = W^t(\theta_t)$ for a white spectrum for the transverse waves. Of course, these results hold outside the generation region and the region of possible cyclotron absorption of the transverse waves.

When isotropically distributed the coalescence of two h-waves gives a total emitted power of [270]

$$\left. \begin{aligned} Q &= \frac{32\pi \omega_{pe}^2}{5n_0 m_e c^5 \omega_{He}^{-2}} \int^\circ \frac{(W_k^h)^2 \, dk}{k^2}, \\ \int W_k^h \, dk &= W^h. \end{aligned} \right\}$$

(6.135)

We must conclude by discussing a final mechanism by which a magnetoactive plasma radiates, one which is connected with the presence of a boundary. Generally speaking, the plasma's dimensions transverse to the field are finite. During the spectral transfer process, the wavelength of the fluctuations can become comparable to the transverse dimensions, and the very first vibrational modes of the plasma column will be excited. These oscillations, being dipole oscillations for example, are easily emitted from the plasma. The essence of this effect is that the length of the emitted waves is much greater than the plasma dimensions. This situation can arise for Langmuir oscillations; it is especially possible for waves propagating only transverse to the magnetic field (cyclotron waves are an example [271]). It is also possible to increase the cross section for mixing two Langmuir waves (whose wavelengths are much smaller than the plasma dimensions) to give a transverse wave with frequency $2\omega_{pe}$ and wavelength greater than the plasma dimensions [272].

6.5. Radiation from Epithermal and Relativistic Particles in a Turbulent Plasma

As shown in Chapter 4, turbulence cannot exist without exciting fast particles whose maximum energy is ultrarelativistic for a stationary turbulence. The radiation from such particles can be at frequencies much higher than the Langmuir frequency, and these waves freely leave the plasma if it is optically thin. The optical thickness itself is determined by the reabsorption by fast particles. This indicates that the emission of electromagnetic waves in an optically thick plasma must significantly alter the fast particle spectrum. Let us examine the effects of radiation from fast particles on Langmuir turbulence, which is the most effective for accelerating particles and the radiation produced by those particles. A treatment of emission from other modes of a turbulent plasma can be found in [261]. For Langmuir turbulence the fast particles emit at frequencies $\omega \gg \omega_{pe}$. The emission processes at such high frequencies admit to a simple, clear interpretation, which will be the starting point for our discussion [66]. In this case the distribution of electric fields in the turbulent fluctuations changes very slowly during one period of the high-frequency waves.

Assume that the electric field for the plasma wave is of the form

$$E^l = E_0 \cos(\omega_1 t - k_1 r), \quad E_0 \| k_1. \tag{6.136}$$

We will use the method of successive approximations to find the interaction between the electron and the waves. In the zeroth approximation we assume that the electron moves uniformly in a straight line at a velocity v. A force $eE_0 \cos(\omega_1 t - k_1 vt)$ acts on it because of the wave. In the first approximation small oscillations are added to the uniform motion, so that we now have $r = vt + \varrho \cos \Omega t$ with

$$\varrho = -\frac{eE_0}{e_p \Omega^2 k_1}(k_1 - v(k_1 v)), \quad \Omega = \omega_1 - k_1 v = \omega_1 - k_1 v \cos \theta_1.$$

If the electron moves in a vacuum we can find the resulting radiation by determining the oscillating part of the dipole moment $e\rho$ and calculating the intensity of the dipole radiation. But the electron moves instead in a rather inhomogeneous plasma. The plasma inhomogeneity is related to the oscillations of the electron density n in a wave which is related to the field by the equation

$$\text{div } E = 4\pi e(n - n_0) \tag{6.137}$$

(n_0 is the mean electron density). As the density changes, the dielectric constant, being a function of n, also changes:

$$\varepsilon(\omega) = 1 - \frac{4\pi e^2 n}{m\omega^2} = 1 + \frac{ek_1 E_0}{m\omega^2} \sin(k_1 r - \omega_1 t). \tag{6.138}$$

The polarization produced by the electrons moves along with them. Because of the nonuniformity in ε a dipole moment arises which partially compensates (or in the case of a nonrelativistic electron, completely compensates) the dipole moment due to the electron oscillations.

A charge which moves in a medium with a dielectric constant ε which varies in both time and space radiates transverse waves. We remark that the corresponding radiation mechanism has a well-known analogy with the radiation from a charge moving in a layered medium (see [64, 65]). This approach becomes possible when the frequencies of the radiated transverse waves are much larger than the frequency of the longitudinal waves creating the density inhomogeneity.

It is very significant that the phases of the electron vibrations and the changes in ε are not independent. The total radiation is not the sum of the radiation due to the electron vibrations and the emission arising from the medium inhomogeneity.

It is easy to find the emitted power from an electron moving in a medium having a variable ε and vibrating under the influence of a wave. We need only find the work done by the field created by the charge on the charge itself:

$$\frac{1}{T} \int_{T \to \infty}^{T} (jE)\, dr\, dt = \frac{(2\pi)^4}{T} \int dk\, (j_k^* E_k). \qquad (6.139)$$

We must use Maxwell's equations to find the field E_k:

$$\Delta E + \operatorname{grad} \operatorname{div} (\varepsilon - 1)\, E - \frac{\partial^2}{c^2 \partial t^2}\, \varepsilon E = 4\pi \left(\frac{\partial j}{c^2 \partial t} + \operatorname{grad} \rho \right). \qquad (6.140)$$

Both ε and j are known in this equation. Solving (6.140) by using successive approximations in E_0 we find $(v_p \ll c)$

$$Q_p = \frac{e^4 E_0^2 k_1^2}{4\pi m_e^2} \int dk\, |\omega|\, \delta\, (k^2 - \omega^2)\, \delta\, (\omega - (k - k_1)\, v)\, |\Lambda^t|^2;$$
$$\Lambda^t = \Lambda' - \frac{k\, (k\Lambda')}{k^2}; \qquad (6.141)$$

$$\Lambda_e' = \frac{k_1}{\omega} \left[\frac{\sqrt{1 - v^2/c^2}}{\left(1 - \frac{v}{c} \cos \theta\right) k_1^2} - \frac{1}{(k - k_1)^2 - k^2} \right] +$$

$$\div v \left[\sqrt{1 - \frac{v^2}{c^2}}\, \frac{(kk_1 - \omega c^{-2}(k_1 v))}{k_1^2\, (k_1\, v \cos \theta)^2} - \frac{1}{(k - k_1)^2 - k^2} \right]; \qquad (6.142)$$

$$\Lambda_i' = \left(\frac{k_1}{\omega} + v \right) ((k_1 - k)^2 - k^2)^{-1}. \qquad (6.143)$$

The first term in Λ_e' corresponds to the Compton scattering, and the second term is the nonlinear scattering. We can arrive at this same result by using the general expressions for the emission probability for scattering on fast particles if we use the form of the inverse Maxwell operator Π_{ij} appropriate for $\omega \gg \omega_{pe}$:

$$\Pi_{ij} = -\frac{k_i k_j c^2}{k^2 \omega^2 \varepsilon(\omega)} + \frac{\delta_{ij} - \dfrac{k_i k_j}{k^2}}{k^2 - \omega^2 \varepsilon(\omega) c^{-2}}, \qquad \varepsilon(\omega) = 1 - \frac{\omega_{pe}^2}{\omega^2}, \qquad (6.144)$$

and the expression valid at high frequencies for the nonlinear currents:

$$S_{ijl} = -\frac{e\omega_{pe}^2}{8\pi m_e \, \omega_1 \omega_2} \left(\delta_{ij} \frac{k_{2l}}{\omega_2} + \delta_{il} \frac{k_{1j}}{\omega_1} + \delta_{jl} \frac{k_i}{\omega} \right). \qquad (6.145)$$

Finally, for Compton scattering $\Lambda = [iek_1/m_e(2\pi)^3]\Lambda'$ [see (2.225)] we have

$$\Lambda = \frac{ie^2}{(2\pi)^3 m_e k_1} \sqrt{1 - \frac{v^2}{c^2}} \, \frac{1}{(\omega - kv)^2} \times$$

$$\times \left\{ v \left(kk_1 - \frac{\omega}{c^2}(k_1 v) \right) + k_1 (\omega - kv) \right\}. \qquad (6.146)$$

The result found from (6.144) and (6.146) corresponds to replacing the energy $E_0^2/4\pi$ of an individual wave in (6.141) by $W_{k_1}^l \, dk_1$ and integrating over k_1. Thus, the radiation of high-frequency waves, $\omega \gg \omega_{pe}$, is only weakly dependent on the nature of the plasma oscillations, whether it be an individual wave or a broad turbulence spectrum. This result is easily interpreted, for high-frequency radiation at ω is automatically averaged over a period, and therefore over the phases of the turbulent fluctuations.

In the limit $v \ll c$ Eq. (6.141) gives the following expression for the total emitted intensity of the individual electrons and ions:

$$Q_p^{(e)} = \frac{e^4 E_0^2}{15 m_e^2} \cdot \frac{v^2}{c^5} (3 + 13 \cos^2 \theta_1), \qquad (6.147)$$

$$Q_p^{(i)} = \frac{e \, E_0^2}{3 m_e^2 c^3}. \qquad (6.148)$$

When $v \ll c$ both the ions and electrons radiate at a frequency $\omega = k_1 v |\cos \theta_1|$. For fast (in general relativistic) particles and an isotropic distribution of Langmuir waves, we average Q over all the angles between the plasma wave and the particle momentum:

$$\bar{Q}_p = \frac{1}{2} \int_{-1}^{1} d\cos\theta_1 \, Q = \int_0^\infty Q_\omega^p \, d\omega, \qquad (6.149)$$

$$Q_\omega^p = \frac{e^4 E_0^2 k_1^2 \omega^2}{16 m_e^2 \pi} \int d\cos\theta_1 \, d\cos\theta \, d\varphi \left\{ \Lambda'^2 - \frac{(k\Lambda')^2}{k^2} \right\} \times$$

$$\times \left[\delta\left(\omega\left(1 - v\cos\theta\right) + k_1 v\cos\theta_1\right) + \delta\left(\omega\left(1 - v\cos\theta\right) - k_1 v\cos\theta_1\right) \right]$$

$$(6.150)$$

(φ is the angle between the planes containing the vectors $\mathbf{k_1}$, $\mathbf{v}$ and $\mathbf{k}$, $\mathbf{v}$). Evaluation of the integral gives [66]

$$Q_\omega^p = \frac{e^4 E_0^2 \varepsilon_p^2}{4 m_e^4 \omega c^7} \, \Phi\left(\frac{\varepsilon_p}{mc^2}, \, q\right), \qquad (6.151)$$

$$q = \frac{\omega\,(1-v/c)}{k_1 v}, \quad \Phi\left(\frac{\varepsilon}{mc^2}, \, q\right) = 0 \text{ for } q > 1, \qquad (6.152)$$

where q is the ratio of the frequency of the emitted wave to the maximum frequency possible for a given k_1 and θ. The function $\Phi(\varepsilon/mc^2, q)$ is given by a very complicated expression; therefore we will present just graphs of $\Phi(\varepsilon_p/mc^2, q)$.

The presence of two maxima in the curves shown in Fig. 6.1 results from the presence of two scattering mechanisms in our

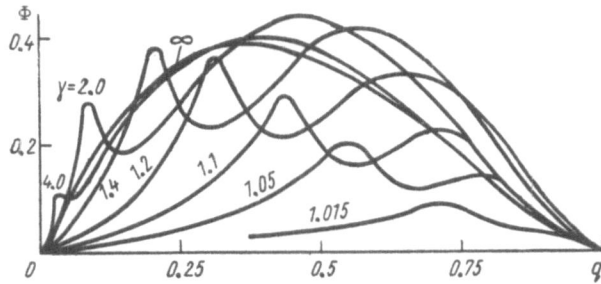

Fig. 6.1. Frequency dependence of the intensity of spontaneous emission by electrons for scattering on isotropic Langmuir fluctuations: $q = (\omega/k_1 v)[1 - (v/c)] < 1$ is the ratio of the emitted frequency to the maximum possible frequency. The curves are labeled with the value of $\gamma = \varepsilon/mc^2$.

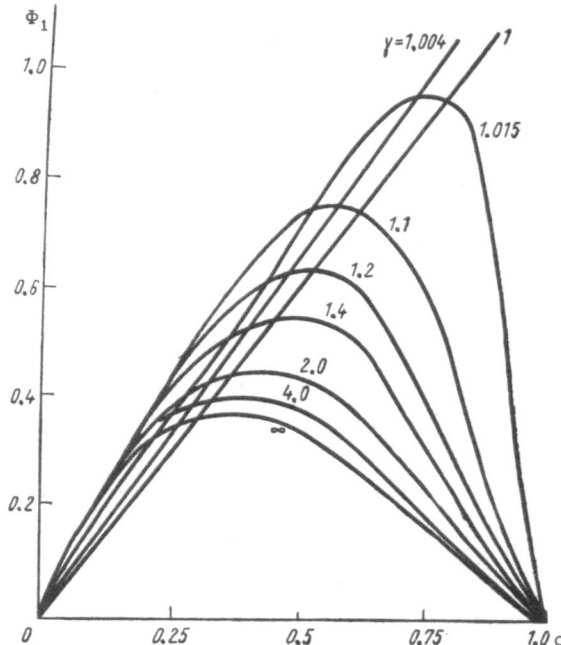

Fig. 6.2. Frequency dependence of the intensity of spontaneous emission of electrons for scattering on Langmuir fluctuations when the nonlinear scattering is not included   (γ =
ε/mc^2).

treatment. At ultrarelativistic energies $(\varepsilon_p/mc^2 \gg 1)$ the primary
role goes to radiation due to electron oscillations which create a
broad, smooth maximum. The narrow maximum at small q arises
from a mechanism which is similar to that of the transition radiation (radiation from the density inhomogeneities created by the
plasma wave). Since the latter type of radiation has a constant
intensity and constant average frequency in the limit $\varepsilon/mc^2 \to \infty$
and does not grow with ε/mc^2, while the radiation due to the electron oscillations has an intensity and frequency which increase as
ε/mc^2 grows [increasing proportional to $(\varepsilon/mc^2)^2$], the left maximum decreases in size and shifts to the left as $\varepsilon/mc^2 \to \infty$. The
limiting value Φ for $\varepsilon/mc^2 \to \infty$ is easily calculated if one takes
into account just the Compton scattering:

$$\Phi(\infty, q) = \frac{8}{3} \, q \, [(1-q)^3 - 3q^2(1-q+\ln q)]. \tag{6.153}$$

In the nonrelativistic limit $(\varepsilon/mc^2 \to 1)$ the two maxima merge and the two mechanisms tend to inhibit each other. For comparison Fig. 6.2 shows curves of $\Phi_1(\varepsilon/mc^2, q)$ which do not take into account the density fluctuations of the plasma wave. These curves all have the same limit when $\varepsilon/mc^2 \to \infty$, which is given by (6.153), but they behave rather differently when $\varepsilon/mc^2 \to 1$.

When transverse radiation appears during the scattering of plasma waves by ions, Eq. (6.143) holds, with $\Phi_2(\varepsilon/mc^2, q)$ having the form shown in Fig. 6.3.

The total power emitted by an electron can be found by numerically integrating Eq. (6.150):

$$Q_p = \frac{2e^4 E_0^2 e^2}{9 m_e^4 c^7} \Pi\left(\frac{v^2}{c^2}\right). \tag{6.154}$$

The function $\Pi(v)$ is shown in Fig. 6.4. The coefficient of $\Pi(v)$ in (6.154) is selected from the condition that $\Pi(v) \to 1$ when $v \to 1$.

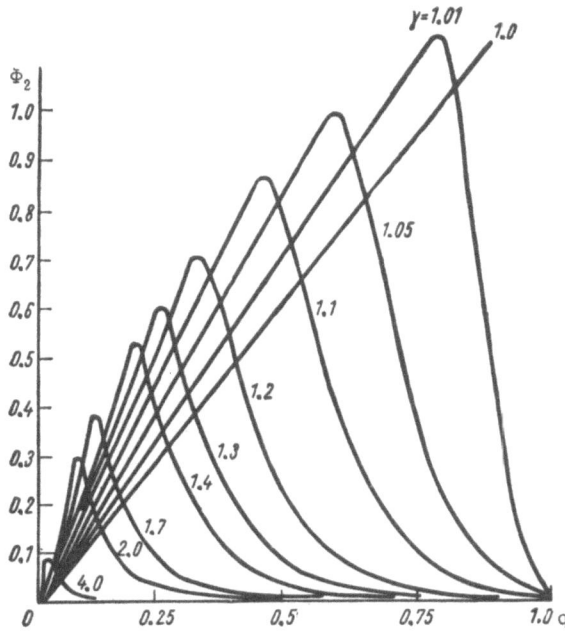

Fig. 6.3. Frequency dependence of the spontaneous emission intensity for ions for scattering on isotropic Langmuir fluctuations ($\gamma = \varepsilon/mc^2$).

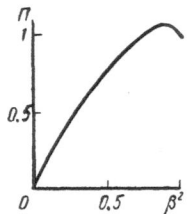

Fig. 6.4. Total intensity of the electron radiation as a function of $\beta^2 \approx \dfrac{v^2}{c^2} \left(\Pi(v) = \dfrac{9 m^2 Q c^3}{2 e^4 E_0^2 \gamma^2} \right)$.

According to (6.153) the intensity of the spontaneous emission from a single ultrarelativistic electron can be written in the form $Q_\omega^p = 4 \pi k^2 Q_{p,k} c^{-1}$, where

$$Q_{p,k} = \frac{e^4}{3 \omega^3} \int\limits_{\frac{\omega m_e^2 c^3}{2 \varepsilon^2}} W_{k_1}^l \frac{\omega d k_1}{k_1} \left\{ \left(1 - \frac{\omega m_e^2 c^3}{k_1 \varepsilon^2} \right)^3 - 3 \frac{\omega^2}{k_1^2} \frac{m_e^4 c^6}{4 \varepsilon^2} \times \right.$$

$$\left. \times \left(1 - \frac{\omega m_e^2 c^3}{k_1 2 \varepsilon^2} + \ln \frac{\omega m_e^2 c^3}{2 \varepsilon^2 k_1} \right) \right\} \frac{1}{m_e^2 c}. \qquad (6.155)$$

The intensity of the radiation from all the ultrarelativistic electrons is

$$Q_\omega^l = \int Q_\omega^p f_\varepsilon d\varepsilon = \frac{\omega_{pe}^4}{12 \pi n_0^2} \int \frac{W_{k_1}^l d k_1}{k_1} \int\limits_{mc^2 \sqrt{\frac{\omega}{2 k_1 c}}} f_{\tilde{\varepsilon}} d\varepsilon \times$$

$$\times \left\{ (1 - q)^3 - 3 q^2 (1 - q + \ln q) \right\} c^{-4}. \qquad (6.156)$$

Here $f_\varepsilon d\varepsilon$ is the number of particles in the energy interval $d\varepsilon$, and $k_1 \gg \omega_{pe}/c$. By using $W_{k_1}^l = \sqrt{\dfrac{2Q}{\alpha}} = \text{const}$, we can replace the integration over k with an integration over $q = \dfrac{\omega}{2 k_1 c} \left(\dfrac{mc^2}{\varepsilon} \right)^2$. We set

$$q_{max} = \frac{\omega}{2 \omega_{pe}} \left(\frac{mc^2}{\varepsilon} \right)^2, \qquad q_{min} = \frac{\omega}{\omega_{pe}} \cdot \frac{v_{Te}}{c} \left(\frac{m_i}{m_e} \right)^{1/5} \left(\frac{mc}{\varepsilon} \right)^2.$$

These values of q_{max} and q_{min} correspond to the limiting wave numbers for the turbulent fluctuations for which Eq. (6.156) and the spectrum $W_{k_1}^l = \sqrt{\dfrac{2Q}{\alpha}}$ are valid. By definition, $q < 1$. For those particle energies for which $q_{max} > 1$,

$$Q_\omega^t = \frac{m_e \sqrt{6}\, \omega_{pe}^3 v_{Te}^3}{4\pi n_0 \sqrt{\pi} c^4} \left(1 + \frac{T_e}{T_i}\right) \sqrt{\frac{m_i}{m_e}} \int\limits_{q_{min} < 1}^{mc^2 \sqrt{\frac{\omega}{2\omega_{pe}}}} f_\varepsilon \, d\varepsilon \sqrt{\frac{Q}{Q_0}} \times$$

$$\times \left[\ln \frac{2\omega_{pe}\, c}{\omega v_{Te}} \sqrt{\frac{m_e}{m_i}} \left(\frac{\varepsilon}{mc^2}\right)^2 - \frac{3}{2}\right]. \tag{6.157}$$

The expression in the logarithm in (6.157) is approximate, just as is the upper limit on the energy. When $q_{max} < 1$ we obtain

$$Q_\omega^t = \frac{m_e \sqrt{6}\, v_{Te}^{3\prime} \omega_{pe}^3}{4\pi^{3/2} n_0 c^4} \sqrt{\frac{m_i}{m_e}} \left(1 + \frac{T_e}{T_i}\right) \int\limits_{mc^2 \sqrt{\frac{\omega}{2\omega_{pe}}}} f_\varepsilon \, d\varepsilon \sqrt{\frac{Q}{Q_0}} \times$$

$$\times \left\{\ln \frac{c}{v_{Te}} \left(\frac{m_e}{m_i}\right)^{1/5} - 3 \frac{\omega}{2\omega_{pe}} \left(\frac{mc^2}{\varepsilon}\right)^2 + \frac{2}{3} \frac{\omega^3}{(2\omega_{pe})^3} \left(\frac{mc^2}{\varepsilon}\right)^6 - \frac{3}{2} \frac{\omega^3}{(2\omega_{pe})^2} \times\right.$$

$$\left.\times \left(\frac{mc^2}{\varepsilon}\right)^4 \ln \frac{\omega}{2\omega_{pe}} \cdot \frac{mc^2}{\varepsilon} + 3\omega^2 (2\omega_{pe})^{-2} \left(\frac{mc^2}{\varepsilon}\right)^4\right\}. \tag{6.158}$$

If f_ε is a decreasing function of the particle energy, such as

$$f_\varepsilon = \frac{(\gamma - 1)\, n_1}{m} \left(\frac{m}{\varepsilon}\right)^\gamma c^{2(\gamma-1)}, \tag{6.159}$$

the upper limit in (6.157) is not very important; the emission given by (6.157) is much greater than that of (6.158), and is given by

$$Q_\omega^t = a(\gamma) \left(\frac{2\omega_{pe}}{\omega}\right)^{\frac{\gamma-1}{2}} T_e \frac{n_1}{n_0} \left(\frac{c}{v_{Te}}\right)^{\frac{\gamma-3}{2}} \left(\frac{m_e}{m_i}\right)^{\frac{\gamma-6}{10}} \sqrt{\frac{Q}{Q_0}} \frac{\omega_{pe}^3}{c^3}, \tag{6.160}$$

where $\alpha(\gamma)$ is a numerical coefficient of order unity. For the power spectrum of (6.159) one can obtain an expression for the intensity of the radiation which will be valid for an arbitrary turbulence spectrum [66]:

$$Q_\omega^t = b(\gamma) \frac{\omega_{pe}^4}{2c^4 n_0} \cdot \frac{n_1}{n_0} \left(\frac{2c}{\omega}\right)^{\frac{\gamma-1}{2}} \int W_{k_1}^l \frac{dk_1}{k_1^{\frac{3-\gamma}{2}}}, \tag{6.161}$$

$$b(\gamma) = \frac{\gamma-1}{6\pi} \left[\frac{1}{\gamma-1} - \frac{3}{\gamma+1} + \frac{2}{\gamma+5} + \frac{6}{(\gamma+3)^2}\right]. \tag{6.162}$$

For the spectrum $(2Q/\alpha)^{1/2}$ Eq. (6.161) gives us Eq. (6.160) when $\gamma > 1$, where $a(\gamma) = b(\gamma) 3\sqrt{6}/(\gamma-1)\sqrt{\pi}$.

In order to evaluate the total radiation intensity, we must have some idea of the optical thickness. From the general expression for the growth rate, it is easy to establish that

$$\gamma_k = \int \frac{d\Omega_k}{4\pi} \gamma_k = (2\pi)^3 \int Q_{p, k} \varepsilon^2 \left(\frac{\partial}{\partial\varepsilon} \cdot \frac{f_\varepsilon}{\varepsilon^2}\right) d\varepsilon \tag{6.163}$$

in the case of isotropic turbulence and an isotropic distribution of the radiating particles. Using Eq. (6.155) we find

$$\gamma_k = -\frac{\pi\omega_{pe}^4}{8\omega^2 n_0^2 c} \int \frac{f_\varepsilon d\varepsilon}{\varepsilon} \Phi_*(\infty, q) W_{k_1}^l \frac{dk_1}{k_1}, \tag{6.164}$$

where

$$\Phi_*(\infty, q) = \frac{8}{3} [(1-q)^3 + 3q(1-q^2+q \ln q)]. \tag{6.165}$$

Using the power spectrum of (6.159) for the accelerated particles, we find

$$\gamma_\omega = -\frac{\pi\omega_{pe}^4 n_1}{3n_0^2 c^5 m_e} \left(\frac{2c}{\omega}\right)^{\frac{\gamma}{2}+2} c(\gamma) \int dk_1 k_1^{\frac{\gamma-2}{2}} W_{k_1}, \tag{6.166}$$

where

$$c(\gamma) = (\gamma-1) \frac{1}{4} \left[\frac{1}{\gamma} + \frac{3}{\gamma+4} - \frac{4}{\gamma+6} - \frac{6}{(\gamma+4)^2}\right]. \tag{6.167}$$

As is clear from the result given by (6.116), relative to self-absorption the optical thickness $L_* = c/\gamma_\omega$ is very strongly dependent on frequency, being larger the greater the frequency. For the turbulence spectrum $(2Q/\alpha)^{1/2}$ we have

$$\gamma_\omega = -d\left(\gamma\right)\frac{n_1}{n_0}\cdot\frac{v_{Te}^3}{c^3}\sqrt{\frac{Q}{Q_0}}\,\omega_{pe}\left(\frac{2\omega_{pe}}{\omega}\right)^{\frac{\gamma}{2}+2}\left(\frac{c}{v_{Te}}\right)^{\frac{\gamma}{2}}\left(\frac{m_e}{m_i}\right)^{\frac{\gamma-1}{2}},$$

$$d\left(\gamma\right) = c\left(\gamma\right)\sqrt{\frac{3\pi}{2}\cdot\frac{8}{\gamma}}\,. \tag{6.168}$$

Let us assume that the plasma layer is so thin that the plasma can be assumed to be optically transparent to frequencies ω of the order of ω_{pe}. Then all frequencies from ω_{pe} to the maximum frequencies will be emitted, where the highest frequencies are those at which the particle spectrum is a power spectrum (more precisely, the spectrum does not fall off rapidly with energy). We will assume that the spectrum is a power function up to an energy $\varepsilon = \varepsilon_*$. According to (6.160), when $\gamma < 3$ the bulk of the emitted power is at the maximum frequencies. From the condition that $q < 1$, we have

$$\omega_{max} = \left(\frac{\varepsilon_*}{mc^2}\right)^2 2\,\frac{\omega_{pe}\,c}{v_{Te}}\left(\frac{m_e}{m_i}\right)^{1/5},$$

$$Q^t = \int Q_\omega^t\,d\omega = \frac{4a\left(\gamma\right)}{3-\gamma}\,Q_0\left(\frac{\varepsilon_*}{mc^2}\right)^{3-\gamma}\left(\frac{m_i}{m_e}\right)^{3/10}\frac{n_1}{n_0}\sqrt{\frac{Q}{Q_0}}\;\omega_{pe}^3/n_0\,c^3,$$

$$Q_0 = \omega_{pe}\,n_0\,T_e. \tag{6.169}$$

Actually, Eq. (6.169) is applicable when the low frequencies do not satisfy the transparency condition. It is only essential that the frequencies be of the order of ω_{max}, while frequencies an order of magnitude smaller would have satisfied the condition for optical transparency.

In the previous chapter it was shown that conditions can result in which the energy of high-frequency turbulence can be almost entirely dissipated on fast particles which are accelerated by the turbulence. It is essential to determine the conditions under which the energy of the fast particles will be dissipated as radiation. For the processes considered here the turbulence energy is essentially undistributed. It is easy to show that when $n_1 < n_0$ the damping of Langmuir turbulence due to conversion processes is always smaller than that due to pair collisions. This

is the means by which the dissipation of turbulence into radiation
can be realized through the use of accelerated particles. Evidently, the radiation from electrons (if they are injected in sufficient
numbers, and $n_1^{(e)}$ is not very small) plays a rather significant
role. In order to make an estimate, we can use Eq. (6.169), but
we must know the value of ε_* and its dependence on Q, and we
must also know γ (it is shown in §6.6 and Chapter 7 that the effects of scattering in the turbulent plasma can increase ε_*).

Before discussing these problems, we must consider conversion on fluctuations with $v_p \gg c$. As a rule the optical thickness
for radiative processes at frequencies of ω_{pe} and $2\omega_{pe}$ is much
smaller than the thickness for the radiative processes involving
accelerated particles. Therefore, if it is assumed that the plasma is optically thick relative to the latter processes, at least for
some $\omega_* \gg \omega_{max}$, it is then optically thick relative to processes
at frequencies of ω_{pe} and $2\omega_{pe}$, and, from the previous discussion,
the energy in these frequencies is small and the maximum in the
turbulence spectrum lies in the region where $v_p \gg c$ for large Q.
Thus, the emission from fluctuations with $v_p \gg c$ can be more
important, not only with respect to the power involved but also
relative to the reverse effect on the spectra of the radiating particles. In accord with our previous assumptions, we will postulate
the presence of both Langmuir waves, with $\omega_k^l = \omega_{pe} + \dfrac{3}{2}\dfrac{k^2 v_{Te}^2}{\omega_{pe}}$,

and transverse plasmons (labeled by p), with $\omega_k^p = \omega_{pe} + \dfrac{k^2 c^2}{2\omega_{pe}}$.

The probabilities of $l \to t$ and $p \to t$ scatterings by relativistic
particles, averaged over the directions of the pulsations and the
directions of the particle velocities, are [251]

$$w_l^t = \frac{e^4 (2\pi)^3 c^4}{6\omega^2 \varepsilon_p^2 \omega_{pe}^2} \left(\omega_{pe} + \frac{\omega}{2\omega_{pe}} \cdot \frac{m^2 c^4}{\varepsilon_p^2} \left(\omega \frac{m^2 c^4}{\varepsilon_p^2} - 2\omega_{pe} \right) \right), \quad (6.170)$$

$$w_p^t \simeq w_l^t. \quad (6.171)$$

When $\omega > 2\omega_{pe} (\varepsilon_p^2 / m^2 c^4)$ these probabilities are zero. Thus, the
maximum radiated frequency cannot depend on the wave number
of the turbulent fluctuations if $k_1 \ll \omega_{pe}/c$. This result has important consequences. One is that the radiated power and the
optical thickness of the pulsation are independent of the turbulence

spectra, that is,

$$Q_\omega = \frac{\omega^3}{c^3 \pi^2 \, \omega_{pe}} \, (W^l + W^l) \int_{mc^2 \sqrt{\frac{\omega}{2\omega_{pe}}}}^{\infty} w_l^t f_\varepsilon \, d\varepsilon, \tag{6.172}$$

$$\gamma_\omega = \frac{\omega}{\omega_{pe}} \, (W^l + W^p) \int_{mc^2 \sqrt{\frac{\omega}{2\omega_{pe}}}}^{\infty} w_l^t \, \varepsilon^2 \, d\varepsilon \, \frac{d}{d\varepsilon} \cdot \frac{f_\varepsilon}{\varepsilon^2}. \tag{6.173}$$

Only the total energies of the oscillations, W^l and W^p, figure in these expressions. For the power spectrum given in (6.159) we have

$$Q_\omega = \frac{\omega_{pe}^3}{6\pi} \left(\frac{2\omega_{pe}}{\omega} \right)^{\frac{\gamma-1}{2}} \frac{n_1 \, (\gamma-1) \, [\, (\gamma^2 + 4\gamma + 11) \, (W^l + W^p) \,]}{n_0^2 \, (\gamma+1) \, (\gamma+3) \, (\gamma+5) c^3}. \tag{6.174}$$

Assuming that $W^l = Q/\nu_e$, we obtain the estimate

$$Q_\omega = \frac{n_1}{n_0} \, Q \, \frac{\omega_{pe}^3}{12\pi\nu_e \, n_0 c^3} \left(\frac{2\omega_{pe}}{\omega} \right)^{\frac{\gamma-1}{2}} \tag{6.175}$$

By comparing with (6.160) we see that even when

$$Q > \frac{\nu_e}{\omega_{pe}} \, Q_{\max} \left(\frac{m_e}{m_i} \right)^{\frac{\gamma-6}{5}} \left(\frac{c}{v_{Te}} \right)^{\gamma-3} \tag{6.176}$$

the radiation given in (6.175) dominates. Thus, when (6.176) is fulfilled, one must take into account, at the very least, the radiation for which $v_p \gg c$. However, the $\omega_{\max}$ in (6.175) is smaller than its counterpart in (6.160) by a factor of $(c/v_{Te})(m_e/m_i)^{1/5}$ because here we have $\omega_{\max} = 2\omega_{pe}(\varepsilon_\ast/mc^2)^2$.

If the power spectrum is valid only up to those energies for which the medium is optically thick relative to (6.175), then when $\omega = \omega_{\max}$ the radiation for (6.175) is still optically opaque and does not act as a source of radiation losses; thus only (6.169) will give radiative losses because the maximum frequencies for it will be transparent.† The coefficient for reabsorption by waves with $v_p \gg c$ and by relativistic particles having a power function spectrum such as (6.159) is easily found from (6.173):

$$-\gamma_\omega = \frac{\pi}{24} \left(\frac{2\omega_{pe}}{\omega} \right)^{\frac{\gamma}{2}+2} \omega_{pe} \frac{n_1}{n_0^2 \, mc^2} (\gamma - 1) \frac{[\,(\gamma^2+6\gamma+16)\,(W^l+W^p)\,]}{(\gamma+4)\,(\gamma+6)}.$$

(6.177)

We note that (6.168) is applicable at frequencies for which (6.177) does not hold.

At optically thick frequencies the radiation spectrum is found from (6.174) and (6.177):

$$I_\omega = \frac{Q_\omega}{|\gamma_\omega|} = m\omega_{pe}^2 \frac{4}{\pi^2 c} \left(\frac{\omega}{2\omega_{pe}} \right)^{5/2} \frac{(\gamma+4)\,(\gamma+6)}{(\gamma+1)\,(\gamma+3)\,(\gamma+5)} \frac{[\,(\gamma^2+4\gamma+11)\,(W^l+W^p)\,]}{[\,(\gamma^2+6\gamma+16)\,(W^l+W^p)\,]}.$$

(6.178)

The spectrum with $I_\omega - \omega^{5/2}$ is also characteristic of synchrotron radiation in the reabsorption region.

6.6. The Effect of Radiation from a Turbulent Plasma on the Spectra of Fast Particles

The reaction of the radiation on the particles which radiate leads not only to energy losses by these particles, but to energy gains through the actions of the turbulence and the radiation. The energy losses correspond to spontaneous scattering of fast particles by turbulent fluctuations with the emission of electromagnetic waves; the acceleration processes are due to the induced scattering processes. Assuming the turbulence is sufficiently strong, we will take into account only those effects which arise when $v_p \gg c$. Then the equation for the relativistic particles takes the form [251]

$$\frac{\partial f_\varepsilon}{\partial t} = \frac{\partial}{\partial \varepsilon} \, D_\varepsilon^p \, \varepsilon^2 \frac{\partial}{\partial \varepsilon} \cdot \frac{f_\varepsilon}{\varepsilon^2} + \frac{\partial}{\partial \varepsilon} \, A_\varepsilon^p f_\varepsilon,$$

(6.179)

---

†One must bear in mind the possibility of "detuning" the frequencies of the optically transparent and radiating mechanisms of (6.175) due to the effect of synchrotron radiation (Sec. 6.6) and scattering (Chapter 7).

where $A_\varepsilon^p = Q_\varepsilon^p$ describes the spontaneous scattering (p labels a plasma mechanism):

$$A_\varepsilon^p = \frac{1}{\pi^2} \left(W^l + W^p \right) \int_0^{2\omega_{pe} \frac{e^2}{m^2 c^4}} w_l^t \frac{\omega^3 \, d\omega}{\omega_{pe} c^3} , \qquad (6.180)$$

and D_ε^p describes the induced scattering:

$$D_\varepsilon^p = \int_0^{2\omega_{pe} \frac{e^2}{m^2 c^4}} \left(W^l + W^p \right) w_l^t \frac{\omega}{\omega_{pe}} I_\omega \, d\omega. \qquad (6.181)$$

Furthermore, we will take into account the synchrotron acceleration, treated in Chapter 5, and the synchrotron deceleration, which are related to the presence of magnetic fields (H is the synchrotron mechanism).

Then

$$A_\varepsilon = A_\varepsilon^p + A_\varepsilon^H = \frac{16\pi \, e^4}{9m^2 \, c^3} \cdot \frac{e^2}{m^2 \, c^4} \left(W^l + W^p + 3 \frac{H^2}{8\pi} \right), \qquad (6.182)$$

where A_ε^H is the synchrotron deceleration [46]. In order to calculate the acceleration, one must know I_ω. We will consider frequencies lower than those at which the plasma becomes transparent, where both the synchrotron radiation and the plasma radiation give spectra of the form $\sim \omega^{5/2}$. Including the synchrotron radiation means that we must change at most only a few numerical factors in (6.174), (6.177), and (6.178), because the frequency dependence of the intensities and reabsorption coefficients are the same for both synchrotron radiation and plasma radiation. In particular, the bracket in (6.174) and the numerator of (6.178) must be replaced with

$$\left(W^l + W^p \right) \left(\gamma^2 + 4\gamma + 11 \right) + a_1 \left(\gamma \right) \xi^{\frac{\gamma - 3}{3}} \frac{H^2}{4\pi}, \qquad (6.183)$$

$$a_1 \left(\gamma \right) = \left(\gamma + 3 \right) \left(\gamma + 5 \right) \left(\frac{3}{2} \right)^{\frac{\gamma}{2} + 1} \frac{\Gamma \left(\frac{3\gamma - 1}{12} \right) \Gamma \left(\frac{3\gamma + 19}{12} \right) \Gamma \left(\frac{\gamma + 5}{4} \right)}{\sqrt{2\pi} \, \Gamma \left(\frac{\gamma + 7}{4} \right)},$$

$$\qquad (6.184)$$

$$\xi = \frac{eH}{mc\omega_{pe}},$$

while the bracket in (6.177) and the denominator of (6.178) are changed into

$$(W^l + W^p)(\gamma^2 + 6\gamma + 16) + a_2(\gamma)\,\xi^{\frac{\gamma}{2}-1}\,\frac{H^2}{4\pi}, \qquad (6.185)$$

$$a_2(\gamma) = \frac{\gamma(\gamma+4)(\gamma+6)}{\gamma+2}\left(\frac{3}{2}\right)^{\frac{\gamma+3}{2}}\frac{\Gamma\left(\frac{\gamma+6}{4}\right)\Gamma\left(\frac{3\gamma+2}{12}\right)\Gamma\left(\frac{3\gamma+22}{12}\right)}{\sqrt{2\pi}\,\Gamma\left(\frac{\gamma+1}{4}\right)}. \qquad (6.186)$$

Expressions for the synchrotron reabsorption and emission coefficients were obtained in [46].

Finally,

$$D_\varepsilon = D_\varepsilon^p + D_\varepsilon^H, \qquad (6.187)$$

where D_ε^H is the diffusion coefficient for synchrotron radiation, as calculated in Chapter 5. In accord with (6.178), we can approximate I_ω by the equations

$$I_\omega = \begin{cases} \dfrac{I}{\omega_*}\left(\dfrac{\omega}{\omega_*}\right)^{\nu'} & \text{for} \quad \omega < \omega_*, \\[3mm] \dfrac{I}{\omega_*}\left(\dfrac{\omega_*}{\omega}\right)^{\nu} & \text{for} \quad \omega > \omega_*, \end{cases} \qquad (6.188)$$

where ω_* is the frequency at which the plasma becomes optically transparent; for (6.178) we have $\nu' = 5/2$. The coefficient D_ε is a rapidly increasing function of the particle energy until $\varepsilon_* = m(\omega_*/2\omega_{pe})$. When $\varepsilon < \varepsilon_*$,

$$D_\varepsilon = \frac{\pi}{24}\cdot\frac{I}{n_0^2}\,\omega_{pe}\left(\frac{2\omega_{pe}}{\omega_*}\right)^{\nu'+1}\left(\frac{\varepsilon}{mc^2}\right)^{2(\nu'-1)}\frac{1}{\nu'(\nu'+1)(\nu'+2)}\times$$

$$\times\left[(W^l + W^p)(\nu'^2 + \nu' + 2) + a_3(\nu')\,\xi^{\nu'-2}\,\frac{H^2}{8\pi}\right], \qquad (6.189)$$

$$a_3(\nu') = \frac{12\sqrt{2}\,\nu'(\nu'+1)(\nu'+2)}{8\sqrt{\pi}\,\Gamma\left(\frac{\nu'+3}{2}\right)}\left(\frac{3}{2}\right)^{\nu'-\frac{1}{2}}\Gamma\left(\frac{\nu'}{2}\right)\Gamma\left(\frac{\nu'}{2}+\frac{4}{3}\right)\Gamma\left(\frac{\nu'}{2}-\frac{1}{3}\right).$$

$$(6.190)$$

This makes it easy to find average increase in particle energy:

$$\dot{\varepsilon}_p = \alpha_T \, mc^2 \left(\frac{\overline{\varepsilon}}{mc^2} \right)^2, \quad v' = \frac{5}{2} ; \tag{6.191}$$

$$\alpha_T = \frac{1}{315\pi} \cdot \frac{\omega_{pe}^4 \, (\gamma+4)\,(\gamma+6)}{n_0^2 \, m \,(\gamma+1)\,(\gamma+2)\,(\gamma+3)c^5} \left[43\,(W^l + W^p) + a_1\,(\gamma)\,\xi^{\frac{\gamma-3}{2}} \frac{H^2}{4\pi} \right] \times$$

$$\times \left[(W^l + W^p)\,(\gamma^2 + 6\gamma + 16) + a_2\,(\gamma)\,\xi^{\frac{\gamma-2}{2}} \frac{H^2}{4\pi} \right]^{-1}. \tag{6.192}$$

Here we have used the explicit form of I as given in (6.178), with the synchrotron effects included. The acceleration of (6.191) is very strong and can completely compensate for the synchrotron losses.

Equation (6.179) can serve to define the equilibrium spectrum

$$D_\varepsilon \, \varepsilon^2 \, \frac{\partial}{\partial \varepsilon} \cdot \frac{f_\varepsilon}{\varepsilon^2} + A_\varepsilon \, f_\varepsilon = \text{const} = 0 \tag{6.193}$$

or with $A_\varepsilon = \beta \varepsilon^2$,

$$\frac{\varepsilon^3 \, \alpha_T}{5} \cdot \frac{\partial}{\partial \varepsilon} \cdot \frac{f_\varepsilon}{\varepsilon^2} + \beta f_\varepsilon = 0, \tag{6.194}$$

where α_T is given by (6.192). Then, from (6.194)

$$f_\varepsilon = \frac{\text{const}}{\varepsilon^\gamma},$$

where

$$\gamma + 2 = \frac{5\beta}{\alpha_T}. \tag{6.195}$$

This last equation will define γ. Thus, it has been shown that the particle spectrum becomes a power function until ε_* reaches the value $mc^2(\omega_*/2\omega_{pe})^{1/2}$. Incidentally, it is this value of ε_* which must enter into (6.169).

The power function spectrum for the energy distribution of relativistic particles corresponds to the spectrum observed for

cosmic-ray electrons and ions. The solutions to (6.195) give us
the value of γ. When γ is known, $\nu = (\gamma - 1)/2$. If one considers only
the scattering on the plasma waves together with the synchrotron
processes in the case of homogeneous fields and homogeneous
relativistic particles, the solution of (6.195) can be proved to be
unique and is $\gamma = 3$ [362]. This result can be proved to be more
general, i.e., the solution $\gamma = 3$ holds even for anisotropic distributions of relativistic particles in the case when their distribution
function can be considered as slowly varying in the narrow angular
interval mc^2/ε , and the same solution is found for other types of
plasma waves, which can have any phase velocity (greater or
smaller than the velocity of light) as well as have to be anisotropically distributed. But the presence of an anisotropy leads to the
excitation of Alfven waves by relativistic particles. Then it is
necessary to take the scattering of particles on the Alfven waves
into account. This effect is of essential importance in the case of
inhomogeneous distributions of relativistic particles, and also
inhomogeneity of the external magnetic field. It is simple to show
that the distribution $1/\varepsilon^{\gamma}$ is again found if the Alfven wave spectrum is proportional to $1/\omega$, as it must be according to (4.307).
But now γ is less than 3. A similar effect of the diminishing of
γ values is found by taking into account the small anisotropy of
the particle distribution function in the above-mentioned narrow
angular interval mc^2/ε. In the limiting case of highly anisotropic
distribution functions of relativistic particles, when $\Delta\theta \ll mc^2/\varepsilon$,
the solution for γ is $\gamma = 2$ [363]. Thus one would expect to find
γ in the interval $2 < \gamma < 3$.

From γ we can find $\nu = (\gamma - 1)/2$, and it turns out that the
spectra of most of the observed sources of cosmic radio waves
are power functions with values of ν which fall within the theoretical interval. Finally, the value of γ observed for cosmic rays
falls in the predicted interval also. We want to emphasize an
important deduction from the general theory of plasma turbulence:
In a plasma object of sufficient size (that is, optically thick relative to the emissions of fast relativistic particles accelerated by
the turbulence), a power function spectrum for the particles must
be automatically set up, one which extends to high energies. The
maximum value of the energy is determined by the radiation cutoff. The cutoff condition can depend on both specific properties
of the plasma object and on a number of effects which arise in a
turbulent plasma. In particular the radiation can be cut off auto-

matically because the radiation emitted from any object becomes locally anisotropic; as a consequence intense turbulence is excited, according to Chapter 3. The radiation cutoff aids scattering (Chapter 7).

Thus, the turbulent plasma accelerates particles up to relativistic energies ($T_{eff} \gg mc^2$, Chapter 5), their radiation then distributes the particles so that their spectrum becomes a power function when $\varepsilon > \varepsilon_{cr} = \left(\dfrac{5 D_k^l > \omega_{pe/c}}{\alpha_T} \right)^{1/3}$, where $D_{k > \omega_{pe/c}}^l$ is the diffusion coefficient for resonant plasmons. This criterion follows from Eq. (6.189) for $D_{k > \omega_{pe/c}}^l$, which was found in Chapter 5. Cosmic rays are characterized by a spectrum of the same qualitative form. Therefore, the theory of turbulent plasma predicts the existence of ultrarelativistic particles in a turbulent plasma with spectra which agree with observed spectra.

Chapter 7

Propagation of Electromagnetic Waves through Turbulent Plasma

7.1. The Problem

Chapters 7 and 8 will be devoted to the propagation of electromagnetic waves and the development of turbulence in a turbulent plasma. This chapter will cover the effects which arise in a turbulent plasma for waves which can propagate freely in the plasma in the absence of turbulence (propagating waves); Chapter 8 will deal with low-frequency waves and excitations.

There are significant differences between the problems encountered with the propagation of high-frequency waves and that of low-frequency waves. The differences arise because for the high-frequency waves the turbulent plasma is not able to change the collective modes of oscillations and waves for the plasma, whereas in the low-frequency region new modes do appear, such as the second-sound type vibrations, along with new specific instabilities and the stabilization of those instabilities which were present in the absence of turbulence. In addition, it will be assumed for both high and low frequencies that the electromagnetic disturbance or wave is weak, which means that the disturbance is not capable of altering the turbulence spectrum. In the present chapter we shall consider only high-frequency waves propagating through a turbulent plasma, with the assumption that $\omega \gg \omega_{He}$, ω_{pe}, so that ω is much greater than all the possible frequencies of turbulent fluctuations. We will not be interested in the origin of these high-frequency waves; they can either be waves due to some turbulent process within the plasma, or they can be waves incident upon the plasma from outside it. There are three processes which are most important from an applications point of view: 1) the scattering of waves by turbulent fluctuations; 2) the amplification of the electromagnetic waves because of turbulence anisotropy or fast particles, and 3) fluctuations in the intensities of the electromagnetic waves.

Of course, scattering does not play a role when the radiation, the fluctuations, and the particles are isotropically distributed; therefore it was neglected in the preceding chapter. However, anisotropy automatically arises when the radiation leaves the plasma; at least it arises close to the surface. Scattering increases the path of a ray in the plasma, thereby increasing the optical thickness. Thus scattering must be taken into account when calculating the optical thickness.

Scattering also has an effect on the amplification of electromagnetic waves, again by increasing the path of the ray. In a single scattering act the change in the direction of the electromagnetic wave can be large, and since only waves in particular directions can be amplified, the amplification of scattered waves becomes possible. Thus, the amplification of waves can increase the scattering.

If the spectrum of the incident radiation is sufficiently narrow, the spectrum of the waves scattered by turbulence will be displaced above and below the frequency of the incident radiation by an amount equal to the frequency of the turbulent fluctuations (combination scattering). In fact, in the absence of fast particles the high frequencies interact with turbulent fluctuations σ only as the result of decay processes such as

$$t \rightleftarrows t' + \sigma, \tag{7.1}$$

which are nothing more than the induced combination scattering. When the width of the spectrum for the scattered wave is much smaller than ω_σ, the scattering leads to combination frequencies $\omega_t + \nu\omega_\sigma$, where ν is an integer. As a rule $\nu = \pm 1$ for a weak wave because the next combination frequencies appear only as the result of a second scattering of the wave which has already been scattered.

If $\omega^\sigma \ll \omega^t$, the change in the energy of transverse quanta is quite small, and the scattering process can be considered elastic in the first approximation. For a quasielastic scattering process it is necessary that the change in the momentum of the scattered wave be much greater than the change in its energy. If the change in momentum is equal to the momentum of the turbulent fluctuations, the required condition is

$$\frac{v_p^\sigma}{c} = \frac{\omega^\sigma}{k^\sigma c} \ll 1. \tag{7.2}$$

In other words, the scattering is quasielastic if it takes place on turbulent fluctuations whose phase velocities are much less than the speed of light.

The largest possible scattering angle is 2π. In this case the change in momentum of the scattered wave is $2k_t$, where k_t is the momentum of the incident wave. This change in momentum is equal to the maximum momentum of the turbulent fluctuations on which the wave can be scattered, i.e.,

$$k_\sigma < 2k_t \tag{7.3}$$

or $\lambda_t < 2\lambda_\sigma$. That is, the scattered wave "senses" only those turbulent fluctuations whose wavelengths are greater than half the wavelength of the scattered wave.

The appearance of scattered combination frequencies plays a very important role in plasma diagnostics. The observation of the combination frequencies enables us to identify the type of turbulent fluctuations excited in the plasma, without changing the plasma state (its turbulence) in any significant way. The other possible methods of diagnosis (probes and others) are connected with significant, although perhaps localized, changes in the plasma properties. It is especially difficult to identify the low-frequency turbulent fluctuations by well-known methods, but the effects of combination scattering make it possible, in principle, to measure the frequencies of the turbulent fluctuations. If it is known that turbulent fluctuations of a given type are in fact excited in a plasma, then measurements of v permit us to draw a number of conclusions about the plasma parameters; such things as density and electron and ion temperatures are now available because the frequencies of the turbulent fluctuations depend on all these parameters.

Finally, the intensity and angular distribution of the scattered radiation can tell us something about the intensity and spectrum of the turbulent fluctuations.

The possibility of using the processes causing wave scattering in a turbulent plasma for its diagnostics relative to the spectrum of the satellites was considered in [185, 273, 274]. Scattering theory has been studied in many works [265, 275, 272, 276]. Scattering problems in a bounded plasma were examined in [272].

The first experimental confirmation of the effect of scattering in a turbulent plasma was reported in [277-279] (see also [280-282]). Scattering on large-scale oscillations of the entire plasma column has also been investigated [283, 284].

Scattering in a turbulent plasma plays a very important role in astrophysics [285-287]. It is also evidently observed [288] in radiofrequency emissions from quasars [287], and experiments on radio location of the sun [289] can be explained in terms of scattering within a turbulent plasma [290].

The amplification of electromagnetic waves in a turbulent plasma is very important for using a plasma as an emitter of radiation, for the interpretation of emissions from cosmic objects [291], and for the excitation of high-frequency turbulence by generating intense electromagnetic waves (even in the nonlinear regime). Finally, intensity fluctuations are also important in turbulence diagnostics [292]. The use of fluctuation theory enables us to interpret the many observations of fluctuations in radio emission passing through the plasma in the space around the earth and the plasma of the solar wind [293]. The application of the theory for the propagation of electromagnetic waves through turbulent plasma is therefore quite widespread. We will limit our present treatment to the most important and the primary aspects of the theory, and we will use the turbulence spectra found above for predicting qualitative properties of radiation scattering, amplification, and intensity fluctuations.

7.2.   General Scattering Theory

The general equations of the preceding chapter, which take into account polarization effects, can be used to describe the scattering if we consider just decay processes and assume that the intensity is small. If one averages over the polarizations, it is more convenient to use the balance equations directly for the transverse quanta and plasmons, as obtained in Chapter 2.

The equation for the change in the number of t-quanta can be written as

$$\frac{\partial N_k^t}{\partial t} + \mathbf{v}_g \frac{\partial N_k^t}{\partial \mathbf{r}} = \int \frac{dk_1 dk'}{(2\pi)^6} N_{k_1}^\sigma (N_{k'}^t - N_k^t) \times$$

$$\times [w_t^{t\sigma}(\mathbf{k}_1, \mathbf{k}) \delta(k_1 - k + k') + w_t^{t\sigma}(\mathbf{k}_1, \mathbf{k}') \delta(k_1 - k' + k)], \qquad (7.4)$$

where $w_t^{t\sigma}(\mathbf{k}_1, \mathbf{k}) \delta(k_1 - k + k')$ is the probability of the decay. After integrating over $\mathbf{k}'$ Eq. (7.4) can be written as

$$\frac{\partial N_k^t}{\partial t} + \mathbf{v}_g \frac{\partial N_k^t}{\partial \mathbf{r}} = \int \frac{N_{k_1}^\sigma dk_1}{(2\pi)^6} [(N_{k-k_1}^t - N_k^t) w_t^{t\sigma}(\mathbf{k}_1, \mathbf{k}) \times$$

$$\times \delta(\omega_{k_1}^\sigma - \omega_k + \omega_{k-k_1}) + (N_{k+k_1}^t - N_k^t) w_t^{t\sigma}(\mathbf{k}_1, \mathbf{k}+\mathbf{k}_1) \times$$

$$\times \delta(\omega_{k_1}^\sigma - \omega_{k+k_1} + \omega_k)]. \qquad (7.5)$$

In the form of (7.5) these equations are especially convenient for the case when the largest possible value of $|\mathbf{k}_1|$ for the turbulent fluctuations is much smaller than $|\mathbf{k}|$ for the scattered electromagnetic waves. Then, if the frequency spectrum of the scattered waves is wide enough, $\Delta\omega_k \gg \omega_{k_1}^\sigma$, i.e., the spectrum width is much greater than the frequency of the turbulent fluctuations, we can expand (7.5) in terms of k_1, and obtain in the process a diffusion-like equation:

$$\frac{\partial N_k^t}{\partial t} + \mathbf{v}_g \frac{\partial N_k^t}{\partial \mathbf{r}} = \frac{\partial}{\partial k_i} D_{ij} \frac{\partial N_k^t}{\partial k_j},$$

$$D_{ij} = \int k_{1i} k_{1j} \frac{dk_1}{(2\pi)^6} w_t^{t\sigma}(\mathbf{k}_1, \mathbf{k}) \delta(\omega_{k_1}^\sigma - \mathbf{k}_1 \mathbf{v}_{gk}) N_{k_1}^\sigma. \qquad (7.6)$$

This method of describing the scattering is useless even when

$\Delta \omega_k \gg \omega_{k_1}^\sigma$ if $|k_1|$ can be comparable with $|k|$. For then the scattering takes place over a significant angle and the diffusion approximation does not describe the process. In a number of cases the change in the frequency of the scattered photon is small compared with the change in the direction of motion, i.e., the scattering is quasielastic.

In the quasielastic scattering approximation (7.4) can be simplified. We note that to obtain such equations it is convenient to use $\delta(k_1 - k + k')$ in (7.4) for the integration over k_1, and $\delta(\omega_{k_1}^\sigma - \omega_k + \omega_{k'})$ for the integration over the modulus of k'. Then Eq. (7.5) takes the form

$$\frac{\partial N_{kn}^t}{\partial t} + v_{g,\,k}\,\frac{\partial N_{kn}^t}{\partial r} = \int \frac{k'^2 N_{kn-k'n'}^\sigma}{v_{gk'}\,(2\pi)^6}\,(N_{k'n'}^t - N_{kn}^t)\,dn' \times$$

$$\times\, w_t^{t\sigma}\,(kn - k'n',\ kn) + \int \frac{k''^2}{v_{gk''}\,(2\pi)^6}\,(N_{k''n'}^t - N_{kn}^t)\,N_{k''n'-kn}^\sigma \times$$

$$\times\, w_t^{t\sigma}\,(k''n' - kn,\ k''n')\,dn'. \tag{7.7}$$

Here dn' is the integral over the solid angle of the scattered photons, n' is a unit vector in the direction of the incident photon, and k' and k" are functions of k, n, n' determined by the solutions of the equations

$$\omega_{k_1}^\sigma = \omega_{kn-k'n'}^\sigma = \omega_k - \omega_{k'}, \tag{7.8}$$

$$\omega_{k_1}^\sigma = \omega_{k''n'-kn}^\sigma = \omega_{k''} - \omega_k. \tag{7.9}$$

Equation (7.7) is a strict consequence of Eq. (7.4) and is valid even for inelastic scattering. The elastic scattering case corresponds to the approximation in which

$$k' \approx k'' \approx k, \quad \omega_{k'} \approx \omega_{k''} \approx \omega_k. \tag{7.10}$$

In this case (7.7) transforms into

$$\frac{\partial N_n^t}{\partial t} + v_g\,\frac{\partial N_n^t}{\partial r} = -N_n^t \int \gamma_{n,\,n'}\,dn' + \int N_{n'}^t\,\gamma_{n,\,n'}\,dn', \tag{7.11}$$

where

$$\gamma_{n,\,n'} = \frac{k^2}{v_{g,\,k}}\,\big(N_{k\,(n-n')}^\sigma\,w_t^{t\sigma}\,(k\,(n-n'),\ kn) +$$

$$+\, N_{k\,(n'-n)}^\sigma\,w_t^{t\sigma}\,(k\,(n'-n),\ kn'))\,(2\pi)^{-6}. \tag{7.12}$$

An equation written in the form of (7.11) is common in the theory
of radiation transport [9]. It is important that in a turbulent plasma the scattering of waves can be much greater than the Thomson
scattering; therefore the scattering processes indicated here play
a very important role. Finally, we must present one more form
for the equations describing the scattering of electromagnetic
waves in turbulent plasma in the limit where (7.11) and (7.6) are
not applicable, and that is when $\Delta\omega_{k_1}\ll\omega_{k_1}^{\sigma}$. In this case the
scattered waves have frequencies which differ from the incident
frequencies by an amount $\nu\omega_{k_1}^{\sigma}$, $\nu = \pm1$, ±2, ..., that is, we have the
satellite spectral lines which are characteristic of induced com-
bination scattering. In the case indicated the line frequencies are
completely determined, and one can introduce a distribution func-
tion for the ν-th satellite, $N_{\nu,n}^{t}$, whose frequency is $\omega_k + \nu\omega_{k_1}^{\sigma}$,
where ω_k is the frequency of the incident photon.

The general equation of this type is

$$\frac{\partial N_{\nu,\,n}^{t}}{\partial t} + v_{g,\,\nu}\,\frac{\partial N_{\nu,\,n}^{t}}{\partial r} = -N_{\nu,\,n}^{t}\int \gamma_{\nu,\,n,\,n'}\,dn' +$$

$$+\int \gamma_{\nu+1,\,n,\,n'}'\,N_{\nu+1,\,n'}^{t}\,dn' +\int \gamma_{\nu-1,\,n,\,n'}''\,N_{\nu-1,\,n'}^{t}\,dn', \qquad (7.13)$$

$$\gamma_{\nu,\,n,\,n'} = \gamma_{\nu+1,\,n,\,n'}' + \gamma_{\nu-1,\,n,\,n'}''\,,$$

$$\left.\begin{aligned}
\gamma_{\nu+1,\,n,\,n'}' &= \frac{k''^{2}}{v_{gk''}(2\pi^{6})}\,N_{k''n'-kn}^{\sigma}w_{t}^{t\sigma}(k''n'-kn,\,k''n')\big|_{\omega=\omega_k+(\nu+1)\omega_{k_1}^{\sigma}},\\[6pt]
\gamma_{\nu-1,\,n,\,n'}'' &= \frac{k'^{2}}{v_{gk''}(2\pi)^{6}}\,N_{kn-k'n'}^{\sigma}w_{t}^{t\sigma}(kn-k'n',\,kn)\big|_{\omega=\omega_k+(\nu-1)\omega_{k_1}^{\sigma}}.
\end{aligned}\right\} \quad (7.14)$$

In plasma diagnostics problems, when the intensity of the scattered
waves is rather small, we can limit our attention to just the first
two satellites $\nu = 0$, ±1:

$$\left.\begin{aligned}
\frac{\partial N_{0,\,n}^{t}}{\partial t} + v_{g,\,0}\,\frac{\partial N_{0,\,n}^{t}}{\partial r} &= -N_{0,\,n}^{t}\int \gamma_{0,\,n,\,n'}\,dn',\\[10pt]
\frac{\partial N_{1,\,n}^{t}}{\partial t} + v_{g,\,1}\,\frac{\partial N_{1,\,n}^{t}}{\partial r} &\approx \int N_{0,\,n'}^{t}\,\gamma_{0,\,n,\,n'}''\,dn',\\[10pt]
\frac{\partial N_{-1,\,n}^{t}}{\partial t} + v_{g,\,-1}\,\frac{\partial N_{-1,\,n}^{t}}{\partial r} &= \int N_{0,\,n'}^{t}\,\gamma_{0,\,n,\,n'}'\,dn'.
\end{aligned}\right\} \qquad (7.15)$$

In the general case of a magnetoactive plasma, the probabilities $w_t^{t\sigma}(\mathbf{k}_1, \mathbf{k})$ can be found from the nonlinear current obtained in §2.10. Assuming that the frequencies of the scattering (and scattered) waves are much larger than all the characteristic plasma frequencies (ω_{pe}, ω_{He}), we find

$$S_{ijl} = \delta_{ij} \frac{e}{4\pi m_e \omega} (k_{1s} \varepsilon_{si}^{(e)}(k_1) - k_{1l}) = \delta_{ij} S_l \frac{e}{4\pi m_e \omega}. \qquad (7.16)$$

Here $\varepsilon_{ij}^{(e)}$ is the electron part of the linear plasma dielectric tensor. The scattering probability is determined by $|S_{k_1}^{\sigma}|^2$:

$$S_{k_1}^{\sigma} = S_l e_{k_1, l}^{\sigma}, \qquad (7.16a)$$

where $e_{k_1}^{\sigma}$ is a unit vector which characterizes the polarization of the turbulent fluctuations of the type under consideration [see (2.205)]:

$$w_t^{t\sigma}(\mathbf{k}_1, \mathbf{k}) = \frac{(2\pi)^6 e^2}{8\pi \omega^2 m_e^2} (1 + (\mathbf{nn'})^2) \frac{\omega_1^2 |S_{k_1}^{\sigma}|^2}{\frac{\partial}{\partial \omega_1} \omega_1^2 \varepsilon_{k_1}^{\sigma}\Big|_{\omega_1 = \omega_{k_1}^{\sigma}}}. \qquad (7.17)$$

We have averaged over the polarizations of the scattered and scattering waves in (7.17). $S_{k_1}^l = k_1 (\varepsilon_{k_1}^{(e)\,l} - 1)$ for scattering on longitudinal oscillations. For high-frequency oscillations $\varepsilon_{k_1}^{(e)\,l} = 0$, $S_{k_1}^l = -k_1$, but in the region where $\omega_1 \ll k_1 v_{Te}$

$$\varepsilon_{k_1}^{l\,(e)} = \frac{\omega_{pe}^2}{k_1^2 v_{Te}^2} \gg 1, \quad S_{k_1}^l = \frac{\omega_{pe}^2}{k_1 v_{Te}^2}. \qquad (7.18)$$

The exact probability of scattering on Langmuir fluctuations in an isotropic plasma for any ω is

$$w_t^{tl}(\mathbf{k}_1, \mathbf{k}) = \frac{(2\pi)^6 e^2 \omega_{pe} k_1^2}{16\pi m_e^2 \omega_k \omega_{k-k_1}} (1 + (\mathbf{nn'})^2), \qquad (7.19)$$

and for ion-acoustic waves it is

$$w_t^{ts}(\mathbf{k}_1, \mathbf{k}) = \frac{e^2 (\omega_{k_1}^s)^3 m_i \omega_{pe}^2 (2\pi)^6}{16\pi m_e^3 v_{Te}^4 k_1^2 \omega_k \omega_{k-k_1}} (1 + (\mathbf{nn'})^2). \qquad (7.20)$$

It is just as easy to find the probabilities for other processes which describe scattering on h, w, A, M, Ms, and drift waves d (see [294]). These are presented below. In order to simplify our future discussion of specific problems related to the scattering of electromagnetic waves by various types of fluctuations, we will treat in some detail the case of isotropic turbulence. The diffusion equation (7.6) differs from Eq. (5.2), which describes the scattering and acceleration of charged particles in a turbulent plasma, only in notation (instead of f_p, the particle distribution function, we have N_k^t, the photon distribution function, and in place of the velocity v one finds the wave group velocity). From this analogy it is clear that the photon scattering must be accompanied by changes in both the average frequency and in the frequency distribution (similar to the systematic and fluctuation acceleration for particles). For an isotropic turbulence one can write

$$D_{ij} = D^l \frac{k_i k_j}{k^2} + D^t \left(\delta_{ij} - \frac{k_i k_j}{k^2} \right). \tag{7.21}$$

The coefficient D^l describes the changes in the photon frequencies and D^t gives their angular scattering. Using (7.6) and (7.17) we can write

$$D^l = \frac{\pi}{2} \int \frac{(\omega_{k_1}^\sigma)^3 \, \omega_{pe}^2 \, |S_{k_1}^\sigma|^2}{k^2 m_e n_0 \left| \frac{\partial}{\partial \omega_1} \omega_1^2 \varepsilon_{k_1}^\sigma \right|_{\omega_1 = \omega_{k_1}^\sigma}} c^4 \, \delta \left(\omega_{k_1}^\sigma - \frac{k k_1 c}{k} \right) W_{k_1}^\sigma \, dk_1, \tag{7.22}$$

$$D^t = \frac{\pi}{8} \int \frac{[k k_1]^2 \, \omega_{pe}^2 \, |S_{k_1}^\sigma|^2}{k^4 n_0 m_e \left| \frac{\partial}{\partial \omega_1^2} \omega_1^2 \varepsilon_{k_1}^\sigma \right|_{\omega_1 = \omega_{k_1}^\sigma}} c^2 \, \delta \left(\omega_{k_1}^\sigma - \frac{k k_1 c}{k} \right) W_{k_1}^\sigma \, dk_1. \tag{7.23}$$

When $v_p^\sigma \ll c$,

$$D^t \approx \frac{c^2}{(v_p^\sigma)^2} D^l \gg D^l.$$

The change in the spectra is especially important for narrow lines; it is less important for broad spectra. This type of broadening is different from that obtained in Stark broadening, which is connected with the interactions of the electric fields due to the turbulent fluctuations with the radiating atom.

When $v_p^\sigma \ll c$ one can throw away the term $k_i k_j / k^2$ multiplying D^t in (7.21) in analyzing the scattering. Assuming that we can neglect frequency changes in treating the scattering, we find that the scattering is described by the angular part of the Laplacian operator. By introducing the electromagnetic field intensity

$$I_{\omega, \mathbf{n}} = I_{\omega, \Omega} = \frac{N_k^t \omega_k k^2}{(2\pi)^3} \cdot \frac{dk}{d\omega} \tag{7.24}$$

for the waves we arrive at the following for the diffuse scattering:

$$\frac{\partial I_{\omega, \mathbf{n}}}{\partial t} + c\mathbf{n} \frac{\partial I_{\omega, \mathbf{n}}}{\partial \mathbf{r}} = \frac{\sigma_d}{2} \Delta_{\theta, \varphi} I_{\omega, \mathbf{n}}, \tag{7.25}$$

$$\sigma_d = 2D^t \frac{c^2}{\omega^2}; \quad \Delta_{\theta, \varphi} = \frac{1}{\sin\theta} \cdot \frac{\partial}{\partial\theta} \sin\theta \frac{\partial}{\partial\theta} + \frac{\frac{\partial^2}{\partial\varphi^2}}{\sin^2\theta}. \tag{7.26}$$

According to (7.22), for large frequencies we have

$$D^l = \text{const} \frac{1}{\omega^2} = \frac{\mu}{c^2 \omega^2}, \tag{7.27}$$

so that the fluctuation smearing of the radiation frequencies and the broadening of the spectral lines are described by the following additional term on the right side of (7.25):

$$\mu\omega \frac{\partial^2}{\partial\omega^2} \cdot \frac{I_{\omega\mathbf{n}}}{\omega^3} = \hat{\mu} I_{\omega, \mathbf{n}}. \tag{7.28}$$

Near the frequency ω_*, which corresponds to the center of the spectral line, the frequency dependence of $I_{\omega\mathbf{n}}$ is the strongest, and we can use an approximate expression for the operator $\hat{\mu}$:

$$\hat{\mu} = \frac{\mu}{\omega_*^2} \cdot \frac{\partial^2}{\partial\omega^2}. \tag{7.29}$$

It is convenient to measure the angle θ in (7.25) from the initial direction of diffusion of the electromagnetic waves. Then, if the initial conditions are independent of φ we have

$$\frac{\partial I_{\omega,n}}{\partial t} + c\cos\theta\,\frac{\partial I_{\omega,n}}{\partial z} = \frac{\sigma_d}{2}\,\frac{1}{\sin\theta}\cdot\frac{\partial}{\partial\theta}\sin\theta\,\frac{\partial I_{\omega,n}}{\partial\theta} +$$

$$+\,Q_{\omega,n} + \gamma_\omega I_{\omega,n} + \mu\omega\,\frac{\partial^2}{\partial\omega^2}\cdot\frac{I_{\omega,n}}{\omega^3}. \tag{7.30}$$

This equation differs from the radiation transport equation of the preceding chapter in that it includes both scattering and line broadening. It should be clear from (7.30) that it is incorrect, in general, to neglect scattering when the intensity distribution is anisotropic. As a qualitative estimate we have, for small scattering angles,

$$\langle\theta^2\rangle = \frac{\sigma_d}{2c}\,L, \tag{7.31}$$

where L is the path of a ray in the plasma, and σ_d is a quantity which acts as a scattering cross section. Equation (7.11), which describes elastic scattering at an angle of order unity, can be written as

$$\frac{\partial I_{\omega,n}}{\partial t} + c n\,\frac{\partial I_{\omega,n}}{\partial r} = -\sigma I_{\omega,n} + \sigma\int \varkappa_{nn'}\,I_{\omega n'}\,dn', \tag{7.32}$$

where

$$\sigma = \int\gamma_{nn'}\,dn', \quad \varkappa_{nn'} = \frac{1}{\sigma}\,\gamma_{nn'}. \tag{7.33}$$

The first term gives the decrease in the number of photons in the direction n due to scattering into other directions n'; the second term describes the reverse process, that of scattering from n' into n. If the scattering is at some small angle and n' is close to n, then (7.32) can be expanded in terms of $n' - n = \Delta n$. We obtain

$$\frac{\partial I_{\omega n}}{\partial t} + c n\,\frac{\partial I_{\omega n}}{\partial r} = \frac{\partial}{\partial n_i}\,D_{ij}^{(n)}\,\frac{\partial I_{\omega n}}{\partial n_j}, \tag{7.34}$$

$$D_{ij}^{(n)} = \int\frac{k^2}{v_g}\,\Delta n_i\Delta n_j N_{k\Delta n}^\sigma\,w_t^{t\sigma}\,(k\Delta n,\ kn)\,d\Delta n, \tag{7.35}$$

which can be transformed into the form of (7.25).

Let us now give qualitative consideration to those effects which arise when radiation leaves the plasma. We will assume that, because of intense scattering and the isotropy and uniformity of the sources generating the radiation $Q_{\omega n} = Q_\omega$, the radiation inside the plasma is completely isotropic, with $I_{\omega n} = I_\omega$.

Consider what takes place as one approaches the plasma boundary. If the distance to the boundary is still large enough (see below for the appropriate criterion), the radiation will be weakly anisotropic and $I_{\omega n}$ can be expanded in terms of Legendre polynomials. We keep the first two terms:

$$I_{\omega n} = A_\omega + B_\omega \cos \theta. \tag{7.36}$$

Substituting (7.36) into (7.30) and averaging over angles, we find that if the spectrum width is neglected,

$$c^2 \frac{\partial^2 A_\omega}{\partial z^2} = 3 \left(\gamma_\omega + \sigma_d \right) \gamma_\omega A_\omega + 3 \left(\gamma_\omega + \sigma_d \right) Q_\omega, \tag{7.37}$$

i.e., the characteristic length over which there is a marked change in intensity,

$$L_{\text{eff}} = \frac{c}{\sqrt{3 \gamma_\omega \left(\gamma_\omega + \sigma_d \right)}}, \tag{7.38}$$

is visibly changed due to scattering if

$$\sigma_d \gg \gamma_\omega, \tag{7.39}$$

while the equilibrium value of the intensity

$$A_\omega = -\frac{Q_\omega}{\gamma_\omega} \tag{7.40}$$

stays the same. Therefore, when $\sigma_d \gg \gamma_\omega$ the optical thickness of the plasma is significantly reduced. The optical thickness is also reduced when there is a buildup of radiation, with $\gamma > 0$ [286]. This result follows from (7.32) for scattering at an angle of order unity, but in (7.38) one must replace σ_d with the cross section σ in this case [286]. Following the procedure used with (7.32) and (7.36), by setting up one equation like (7.38) for A_ω, it would be easy to estimate the effects of spectral line broadening. We obtain the following estimates of the broadening:

1) when the ray path L is much shorter than the scattering length c/σ,

$$(\Delta\omega)^2 \simeq 3\frac{\mu}{\omega_*^2} L c^{-1}; \tag{7.41}$$

2) when the ray path L is much longer than the scattering length c/σ,

$$(\Delta\omega)^2 \simeq 3\frac{\mu}{\omega_*^2} \sigma L^2 c^{-2}, \tag{7.42}$$

i.e., the broadening increases by a factor of $\sigma L c^{-1} \gg 1$.

7.3. Scattering of Electromagnetic Waves in Turbulent Plasma

We will examine the effects of scattering on the Langmuir fluctuations. It follows from the conservation laws pertaining to a decay that only fluctuations with $v_p < c$ interact with transverse waves. If $3v_{Te}^2/v_{Ti} > c$, then for practical purposes the turbulence spectrum can be treated as flat up to $v_p \approx c$. Scattering at angles of order unity is possible if $k \sim k^l$:

$$\omega < \omega_{pe}\frac{c}{v_{Te}} . \tag{7.43}$$

When $\omega \gg \omega_{pe}$ it is impossible to have scattering at an angle of order unity with $v_p \sim c$. Thus, when using (7.12) one can assume that $W_k = \mathrm{const} = (2Q/\alpha)^{1/2}$. Using the probabilities given in (7.19) and (7.12) we find

$$\gamma_{n,\,n'} = \frac{\omega_{pe}\, 3\sqrt{6}\left(1+\dfrac{T_e}{T_i}\right)}{16\sqrt{\pi}}\sqrt{\frac{Q}{Q_0}}\sqrt{\frac{m_i}{m_e}}\cdot\frac{v_{Te}^3}{c^3}(1+(nn')^2), \tag{7.44}$$

where $\gamma_{n,n'}$ is frequency-independent only up to $\omega \simeq \omega_{pe}(c/v_{Te}) \times (m_e/m_i)^{1/5}$. At large frequencies scattering at angles of order unity is possible, but drops off as $1/\omega^{5/2}$ with the frequency; when $\omega \gg \omega_{pe}(c/v_{Te})$ the scattering always occurs at small angles and the diffusion approximation is valid. Equation (7.44) gives

$$\sigma = \sqrt{6\pi}\,\sqrt{\frac{m_i}{m_e}}\cdot\frac{v_{Te}^3}{c^3}\left(1+\frac{T_e}{T_i}\right)\omega_{pe}\sqrt{\frac{Q}{Q_0}}, \tag{7.45}$$

$$Q_0 = \omega_{pe}\, n_0 T_e.$$

When $T_e \gg T_i$ the factor $(1 + T_e/T_i)$ must be dropped. For diffusion scattering we have

$$\sigma_d = \frac{\pi\omega_{pe}^2}{8m_e} \int \frac{k_1^4}{\omega^4} W_{k_1}^l \frac{dk_1 dn_1}{4\pi n_0} \delta(\omega_{pe} - k_1 n_1 c) c^2. \qquad (7.46)$$

Assuming that, since $W_k = $ const, the main contribution to the integral in (7.46) comes from $k_1 \gg \omega_{pe}/c$, we obtain

$$\sigma_d = \frac{3\sqrt{6\pi}}{4} \sqrt{\frac{m_i}{m_e}} \cdot \frac{v_{Te}^3}{c^3}\left(1 + \frac{T_e}{T_i}\right)\sqrt{\frac{Q}{Q_0}} \times$$
$$\times \left(\frac{\omega_{pe}c}{2v_{Te}\omega}\left(\frac{m_e}{m_i}\right)^{1/5}\right)^4 \omega_{pe}. \qquad (7.47)$$

Here $k_{1\,max} = \omega_{pe}(m_e/m_i)^{1/5}/v_{Te}$; that is, the scattering drops off as $1/\omega^4$. It is interesting to compare (7.45) with the Thomson scattering, for which $\sigma = \sigma_T$:

$$\sigma_T = \frac{4\omega_{pe}}{3\pi^2} \cdot \frac{\omega_{pe}^3}{n_0 c^3}. \qquad (7.48)$$

The critical value Q_{cr} for which (7.45) is larger than (7.48) is very small:

$$Q_{cr} = Q_0 \frac{m_e}{m_i}\left(\frac{2\sqrt{2}}{(3\pi)^{5/2}} \cdot \frac{\omega_{pe}^3}{n_0 v_{Te}^3}\right)^2 \sim Q_0 \frac{m_e}{m_i N_D^2} \sim Q_0 \left(\frac{v_e}{\omega_{pe}}\right)^2 \frac{m_e}{m_i}. \qquad (7.49)$$

Thus, in turbulent plasma the scattering is always anomalously large, and is usually many orders of magnitude greater than the Thomson scattering.

If ion-acoustic waves are excited in the plasma, then although the scattering is always elastic (since $\omega^s \ll \omega$), it is possible for it to occur at angles of order unity. Then, from (7.20) we obtain

$$\gamma_{nn'} = \frac{\pi\omega_{pe}^4}{8cn_0 T_e \omega^2} \cdot \frac{(1+(nn')^2)\left(W_{k\,(n-n')}^s + W_{k\,(n'-n)}^s\right)}{(n-n')^2\left(1 + 4\omega^2 v_{Te}^2(1 - nn')/\omega_{pe}^2 c^2\right)}. \qquad (7.50)$$

When $n \to n'$, $W^s \to 0$. On the other hand, when $n' \to n$ one must use the diffusion approximation. If the spectrum given in (4.231) is used, then

$$\gamma_{nn'} \simeq \frac{\pi\omega_{pe}}{16\sqrt{2\pi}} \frac{15 T_e \,\omega_{pe}^3}{T_i \omega^3} \cdot \frac{(1+(nn')^2)}{|n-n'|^3} \sqrt{\frac{m_e}{m_i}} \ln\frac{\omega|n-n'|}{k_0 c}. \tag{7.51}$$

When $|n-n'| \sim 1$ this scattering can, even though it falls off rapidly with frequency, exceed the scattering by Langmuir fluctuations (7.45) over a wide range of the parameters. Langmuir scattering occurs when

$$Q < Q_0 \left(\frac{m_e}{m_i}\right)^2 \left(\frac{6\omega_{pe}}{v_{Te}\omega}\right)^6. \tag{7.52}$$

In the diffusion approximation, which is always valid when $\omega \gg \omega_{pe}c/v_{Te}$, we have

$$\sigma_d = \frac{\pi}{16} \cdot \frac{\omega_{pe}^4}{\omega^4} \int \frac{ck_1 dk_1 W_{k_1}^s}{n_0 T_e} = \frac{15\pi T_e}{64\sqrt{2\pi}T_i} \sqrt{\frac{m_e}{m_i}} \, \omega_{pe} \frac{\omega_{pe}^3}{\omega^3} \cdot \frac{\omega_{pe} c}{v_{Te}\,\omega} \left(\ln\frac{\omega_{Te}}{v_{Te}k_0} - 1\right). \tag{7.53}$$

In the last equality of (7.53) we have $k_{max} \simeq \omega_{pe}/v_{Te}$ approximately. Thus, when comparing with (7.51) there is an additional small factor $\omega_{pe}c/v_{Te}\omega \ll 1$.

We will now consider a plasma situated in a magnetic field. The scattering on Langmuir fluctuations and on ion-acoustic waves has about the same frequency dependence as it does when there is no magnetic field. In a strong field such that $\omega_{He} \gg \omega_{pe}$ a branch with $\omega_k^h \approx \omega_{He}$ appears. Calculation of $w_t^{th}(k_1, k)$ for this branch from Eq. (7.17) gives

$$w_t^{th}(k_1,\ k) = \frac{(2\pi)^6 \,\omega_{pe}^4 \,(k-k')^2}{4n_0\,\omega^2\,|\omega_{He}|(4\pi)^2\,m_e} \,(1+(nn')^2) \left[1 - \frac{(h,\ k-k')^2}{(k-k')^2}\right], \tag{7.54}$$

where $h = H_0/H_0$. It follows from 7.54 that

$$\gamma_{nn'} = \frac{\omega_{pe}^4 \,[1+(nn')^2]}{32\omega_{He}^2 \,n_0 \,m_e c^2} \left(1 - \frac{(h(n-n'))^2}{4(1-nn')}\right) \,(W_{k\,(n-n')}^h + W_{k(n'-n)}^h). \tag{7.55}$$

There is no scattering on transverse waves with frequencies $\omega \sim \omega_{He}$ (into which h-waves turn) because the transverse waves do not decay. For scattering on magnetized Langmuir fluctuations with $\omega = \omega_{pe}|\cos\theta|$, the only difference from the case considered above with $H = 0$ will be the appearance of the factor $\frac{(h,n-n')}{(n-n')}$ in Eq. (7.19).

In a weak magnetic field $(\omega_{He} \ll \omega_{pe})$ new branches of whistlers and $\omega_{He}|\cos\theta|$ appear. For the whistlers we have

$$w_i^{tw}(\mathbf{k_1},\ \mathbf{k}) = \frac{(2\pi)^6 e^2}{\omega^2\,16\pi m_e^2}(1+(nn')^2)\,\frac{\omega_{pe}^2\,k_1^2\,\sin^2\theta_1\,|\cos\theta_1|}{\omega_{He}}. \qquad (7.56)$$

Since $\omega \gg \omega_{pe}$, $k_1 \ll \omega_{pe}/c$, and $\omega \gg k_1 c$, the diffusion approximation is always valid:

$$D_{ij} = \int\limits_{x_1^2 < 1-x^2} k_{1i}k_{1j}\,\frac{d\omega_1 W_{\omega_1,\,\Omega_1}^w\,\pi d\Omega_1\,(1-x_1^2)\omega_{pe}^6\,c^{-3}}{4\,\omega^2\omega_{He}^2\,n_0\,m_e\,k_1\,\sqrt{1-x^2-x_1^2}}. \qquad (7.57)$$

Here $x_1 = \cos\theta_1$, $x = \cos\theta$; W_{ω_1,Ω_1}^w is the turbulent energy of the whistlers normalized to $d\omega_1 dx_1$ (as introduced in Chapter 4). According to Chapter 4 $W_{\omega_1 x_1} \simeq \frac{W(x_1)}{\omega^\nu}$ with $\nu = 1/2$ in the absence of ion-acoustic waves and $\nu = 0$ in the presence of intense ion-acoustic waves. The integrand in (7.57) is proportional to $-(\nu - 1/2)$ so that for any of the ν given above the main contribution comes from the largest possible frequencies, which are $\sim\omega_{He}$. This gives us this estimate

$$\sigma_d \simeq \frac{2D}{\omega^2} \approx \omega_{pe}\frac{W^w}{n_0 m_e c^2}\left(\frac{\omega_{pe}}{\omega}\right)^4\left(\frac{\omega_{pe}}{\omega_{He}}\right)^2. \qquad (7.58)$$

Turning to the low-frequency vibrations, we must point out that the scattering on the magnetized sound waves with $\omega = kv_s|\cos\theta|$ $(v_A \gg v_S)$ is qualitatively little different from the scattering examined earlier for the unmagnetized sound waves [one must add the additional factor $\left|\frac{h(n-n')}{n-n'}\right|$ to the probability given in (7.20)]. The probability of scattering on the longitudinal ion waves with frequency $\omega_{Hi}|\cos\theta_1|$ is

$$w_i^{tl}(\mathbf{k_1},\ \mathbf{k}) = \frac{(2\pi)^6\,\omega_{pe}^4\omega_{He}\,\omega_{Hi}^2\sin^2\theta_1|\cos\theta_1|}{64\pi^2\,k_1^2\,v_{Te}^2\,n_0\,T_e\,\omega^2}(1+(nn')^2). \qquad (7.59)$$

Assuming that $k_1 \sim \omega_{Hi}/v_s$ we find

$$\sigma_d \approx \omega_{Hi}\, \frac{c}{v_s} \left(\frac{\omega_{pe}}{\omega} \right)^4 \frac{W}{n_0 T_e}\,. \tag{7.60}$$

For Alfven waves with $\omega^A = k v_A |\cos \theta_1|$, we have

$$w_t^{tA}(\mathbf{k}_1,\ \mathbf{k}) = \frac{(2\pi)^6 \omega_{pe}^2 k_1^3 |\cos \theta_1| \sin^2 \theta_1 c^2}{64 \pi^2 n_0 m_e v_A\, \omega^2} \cdot (1 + (\mathbf{n}\mathbf{n}')^2). \tag{7.61}$$

The diffusion coefficient can be written as

$$D_{ij}^A = \int \frac{k_{1i}\, k_{1j}\, d\omega_1\, d\Omega_1\, W_{\omega_1,\,\Omega_1}^A \pi k_1 \omega_{pe}^2\, c\, (1 - x_1^2)}{4 n_0\, m_e\, v_A^2 \omega^2\, \sqrt{1 - x^2 - x_1^2}}\,, \tag{7.62}$$

where $x_1 = \cos \theta_1$, $x = \cos \theta$. With stationary, one-dimensional conditions the quantity $W_{\omega_1}^A$ introduced in Chapter 4 has the form const/ω_1 and the main contribution to the integral in (7.62) comes from the largest frequencies ω_{Hi}. This enables us to arrive at the following estimate:

$$\sigma_d = \frac{2Dc^2}{\omega^2} \approx \omega_{pi}\, \frac{4\pi W}{H_0^2} \left(\frac{\omega_{pe}}{\omega} \right)^4. \tag{7.63}$$

For magnetoacoustic waves with $\omega^M = k v_A$ we have

$$w_t^{tM}(\mathbf{k}_1,\ \mathbf{k}) = \frac{(2\pi)^6\, \omega_{pe}^2\, \omega_{pi}^2\, k_1 \sin^2 \theta_1}{64 \pi^2 m_e\, \omega^2 v_A n_0} (1 + (\mathbf{n}\mathbf{n}')^2),$$

and the diffusion coefficient is

$$D_{ij}^M = \int \frac{c^{-1} k_{1i}\, k_{1j}\, \pi \omega_{pe}^2 \omega_{pi}^2\, (1 - x_1^2)}{4 n_0\, m_e\, v_A^2 k_1 \omega^2\, \sqrt{1 - x^2 - x_1^2}}\, W_{\omega_1,\,\Omega_1}^M\, d\omega_1\, d\Omega_1. \tag{7.64}$$

If the spectrum has the form $W_{\omega_1,\,\Omega_1}^M = \dfrac{1}{\omega^\nu} W_{x_1}^M$, where $\nu = 2$, then

$$2\pi\, W_{\omega_1,\,\Omega_1}^M = \frac{W^M}{\omega_{\min}} \left(\frac{\omega_{\min}}{\omega} \right)^2 W^M(x_1),\quad \int W^M(x_1)\, dx_1 = 1,$$

where W^M is the total energy in the magnetoacoustic waves; $\omega_{\min} = k_0 v_A$ is the minimum frequency and $2\pi/k_0 \simeq L_0$ is the fundamental scale of the turbulence. Under these conditions we estimate σ_d to be of the form

$$\sigma_d = k_0 c \frac{4\pi W^M}{H_0^2} \ln \frac{\omega_{pi}}{k_0 c} \left(\frac{\omega_{pe}}{\omega} \right)^4. \qquad (7.65)$$

7.4. Amplification of Electromagnetic Waves Propagating in Turbulent Plasma

Amplification arises either because of an anisotropy in the fluctuation distribution or because of an anisotropy in the distribution of the fast particles which are accelerated by the turbulence.

We will consider first the effects connected with anisotropies in the turbulence. We will limit ourselves to high frequencies, much higher than any of the characteristic frequencies of the plasma, such as ω_{He}, ω_{pe}. In the presence of turbulent fluctuations, effective collisions between thermal plasma particles and turbulent fluctuations arise. The high-frequency wave sets the plasma particles into oscillation. Their collisions with the turbulent fluctuations should be able to dissipate the vibrational energy, and therefore the energy of the high-frequency wave. In reality, however, the high-frequency wave can destroy the resonance condition which set up the effective interaction of particle and wave, and thereby eliminate the above-mentioned dissipation. Even from the general quasilinear equation (2.130) we can see that there may be high-frequency electromagnetic waves. In fact, taking account of the $\delta(k - k_1 - k_2)$, one can write the resonance denominator of (2.130) as

$$\frac{1}{\omega_1 + \omega_2 - (k_1 + k_2)\, v + i\delta} \cdot \cdot$$

For slow, low-frequency motions the frequency ω_2, which characterizes the change in the regular component of the distribution function, is small and can satisfy the resonance condition $\omega_1 = k_1 v$, whereas small high-frequency oscillations give a nonresonant denominator $1/(\omega_2 - k_2 v)$ (because $\omega_2/k_2 = v_p^t \simeq c$ and $v \ll c$).

The influence of turbulent collisions on the propagation of electromagnetic waves is described by a correction to the collision integral which is related to a perturbation of the particle distribution by the electromagnetic wave. Thus, the effective turbulent frequencies for the high-frequency radiation have a much different character than the perturbed low frequencies. Since the turbulence is weak and the frequencies are high, we must find the nonlinear γ_k^N for the transverse waves in the presence of low-frequency waves of type σ. These waves will be assumed to be resonant with the plasma particles, i.e., when H = 0, $\omega_{k_1}^\sigma = k_1 v$ and when H $\neq$ 0, $\omega_{k_1}^\sigma - k_z v_z - \nu \omega_{Ha} = 0$. We will use an equation similar to (2.124), dropping the terms linear in the intensity of the transverse waves:

$$(k^2 c^2 - \omega^2 \, \varepsilon_h) \, I_k^t = 8\pi i \omega I_k^t \int \Sigma_{kk_1}^{\text{eff}} \, I_{k_1}^\sigma \, dk_1 +$$

$$+ \frac{32\pi^2 \omega^2}{k^2 c^2 - \omega^2 \varepsilon_{-k}} \int | \, S_{k,\,k_1,\,k_2} \, |^2 \, I_{k_1}^\sigma \, I_{k_2}^t \, \delta \, (k - k_1 - k_2) \, dk_1 \, dk_2. \qquad (7.66)$$

Here

$$S_{k,\,k_1,\,k_2} = S_{ijl} \, (k, \, k_1, \, k_2) \, e_{k_1,\,j}^\sigma \, e_{k,\,i}^{t*} \, e_{k_2,\,l}^t,$$
$$\Sigma_{k,\,k_1}^{\text{eff}} = \Sigma_{ijls}^{\text{eff}} \, (k, \, k_1, -k_1, \, k) \, e_{k,\,i}^{t*} \, e_{k,\,s}^t \, e_{k_1,\,j}^\sigma \, e_{k_1,\,l}^{\sigma*}, \qquad (7.67)$$
$$\Sigma_{ijls}^{\text{eff}} \, (k, \, k_1, -k_1, \, k) = \tfrac{1}{2} \, (\Sigma_{ijls} \, (k, \, k_1, -k_1, \, k) +$$
$$+ \Sigma_{ijsl} \, (k, \, k_1, \, k, -k_1)) + S_{inj} \, (k, \, k - k_1, \, k_1) 8\pi i \, (\omega - \omega_1) \times$$
$$\times \Pi_{nm} \, (k - k_1) \, S_{msl} \, (k - k_1, \, k, -k_1). \qquad (7.68)$$

Here Π_{nm} (k) is the inverse Maxwell operator:

$$(k^2 \, \delta_{ij} - k_i \, k_j - \tfrac{1}{c^2} \omega^2 \, \varepsilon_{ij}) \, \Pi_{jl} \, (k) = \delta_{il}. \qquad (7.69)$$

Equation (7.66) is similar to (2.124); it differs from it only in the types of interacting waves, transverse t and turbulent σ, which is reflected in the unit vectors $\mathbf{e}^t$ and $\mathbf{e}^\sigma$ with which the coefficients of second S_{ijl} and third Σ_{ijls} order for the nonlinear currents are contracted. We are limited to the high-frequency region $\omega \gg \omega_{pe}$, ω_{He} in (7.66). Then, in the operator in (7.69), with $\varepsilon_{ij} \approx \varepsilon \, (\omega) \delta_{ij} \simeq \delta_{ij}$, we have

$$\Pi_{ij} \, (k) \simeq \frac{1}{k^2 - \omega^2/c^2} \left(\delta_{ij} - \frac{k_i \, k_j}{k^2} \right). \qquad (7.70)$$

In the following it is important that $\Pi_{ij}(k)$ contains a pole at $k^2 = \omega^2/c^2$ which corresponds to the characteristic mode of the transverse wave. By using the explicit expression for S_{inj} in (7.16) it is easy to see that the second term of (7.68) contains just $|S_k^{\sigma}|^2$, where S_k^{σ} is defined by (7.16).[†] But this means that in the balance equations, which are obtained from (7.66) by taking the imaginary part and integrating over ω, an imaginary part appears only because of the denominator of (7.70), $\frac{1}{k^2 - \omega^2/c^2} = i\pi\delta(k^2 - \omega^2/c^2)$. A similar result is obtained for the last term in (7.66). From this it follows that the last term in (7.68) and the last term of (7.66) describe just scattering, and with any anisotropy in the distribution of turbulent fluctuations they do not give instabilities.

Let us consider Σ_{ijls} in more detail. Intrinsically, this term takes into account both the fluctuation anisotropy and the anisotropy of the fast particles (this latter feature arises because the third-order nonlinear current describes Compton scattering, which is only important for relativistic particles). Let us examine the effect of the thermal particles. As shown by the results of Chapter 2, Σ_{ijls} contains a denominator of the type $1/(\omega - \mathbf{k}\mathbf{v})$ for the first argument, the difference between the second and third arguments, and for the fourth argument as well. The resonance denominator, which can lead to buildup, is present only in the second term of (7.68). The first two denominators of this Σ contain the high frequencies ω and $\omega - \omega_1 \approx \omega$, and can be expanded rather easily in terms of $\mathbf{k}\mathbf{v}/\omega$ and ω_{He}/ω. Thus, in the solutions to an equation like

$$\left(i(\omega - \mathbf{k}\mathbf{v}) + \omega_{He}\frac{\partial}{\partial\varphi}\right)f_k^{(3)} = e\int F_{k_1}\frac{\partial f_{k_2}^{(2)}}{\partial\mathbf{p}}\delta(k - k_1 - k_2)\,dk_1\,dk_2 \qquad (7.71)$$

one can neglect $\mathbf{k}\mathbf{v}$ and $\omega_{He}\frac{\partial}{\partial\varphi}$ on the left side in the first approximation and then find the contributions from these terms in the first approximation by means of perturbation theory based on $\mathbf{k}\mathbf{v}/\omega$ and ω_{He}/ω. In the same manner, $f^{(2)}$ is expressed in terms of $f^{(1)}$. The integral of $f^{(1)}$ can be expressed in terms of the

---

[†]This follows from the fact that $\varepsilon_{-k} = \varepsilon_k^*$.

linear dielectric constant without any expansions. We obtain

$$\sum_{iijls}(k, k_1, k, -k_1) \approx \frac{e^2}{m_e^2\,\omega^2\,4\pi i}\left\{\delta_{ij}\frac{k_{1l}}{\omega_1} - \delta_{lj}\frac{k_{1i}}{\omega_1} - \right.$$

$$\left. - \delta_{ij}\frac{k_l}{\omega} - \delta_{li}\frac{k_j}{\omega}\right\}\left(k_{1m}\,\varepsilon_{ms}^{(e)}(-k_1) - k_{1s}\right). \qquad (7.72)$$

For longitudinal waves the large terms c/v_p in (7.72) are retained and

$$\sum_{k, k_1}^{\text{eff}} = -\frac{e^2\,(k\,k_1)}{m_e^2\,\omega^3\,8\pi_e}\left(\varepsilon_{-k_1}^{l(e)} - 1\right). \qquad (7.73)$$

The nonlinear dielectric constant will take the form

$$\varepsilon_k^N = -\frac{8\pi i}{\omega}\int\sum_{k, k_1}^{\text{eff}} I_{k_1}^\sigma\,dk_1 = \frac{\omega_{pe}^2}{4\pi n_0\,m_e\,\omega^4}\int(k k_1)\,I_{k_1}^l\,dk_1\left(\varepsilon_{-k_1}^{l(e)} - 1\right). \qquad (7.74)$$

The last equality is written for longitudinal turbulent fluctuations. Since $(k k_1)$ has odd parity, only the imaginary part of $\varepsilon_{k_1}^{l(e)}$ remains in (7.74). Equation (7.74) is also valid when there is no unique relationship between ω and k for the fluctuations. But if there is such a relationship, then

$$I_{k_1}^l = \frac{4\pi}{\omega_1\left.\dfrac{\partial\varepsilon_{k_1}^l}{\partial\omega_1}\right|_{\omega_1=\omega_{k_1}^l}}\left(W_{k_1}^l\,\delta\left(\omega - \omega_{k_1}^l\right) + W_{-k_1}^l\,\delta\left(\omega + \omega_{k_1}^l\right)\right), \qquad (7.75)$$

or

$$\text{Im}\,\varepsilon_k^N = -\frac{2\omega_{pe}^2}{n_0\,m_e\,\omega^4}\int\frac{(k k_1)\,\text{Im}\varepsilon_{k_1}^l\,dk_1}{\omega_1\left.\dfrac{\partial\varepsilon_{k_1}^l}{\partial\omega_1}\right|_{\omega_1=\omega_{k_1}^l}}\,W_{k_1}^l, \qquad (7.76)$$

$$W^l = \int W_{k_1}^l\,dk_1. \qquad (7.77)$$

Thus the imaginary part of ε_k^N is opposite to the imaginary part of $\varepsilon_{k_1}^l$ if k and k_1 are aligned. This means that if there is a preferred direction for the propagation of the turbulent fluctuations, waves making an acute angle with this direction will build up. This buildup is absent for an isotropic turbulence. The damping which arises is another order of $\mathbf{kv}/\omega \sim v_{Te}/c$ smaller. Since damping of the electromagnetic waves appears only in the next order, only a small amount of anisotropy, of the order of $\sim v_{Te}/c$, is needed to have buildup. For Langmuir turbulence

$$\text{Im } \varepsilon_k^l \approx 0 \tag{7.78}$$

and only fast particles, which create effective Landau absorption, can lead to instability. For ion-acoustic turbulence we have

$$\gamma_k^N = -\frac{\omega^2 \text{ Im } \varepsilon_k^N}{\frac{\partial}{\partial \omega} \omega^2 \varepsilon_k^l} \approx -\frac{\omega}{2}\text{Im } \varepsilon_k^N = \frac{\omega_{pe}^2}{\omega^2} v_{\text{eff}}, \tag{7.79}$$

$$v_{\text{eff}} = \frac{v_{Te}}{2c} \sqrt{\frac{\pi}{2}} \int \frac{W_{k_1}^s}{n_0 T_e} \omega_{k_1}^s \frac{(\mathbf{kk_1})}{kk_1} \, d\mathbf{k_1}. \tag{7.80}$$

The magnitude of v_{eff} is $(v_{Te}/c)(m_e/m_i)^{1/2}$ times smaller than the effective frequency of the quasilinear collisions between thermal particles and the turbulent fluctuations. If Imε in (7.73) is connected with synchrotron radiation from the relativistic particles, then σ describes electromagnetic waves. This process of building up high-frequency radiation on anisotropic low-frequency fluctuations is qualitatively different from induced Compton scattering.

Induced Compton scattering of turbulent fluctuations by electromagnetic radiation is described by another resonance in the expression for Σ, one which resonates at the frequency $\omega - \omega_1$. The possibility of amplifying the radiation in this case is connected with the anisotropic distribution of relativistic particles. This type of effect was studied first in [68] for a sharply anisotropic distribution in a beam of relativistic electrons in a turbulent plasma; it was studied in [261] for the case of a slightly anisotropic distribution. We will present just the latter case here.

Assume that the electron distribution is of the form

$$f_e = f_e^{(0)}(1 + \alpha \cos \theta), \tag{7.81}$$

where θ is the angle between the particle velocity and some selected direction. To find the growth rate we must use the general expression for the growth rate in terms of the scattering probability [see (6.146) and (2.233)]. We then obtain Eq. (6.164), in which $\Phi_\bullet$ will be replaced by

$$\Phi_{\bullet a} = \frac{4}{3}\{(2+3\alpha\cos\theta)(1-q)^3+6q(1-q^2)-3q^2(1-q)\times$$
$$\times\,\alpha\cos\theta+3q^2\ln q(2+\alpha\cos\theta)\}, \qquad (7.82)$$

where θ is the angle between k and H. For the spectrum given in (6.159) we find

$$\gamma_\omega = -\frac{\pi\omega_{pe}^4\,n_1}{3\omega^2\,n_0^2\,c^5 m_e}\left(\frac{2c}{\omega}\right)^{\gamma/2}c_*(\gamma)\int k_1^{\frac{\gamma-2}{2}}W_{k_1}^l\,dk_1, \qquad (7.83)$$

$$c_*(\gamma)=c(\gamma)+\frac{3}{2}\alpha\cos\theta\,(\gamma-1)\left(\frac{1}{\gamma}-\frac{3}{\gamma+2}-\frac{2}{\gamma+4}-\frac{2}{(\gamma+4)^2}\right). \qquad (7.84)$$

When α is of order 1, γ_ω can become positive, signaling a buildup in the oscillations. A buildup in the radiation is possible without the plasma because of the induced synchrotron radiation [269]; however, this situation requires a different degree of anisotropy $\alpha \sim \varepsilon/mc^2$.

7.5. Fluctuations in Intensities

of Electromagnetic Waves when

Propagating through Turbulent Plasma

In addition to the scattering effects and the amplification of radiation in a turbulent plasma, intensity fluctuations are also possible. According to [295], these fluctuations can be interpreted as resulting from interference between waves which have acquired different phase shifts during their passage through a medium containing random electromagnetic inhomogeneities (see [296–298] for discussions of fluctuations during propagation through turbulent fluids and gases). In a turbulent plasma the random inhomogeneities and random electromagnetic characteristics are produced by the turbulent fluctuations, and the distribution of the inhomogeneities over the various scales can be found from the turbulence spectra (see Chapter 4). In principle a plasma is not required to be present in order to have wave interference; the plasma can create a random distribution of phases in the electromagnetic

waves leaving the plasma, and the interference will then take place
in the vacuum outside the plasma. Therefore, the notion of a
phase screen is often used, in which one assumes that the plasma
layer is narrow and serves only to modulate the phase of the electromagnetic wave passing through it. In reality, of course, this
assumption is much too crude, because the change in the phase
relations and the interference both take place simultaneously.
Moreover, as we have seen in the example of the scattering effects, the wavelengths of the turbulent fluctuations and the radiation passing through the plasma can be comparable ($k \sim k_1$) and
the graphic geometrical-optics picture with a phase advance at
the inhomogeneities is not applicable in this instance.

Assume that a high-frequency monochromatic wave is transmitted in the plasma:

$$E^t = E^t_{k,\omega} e^{-i\omega t + i kr}. \tag{7.85}$$

The wave which will interact with the turbulent plasma will be
modulated in both time and space:

$$E^t = E^t_{k\omega}(r, t) e^{-i\omega t + i\, kr}. \tag{7.86}$$

The intensity of the wave is $|E^t_{k,\omega}(r, t)|^2$, and the intensity fluctua-
tion is

$$\langle \Delta I^2 \rangle = \langle |E^t_{k,\omega}(r, t)|^4 \rangle - (\langle |E_{k\omega}(r, t)|^2 \rangle)^2. \tag{7.87}$$

The last term in (7.87), the square of the intensity, can be found
from the scattering equation; therefore, we will study the first
term in greater detail. Expand $E_{k,\omega}(r, t)$ in a Fourier series:

$$E^t_{k\omega}(r, t) = \int E_{k,\omega,k_1,\omega_1} e^{-i\omega_1 t + i k_1 r}\, d\omega_1\, dk_1. \tag{7.88}$$

If $E^t_{k,\omega}$ is the complete Fourier component of the field, then evidently

$$E^t_{k,\omega,k_1,\omega_1} = E^t_{k+k_1,\, \omega+\omega_1}. \tag{7.89}$$

The mean value of the fourth power of the field in (7.87) can be written as

$$\langle \, | E_{k\omega}^{t}\,(\mathbf{r},\, t) \, |^{4} \rangle = \int \langle E_{k+k_1,\,\omega+\omega_1}^{t} E_{k+k_2,\,\omega+\omega_2}^{t} \times$$

$$\times E_{k+k_3,\,\omega+\omega_3}^{t*} E_{k+k_4,\,\omega+\omega_4}^{t*} \rangle \, dk_1 \, dk_2 \, dk_3 \, dk_4 \times$$

$$\times \exp \{ - i\, (\omega_1 + \omega_2 - \omega_3 - \omega_4)\, t + i\, (\mathbf{k}_1 + \mathbf{k}_2 - \mathbf{k}_3 - \mathbf{k}_4) \, \mathbf{r} \}. \quad (7.90)$$

Changing the signs of dk_3, dk_4 and using $E_{k,\,\omega}^{*} = E_{-k,\,-\omega}$, we find

$$\langle \, | E_{k\omega}^{t}\,(\mathbf{r},\, t) \, |^{4} \rangle = \int \langle E_{k_1'\omega_1'}^{t} E_{k_2',\,\omega_2'}^{t} E_{k_3'\omega_3'}^{t} E_{k_4'\,\omega_4'}^{t} \rangle \times$$

$$\times \exp \{ - i(\omega_1' + \omega_2' + \omega_3' + \omega_4')\, t + i\, (\mathbf{k}_1' + \mathbf{k}_2' + \mathbf{k}_3' + \mathbf{k}_4') \, \mathbf{r} \}, \quad (7.91)$$

where

$$k_1' = k_1 + k, \quad k_2' = k_2 + k, \quad k_3' = k_3 - k, \quad k_4' = k_4 - k. \quad (7.92)$$

Thus, the problem reduces to computing the average values of four transverse fields, the first two of which have values of their four-dimensional wave vectors close to $+k$, and the last two close to $-k$. When the fluctuations are stationary we have

$$\langle E_k^{t} E_{k_1}^{t} | E_{b}^{t}{}' E_{k_1'}^{t} \rangle = \langle kk_1 | k', k_1' \rangle = I_{k,k_1,\,k',\,k_1'} \delta \, (k + k' + k_1 + k_1'). \quad (7.93)$$

Therefore, only three vectors in (7.93) are independent.

If we introduce

$$\varkappa_1 = k_1 - k, \quad \varkappa' = k' + k, \quad \varkappa_1' = k_1' + k, \quad (7.94)$$

then $\varkappa_1 = -\varkappa - \varkappa'$. We will assume that $I_{k,\,k_1,\,k',\,k_1'}$ is a function of k, $\varkappa_1'$, $\varkappa'$. Let us put together an equation for $\langle kk_1 | k, k_1' \rangle$. Assuming that the field t is weak and high-frequency ($\omega \gg \omega_{\text{He}}$, ω_{pe}), we keep only the term which is linear in this field in the nonlinear equations:

$$(k^2 - \tfrac{1}{c^2} \, \omega^2) \, E_k^{t} = 8\pi i\omega \int S_{k,\,k^\sigma,\,k_1} E_{k^\sigma}^{\sigma} E_{k_1}^{t} \, \delta \, (k - k^\sigma - k_1) \, dk^\sigma dk_1 \, c^{-2}, \quad (7.95)$$

$S_{k,\,k^\sigma,\,k_1}$ is the component of the nonlinear current of second order, contracted with the transverse unit vectors and the unit vectors of the turbulent σ-waves and then averaged over the polarizations. The complete field is involved in (7.95). It is inconvenient to divide it into random and nonrandom components in this case.

After multiplying (7.95) by three fields, and averaging over the ensemble, one can determine with greater accuracy the transverse fields on the right side by using (7.95), and thereby obtain an equation for $\langle kk_1|k'k_1'\rangle$:

$$(k^2 - \frac{1}{c^2}\omega^2)\langle k, k_1 | k', k_1' \rangle = 8\pi i\omega \langle k, k_1 | k', k_1'\rangle \int S_{k, k^\sigma, k-k^\sigma} \times$$

$$\times \frac{8\pi i\,(\omega-\omega^\sigma)\,c^{-4}}{(k-k^\sigma)-(\omega-\omega^\sigma)^2 c^{-2}} S_{k-k^\sigma,-k^\sigma,\,k}\, I_{k^\sigma}^\sigma\, dk^\sigma - \frac{64\omega\omega_1\pi^2 c^{-4}}{k_1^2-\omega_1^2 c^{-2}} \times$$

$$\times \int S_{k, k^\sigma,\ k-k^\sigma} S_{k_1,\ -k^\sigma,\ k_1+k^\sigma}\langle k-k^\sigma, k_1+k^\sigma | k', k_1'\rangle\, I_{k^\sigma}^\sigma\, dk^\sigma -$$

$$- \frac{64\pi^2\,\omega\omega'c^{-4}}{k'^2-\omega'^2 c^{-2}} \int dk^\sigma\, S_{k, k^\sigma, k-k^\sigma} S_{k',\,-k^\sigma,\ k'+k^\sigma} \times$$

$$\times \langle k-k^\sigma, k_1 | k'+k^\sigma, k_1'\rangle\, I_{k^\sigma}^\sigma - \frac{64\pi^2\,\omega\omega_1'c^{-4}}{k_1'^2-\omega_1'^2 c^{-2}} \int dk^\sigma\, S_{k, k^\sigma,\ k-k^\sigma} \times$$

$$\times S_{k_1',\,-k^\sigma,\ k_1'+k^\sigma}\langle k-k^\sigma, k_1 | k', k'+k^\sigma\rangle\, I_{k^\sigma}^\sigma. \qquad (7.96)$$

In a similar fashion, equations can be obtained whose left sides are

$$(k_1^2-\omega_1^2 c^{-2})\langle k, k_1 | k', k_1'\rangle, \quad (k'^2-\omega'^2 c^{-2})\langle k, k_1 | k', k_1'\rangle,$$

and

$$(k_1'^2-\omega_1'^2 c^{-2})\langle k, k_1 | k', k_1'\rangle.$$

We will now assume that the differences of the four-dimensional vectors k_1, k', and k_1' are small compared with k; i.e.,

$$(k_1^2-\omega_1^2 c^{-2}) = (k+\varkappa_1)^2 - c^{-2}(\omega+v_1)^2 \approx 2k\varkappa_1 - 2\omega v_1 c^{-2} + k^2-\omega^2 c^{-2},$$

$$k'^2-\omega'^2 c^{-2} = -2k\varkappa + 2\omega v c^{-2} + k^2-\omega^2 c^{-2},$$

$$k_1'^2-\omega_1'^2 c^{-2} = -2k\varkappa_1' + 2\omega v_1' c^{-2} + k^2-\omega^2 c^{-2}, \qquad \varkappa = \{\varkappa, v\}. \qquad (7.97)$$

Combining (7.96) with the equation for $(k_1^2 - c^{-2}\omega_1^2) \langle k, k_1 | k', k_1' \rangle$ and taking into account the two equations containing $k'^2 - c^{-2}\omega'^2$ and $k_1'^2 - c^{-2}\omega_1'^2$, we obtain $2k(\varkappa_1 + \varkappa' + \varkappa_1') - 2\omega(\nu_1 + \nu' + \nu_1')c^{-2}$ on the left. We multiply this result by

$$\exp\left[i\left(\varkappa_1 + \varkappa' + \varkappa_1'\right)\mathbf{r} - i\omega\left(\nu_1 + \nu_1' + \nu'\right)t\right] \tag{7.98}$$

and integrate over $\varkappa_1, \nu_1$. We neglect the weak instability on the right side, that is, we are assuming that (7.93) is satisfied. Then, with $I_{k,\varkappa,\varkappa_1'}(\mathbf{r},t)$ given by

$$I_{k,\varkappa',\varkappa_1'}(\mathbf{r},t) = \int I_{k,\varkappa_1,\varkappa'\varkappa_1'} e^{i(\varkappa_1 + \varkappa_1' + \varkappa')\mathbf{r} - i\omega(\nu_1 + \nu' + \nu_1')t} d\varkappa_1 d\nu_1, \tag{7.99}$$

we obtain the equation

$$\left(\frac{\partial}{\partial t} + \frac{k}{k} \cdot \frac{\partial}{\partial \mathbf{r}}\right) I_{k,\varkappa'\varkappa_1'} = \int |S_{k,k\sigma}|^2 I_{k\sigma}^\sigma M_{k,k\sigma,\varkappa_1'\varkappa'} dk\sigma. \tag{7.100}$$

Here

$$|S_{k,k\sigma}|^2 = \frac{e^2\pi}{8m_e\omega_k^2} |(k_j^\sigma e_{sl}^{(e)}(k^\sigma) - k_l^\sigma) e_{lk\sigma}^\sigma|^2 \left(1 + \frac{(k, k-k^\sigma)^2}{k^2(k-k^\sigma)^2}\right). \tag{7.101}$$

Equation (7.100) includes the first term in the expansion of the nonlinear currents in (7.96) in terms of the small parameters $\varkappa'/k$ and is averaged over the polarizations of the transverse waves. Furthermore,

$$M_{k,k\sigma\varkappa'\varkappa_1'} = -I_{k,\varkappa',\varkappa_1'} \{\delta(\omega_k - \omega^\sigma - \omega_{k-k\sigma}) + \delta(\omega_k - \nu' - \nu_1' - \omega^\sigma -$$

$$-\omega_{k-\varkappa'-\varkappa_1'-k\sigma}) + \delta(\omega_k - \nu - \omega^\sigma - \omega_{k-\varkappa'-k\sigma}) + \delta(\omega_k - \nu_1' -$$

$$-\omega^\sigma - \omega_{k-\varkappa_1'-k\sigma})\} + I_{k-k\sigma,\varkappa',\varkappa_1'-k\sigma} \{\delta(\omega_{k-k\sigma} + \omega^\sigma -$$

$$-\nu' - \omega_{k-\varkappa'}) + \delta(\omega^\sigma + \omega_{k-k\sigma} - \omega_k)\} + I_{k-k\sigma\varkappa'-k\sigma,\varkappa_1'} \times$$

$$\times \{\delta(\omega_{k-k\sigma} + \omega^\sigma - \nu_1' - \omega_{k-\varkappa_1'}) + \delta(\omega_{k-k\sigma} + \omega^\sigma - \omega_k)\} +$$

$$+ I_{k,\varkappa',\varkappa_1'-k\sigma} \{\delta(\omega_k - \nu_1' - \omega_{k-\varkappa_1'}) + \delta(\omega_k - \nu_1' - \nu' - \omega_{k-\varkappa'-\varkappa_1'})\} +$$

$$+ I_{k, \varkappa' - k^\sigma, \varkappa_1'} \left\{ \delta \left(\omega_k - v_1' - \omega_{k - \varkappa'} \right) + \delta \left(\omega_k - v_1' - v' - \omega_{k - \varkappa' - \varkappa_1'} \right) \right\} -$$

$$- I_{k - k^\sigma, \varkappa' - k^\sigma, \varkappa_1' - k^\sigma} \left\{ \delta \left(\omega_{k - k^\sigma} + \omega^\sigma - v' - v_1' - \omega_{k - \varkappa' - \varkappa_1'} \right) +$$

$$+ \delta \left(\omega_{k - k^\sigma} + \omega^\sigma - \omega_k \right) \right\} - I_{k, \varkappa' - k^\sigma, \varkappa_1' + k^\sigma} \times$$

$$\times \left\{ \delta \left(\omega_k - v_1' - \omega_{k - \varkappa_1'} \right) + \delta \left(\omega_k - v' - \omega_{k - \varkappa'} \right) \right\}. \qquad (7.102)$$

If k^σ is of order k, and $\varkappa'$, $\varkappa_1' \ll k$ by necessity, then all the terms of (7.102) except the first contain I in which all but one $\varkappa'$ is k^σ. They are equal to zero, and since the first term is negative the fluctuations are damped. If $k^\sigma \ll k$ it is possible to have both $\varkappa \ll k_\sigma$ and $\varkappa \gg k^\sigma$. When evaluating these or other terms one must take into account the fact that those δ-functions not containing k^σ are of order $1/\varkappa$, while those containing k^σ are of order $1/k^\sigma$. Therefore, when $\varkappa' \varkappa_1' \ll k^\sigma$, we arrive at the following approximation:

$$M_{k, k^\sigma, \varkappa', \varkappa_1'} \simeq I_{k, \varkappa', -k^\sigma} \left\{ \delta \left(\frac{k}{k} \varkappa_1' \right) + \delta \left(\frac{k}{k} (\varkappa' + \varkappa_1') \right) \right\} +$$

$$+ I_{k, -k^\sigma, \varkappa_1'} \left\{ \delta \left(\frac{k}{k} \varkappa' \right) + \delta \left(\frac{k}{k} (\varkappa' + \varkappa_1') \right) \right\} - I_{k, -k^\sigma, k^\sigma} \times$$

$$\times \left\{ \delta \left(\frac{k}{k} \varkappa' \right) + \delta \left(\frac{k}{k} \varkappa_1' \right) \right\}. \qquad (7.103)$$

Here it has been assumed that $v'/|\varkappa'|$, $v_1'/|\varkappa_1'| \ll 1$. When $\varkappa'$, $\varkappa_1' \gg k^\sigma$ we have

$$M_{k, k^\sigma, \varkappa', \varkappa_1'} = \left\{ - I_{k, \varkappa', \varkappa_1'} + I_{k - k^\sigma, \varkappa', \varkappa_1' - k^\sigma} +$$

$$+ I_{k - k^\sigma, \varkappa' - k^\sigma, \varkappa_1'} - I_{k - k^\sigma, \varkappa' - k^\upsilon, \varkappa_1' - k^\sigma} \right\} \delta \left(\frac{k}{k} k^\sigma \right). \qquad (7.104)$$

Because of the subtraction in (7.104) the fluctuations are much reduced in magnitude. Therefore, the most important values of $\varkappa'$ and $\varkappa_1'$ are those close to k^σ.

However, for the total fluctuation $\langle I^2 \rangle - (\langle I \rangle)^2 \simeq \langle I^2 \rangle = \int I_{k, \varkappa, \varkappa'} \times$ $d\varkappa d\varkappa' dv dv'$ the phase volume is $d\varkappa d\varkappa' \sim \varkappa^2 d\varkappa d\varkappa' \varkappa'^2$, so that $\varkappa'$ of order k^σ make the main contributions. This enables us to obtain

an estimate of the intensity fluctuations. For example, from
(7.102) and (7.100) we find for Langmuir fluctuations

$$\frac{\langle \Delta I^2 \rangle}{(\langle I \rangle)^2} \simeq \left(\frac{\omega_{pe}}{\omega} \right)^2 \frac{L}{n_0 m_e c^2} \int k W_k^l \, dk, \qquad (7.105)$$

where L is the path of the ray in the plasma. Equation (7.105) is
valid for $k^2 \gg \omega/Lc$. With the spectrum $W_k = (2Q/\alpha)^{1/2}$ we find
that this last condition is satisfied if

$$\omega \ll \omega_{pe} \frac{\omega_{pe} Lc}{v_{Te}^2} \left(\frac{m_e}{m_i} \right)^{2/5} .$$

In cases of practical interest L is so large that (7.105) is a good
estimate.

We have treated here just a few cases; readers interested in
the details of the analysis of these equations are referred to the
original sources [295, 297].

Chapter 8

Electromagnetic Properties of Turbulent Plasma

8.1. General Statement

of the Problem

Assume that stationary turbulence has been established in a plasma. Let us follow the behavior of a weak electromagnetic field E^R which appears in the plasma either because of fluctuations or because of some external influence. The action of the field E^R on the turbulent plasma changes the flow of turbulent energy and the plasmon distribution, along with the distribution of charged plasma particles. These changes in the plasmons and particles produce an additional flow, which then alters the field. This effect is described by the dielectric constant of the turbulent plasma, which is the response of the plasma to the weak field E^R. An important difference between a turbulent plasma and a nonturbulent one is that in a nonturbulent plasma the field E^R can change just the particle distribution, while in a turbulent plasma the plasmon state can be significantly altered. The frequency ω of the perturbing field is a very important quantity. If this frequency is sufficiently low, for example much lower than the characteristic times for the nonlinear transfer of turbulent energy across the spectrum, one can expect marked differences between the electromagnetic properties of turbulent and nonturbulent plasmas. This chapter will consider just the low-frequency region, in which new electromagnetic properties appear which are related to the plasma's turbulence. It is useful to introduce the concept of effective turbulent collisions ν^T as effective frequencies for collisions among the fluctuations themselves and with the plasma particles. They will be dependent on the turbulence energy.

417

In the case $W/n_0 T_e \ll 1$, which will be the only subject of consideration, the effective turbulent collisions, which are proportional to the higher powers of the turbulence energy, have a lower frequency. Turbulent collisions proportional to the first power of the turbulence energy have the highest frequency. They describe quasilinear relaxation processes, decay interactions, and induced scattering. In the region of perturbing frequencies lower than the effective turbulence frequencies, the dielectric tensor cannot be expanded in terms of the effective turbulent frequencies, or the turbulence energy. This seems to indicate that the electromagnetic properties of the plasma in this frequency region are changed in a fundamental way.

For frequencies larger than the effective turbulent frequencies the dielectric tensor is expandable in terms of the turbulent frequencies or the turbulence energy. The oscillational modes of the plasma then stay about the same as they are in the absence of turbulence, but the corrections proportional to the turbulence energy describe a weak change in the frequency and a small linear buildup or damping of the oscillations, and the scattering of those oscillations (see Chapter 7). When low frequencies are excited in a turbulent plasma, one cannot speak of the nonlinear interactions of modes any longer; rather, we must now consider fundamental changes in the modes, particularly the disappearance of old modes and the production of new ones. The change in the low-frequency plasma properties in the presence of turbulence is of special interest because the magnetohydrodynamic and drift instabilities most inimical to plasma confinement are found in the low-frequency region. Changes in the conditions for the excitation and dissipation of high-frequency turbulence, its intensity, and the types of turbulent fluctuations can regulate the low-frequency properties and plasma instabilities over wide limits.

The problem is to find an analytic form for the low-frequency dielectric tensor $\varepsilon_{ij}(\omega,\ k,\ W_{k_1})$ for the plasma, which is functionally dependent on the spectral turbulence energy W_{k_1}:

$$W = \int W_{k_1}\, dk_1. \qquad (8.1)$$

Here W is the energy of the turbulent fluctuations in 1 cm^3, ω is the frequency, and k the wave vector of the perturbation. In order to

construct a theory we can use physical concepts which are similar
to those required for describing weak turbulence. That is, we
must expand in terms of the turbulence energy in the kernels of
the collision integrals, and take into account processes with an
ever-increasing number of external plasmon lines. The general
calculation scheme generalizes the method used in Chapter 2. The
particle distribution function f^α and the electric field E are split
into the sums of turbulent $f^{T\alpha}$, E^T and regular $f^{R\alpha}$, E^R components, giving

$$f^\alpha = f^{T\alpha} + f^{R\alpha} : E = E^T + E^R, \qquad (8.2)$$

where $\langle f^{T\alpha} \rangle = 0$, $\langle E^T \rangle = 0$; the averaging is over a statistical ensemble. By averaging the equations of motion and Maxwell's
equations over a statistical ensemble and then subtracting the
averaged equations from the original equations, we can obtain a
system of equations for the regular and random components.
Further, one must isolate those quantities which characterize the
stationary turbulent state; these quantities will be distinguished
by the index (0) thus: $f^{R(0)\alpha}$, $f^{T(0)\alpha}$, $E^{R(0)}$, $E^{T(0)}$. Consider now
the perturbation of the turbulent field and the field $E^{R(1)}$ connected
with it [we use the index (1) for the perturbation]. All the quantities (E^T, E^R, f^T, f^R) are expanded in terms of $E^{R(1)}$, and only the
terms linear in $E^{R(1)}$ are retained.[†] Thus we obtain two systems
of equations for the fundamental turbulent state (0) and the deviations from it (1). These equations are general in nature and can
in principle be used for turbulence which is not weak. In order to
obtain the truncated equations describing the weakly turbulent
state, one can, as usual, expand in terms of the turbulent field
$E^{T(0)}$ in the kernels of the integral equation for the original turbulent state. One can also use the expansions in terms of $E^{T(0)}$ in
the kernels of the equations which give the deviations from the
fundamental turbulent state. In view of the fact that we are retain-
ing a finite number of terms in the expansions, one can hope to
take into account only the first few turbulent collision frequencies.

This allows us to describe the electromagnetic properties of
the plasma at frequencies smaller than these turbulent frequencies
but larger than the turbulent frequencies not taken into account,

---

[†] If $E^{R(0)} = 0$ then $E^{R(1)} = E^R$.

which are of higher order in the turbulence energy. This approach to the problem is dictated by the fact that the effective turbulent frequencies are, in a number of very important practical cases, dependent not only on the turbulence energy but on the frequency ω and the perturbation wave number k. Therefore the condition that the turbulent collisions ($\omega \ll \nu^T$) be dominant is in fact usually a condition which includes the values of ω, k, and the turbulence energy.

The limitations on the applicability of an approach based on the expansions of the particle–turbulent fluctuation collision integrals in terms of the turbulence energy are quite strict in a number of cases. These limitations are closely related to the effects of nonlinear changes in the dispersion properties of the high-frequency oscillations. Although the changes in the frequencies of the high-frequency waves are always small for weak turbulence, as we have already mentioned, they can be sensed by the low-frequency perturbations at frequencies which are, roughly speaking, smaller than these nonlinear shifts. Inclusion of these effects leads to the renormalization of the high-frequency plasmon propagators and the particle charges. This renormalization can be taken into account by setting up an equation for the integrals of the particle and turbulent fluctuation collisions (compare this with the corelation effects in Chapter 2). These effects turn out to be very important for Langmuir turbulence. The instability of a turbulent plasma, corresponding to the instability of a gas of cold Langmuir plasmons, was first examined in [299][†]; it arises in the approach which uses expansions of the collision integrals for particles and turbulent fluctuations in terms of the turbulence energy. This type of instability is possible in a narrow interval of plasma parameters, and for turbulence located at small plasmon phase velocities. The use of the dielectric constant, which is found by summing the perturbation series with respect to the turbulence energy in the particle–turbulent-fluctuation collision integrals, reveals new instabilities in a turbulent plasma [301] which are developed at large plasmon phase velocities. Such instabilities are of interest because the turbulence energy is concentrated at large phase velocities due to the spectral transfers. Furthermore, in an inhomogeneous plasma the high-frequency

---

[†] The instability criterion [299] was also obtained in [300] using the energy principle.

turbulence and stochastic high-frequency fields produce stabilization effects for the drift oscillations [302]. It is also possible
to excite magnetic fields by turbulent fluctuations of the potential
type, such as Langmuir waves [303]. This is similar to the spontaneous excitation of magnetic fields in a turbulent fluid [304].

It is known that low-frequency electromagnetic waves do not
penetrate a plasma (the skin effect). Usually the thickness of the
skin layer in a nonturbulent plasma is very dependent on the frequency of the pairwise particle collisions. In a turbulent plasma
the role of those collisions is given to the nonlinear effects in the
interactions of turbulent fluctuations among themselves (plasmon—
plasmon collisions) and the interactions of particles with turbulent fluctuations (plasmon—particle collisions, collisions between
two plasmons and a particle, and so on). We must assume that it
is the frequencies of these processes which determine the skin
effect in a turbulent plasma. In particular, the skin layer thickness must depend on the characteristic spectral transfers, the
time needed to establish a stationary turbulence, and so on.

Therefore, study of the skin effect will enable us to detect the
processes which influence it. Since the frequencies of these effective collisions are much higher than the frequencies of particle
collisions, the skin-layer thickness in a turbulent plasma should
be much greater than the thickness for a nonturbulent plasma.
Although in general the low-frequency perturbations affect both
the plasma particles and the plasmons, there are instances in
which the plasmon displacements are dominant. The excitation of
these interactions is revealed only in the plasmon distribution.
Under these conditions we can speak of perturbations in a gas of
plasmons, or, in the terminology of Landau [305], second sound in
a turbulent plasma. The problem of second sound in this aspect
has been examined in [306].

In order to simplify our discussion of the main problems, we
will consider just magnetized electrons and ions which are located
in an infinitely strong magnetic field. Under these conditions the
particle motion (but not the plasmon motion) is in one direction
along the strong field lines, which are assumed to be straight lines.
Taking a finite field magnitude into account enables us to rather
simply examine the drift instabilities, but the generalizations to
the case where $H \neq 0$ and to an isotropic turbulence do not repre-

sent any great problem, so we will merely present some final expressions for these cases.

8.2. Expansions of Collision Integrals in Terms of the Turbulence Energy

The drift kinetic equation which describes the distribution of the centers of the Larmor circles in the limit $H \to \infty$ (i.e., when drift effects are neglected in an inhomogeneous plasma) takes the form [307]

$$\frac{\partial f^a}{\partial t} + v_z \frac{\partial f^a}{\partial z} + \frac{e_a}{m_a} E \frac{\partial f^a}{\partial v_z} = 0, \tag{8.3}$$

where v_z is the particle velocity and E is the component of the electric field along H. Using the method described above, we obtain a system of equations for the fundamental turbulent state [indexed as (0), $\mathbf{E}^{R(0)} = 0$] and the perturbations on this state [labeled (1)]:

$$-i(\omega - k_z v_z) f_k^{T\,(0)\,a} + \frac{e_a}{m_a} E_k^{T\,(0)} \frac{\partial f^{R\,(0)\,a}}{\partial v_z} + \frac{e_a \partial}{m_a \partial v_z} \times$$

$$\times \int dk_1 dk_2 \, \delta(k - k_1 - k_2)\, (E_{k_1}^{T\,(0)} \, f_{k_2}^{T\,(0)\,a} - \langle E_{k_1}^{T\,(0)} \, f_{k_2}^{T\,(0)} \rangle) = 0; \tag{8.4}$$

$$\left(1 + \frac{k_\perp^2 c^2}{c^2 k_z^2 - \omega^2}\right) E_k^{T\,(0)} = -\frac{4\pi i}{k_z} \sum_a e_a \int f_k^{T\,(0)\,a} \, dv_z; \tag{8.5}$$

$$-i(\omega - k_z v_z) f_k^{R\,(1)\,a} + \frac{e_a}{m_a} E_k^R \frac{\partial f^{R\,(0)\,a}}{\partial v_z} = -\frac{\partial}{\partial v_z} \cdot \frac{e_a}{m_a} \int dk_1 dk_2 \times$$

$$\times \delta(k - k_1 - k_2)\, (\langle E_{k_1}^{T\,(1)} \, f_{k_2}^{T\,(0)\,a} + E_{k_1}^{T\,(0)} \, f_{k_2}^{T\,(1)\,a} \rangle); \tag{8.6}$$

$$-i(\omega - k_z v_z) f_k^{T\,(1)\,a} + \frac{e_a}{m_a} E_k^{T\,(1)} \frac{\partial f^{R\,(0)\,a}}{\partial v_z} \doteq -$$

$$-\frac{e_a}{m_a} \cdot \frac{\partial}{\partial v_z} \int dk_1 dk_2 \, \delta(k - k_1 - k_2)\, (E_{k_1}^{T\,(0)} \, f_{k_2}^{R\,(1)\,a} + E_{k_1}^R \, f_{k_2}^{T\,(0)\,a} +$$

$$+ E_{k_1}^{T\,(1)} \, f_{k_2}^{T\,(0)\,a} + E_{k_1}^{T\,(0)} \, f_{k_2}^{T\,(1)\,a} - \langle E_{k_1}^{T\,(1)} \, f_{k_2}^{T\,(0)\,a} + E_{k_1}^{T\,(0)} \, f_{k_2}^{T\,(1)\,a} \rangle); \tag{8.7}$$

$$\left(1 + \frac{k_\perp^2 c^2}{k_z^2 c^2 - \omega^2}\right) E_k^{T\,(1)} = -\frac{4\pi i}{k_z} \sum_a e_a \int dv_z f_k^{T\,(1)\,a}; \tag{8.8}$$

$$\left(1 + \frac{k_\perp^2 c^2}{k_z^2 c^2 - \omega^2}\right) E_k^R = -\frac{4\pi i}{k_z} \sum_\alpha e_\alpha \int dv_z f_k^{R(1)\alpha}. \tag{8.9}$$

The fundamental turbulent state is stationary, i.e., $f_k^{R(0)\alpha} = \Phi^\alpha \delta(k)$ and the spectrum I_k of the stationary turbulence is determined by the equation

$$\langle E_{k_1}^{T(0)} E_{k_2}^{T(0)} \rangle = I_{k_1} \delta(k_1 + k_2), \quad n_\alpha = \int \Phi^\alpha dv_z. \tag{8.10}$$

The frequency of the linear turbulent fluctuations is determined by the following dispersion relation:

$$\Pi(k) = \Pi(\omega, \mathbf{k}) = \varepsilon_0^{(e)}(\omega, \mathbf{k}) + \varepsilon_0^{(i)}(\omega, \mathbf{k}) - 1 + \frac{k_\perp^2 c^2}{k_z^2 c^2 - \omega^2} = 0, \tag{8.11}$$

$$\varepsilon_0^{(\alpha)}(\omega, \mathbf{k}) = 1 + \frac{4\pi e_\alpha^2}{m_\alpha k_z} \int dv_z \frac{\partial \Phi^\alpha}{\partial v_z} \cdot \frac{1}{(\omega - k_z v_z + i\delta)}. \tag{8.12}$$

For high-frequency waves, whose phase velocities are much higher than the average thermal velocities of the electrons, we find upon neglecting the spatial dispersion that (8.11) gives

$$\omega^2 = \omega_{k_1, \pm}^2 = \frac{1}{2}\left[k_1^2 c^2 + \omega_{pe}^2 \pm \sqrt{(k_1^2 c^2 + \omega_{pe}^2)^2 - 4\omega_{pe}^2 k_{1z}^2 c^2}\right], \tag{8.13}$$

and I_{k_1} is related to the spectral turbulence density W_{k_1} introduced above by the equation

$$I_{k_1} = 2\pi \sum_{s=\pm 1}\left[1 + \frac{k_{1\perp}^2 k_{1z}^2 c^4}{(k_{1z}^2 c^2 - \omega_{k_1 s}^2)^2}\right]^{-1}\{W_{k_1}^s \delta(\omega - \omega_{k_1, s}) + W_{-k_1}^s \delta(\omega + \omega_{k_1, s})\}. \tag{8.14}$$

For Langmuir fluctuations along H we have $\omega_{k_1} = \omega_{pe} + 3k_1^2 v_{Te}^2 / 2\omega_{pe}$. Confining attention to high-frequency oscillations, we can treat the ions in a linear manner; in particular

$$\left.\begin{array}{r}f_k^{T(1)i} \\ f_k^{R(1)i}\end{array}\right\} = \frac{e_i}{im_i(\omega - k_z v_z + i\delta)} \cdot \frac{\partial \Phi^i}{\partial v_z} \left\{\begin{array}{l}E_k^{T(1)} \\ E_k^R\end{array}\right. \tag{8.15}$$

Now consider the integral for collisions between electrons and turbulent fluctuations, which appears on the right side of (8.6). We will expand this integral in terms of the turbulence energy I_{k_1}, or what is the same thing, in terms of $E_{k_1}^{T(0)}$, assuming that $\langle E_{k_1}^{T(0)} \rangle = 0$. To begin with we will keep the terms linear in I_{k_1}. Then, in accord with our discussion above, we can obtain

$$-i\,(\omega - k_z\,v_z)\,f_k^{R\,(1)\,e} - \frac{e}{m_e}\,E_k^R\,\frac{\partial \Phi^e}{\partial v_z} = \frac{\partial}{\partial v_z}\,D_0\,\frac{\partial f_k^{R\,(1)\,e}}{\partial v_z} +$$

$$+ \frac{\partial}{\partial v_z}\,(E_k^R\,\hat{D} + D_1 + E_k^R\,D_2)\,\frac{\partial \Phi^e}{\partial v_z}\,; \qquad (8.16)$$

$$D_0 = i\,\frac{e^2}{m_e^2}\int I_{k_1}\,dk_1\,\frac{1}{(\omega + \omega_1 - (k_z + k_{1z})\,v_z + i\delta)}\,; \qquad (8.17)$$

$$\hat{D} = \frac{e^2}{m_e^2}\int I_{k_1}\,dk_1\,\frac{1}{(\omega + \omega_1 - (k_z + k_{1z})\,v_z + i\delta)}\,\frac{\partial}{\partial v_z}\cdot\frac{1}{(\omega_1 - k_{1z}\,v_z + i\delta)}\,; \qquad (8.18)$$

$$D_1 = i\,\frac{e^2\,\omega_{pe}^2}{m_e^2\,n_0}\int \frac{I_{k_1}\,dk_1\,(\omega - k_z\,v_z)}{\Pi\,(k + k_1)\,(k_z + k_{1z})\,(\omega_1 - k_{1z}\,v_z + i\delta)} \times$$

$$\times \frac{1}{(\omega_1 + \omega_1 - (k_z + k_{1z})\,v_z + i\delta)}\int dv_z'\,\frac{1}{(\omega + \omega_1 - (k_z + k_{1z})\,v_z' + i\delta)}\cdot\frac{\partial f_k^{R\,(1)\,e}}{\partial v_z'}\,; \qquad (8.19)$$

$$D_2 = -\frac{e^2\,\omega_{pe}^2}{m_e^2\,n_0}\int I_{k_1}\,dk_1\,\frac{(\omega - k_z\,v_z)}{\Pi\,(k + k_1)\,(k_z + k_{1z})} \times$$

$$\times \frac{1}{(\omega + \omega_1 - (k_z + k_{1z})\,v_z + i\delta)}\int dv_z'\,\frac{1}{(\omega + \omega_1 - (k_z + k_{1z})\,v_z' + i\delta)} \times$$

$$\times \frac{\partial}{\partial v_z'}\cdot\frac{1}{(\omega_1 - k_{1z}\,v_z' - i\delta)}\cdot\frac{\partial}{\partial v_z'}\,\Phi^e\,(v_z')\,. \qquad (8.20)$$

The various terms in the collision integral (8.16) have simple physical interpretations. The diffusion coefficient D_0 describes the change in quasilinear effects of relaxation of resonant particles connected with deviations in their distribution from the equilibrium distribution Φ. This is true for $\omega \ll \omega_1$, $k \ll k_1$. In the absence of resonant particles, which is the subject of our treatment when $\omega \ll \omega_1$, $k \ll k_1$, the two terms in (8.14) cancel each other almost completely, and

$$D_0 = -i\,\frac{4\pi e^2}{m_e^2}\,(\omega - k_z\,v_z)\sum_{s=\pm 1}\int \frac{dk_1\,W_{k_1}^s}{\omega_{k_1,\,s}^2}\,. \qquad (8.21)$$

By comparing (8.21) with the first term on the left side of (8.16), it is easy to see that (8.21) is smaller by a factor of $W/n_0 T_e \ll 1$ and can therefore be dropped. The term in (8.16) containing D, which describes changes in the effects of induced Compton scattering under the same conditions as before, is a factor of $W/n_0 T_e$ smaller than the second term on the left side of (8.16). The diffusion coefficient D_2 describes the nonlinear induced scattering and D_1 gives the decay interaction.† If, in addition to the stipulations $\omega \ll \omega_1$, $k \ll k_1$, $\omega_1/k_1 \gg v_{Te}$ we also have $\omega/k \ll \omega_1/k_1$, then D_2 will be small compared with D_1, which has the approximate form

$$D_1 = -i(\omega - k_z v_z) n_k^{(1)\, e} d_1,$$

$$n_k^{(1)\, e} = \int f_k^{R\,(1)\, e}\, dv_z, \tag{8.22}$$

$$d_1 = -\frac{\omega_{pe}^2}{n_0} \sum_{s=\pm 1} \frac{4\pi e^2}{m_e^2} \int \frac{dk_1}{(\omega - kv_{g,\, k_1}^s + i\delta)} \left(k \frac{\partial}{\partial k_1}\right) \frac{W_{k_1}^s}{\omega_{k_1,\, s}} \left(\frac{\partial}{\partial \omega_1} \omega_1^2 \Pi(k_1)\right)^{-2}_{\omega_1 = \omega_{k_1,\, s}},$$

$$\tag{8.23}$$

$s = \pm 1$ corresponds to the two signs in (8.13), and $v_{g,\, k_1}^s = \frac{\partial}{\partial k_1} \omega_{k_1,\, s}$ is the group velocity of the linear spectra in (8.13). Note the small factor $\Pi(k + k_1)$ in the denominator of (8.19). In obtaining (8.23) it was assumed that $\Pi(k_1) = 0$ for the linear spectra in (8.13), so that

$$\Pi(k_1 + k) \approx \left.\frac{\partial \Pi(k_1)}{\partial \omega_1}\right|_{\omega_1 = \omega_{k_1}} (\omega - kv_{g,\, k_1}).$$

We will first show how to obtain the decay instabilities. They arise if $\omega \gg \nu^T$, when the turbulent collisions can be treated using perturbation theory. In the equation

$$i(\omega - k_z v_z) f_k^{R\,(1)\, e} = -\frac{e}{m_e} E_k^R \frac{\partial \Phi^e}{\partial v_z} + i n_k^{(1)\, e} d_1 \frac{\partial}{\partial v_z}(\omega - k_z v_z)\frac{\partial \Phi^e}{\partial v_z} \tag{8.24}$$

one can neglect the term containing d_1 in the first approximation, find $f_k^{R\,(1)\, e}$ and then substitute it into $n_k^{(1)}$. Then when $k_z v_{Ti} \ll$

---

† More precisely, D_1 describes those turbulent collisions which are decay processes when $\omega \gg \nu^T$.

$\omega \ll k_z v_{Te}$, it is easy to find $\Pi(k) = \varepsilon(k) + \dfrac{k_\perp^2 c^2}{k_z^2 c^2 - \omega^2} = 0$, where

$$\varepsilon(k) = 1 + \frac{\omega_{pe}^2}{k_z^2 v_{Te}^2} - \frac{\omega_{pi}^2}{\omega^2} - \frac{\omega_{pe}^2}{k_z^2 v_{Te}^4} n_0 d_1, \tag{8.25}$$

which agrees with the nonlinear dielectric function describing the decay instabilities. Equation (8.24) has an exact solution[†]:

$$\varepsilon(k) = \varepsilon_0^{(i)}(k) + \frac{\varepsilon_0^{(e)}(k) - 1}{1 + k_z^2 n_0 \left(\varepsilon_0^{(e)}(k) - 1\right) d_1 / \omega_{pe}^2}. \tag{8.26}$$

As an example illustrating the radical change in the dispersion properties of the plasma at low frequencies, let us consider the one-dimensional Langmuir turbulence with $\omega_{k_1} = \omega_{pe} + {}^3/_2 \times k_{1z}^2 v_{Te}^2 / \omega_{pe}$, where all the fluctuations are directed along H. We have

$$d_1 = -\frac{\omega_{pe}}{4 n_0^2 m_e} \int_0^\infty dk_{1z} \, \frac{k_z \dfrac{\partial}{\partial k_{1z}} W_{k_{1z}}}{\omega - \dfrac{3 v_{Te}^2 k_{1z} k_z}{\omega_{pe}} + i\delta}. \tag{8.27}$$

If $\omega \gg k_z v_g$, then

$$d_1 \approx \frac{3 k_z^2 v_{Te}^2}{4 n_0^2 m_e \omega^2} W, \tag{8.28}$$

and when $k_z v_{Ti} \ll \omega \ll k_z v_{Te}$, $\varepsilon_0^{(i)}(k)$, $\varepsilon_0^{(e)}(k) \gg 1$, we have

$$\varepsilon(k) \approx -\frac{\omega_{pi}^2}{\omega^2} + \frac{\omega_{pe}^2}{k_z^2 v_{Te}^2} \cdot \frac{1}{\left(1 + \dfrac{3 W k_z^2}{4 n_0 m_e \omega^2}\right)}, \tag{8.29}$$

$$\omega^2 = k_z^2 v_\pm^2;$$

$$v_\pm^2 = \frac{1}{2} v_s^2 \pm \sqrt{\frac{1}{4} v_s^4 + \frac{3W}{4 n_0 m_e} v_s^2}, \tag{8.30}$$

$$v_s = v_{Te} \sqrt{\frac{m_e}{m_i}}.$$

[†] It is found by formally solving (8.24) relative to $f_k^{R\,(1)\,e}$ and then creating a linear equation for $n_k^{(1)\,e}$.

The solution with the minus sign is aperiodically unstable. This instability is similar to that instability found in [299] for cold isotropic plasmons.

If $\omega \ll k_z v_g$ for the entire turbulence spectrum (including the smallest v_g in the spectrum), then

$$d_1 = \frac{\omega_{pe}^2}{12 n_0^2 m_e v_{Te}^2} \int \frac{W_{k_{1z}} dk_{1z}}{k_{1z}^2},$$ (8.31)

and under the same conditions as stipulated for (8.29) we find

$$\omega^2 = k_z^2 v_s^2 \left(1 + \frac{\omega_{pe}^2}{12 n_0 T_e} \int \frac{W_{k_{1z}} dk_{1z}}{k_{1z}^2 v_{Te}^2}\right).$$ (8.32)

The solution given in (8.32) indicates the possibility that there exist sound vibrations in an isothermal plasma which are strongly damped because of Landau absorption by ions in the absence of turbulence. If $\frac{W}{n_0 T_e} \gg 12 \frac{v_{Te}^2}{v_p^2}$, where $v_p = \frac{\omega_{pe}}{k_{1z}}$ is the phase velocity of the fluctuations, then the speed of sound increases according to $\omega^2 = k^2 \tilde{v}_s^2$, where $\tilde{v}_s^2 \approx \frac{m_e}{12 m_i} v_p^2 W / n_0 T_e$, and the damping by ions becomes exponentially small, $\sim \exp\{-W v_p^2 / 12 v_{Te}^2 n_0 T_e\}$. The observed effect has some similarity with the possible existence of sound in a plasma situated in an intense high-frequency field [183]. The possibility that the isothermal sound of (8.32) will appear has a threshold connected with W. Since $\omega \ll k_z v_g$ we have $v_p^2 / v_{Te}^2 \ll 9 m_i / m_e$, so that $W / n_0 T_e \gg 4 m_e / 3 m_i$.

In a turbulent plasma nonlinear corrections to the plasmon propagator $\Pi(k)$ [the Green's function $\Pi^{-1}(k)$] appear. These corrections are proportional to the turbulence energy I_k in the first approximation and describe the electromagnetic "fur coat" of the plasmon. Unlike the usual renormalization, the dependence of $\Pi(k)$ on I_k is a real effect. After expanding all the quantities in (8.4) and (8.5) for the collisions of particles with turbulent fluctuations in terms of the turbulent field $E_k^{T(0)}$ we find

$$\Pi(k_1) I_{k_1} = \frac{e^2}{m_e^2} I_{k_1} \int \widetilde{\sum}_{k_1, k_2, k_1, -k_2} I_{k_2} dk_2 +$$

$$+ \frac{e^2}{2 m_e^2} \int dk_2 dk_3 \, \delta(k_1 - k_2 - k_3) I_{k_2} I_{k_3} |S_{k_1, k_2, k_3}|^2 \Pi^{-1}(-k_2 - k_3).$$ (8.33)

This equation is analogous to (2.113). Here we have

$$\widetilde{\Sigma}_{k,\,k_1,\,k_2,\,k_3} = \Sigma_{k,\,k_1,\,k_2,\,k_3} - S_{k,\,k_1,\,k-k_1} \times$$
$$\times \Pi^{-1}\,(k-k_1)\,S_{k-k_1,\,k_2,\,k_3}, \qquad (8.34)$$

$$S_{k,\,k_1,\,k_2} = \frac{\omega_{pe}^2}{n_0} \int dv_z \frac{d\Phi^e}{dv_z} (\omega - k_z\,v_z + i\delta)^{-1} \times$$
$$\times (\omega_1 - k_{1z}\,v_z + i\delta)^{-1}\,(\omega_2 - k_{2z}\,v_z + i\delta)^{-1},$$

$$\Sigma_{k,\,k_1,\,k_2,\,k_3} = \frac{\omega_{pe}^2}{n_0} \int \frac{dv_z}{(\omega - k_z\,v_z + i\delta)^2} \left[\frac{2k_z}{(\omega - k_z\,v_z + i\delta)} + \right.$$
$$\left. + \frac{k_{2z} + k_{3z}}{(\omega_2 + \omega_3 - (k_{2z} + k_{3z})\,v_z + i\delta)} \right] (\omega_2 - k_{2z}\,v_z + i\delta)^{-1} \times$$
$$\times (\omega_3 - k_{3z}\,v_z + i\delta)^{-1} \frac{\partial \Phi^e}{\partial v_z}.$$

For a nondecaying turbulence, such as the Langmuir turbulence, we have instead of $\Pi\,(k) = 0$

$$\widetilde{\Pi}\,(k_1)\,I_{k_1} = 0,$$

$$\widetilde{\Pi}\,(k_1) = \Pi\,(k_1) - \frac{e^2}{m_e^2} \int I_{k_2} \widetilde{\Sigma}_{k_1,\,k_2,\,k_1,\,-k_2}\,dk_2. \qquad (8.35)$$

Note that from (8.35) $\Pi\,(\omega_1,\,k_1) \approx (\omega_1 - \omega_{k_1}) \frac{\partial \Pi\,(k_1)}{\partial \omega_1}\Big|_{\omega_1 = \omega_{k_1}} \neq 0$, indicating that the factor $1/\Pi\,(k + k_1)$ has no resonant properties when $k \to 0$. We therefore conclude that these results are applicable when $\Pi\,(k_1) \ll \frac{\partial \Pi\,(k_1)}{\partial \omega_1}\Big|_{\omega_1 = \omega_{k_1}} |\omega - kv_{gk_1}|$, or more precisely, if the compensation of the positive frequency and negative frequency parts in (8.23) is taken into account (the small factor k/k_1),

$$\frac{k}{k_1} \max(\omega,\,kv_g) \gg |\omega_1 - \omega_{k_1}|. \qquad (8.36)$$

This same criterion can be obtained if we consider the terms of next higher order in the turbulence energy $(\sim I_k^2)$ in the collision integrals for turbulent fluctuations [308].

Applying the criterion of (8.36) to the turbulent sound oscillations of (8.32), we find

$$\frac{W}{n_0 T_e} \cdot \frac{m_e}{54 m_i} \cdot \frac{v_p^4}{v_{Te}^4} \ll \frac{k^2}{k_1^2} \ll 1.$$ (8.37)

This corresponds to neglecting the nonlinear dispersion and the condition $\omega \ll k v_g$, for which such sound oscillations are possible. Thus, the turbulent sound vibrations of (8.32) can exist in a relatively narrow range of parameters, such as for radiative turbulence (see §6.3).

According to (8.36) the conditions under which the turbulence of (8.30) will appear are very rigid:

$$\frac{12 m_i}{m_e} \ll \frac{v_p^2}{v_{Te}^2} \ll \frac{9 T_e m_i}{T_i m_e}, \qquad T_e \gg T_i, \qquad \frac{T_i m_e}{T_e m_i} \ll \frac{3W}{4 n_0 T_e} \ll \frac{m_e}{m_i}.$$ (8.38)

We note that the limitations imposed by (8.36) are less strict for turbulent oscillations whose frequency differences are greater than the Langmuir frequency. This occurs for nonpotential fluctuations of transverse plasmons.

8.3. The Particle — Turbulent-Fluctuation Collision Integral: Summing the Series

The renormalized plasmon group velocity which includes nonlinear correction to the frequency is a physical quantity; therefore, the collision integral for plasma particles and turbulent fluctuations must contain the complete plasmon Green's function $1/\Pi (k_1 + k)$ and not $1/\Pi (k_1 + k)$ as in (8.19). It is easy to see that when $k \to 0$, $\Pi (k + k_1)$ is of order I_k, so that all terms of this order must be included. In this section we will expand the collision integral in terms of the small parameter $W/n_0 T_e \ll 1$, and take into account just the first term in the expansion in this quantity. However, it will be assumed that $\Pi (k_1 + k)$ is of order I_k to a first approximation.

Let us now set up an equation for the kernel of the particle—turbulent-fluctuation collision integral. We will show that such an equation can be obtained in a natural way with the assumptions of a weak correlation of the fields $E_k^{T\,(0)}$ among themselves and with the perturbation field $E_k^{T\,(1)}$ (i.e., the assumptions usually employed in weak perturbation theory). It is convenient to write Eq. (8.6) in a somewhat different form. By introducing the quantities

$$
\begin{aligned}
\widetilde{f}_k^{T\,(1)\,\alpha} &= f_k^{T\,(1)\,\alpha} - \frac{e_\alpha E_k^{T\,(1)} \dfrac{\partial \Phi^\alpha}{\partial v_z}}{m_\alpha\, i\,(\omega - k_z v_z + i\delta)}, \\[2mm]
\widetilde{f}_k^{T\,(0)\,\alpha} &= f_k^{T\,(0)\,\alpha} - \frac{e_\alpha E_k^{T\,(0)}\, \partial \Phi^\alpha/\partial v_z}{m_\alpha\, i\,(\omega - k_z v_z + i\delta)},
\end{aligned}
\tag{8.39}
$$

we have

$$
- i\,(\omega - k_z v_z)\, f_k^{R\,(1)\,\alpha} + \frac{e_\alpha}{m_\alpha} E_k^R \frac{\partial \Phi^\alpha}{\partial v_z} - \frac{\partial}{\partial v_z} \langle D^* \rangle \frac{\partial \Phi^\alpha}{\partial v_z} =
$$

$$
= - \frac{\partial}{\partial v_z} \frac{e_\alpha}{m_\alpha} \int dk_1\, dk_2\, \delta\,(k - k_1 - k_2) \langle E_{k_1}^{T\,(1)} \widetilde{f}_{k_2}^{T\,(0)\,\alpha} + E_{k_1}^{T\,(0)} \widetilde{f}_{k_2}^{T\,(1)\,\alpha} \rangle;
\tag{8.40}
$$

$$
\Pi\,(k)\, E_k^{T\,(1)} + \frac{4\pi i}{k_z} \sum_\alpha e_\alpha \int \widetilde{f}_k^{T\,(1)\,\alpha}\, dv_z = 0;
\tag{8.41}
$$

$$
- i\,(\omega - k_z v_z)\, \widetilde{f}_k^{T\,(1)\,\alpha} + \frac{e_\alpha}{m_\alpha} \int dk_1\, dk_2\, \delta\,(k - k_1 - k_2) \times
$$

$$
\times\, E_{k_1}^{T\,(0)} \frac{\partial}{\partial v_z} f_{k_2}^{R\,(1)\,\alpha} - \frac{\partial}{\partial v_z} (D^* - \langle D^* \rangle) \frac{\partial \Phi^\alpha}{\partial v_z} + \frac{e_\alpha}{m_\alpha} \cdot \frac{\partial}{\partial v_z} \times
$$

$$
\times \int dk_1\, dk_2\, \delta\,(k - k_1 - k_2)\, \{ E_{k_1}^R f_{k_2}^{T\,(0)\,\alpha} + E_{k_1}^{T\,(1)} \widetilde{f}_{k_2}^{T\,(0)\,\alpha} +
$$

$$
+ E_{k_1}^{T\,(0)} \widetilde{f}_{k_2}^{T\,(1)\,\alpha} - \langle E_{k_1}^{T\,(1)} \widetilde{f}_{k_2}^{T\,(0)\,\alpha} + E_{k_1}^{T\,(0)} \widetilde{f}_{k_2}^{T\,(1)\,\alpha} \} = 0;
\tag{8.42}
$$

$$
- i\,(\omega - k_z v_z)\, \widetilde{f}_k^{T\,(0)\,\alpha} = - \frac{e_\alpha \partial}{m_\alpha \partial v_z} \int dk_1\, dk_2\, \delta\,(k - k_1 - k_2) \times
$$

$$
\times\, (E_{k_1}^{T\,(0)} f_{k_2}^{T\,(0)\,\alpha} - \langle E_{k_1}^{T\,(0)} f_{k_2}^{T\,(0)\,\alpha} \rangle).
\tag{8.42a}
$$

Here

$$
D^* = i\,(\omega - k_z v_z) \frac{e_\alpha^2}{m_\alpha^2} \int \frac{dk_1\, dk_2\, \delta\,(k - k_1 - k_2)\, E_{k_1}^{T\,(0)}\, E_{k_2}^{T\,(1)}}{(\omega_1 - k_{1z} v_z + i\delta)\,(\omega_2 - k_{2z} v_z + i\delta)}.
\tag{8.43}
$$

It is clear from (8.40) that for small frequencies ($k \to 0$) the collision integral containing D^* is much larger than the rest of them, as compared with the right side of (8.40). In fact, when $k \to 0$, $k_1 \to -k_1$ and since $E_{k_1}^{T(0)}$ is nearly a linear field, we find that $E_{k_2}^{T(1)}$, which contains $1/\Pi(k_2)$ in view of (8.41), is a large quantity. The factor $\widetilde{f}_k^{T(1)}$ on the right side of (8.40) does not contain this large factor, and according to (8.42a) $\widetilde{f}_k^{T(0)a}$ is nonlinear with respect to $E_k^{T(0)}$, i.e., $\langle E_{k_1}^{T(1)} \widetilde{f}_{k_2}^{T(0)} \rangle$ is proportional to a higher power of the turbulence energy.

Therefore, we will begin our calculation by obtaining an equation for D^*. The right side of (8.40) is not used below, but we will show that the same methods used to obtain $\langle D^* \rangle$ can be used to compute these other integrals. We can construct expressions for $\langle E_{k_1}^{T(0)} E_{k_2}^{T(1)} \rangle$ from (8.41) and (8.42):

$$\Pi(k_2)\langle E_{k_1}^{T(0)} E_{k_2}^{T(1)}\rangle + i\frac{e}{m_e}\int dk_1' \, dk_2' \, \delta(k_2 - k_1' - k_2') \times$$

$$\times S_{k_2, \, k_1', \, k_2'} \left\langle E_{k_1}^{T(0)} E_{k_1'}^{T(0)} E_{k_2'}^{T(1)}\right\rangle + \frac{\omega_{pe}^2}{n_0 k_{2z}}\int \frac{dv_z \, dk_1' \, dk_2'}{(\omega_2 - k_{2z}v_z)} \times$$

$$\times \delta(k_2 - k_1' - k_2')\frac{\partial}{\partial v_z}\left(\left\langle f_{k_1'}^{T(0)\, e} E_{k_1}^{T(0)} E_{k_2'}^{T(1)}\right\rangle + \right.$$

$$\left. + \left\langle E_{k_1}^{T(0)} E_{k_1'}^{T(0)} f_{k_2'}^{T(1)\, e}\right\rangle\right) = -\frac{\omega_{pe}^2}{n_0}\int I_{k_1} \frac{\delta(k_1 + k_2 - k')\, dk'}{k_{2z}(\omega_2 - k_{2z}v_z + i\delta)} \times$$

$$\times \frac{\partial f_{k'}^{R(1)\, e}}{\partial v_z}\, dv_z - \frac{\omega_{pe}^2}{n_0 k_{2z}}\int \frac{dv_z \, dk_1' \, dk_2'}{(\omega_2 - k_{2z}v_z + i\delta)}\cdot\frac{\partial}{\partial v_z} \times$$

$$\times \left\langle E_{k_1}^{T(0)} f_{k_2}^{T(0)}\right\rangle E_{k_1}^{R}\cdot \delta(k_2 - k_1' - k_2') \equiv G. \tag{8.44}$$

The right side of (8.44) does not contain $E_k^{T(1)}$ and can be calculated in the standard manner by using (8.42a) as the sum over the powers of I_{k_1}. In the first approximation it is

$$G = -\frac{\omega_{pe}^2 I_{k_1}}{n_0 k_{2z}}\int \frac{dv_z \, dk'\, \delta(k_1 + k_2 - k')}{(\omega_2 - k_{2z}v_z + i\delta)}\cdot\times$$

$$\times \left(\frac{\partial f_{k'}^{R(1)\, e}}{\partial v_z} - \frac{e}{m_e}\, iE_{k'}^{R}\cdot\frac{\partial}{\partial v_z}\frac{1}{(\omega_1 - k_{1z}v_z + i\delta)}\cdot\frac{\partial\Phi^e}{\partial v_z}\right). \tag{8.45}$$

In order to transform the average $\left\langle E_{k_1}^{T\,(0)} E_{k_1'}^{T\,(0)} E_{k_2'}^{T\,(1)} \right\rangle$ into an average of four E_k^T we can use the assumption of weak correlation between the fields. To express $E_{k_1}^{T\,(0)}$ in terms of quadratic combinations of fields we must use

$$\Pi(k_1)\, E_{k_1}^{T\,(0)} = \frac{ie}{2m_e} \int dk_2\, dk_3\, \delta(k_1 - k_2 - k_3)\, S_{k_1,\,k_2,\,k_3} \times$$

$$\times \left(E_{k_2}^{T\,(0)} E_{k_3}^{T\,(0)} - \left\langle E_{k_2}^{T\,(0)} E_{k_3}^{T\,(0)} \right\rangle \right), \qquad (8.46)$$

while for $E_{k_1}^{T\,(1)}$ we use the relationship obtained from (8.41) and (8.42) if we are limited to linear and quadratic terms in $E_k^{T\,(1)}$ and $E_k^{T\,(0)}$. The linear terms in this relationship contain just $E_k^{T\,(0)}$, and in $\left\langle E_k^{T\,(0)} E_{k_1}^{T\,(0)} E_{k_2}^{T\,(1)} \right\rangle$ they give terms $\sim I_k^2$ which must be related to the right side of (8.44) (which does not depend on $E_k^{T\,(1)}$); they can be dropped in the approximation of (8.45). Therefore $E_k^{T\,(1)}$ can be refined using the relationship

$$-\Pi(k)\frac{e}{m_e}\, E_k^{T\,(1)} = \frac{\omega_{pe}^2}{n_0\, k_z} \int \frac{dv_z}{(\omega - k_z v_z)} \cdot \frac{\partial}{\partial v_z}\, (D^* - \langle D^* \rangle) \frac{\partial \Phi^e}{\partial v_z}. \qquad (8.47)$$

Similar considerations show that the calculation of the average $\left\langle \tilde{f}_{k_2}^{T\,(1)\,e} E_{k_1}^{T\,(0)} E_{k_1'}^{T\,(0)} \right\rangle$ depends on $\tilde{f}_k^{T\,(1)\,e}$ as obtained from the equation

$$-i(\omega - k_z v_z)\, \tilde{f}_k^{T\,(1)\,e} = \frac{\partial}{\partial v_z}\,(D^* - \langle D^* \rangle) \frac{\partial \Phi^e}{\partial v_z}. \qquad (8.48)$$

Finally, it is sufficient to use the first approximation for $f_k^{T\,(0)}$ of (8.42) in computing $\left\langle \tilde{f}_k^{T\,(0)} E_{k_1}^{T\,(0)} E_{k_1'}^{T\,(1)} \right\rangle$. As a result of these calculations, the left side of (8.44) is transformed into a form which contains only averages of four turbulent fields, which can then be divided into all possible products of the averages of two fields. We obtain

$$G = \tilde{\Pi}(k_2) \left\langle E_{k_1}^{T\,(0)} E_{k_2}^{T\,(1)} \right\rangle - \frac{e^2}{m_2^2}\, I_{k_1} \int dk_2'\, dk_2'' \times$$

$$\times \delta(k_1 + k_2 - k_2' - k_2'')\, \tilde{\sum}_{k_2,\,k_2',\,-k_1,\,k_2''} \left(\left\langle E_{k_2'}^{T\,(0)} E_{k_2''}^{T\,(1)} + E_{k_2''}^{T\,(0)} E_{k_2'}^{T\,(1)} \right\rangle \right) +$$

$$+ \frac{e^2}{m_e^2} \int dk_2' \, dk_2'' \, \delta \left(k_1 + k_2 - k_2' - k_2''\right) S_{k_2, \, k_2'' - k_1, \, k_2'} \times$$

$$\times I_{k_2'' - k_1'} \frac{1}{\Pi(k_1)} S_{k_1, \, -k_2'' + k_1, \, k_2''} \left\langle E_{k_2''}^{T\,(0)} E_{k_2'}^{T\,(1)} \right\rangle . \tag{8.49}$$

Note that the operator $\widetilde{\Pi}(k_2)$ automatically appears in (8.49), together with the additional terms $\sim I_k$, which are of the same order of magnitude as either $\Pi(k_2)$ or $\widetilde{\Pi}(k_2)$. For decay turbulence the last term in (8.49) is zero, and the equation sought for takes the form

$$\widetilde{\Pi}(k - k_1) \langle E_{k_1}^{T\,(0)} E_{k-k_1}^{T\,(1)} \rangle = - \frac{\omega_{pe}^2 I_{k_1}}{n_0(k_z - k_{1z})} \int dv_z \frac{1}{\omega - \omega_1 - (k_z - k_{1z})v_z + i\delta} \times$$

$$\times \left(\frac{\partial f_k^{R\,(1)e}}{\partial v_z} + i \frac{e}{m_e} E_k^R \frac{\partial}{\partial v_z} \frac{1}{(\omega_1 - k_{1z}v_z + i\delta)} \cdot \frac{\partial \Phi^e}{\partial v_z} \right) +$$

$$+ \frac{e^2}{m_e^2} I_{k_1} \int dk_1' \left\langle E_{k_1'}^{T\,(0)} E_{k-k_1}^{T\,(1)} \right\rangle \left[\widetilde{\Sigma}_{k-k_1, \, k_1', \, -k_1, \, k-k_1'} + \right.$$

$$\left. + \widetilde{\Sigma}_{k-k_1, \, k-k_1', \, -k_1, \, k_1'} \right]. \tag{8.50}$$

If the integral equation (8.50) is solved, it is easy to find the collision integral which contains D^*. Finally, the collision integral on the right side of (8.40), containing $\langle E_{k_1}^{T(1)} \widetilde{f}_{k_1}^{T(0)\alpha} \rangle$, is of order $W/n_0 T_e$ relative to the term with D^*.

We will solve this integral equation in the limit $|\omega_1 - \omega_1'| \ll |k_{1z} - k_{1z}'| v_{Ti}$, $\omega_1 \ll k_{1z} v_{Ti}$, which, for the example considered above, corresponds to the one-dimensional Langmuir turbulence with $v_p \gg 3 v_{Te}^2 / v_{Ti}$; that is, it corresponds to the limit in which there is no region wherein the results obtained by expanding the collision integral in terms of I_k can be applied. From (8.34) we have

$$\widetilde{\Sigma}_{k-k_1, \, k_1', \, -k_1, \, k-k_1'} \simeq - \frac{\omega_{pe}^2 T_e}{(T_e + T_i) v_{Te}^2 \, \omega_1 \omega_1' (\omega_1 - \omega)(\omega_1' - \omega)} . \tag{8.51}$$

This result holds if the signs of ω_1 and ω_1' are opposite; when the signs of ω_1 and ω_1' are the same, (8.51) is zero. The quantity $\widetilde{\Sigma}_{k-k_1, \, k-k_1', \, -k_1, \, k_1'}$ is not zero when the signs of ω_1 and ω_1' are the same; it is then equal to (8.51). We divide (8.50) by $\widetilde{\Pi}(k - k_1)$ and set up an equation for

$$S_\pm(k) = \int \frac{\langle E_{k_1}^{T\,(0)} E_{k-k_1}^{T\,(1)} \rangle}{\omega_1(\omega_1 - \omega)} \, dk_{1, \pm}, \tag{8.52}$$

where in S_+ the integration goes over a region of positive frequencies, and in S_- it is over negative frequencies. We obtain a system of linear algebraic equations for $S_{\pm}$, the solution to which is

$$S_+(k)+S_-(k)=-\left[1+\frac{\omega_{pe}^4}{4\pi n_0(T_e+T_i)}\int\frac{I_{k_1}dk_1}{\omega_1^2(\omega_1+\omega)^2\tilde{\Pi}(k_1+k)}\right]^{-1}\times$$

$$\times\frac{\omega_{pe}^2}{n_0}\int\frac{I_{k_1}dk_1}{\tilde{\Pi}(k_1+k)(k_{1z}+k_z)}\int dv_z\frac{1}{(\omega_1+\omega-(k_{1z}+k_z)v_z+i\delta)}\times$$

$$\times\left[\frac{\partial f_k^{R(1)e}}{\partial v_z}+i\frac{e}{m_e}E_k^R\frac{\partial}{\partial v_z}\frac{1}{(\omega_1-k_{1z}v_z+i\delta)}\cdot\frac{\partial\Phi^e}{\partial v_z}\right].\qquad(8.53)$$

Here the integration runs over all frequencies. To within the required accuracy we have

$$\langle D^*\rangle=-i(\omega-k_z v_z)\frac{e^2}{m_e^2}(S_+(k)+S_-(k)),\qquad(8.54)$$

which enables us to find the coefficients D_1 and D_2 in (8.16). For example,

$$D_1=-i\frac{\omega_{pe}^2}{n_0}(\omega-k_z v_z)n_k^{(1)}\frac{e^2}{m_e^2}\int I_{k_1}dk_1\frac{1}{\tilde{\Pi}(k_1+k)\omega_1(\omega_1+\omega)^3}\times$$

$$\times\left[1+\frac{\omega_{pe}^4}{4\pi n_0(T_e+T_i)}\int\frac{I_{k_1}dk_1}{\omega_1^2(\omega_1+\omega)^2\tilde{\Pi}(k_1+k)}\right]^{-1}.\qquad(8.55)$$

This expression differs from that obtained above using the expansion in terms of I_k in that the denominator contains an expression which differs from unity, and in the numerator the plasmon Green's function $1/\Pi(k)$ is replaced with $\tilde{\Pi}^{-1}(k)$. Thus, this result leads to the renormalization of the plasmon propagator and the renormalization of the electron's effective charge. A denominator similar to that in (8.55) appears in the equation for D_2, and the inequality $D_2/D_1\ll 1$ obtained above for $\omega/k\ll\omega_1/k_1$, $\omega\ll\omega_1$, $k\ll k_1$ is preserved in this case too. We can find the dielectric permeability from (8.55) and (8.16):

$$\varepsilon(k)=\varepsilon_0^{(i)}(k)+\left(\varepsilon_0^{(e)}(k)-1\right)\left(1+\frac{n_0 m_e}{T_e+T_i}d_2\right)\times$$

$$\times\left(1+\frac{(\varepsilon_0^{(e)}(k)-1)n_0 k_z^2}{\omega_{pe}^2}d_1+\frac{n_0 m_e}{T_e+T_i}d_2\right)^{-1};\qquad(8.56)$$

$$d_1 = \frac{e^2 \omega_{pe}^2}{m_e^2 n_0} \int \frac{I_{k_1} d k_1}{\widetilde{\Pi} (k + k_1) \omega_1 (\omega_1 + \omega)^3} \; ;$$

$$d_2 = \frac{e^2 \omega_{pe}^2}{m_e^2 n_0} \int \frac{I_{k_1} dk_1}{\widetilde{\Pi} (k_1 + k) \omega_1^2 (\omega_1 + \omega)^2} \, .$$

(8.57)

Consider as an example the one-dimensional Langmuir turbulence. In the region where $v_p \gg 3v_{Te}^2/v_{Ti}$ the plasmon group velocities are not changed. Therefore, Π can be replaced with $\widetilde{\Pi}$ in (8.57). Furthermore, in the limit $\omega \ll \omega_1$, $k \ll k_1$, $\omega/k \ll \omega_1/k_1$ we have $d_2 \approx d_1$ and both are equal to (8.23). In the limit $\omega \gg k_z v_g$, d_1 coincides with (8.28). If in addition $k_z v_{Ti} \ll \omega \ll k_z v_{Te}$ then (8.56) gives

$$\varepsilon (k) = - \frac{\omega_{pi}^2}{\omega^2} + \frac{\omega_{pe}^2}{k_z^2 v_{Te}^2} \left(1 + \frac{3k_z^2 WT_e}{4\omega^2 n_0 m_e (T_e + T_i)} \right) \left(1 + \frac{3k_z^2 W (2T_e + T_i)}{4\omega^2 n_0 m_e (T_e + T_i)} \right)^{-1} ;$$

(8.58)

$$\omega^2 = k_z^2 \widetilde{v}_{\pm}^2 ; \qquad \widetilde{v}_{\pm}^2 = \frac{1}{2} \left(v_s^2 - v_{\sim}^2 \frac{T_e}{T_e + T_i} \right) \pm$$

$$\pm \sqrt{\frac{1}{4} \left(v_s^2 - v_{\sim}^2 \frac{T_e}{T_e + T_i} \right)^2 + v_{\sim}^2 \, v_s^2 \frac{2T_e + T_i}{T_e + T_i}} \, .$$

(8.59)

Here $v_{\sim}^2 = 3W/4n_0 m_e$. The instability in (8.59) is qualitatively different from that in (8.30). When $v_{\sim}^2 \gg v_s^2$ the square of the growth rate for (8.59) is proportional to W. Finally, when $\omega \ll k_z v_{Ti}$, $\omega \ll k_z v_g$ we obtain

$$\omega^2 = - 2k_z^2 v_{\sim}^2 \frac{T_e + T_i}{T_e} \, .$$

(8.60)

8.4. Dielectric Permeability for an Isotropic Turbulent Plasma

With H = 0 we will assume that a stationary turbulence with the Langmuir frequency has been established in the plasma:

$$\omega_k^l = \omega_{pe} + \frac{3k^2 v_{Te}^2}{2\omega_{pe}} \, .$$

(8.61)

The equation for the perturbed distribution function $f_k^{R\,(1)}$ which replaces (8.16) takes the form

$$-i\,(\omega-\mathbf{k}\mathbf{v})f_k^{R\,(1)}+\mathbf{F}_k^R\,\frac{\partial\Phi}{\partial\mathbf{v}}=\frac{\partial}{\partial v_i}\,D_{ij}^{(0)}\,\frac{\partial}{\partial v_j}\,f_k^{R\,(1)}+$$

$$+F_{k,\,i}^R\,\frac{\partial}{\partial v_i}\,\hat{D}_{ijl}\,\frac{\partial}{\partial v_l}\,\Phi+\frac{\partial}{\partial v_i}\,(D_{ij,\,1}+D_{ij,\,2})\,\frac{\partial}{\partial v_j}\,\Phi, \qquad (8.62)$$

$$\mathbf{F}_k^R=\frac{-e}{m_e}\,\big(\mathbf{E}_k^R+\frac{1}{c}\,[\mathbf{v}\mathbf{H}_k^R]\big)=\frac{-e}{m_e}\,\Big(\mathbf{E}_k^R\big(1-\frac{\mathbf{k}\mathbf{v}}{\omega}\big)+\frac{\mathbf{k}}{\omega}\,(\mathbf{v}\mathbf{E}_k^R)\Big). \qquad (8.63)$$

Here the expressions for the coefficients D_{ij} take the form

$$D_{ij,\,1}=i\,\frac{\omega_{pe}^2}{n_0}\int\frac{I_{k_1}\,dk_1}{\varepsilon_{k_1-k}}\left[\frac{-k_{1i}\,(k_j-k_{1j})}{k_1^2\,(\mathbf{k}_1-\mathbf{k})^2}\cdot\frac{1}{(\omega_1-\omega-(\mathbf{k}_1-\mathbf{k})\,\mathbf{v}-i\delta)}+\right.$$

$$\left.+\frac{k_{1j}\,(k_i-k_{1i})}{k_1^2\,(\mathbf{k}_1-\mathbf{k})^2}\cdot\frac{1}{(\omega_1-\mathbf{k}_1\mathbf{v}+i\delta)}\right]\int\frac{d\mathbf{v}'}{(\omega_1-\omega-(\mathbf{k}_1-\mathbf{k})\,\mathbf{v}'-i\delta)}\times$$

$$\times\Big(\mathbf{k}\,\frac{\partial f_k^{R\,(1)}\,(p')}{\partial\mathbf{v}'}\Big)\frac{e^2}{m_e^2}, \qquad n=\int\Phi d\mathbf{v}, \qquad (8.64)$$

$$D_{ij,\,2}=-\frac{\omega_{pe}^2}{n_0}\int\frac{I_{k_1}\,dk_1}{\varepsilon_{k_1-k}}\left[\frac{k_{1i}\,(k_{1j}-k_j)}{\omega_1-\omega-(\mathbf{k}_1-\mathbf{k})\mathbf{v}-i\delta}-\frac{k_{1j}\,(k_{1i}-k_i)}{\omega_1-\mathbf{k}_1\mathbf{v}+i\delta}\right]\times$$

$$\times\frac{1}{k_1^2\,(\mathbf{k}_1-\mathbf{k})^2}\int\frac{d\mathbf{v}'}{(\omega_1-\omega-(\mathbf{k}_1-\mathbf{k})\,\mathbf{v}'-i\delta)}\,\Big(\mathbf{F}_k^R\,\frac{\partial}{\partial\mathbf{v}'}\Big)\times$$

$$\times\frac{1}{\omega_1-\mathbf{k}_1\mathbf{v}'-i\delta}\,\Big(\mathbf{k}_1\,\frac{\partial}{\partial\mathbf{v}'}\Big)\,\Phi\,(v')\frac{e^2}{m_e^2}, \qquad (8.65)$$

$$D_{ij}^{(0)}=-\int\frac{I_{k_1}\,dk_1\,k_{1i}\,(\omega-\mathbf{k}\mathbf{v})\,k_{1j}\,e^2/m_e^2}{k_1^2\,(\omega_1-\omega-(\mathbf{k}-\mathbf{k}_1)\,\mathbf{v}-i\delta)\,(\omega_1+\omega-(\mathbf{k}_1+\mathbf{k})\,\mathbf{v}+i\delta)}, \qquad (8.66)$$

$$\hat{D}_{ijl}=-\int\frac{I_{k_1}\,dk_1\,k_{1i}\,k_{1l}\,e^2/m_e^2}{k_1^2\,(\omega_1-\omega-(\mathbf{k}_1-\mathbf{k})\,\mathbf{v}-i\delta)}\cdot\frac{\partial}{\partial v_j}\cdot\frac{1}{(\omega_1-\mathbf{k}_1\mathbf{v}-i\delta)}, \qquad (8.67)$$

$$\langle E_{k_1}^T\,E_{k_2}^{T\,(0)}\rangle=-I_{k_1}\,\delta\,(k_1+k_2), \qquad \mathbf{E}_k^T=\frac{\mathbf{k}}{k}\,E_k^T. \qquad (8.68)$$

The approximate expressions for the diffusion coefficients $D_{ij,1}$, $D_{ij,2}$ in the regions where $\omega \ll \omega_{k_1}$ and $k \ll k_1$ take the form

$$D_{ij,1} = i(\omega - kv)\frac{n_k^{(1)}}{n_0} \cdot \frac{\pi e^2}{m_e^2}\int dk_1 \frac{k_{1i}k_{1j}}{k_1^2}\frac{1}{\omega - kv_{g,k_1} + i\delta}\left(k\frac{\partial}{\partial k_1}\right)\frac{W_{k_1}}{\omega_{pe}},$$
(8.69)

$$D_{ij,2} = \frac{6\pi e^3}{m_e^3 \omega_{pe}^3}(\omega - kv)\int dk_1 \frac{k_{1i}k_{1j}(k_1 E_k^R)W_{k_1}}{k_1^2(\omega - kv_{gk_1} + i\delta)} +$$

$$+ \frac{\pi e^3}{m_e^3 \omega\omega_{pe}}\int dk_1 \frac{k_{1i}k_{1j}}{k_1^4}k^2(\omega - kv)\frac{(k_1 E_k^{Rt})\left(k_1\frac{\partial}{\partial k_1}\right)\frac{W_{k_1}}{\omega_{pe}}}{\omega - kv_{g,k_1} + i\delta}, \quad (8.70)$$

where

$$E_k^{Rt} = E_k^R - \frac{k(kE_k^R)}{k^2}.$$

We will introduce the following notation:

$$D_{ij,1} = -i(\omega - kv)d_{ij}n_k^{(1)}, \tag{8.71}$$

$$D_{ij,2} = (\omega - kv)\frac{e}{m_e}d_{ijl}E_{k,l}^R. \tag{8.72}$$

We will only take into account $D_{ij,1}$ and $D_{ij,2}$ since the remaining coefficients make only negligible contributions when $W/nT \ll 1$; then

$$f_k^{R(1)} = \frac{1}{i(\omega - kv)}\left(F_k^R\frac{\partial}{\partial v}\right)\Phi + \frac{1}{i(\omega - kv)}\frac{\partial}{\partial v_l}(\omega - kv)\times$$

$$\times\frac{\partial\Phi}{\partial v_j}\left(i\,d_{ij}n_k^{(1)} - \frac{e}{m_e}d_{ijl}E_{k,l}^R\right). \tag{8.73}$$

From this we find $n_k^{(1)}$:

$$n_k^{(1)} = \int f_k^{R(1)}dv = \int\frac{dv}{i(\omega - kv + i\delta)}\left\{\left(F_k^R\frac{\partial\Phi}{\partial v}\right) - \frac{e}{m_e}E_{k,l}^R d_{ijl}\frac{\partial}{\partial v_l}\times\right.$$

$$\left.\times(\omega - kv)\frac{\partial\Phi}{\partial v_j}\right\}\left\{1 - d_{ij}\int\frac{dv}{(\omega - kv + i\delta)}\frac{\partial}{\partial v_i}(\omega - kv)\frac{\partial\Phi}{\partial v_j}\right\}^{-1}. \tag{8.74}$$

Then it is easy to find the current from (8.74):

$$j_{k,i}^{(1),R} = -\int e v_i f_k^{R(1)} \, dv = \frac{(\varepsilon_{ij}^{(e)} - \delta_{ij})\omega}{4\pi i} E_{k,j}^R,$$

$$F_{kl}^{(R)} = \left[\delta_{ls}\left(1 - \frac{kv}{\omega}\right) + \frac{k_l v_s}{\omega} \right] E_{k,s}^R. \tag{8.75}$$

Thus, the expression for the dielectric permeability of the plasma is

$$\varepsilon_{ij}^{(e)} = \delta_{ij} + \frac{4\pi e^2}{\omega m_e} \int \frac{v_i \left[\delta_{lj}\left(1 - \frac{kv}{\omega}\right) + \frac{k_l v_j}{\omega} \right] \frac{\partial \Phi}{\partial v_l} \, dv}{\omega - kv + i\delta} + $$

$$+ \frac{4\pi e^2}{m_e \omega} \int \frac{v_i \, dv}{\omega - kv + i\delta} \cdot \frac{\partial}{\partial v_l}(\omega - kv)\frac{\partial \Phi}{\partial v_s} \left\{ d_{ls} \int \frac{dv'}{\omega - kv' + i\delta} \times \right.$$

$$\times \left[\left(\delta_{mj}\left(1 - \frac{kv'}{\omega}\right) + \frac{k_m v_j'}{\omega}\right)\frac{\partial \Phi(v')}{\partial v_m'} + d_{mnj}\frac{\partial}{\partial v_m'}(\omega - kv')\times \right.$$

$$\times \frac{\partial \Phi(v')}{\partial v_n'} \right] \left[1 - d_{rp}\int \frac{dv''}{\omega - kv'' + i\delta}\cdot\frac{\partial}{\partial v_r''}(\omega - kv'')\frac{\partial \Phi(v'')}{\partial v_p''} \right]^{-1} + d_{lsj} \right\}. \tag{8.76}$$

An especially simple expression is obtained for the longitudinal dielectric function $\varepsilon^l = (k_i k_j / k_2)\varepsilon_{ij}$, which defines the density fluctuations of the plasma:

$$\varepsilon^{l(e)} = 1 + \frac{4\pi e^2}{m_e k^2} \frac{\int \frac{dv}{\omega - kv + i\delta}(k_j - k_i k_l d_{ijl})\frac{\partial \Phi}{\partial v_j}}{1 + \int \frac{dv}{\omega - kv + i\delta}(k_i d_{ij})\frac{\partial \Phi}{\partial v_j}}. \tag{8.77}$$

This dielectric permeability describes the potential oscillations of the turbulent plasma.

When the particle and turbulent fluctuation distribution functions are isotropic, one can speak about transverse perturbations in a turbulent plasma. The dielectric permeability $\varepsilon^{t(e)} = \dfrac{\mathrm{Tr}\,\varepsilon_{ij}^{(e)} - \varepsilon^{l(e)}}{2}$ is given by

$$\varepsilon^{t(e)} = 1 + \frac{2\pi e^2}{m_e \omega}\int \frac{dv}{\omega - kv + i\delta}\left\{ \left(v\frac{\partial \Phi}{\partial v}\right)\left(1 - \frac{kv}{\omega}\right) + \frac{v^2}{\omega}\left(k\frac{\partial \Phi}{\partial v}\right) - \right.$$

$$- k_m v_j d_{msj}\frac{\partial \Phi}{\partial v_s} \right\} - \frac{2\pi e^2}{m_e}\left[1 + \int \frac{dv' k_i d_{ij}}{\omega - kv' + i\delta}\cdot\frac{\partial \Phi(v')}{\partial v_j'} \right]^{-1} \times$$

$$\times \left\{ \frac{1}{k^2} \int \frac{dv \frac{\partial \Phi}{\partial v_j}}{\omega - kv + i\delta}(k_j - k_i k_l d_{ijl}) + \frac{1}{\omega} \int \frac{dv''}{\omega - kv'' + i\delta} \times \right.$$

$$\times \left[\frac{\partial \Phi(v'')}{\partial v''}\left(1 - \frac{kv''}{\omega}\right) + \frac{v''_l}{\omega}\left(k \frac{\partial \Phi(v'')}{\partial v''}\right) - k_m d_{mnj} \frac{\partial \Phi(v'')}{\partial v''_n} \right] \times$$

$$\left. \times \int \frac{v_j k_l\, dv}{\omega - kv + i\delta} \frac{\partial}{\partial v_l} \Phi \right\}. \qquad (8.78)$$

The approximate expression for $\varepsilon^{t(e)}$ when $\omega \ll kv_{Te}$, $\varepsilon^{t(e)} \gg 1$ is

$$\varepsilon^{t(e)} = i\sqrt{\frac{\pi}{2}} \cdot \frac{\omega_{pe}^2}{\omega k v_{Te}}\left\{ 1 - \frac{k_m d_{msj}}{2}\left[\delta_{sj} - \frac{k_s k_j}{k^2} - \left(\delta_{jl} - \frac{k_j k_l}{k^2}\right) \times \right.\right.$$

$$\left.\left. \times \frac{k_s k_i d_{il} n_0}{k^2 v_{Te}^2} \cdot \frac{1}{\left(1 + \frac{k_i k_j d_{ij} n_0}{k^2 v_{Te}^2}\right)} \right]\right\}. \qquad (8.79)$$

When there is isotropy in the distribution of waves, the coefficients d_{ij} and $k_m d_{mij}$ can only be combinations of the tensors δ_{ij} and $k_i k_j/k^2$:

$$d_{ij} = d^t\left(\delta_{ij} - \frac{k_i k_j}{k^2}\right) + d^l \frac{k_i k_j}{k^2}, \qquad (8.80)$$

$$k_m d_{mij} = \Delta^t\left(\delta_{ij} - \frac{k_i k_j}{k^2}\right) + \Delta^l \frac{k_i k_j}{k^2}. \qquad (8.81)$$

We have

$$\varepsilon^{t(e)} \simeq i\sqrt{\frac{\pi}{2}} \frac{\omega_{pe}^2}{\omega k v_{Te}}(1 - \Delta^t). \qquad (8.82)$$

The general expressions for the coefficients Δ^t take the form

$$\Delta^t = \Delta_1^t + \Delta_2^t;$$

$$\Delta_1^t = \frac{1}{4n_0 m_e v_{Te}^2}\int dk_1\left(1 - \frac{(kk_1)^2}{k^2 k_1^2}\right)\frac{(kv_g, k_1)}{\omega - kv_{gk_1} + i\delta} W_{k_1}; \quad (8.83)$$

$$\Delta_1^t \ll 1; \quad \omega \gg kv_g; \quad \omega \ll kv_g;$$

$$\Delta_2^t = \frac{\omega_{pe}}{24 n_0 m_e v_{Te}^2} \int \frac{k^3}{k_1^2} \frac{[\mathbf{kk_1}]^2}{k^2 k_1^2} \cdot \frac{dk_1}{(\omega - kv_{g, k_1} + i\delta)} \left(\mathbf{k} \frac{\partial}{\partial k_1} \right) W_{\mathbf{k_1}} + \Delta_2^{t'}; \qquad (8.84)$$

$$\Delta_2^{t'} = -\frac{\omega_{pe}}{24 n_0 m_e v_{Te}^2} \int \frac{[\mathbf{kk}]^2}{k_1^4 \omega} \left(\mathbf{k} \frac{\partial}{\partial k_1} \right) W_{\mathbf{k_1}} \, dk_1;$$

$$\Delta_2^{t'} = 0 \text{ when } W_{\mathbf{k_1}} = W_{|\mathbf{k_1}|}. \qquad (8.85)$$

Inclusion of the renormalized effects for isotropic turbulence is carried out just as in §8.3.

8.5. Potential Instabilities

of an Isotropic Turbulent Plasma

We will consider the longitudinal potential perturbations $E_k^R = \frac{\mathbf{k}}{k} E_k^R$ in an isotropic plasma. They can be unstable. The physical reasons for such instability are the following. We will assume that in some region of space the intensity of turbulent fluctuations increases in a fluctuating manner relative to other regions. The appearance of this inhomogeneity in the distribution of turbulent fluctuations means that a plasma electron will experience a force which tries to expel it from the region with the increased turbulent fluctuation energy (the Miller force [309]). Because of the plasma's quasineutrality, ions leave with the electrons. As a result the plasma density at this location is decreased. This process is inhibited by the thermal motion of the particles, which strives to equalize the density. If the force expelling the plasma overcomes the force of the gas kinetic pressure, an instability arises [299, 300]. The instability criterion is most easily obtained from energy considerations [300]. Assume that the quasi-neutral change in the plasma density has the form shown in Fig. 8.1:

$$n(x) = \begin{cases} n_0 + \delta n, & x > x_0, \\ n_0 - \delta n, & x < x_0, \end{cases} \delta n \ll n_0. \qquad (8.86)$$

The isothermal production of such an inhomogeneity during the compression of plasma requires the following amount of work:

$$\delta W = \frac{n_0}{2} (T_e + T_i) \left(\frac{\delta n}{n_0} \right)^2. \qquad (8.87)$$

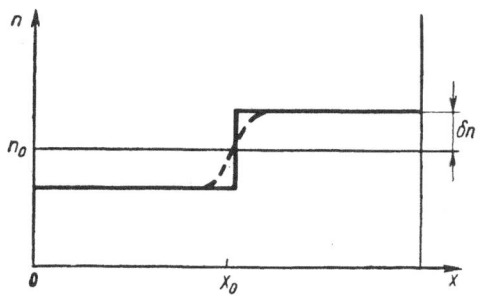

Fig. 8.1. The onset of an inhomogeneity in a turbulent plasma.

But the change in wave energy is [300]

$$\delta W = - \int \frac{\omega_{pe}^3}{24 k_1^2 v_{Te}^2} \cdot \frac{N_k \, dk}{(2\pi)^3} \left(\frac{\delta n}{n_0} \right)^2. \tag{8.88}$$

Therefore, it is more advantageous for a turbulent plasma to go into a state of inhomogeneous density distribution when

$$\int \frac{\omega_{pe}^3 N_{k_1} \, dk_1}{k_1^2 v_{Te}^2 (2\pi)^3} > 12 n_0 \, (T_e + T_i). \tag{8.89}$$

For longitudinal waves Eq. (8.77) describes any potential oscillations, even for an anisotropic plasma. Here we are considering just an isotropic plasma. When $\omega \ll k v_{Te}$ (8.77) gives

$$\varepsilon^{l \, (e)} = 1 + \frac{\omega_{pe}^2}{k^2 v_{Te}^2} \frac{1}{\left(1 + \frac{n_0}{v_{Te}^2} d^l \right)}, \tag{8.90}$$

$$d^l = - \frac{e^2}{4 m_e^2 \omega_{pe} n_0} \int dk_1 \frac{4\pi}{(\omega - k v_{g, k_1} + i\delta)} \left(k \frac{\partial}{\partial k_1} \right) W_{k_1}. \tag{8.91}$$

When $\omega \gg k v_g$,

$$d^l = \frac{3}{4} \frac{k^2 v_{Te}^2}{m_e \omega^2 n_0^2} \int W_{k_1} \, dk_1. \tag{8.92}$$

There is a substantial change in the plasma's dispersion properties when $\omega/k < v_*^l$, where

$$v_*^l = \sqrt{\frac{3}{4} \cdot \frac{W^l}{n_0 m_e}}. \tag{8.93}$$

Then, if $\omega/k \ll v_{Ti}$ we obtain

$$\varepsilon^l = 1 + \frac{\omega_{pi}^2}{k^2 v_{Ti}^2} + \frac{\omega_{pe}^2}{k^2 v_{Te}^2} \cdot \frac{1}{\left(1 + \frac{k^2 (v_*^l)^2}{\omega^2}\right)} \cdot \tag{8.94}$$

Thus $(\varepsilon^l \gg 1)$

$$\omega^2 = -\frac{k^2 (v_*^l)^2}{1 + \frac{T_i}{T_e}} \cdot \tag{8.95}$$

This shows that there is an aperiodic buildup in the oscillations:

$$\omega = \pm ik_*^l v_*^l \left(1 + \frac{T_i}{T_e}\right)^{-1/2} \cdot \tag{8.96}$$

If $kv_{Ti} \ll \omega \ll kv_{Te}$, $\omega \gg kv_g$, then

$$\omega^2 = k^2 \left(\frac{v_s^2}{2} \pm \sqrt{\frac{v_s^4}{4} + v_s^2 (v_*^l)^2}\right). \tag{8.97}$$

This instability is similar to that in (8.30).

Let us now consider isotropic turbulence with a narrow spectrum, such that $k\Delta v_{gk_1} \ll \omega$.[†] In this case all the plasmons can be in resonance with one wave, and in place of ω^2 in the dispersion equation, we have $\omega^2 - k^2 v_g^2$, i.e., when $\omega/k \ll v_{Ti}$

$$\varepsilon^l = 1 + \frac{\omega_{pi}^2}{k^2 v_{Ti}^2} + \frac{\omega_{pe}^2}{k^2 v_{Te}^2 \left(1 + \frac{k^2 (v_*^l)^2}{\omega^2 - k^2 v_g^2}\right)} \cdot \tag{8.98}$$

From this we have

$$\omega^2 = k^2 \left(v_g^2 - \frac{(v_*^l)^2}{1 + \frac{T_i}{T_e}}\right), \tag{8.99}$$

[†]If it is assumed that $\Delta k \sim k_0 = 2\pi/L$, the condition written above means that $\omega/k \gg (k_0/\omega_{pe})v_{Te}^2$.

and the condition for buildup of the oscillations is

$$\frac{W}{n_0(T_e+T_i)} > 12\frac{v_{Te}^2}{v_p^2}.$$ (8.100)

This instability condition was obtained in the above form in [299, 300]. When $\omega/k \gg v_{Ti}$ we have [299]

$$\left(\frac{\omega}{k}\right)^2 = \frac{v_g^2+v_s^2}{2} \pm \sqrt{\frac{(v_g^2+v_s^2)^2}{4}+\left((v_*^l)^2 - v_g^2\right)v_s^2}.$$ (8.101)

If $\omega \ll kv_g$ we obtain

$$d^l = -\frac{\omega_{pe}^2}{12\,n_0} \int \frac{W_{k_1}\,dk_1}{k_1^2\,n_0\,T_e}.$$ (8.102)

The sign of this expression is different from that of (8.31); this is due to the difference in phase volumes in the three-dimensional case versus the one-dimensional case. Therefore, turbulent acoustic waves are not possible in isotropic plasma. When the inequality of (8.100) is fulfilled an instability arises, where

$$\omega^2 = k^2\,v_s^2\left(1 - \frac{\omega_{pe}^2}{12\,v_{Te}^2}\int\frac{W_{k_1}\,dk_1}{k_1^2\,n_0\,T_e}\right).$$ (8.103)

The criterion for the appearance of the instability in (8.97) is very strict and practically coincides with (8.38); W is the total turbulence energy in the region where $v_p \ll 3v_{Te}^2/v_{Ti}$. Therefore, the spectrum must be cut off rather sharply at $v_p \sim 3v_{Te}^2/v_{Ti}$, which can occur for example because of the intense radiation in an optically thin plasma layer (see Chapter 6).

If $v \gg 3v_{Te}^2/v_{Ti}$, renormalization must be taken into account. For transverse plasmons with $\omega_k = \omega_{pe} + k^2c^2/2\omega_{pe}$, the criterion of (8.36) is much easier to satisfy in view of the large frequency difference.

In the presence of transverse plasmons the nonlinear dielectric function for an isotropic plasma with $\omega \ll \omega_{pe}$ is [308]

$$\varepsilon^l{}^{(e)} = 1 + \frac{4\pi e^2}{m_e\,k^2}\int\frac{dv}{\omega-kv+i\delta}\left(k\,\frac{\partial\Phi}{\partial v}\right)\times$$

$$\times\left[1 - \frac{\omega_{pe}^4}{2n_0^2\,m_e}\int\frac{dk_1\left(k\,\frac{\partial}{\partial k_1}\right)\frac{W_{k_1}}{\omega_{k_1}^3}}{(\omega-kv_{g,k_1}+i\delta)}\int\frac{\left(k\,\frac{\partial\Phi}{\partial v}\right)dv}{\omega-kv+i\delta}\right]^{-1}.$$ (8.104)

We will treat a number of examples here. If $\omega \ll kv_{Te}$, $\omega \gg kv_{Ti}$, $\omega \gg kv_g$, then

$$\varepsilon^l = 1 - \frac{\omega_{pi}^2}{\omega^2} + \frac{\omega_{pe}^2}{k^2 v_{Te}^2} \cdot \frac{1 + i\sqrt{\dfrac{\pi}{2}} \cdot \dfrac{\omega}{kv_{Te}}}{1 + \dfrac{k^2 v_*^2}{\omega^2}\left(1 + i\sqrt{\dfrac{\pi}{2}} \cdot \dfrac{\omega}{kv_{Te}}\right)}, \qquad (8.105)$$

where

$$v_*^2 = \frac{W c^2}{4 n_0 T_e}. \qquad (8.106)$$

Neglecting the imaginary contributions we find $(d_e = v_{Te}/\omega_{pe})$

$$\omega_{\pm}^2 = \frac{1}{2}\left\{\omega_s^2 - \frac{k^2 v_*^2 k^2 d_e^2}{1 + k^2 d_e^2} \pm \sqrt{\left[\omega_s^2 - \frac{k^2 v_*^2 k^2 d_e^2}{(1 + k^2 d_e^2)}\right]^2 + 4k^2 v_*^2 \omega_s^2}\right\}. \qquad (8.107)$$

If $kv_{Te} \ll \omega_{pe}$, then $\omega_s = kv_s$ and (8.107) corresponds to an unstable root for $v_* > v_s \omega_{pe}/kv_{Te}$. The growth rates are estimated to be $(\gamma = \mathrm{Im}\,\omega)$

$$\gamma_+ = k^2 d_e v_*; \quad \gamma_- = 0. \qquad (8.108)$$

When $v_* \ll v_s \omega_{pe}/kv_{Te}$

$$\gamma_+ = 0; \quad \gamma_- = k\sqrt{\sqrt{v_*^2 v_s^2 + \frac{v_s^4}{4}} - \frac{v_s^2}{2}}. \qquad (8.109)$$

If $kv_{Te}/\omega_{pe} \gg 1$, $\omega_s = \omega_{pi}$, when $v_* \ll \omega_{pi}/k$

$$\gamma_+ = 0, \quad \gamma_- = kv_*, \qquad (8.110)$$

and when $v_* \gg \omega_{pi}/k$

$$\gamma_- = 0, \quad \gamma_+ = kv_*. \qquad (8.111)$$

As mentioned earlier, these instabilities are of interest in the problem of interactions between intense high-frequency fields and the plasma [310], and for the interactions of lasers with the plasma. Everywhere in the above the frequency is much larger than the particle pair collision frequencies.[†]

[†]The other limit is treated in [311, 312].

8.6. Drift Instabilities in Turbulent Plasma

The radical changes in the plasma's electromagnetic properties in the low-frequency region, resulting from the effect of
high-frequency turbulence, is very important for drift instabilities.
In fact, the drift instabilities most dangerous for plasma confinement
are aperiodic in character [6]. A small change in frequency cannot stabilize such an instability. Only a marked change in the
electromagnetic properties can provide new kinds of drift modes,
which are either stable or at least less unstable. The stabilization of drift oscillations by the turbulence has a number of features in common with stabilization by high-frequency fields [313,
314], especially since, as already pointed out, the condition for
changes in the plasma's electromagnetic properties by transverse
plasmons is more favorable. The essential difference, of course,
is that real high-frequency modes which can propagate in the
plasma are being considered, rather than high-frequency oscillations of the type sin ωt, which are either involved in the skin effect in the plasma or require some limitation on the plasma dimensions (dimension smaller than $1/k$). In either of these cases
the strict statement of the problem encounters significant difficulties in describing the transition layer at the plasma boundary.

In the general discussion of the problem of turbulence presented above, the energy level and its distribution over the spectrum are regulated by internal nonlinear interactions for any
means of excitation (external high-frequency fields or other methods). Since the changes in the plasma's electromagnetic properties are connected with the presence of turbulence, we can assume that when drift instabilities appear, their influence can produce changes in the spectrum and in the spatial distribution of
turbulent fluctuations. It is evident from the analysis in §8.5 that
in the absence of nonuniformity the turbulence even then becomes
nonuniform, dividing into regions which are separated by intervals
in which the intensity of the high-frequency turbulence is increased. But the drift instabilities usually tend to wash out the
density gradients. The possibility of stabilizing the drift instabilities must be considered as opposition to these effects. We will
treat drift instabilities in a magnetized turbulent plasma; that is,
we will begin with the results of §8.2 and include the density and
temperature inhomogeneities of the plasma (see [302]). We will

assume that a localized approach is valid, which means that the wavelengths are much smaller than the characteristic dimensions of the inhomogeneities [6].

Two approximations will be examined:

(1) One-dimensional Langmuir turbulence with $\omega_{k_1} = \omega_{pe} + \dfrac{3k_1^2 v_{Te}^2}{\omega_{pe}}$ ($\omega_{He} \gg \omega_{pe}$). However, because the results depend only on the group velocities of the waves, this case will describe qualitatively the results which arise for isotropic turbulence as well, but with $\omega_{He} \ll \omega_{pe}$.

(2) Isotropic turbulence at a frequency of $\omega_{k_1} = \omega_{pe} + \dfrac{k_{1\perp}^2 c^2}{2\omega_{pe}}$.

For Langmuir turbulence the dielectric permeability has the form [see (8.26)]

$$\varepsilon_k^{l\,(e)} = 1 + \frac{\dfrac{4\pi e^2}{m_e k^2} \displaystyle\int \dfrac{dv_z}{\omega - k_z v_z + i\delta} \left(k_z \dfrac{\partial \Phi}{\partial v_z} - \dfrac{k_x}{\omega_{He}} \cdot \dfrac{\partial \Phi}{\partial y} \right)}{1 + \dfrac{k_z v_{Te}^2}{n_0} d \displaystyle\int \dfrac{dv_z}{\omega - k_z v_z + i\delta} \cdot \dfrac{\partial \Phi}{\partial v_z}}, \tag{8.112}$$

where $\delta \to +0$, with

$$d = -\frac{\pi e^2 k_z}{m_e v_{Te}^2 \omega_{pe}} \int dk_1 \frac{1}{(\omega - k_z v_g + i\delta)} \cdot \frac{\partial}{\partial k_1} W_{k_1}. \tag{8.113}$$

In the frequency region defined by the inequality $\omega \ll k_z v_{Te}$, we have

$$\varepsilon_k^{l\,(e)} = 1 + \frac{\omega_{pe}^2}{k^2 v_{Te}^2} \frac{1 + i \sqrt{\dfrac{\pi}{2}} \dfrac{1}{k_z v_{Te}} \left(\omega - \omega_* + \omega_* \dfrac{\eta}{2} \right)}{1 + d \left(1 + i \sqrt{\dfrac{\pi}{2}} \dfrac{\omega}{k_z v_{Te}} \right)}, \tag{8.114}$$

$$\omega_D = \omega_* = -\frac{T_e k_x}{m_i \omega_{Hi}} \cdot \frac{\partial}{\partial x} \ln n_0, \tag{8.115}$$

$$\eta = \frac{\partial \ln T_e}{\partial \ln n_0}. \tag{8.116}$$

It is clear from (8.114) that the turbulence has its strongest effect on the plasma's electromagnetic properties when $d \gg 1$. Then

$$\varepsilon_k^{l\,(e)} = 1 + \frac{\omega_{pe}^2}{dk^2 v_{Te}^2} \left(1 + i \sqrt{\frac{\pi}{2}} \frac{\omega_*}{k_z v_{Te}} \left(\frac{\eta}{2} - 1 \right) \right). \tag{8.117}$$

Thus, when $\eta > 2$ we have stabilization of the drift oscillations for that branch of the vibrations which corresponds to the fastest oscillating drift wave in the linear approximation [6]. If d depends on ω, new nonlinear branches can appear, and among them can be unstable modes. However, generally the appearance of such instabilities is inhibited. Taking into account the fact that when $\omega \gg k_z v_{Ti}$,

$$\varepsilon^{(i)} = \varepsilon_0^{(i)} = -\frac{\omega_{pi}^2}{\omega^2} - \frac{\omega_{pe}^2}{k^2 v_{Te}^2} \cdot \frac{\omega_*}{\omega} \qquad (8.118)$$

and $\omega \gg k_z^2 v_s^2 / \omega_*$, we can limit our attention to just the second term in (8.118). We find

$$\omega = \frac{\omega_* d}{2} \pm \sqrt{\frac{1}{4} d^2 \omega_*^2 + k_z^2 v_s^2 d}, \qquad (8.119)$$

so that when $k_z v_s \ll \omega_* \sqrt{d}$ one of the roots is

$$\omega = \omega_* d. \qquad (8.120)$$

When $v_p \ll 3 v_{Te}^2 / v_{Ti}$, for example, the condition for the existence of the spectrum in (8.120) imposes a constraint on the size of the density gradient:

$$\frac{W}{nT_e} \gg \frac{m_e}{m_i}, \quad \frac{\omega_*}{\omega_{pe}} \gg \frac{m_e}{m_i}. \qquad (8.121)$$

The condition $\omega \gg k_z v_{Ti}$ is satisfied when $\dfrac{W}{n_0 T_e} \gg \dfrac{k_z v_{Ti}}{\omega_*} 12 \dfrac{v_{Te}^2}{v_p^2}$. When $k_z v_z \ll \omega_* \sqrt{d}$ the slow branch in (8.118) is independent of the turbulence energy.

Let us now consider the case where $\omega \gg k_z v_g$, in which d is now

$$d = \frac{3}{4} \cdot \frac{k_z^2 v_{Te}^2}{\omega^2} \cdot \frac{W}{n_0 T_e}. \qquad (8.122)$$

In this case some new nonlinear drift instabilities can arise. For $\omega \gg k_z v_{Ti}$ and $d \gg 1$ we find

$$\omega = \frac{1 \pm i\sqrt{3}}{2} k_z v_{Te} \left(\frac{3\omega_*}{4 k_z v_{Te}}\right)^{1/3} \left(\frac{W}{n_0 T_e}\right)^{1/3}. \qquad (8.123)$$

The instability in (8.123) will exist when (from $d \gg 1$)

$$1 \gg \frac{W}{n_0 T_e} \gg \left(\frac{\omega_*}{k_z v_{Te}} \right)^2, \tag{8.124}$$

that is, the instability can be generated over small distances which satisfy the inequality $k_z \gg \omega_*/v_{Te}$. Moreover, when both these inequalities are satisfied, $k_z v_{Ti}$, $k_z v_g \ll \omega \ll k_z v_{Te}$, we arrive at the condition

$$1 \gg \frac{W}{n_0 T_e} \gg \frac{k_z v_{Te}}{\omega_*} \max \left\{ \left(\frac{m_e}{m_i} \right)^{3/2}, \left(\frac{3 v_{Te}}{v_p} \right)^3 \right\}, \tag{8.125}$$

which bounds the lengths of the unstable waves from below, such that

$$\frac{\omega_*}{v_{Te}} \min \left\{ \left(\frac{m_i}{m_e} \right)^{3/2}, \left(\frac{v_p}{3 v_{Te}} \right)^3 \right\} \gg k_z \gg \frac{\omega_*}{v_{Te}}. \tag{8.126}$$

Finally, the condition that $\omega \gg (\omega - \omega k_1) k_1/k$ gives (with $v_p \ll 3 v_{Te}^3/v_{Ti}$)

$$1 \gg \frac{k^2}{k_1^2} \gg \frac{1}{9} \left(\frac{v_p}{v_{Te}} \right)^3 \left(\frac{W}{n_0 T_e} \right)^{2/3} \frac{m_e}{m_i} \left(\frac{k_z v_{Te}}{\omega_*} \right)^{1/3}. \tag{8.127}$$

These inequalities give us some idea of the conditions under which these instabilities will arise. We should point out that one result of the development of such instability will be changes in the spatial distribution of the turbulent fluctuations. This is clear from the original equations, which show that the main contribution to ε comes from the effects which alter the turbulence spectrum.

Let us now study drift oscillations when isotropic nonpotential fluctuations at frequency $\omega_{k_1} = \omega_{pe} + k_{1\perp}^2 c^2/2\omega_{pe}$ are present in the plasma. Since the conditions under which the approximations used for the dielectric permeability can be used in this case have been discussed in detail earlier, we will simply consider a number of particular drift spectra here. If $\omega \ll k_z v_{Te}$, $\omega \gg k_g v_\perp$, i.e.,

$v_g \ll v_{Te}$, the dispersion equation is of the form[†]

$$-\frac{\omega_{pi}^2}{\omega^2} \cdot \frac{k_z^2}{k^2} - \frac{\omega_{pe}^2}{k^2 v_{Te}^2} \cdot \frac{\omega_*}{\omega} + \frac{\omega_{pe}^2}{k^2 v_{Te}^2} \left[1 + i \sqrt{\frac{\pi}{2}} \frac{\left(\omega - \omega_* + \omega_* \frac{\eta}{2} \right)}{k_z v_{Te}} \right] \times$$

$$\times \left[1 + \left(1 + i \sqrt{\frac{\pi}{2}} \omega/k_z v_{Te} \right) k_\perp^2 c^2 W/\omega^2 4n_0 T_e \right]^{-1} . \qquad (8.128)$$

If $W/n_0 T_e \gg k_z^2 v_{Te}^6 m_e^3/k_\perp^2 c^2 \omega_*^4 m_i^3$ the last term in (8.128) is small, so that in the low-frequency region we have

$$\omega = - k_z^2 v_{Te}^2 m_e/\omega_* m_i . \qquad (8.129)$$

These oscillations damp out when $\eta > 2$ with a damping rate given by

$$\gamma = 2 \sqrt{2\pi} \frac{n_0 T_e}{W} \cdot \frac{k_z^7 v_{Te}^7}{\omega_*^4 k_\perp^2 c^2} \left(\frac{m_e}{m_i} \right)^4 \left(\frac{\eta}{2} - 1 \right) . \qquad (8.130)$$

If $k_z^2 v_s^2 \ll \omega\omega_*$, then, when a number of inequalities are satisfied (which we will not present here), we have

$$\omega^3 = \omega_* k_\perp^2 c^2 \frac{W}{n_0 T_e} . \qquad (8.131)$$

If $\omega \ll k_\perp v_g$, but $\omega \gg k_z v_{Te}$, the dispersion equation takes the form

$$-\frac{\omega_{pi}^2}{\omega^2} \cdot \frac{k_z^2}{k^2} - \frac{\omega_{Te}^2}{k^2 v_{Te}^2} \cdot \frac{\omega_*}{\omega} - \left[\frac{\omega_{Te}^2 k_z^2}{\omega^2 k^2} \left(1 - \frac{\omega_*}{\omega} \right) + \frac{\omega_{pe}^2}{k^2 v_{Te}^2} \cdot \frac{\omega_*}{\omega} \right] \times$$

$$\times \left[1 + \frac{k_z^2 v_{Te}^2}{\omega^2} \cdot \frac{\omega_{pe}^2}{4n_0 T_e} \int \frac{W_{k_1} dk_1}{k_1^2} \right]^{-1} + \frac{\omega_{pi}^2 k_\perp^2}{\omega_{Hi}^2 k^2} \left(1 + \frac{\omega_*}{\omega} \right) = 0 . \qquad (8.132)$$

When the following restriction is fulfilled:

$$\frac{m_i}{m_e} \gg \frac{k_z^2 v_{Te}^2}{\omega^2} \frac{\omega_{pe}^2}{4n_0 T_e} \int W_{k_1} \frac{dk_1}{k_1^2 c^2} \gg k_\perp^2 \frac{v_{Ti}^2}{\omega_{Hi}^2} , \qquad (8.133)$$

[†]Equation (8.128) was obtained with a Maxwellian particle distribution. In the general case one must consider the influence of transverse waves on the distribution function; these effects are smaller when $k_\perp \ll \omega_{pe}/c$.

the dispersion equation goes into ($\omega \gg k_z^2 v_s^2 / \omega_*$)

$$\frac{\omega_{pe}^2}{\omega^2} \cdot \frac{k_z^2}{k^2} \left(\frac{\omega_*}{\omega} \cdot \frac{\omega_{pe}^2}{4 n_0 T_e} \int W_{k_1} \frac{dk_1}{k_1^2 c^2} + 1 - \frac{\omega_*}{\omega} \right) = 0, \qquad (8.134)$$

i.e., the drift oscillations are hydrodynamically stable. According to [302] it is also possible to stabilize the temperature-drift instabilities.

8.7. Spontaneous Excitation of Magnetic Fields in Turbulent Plasma

It is known [304] that magnetic fields are spontaneously excited in a conducting turbulent fluid; that is, a turbulent conducting fluid is unstable relative to the excitation of perturbations which carry with them magnetic fields. The energy of the magnetic field is drawn from the turbulent hydrodynamic fluctuations of the fluid. Magnetic fields can also be spontaneously excited in turbulent plasma; however, now their energy comes from the high-frequency plasma fluctuations [303].

As an example, we will consider a plasma in which intense Langmuir turbulent oscillations have been excited. These oscillations are potential oscillations, so that no magnetic fields are present. We will show that even with an isotropic distribution of Langmuir fluctuations a turbulent plasma can become unstable relative to perturbations whose ground-state energy is tied up in the energy of a magnetic field.

Let us consider the development of excitations in a turbulent plasma with an isotropic distribution of both particles and turbulent fluctuations. Because of the assumed isotropy, the distribution of the excitations carrying magnetic fields is described by the equation $k^2 c^2 = \omega^2 \varepsilon^{t(e)}$, where $\varepsilon^{t(e)}$ is the transverse dielectric permeability for the turbulent plasma.

In order to show the effects of the instability, it is sufficient to obtain solutions to the dispersion equations in the form of ω as a function of k.

In the limit $\omega \gg k v_{gk_1}$ we find from (8.84) and (8.83) that

$$\Delta_1^t \approx \frac{k^2}{30\, n_0\, m_e\, v_{Te}^2\, \omega^2} \int v_{gk_1}^2\, W_{k_1}\, dk_1 \ll 1, \qquad \Delta_2^t = -\frac{k^4\, v_{Te}^2}{12\omega^2} \int \frac{W_{k_1}\, dk_1}{k_1^2\, n_0\, T_e}.$$

$$(8.135)$$

Note that $\omega \ll k v_{Te}$ in (8.83); therefore the correction Δ_2^t, which is of order $(k^2/12k_1^2)(W/n_0 T_e)(k^2 v_{Te}^2/\omega^2)$, can be larger than unity. When $\Delta_2^t \gg 1$, we have

$$\varepsilon_{\backslash}^{t\,(e)} = i \sqrt{\frac{\pi}{2}} \cdot \frac{\omega_{pe}^2\, k^3}{\omega^3\, v_{Te}} \int \frac{dk_1\, W_{k_1}}{12\, n_0\, m_e\, k_1^2}.$$

$$(8.136)$$

The equation $k^2 = \omega^2 \varepsilon^{t(e)}/c^2$ gives

$$\omega = i\, k v_*^t,$$

$$(8.137)$$

$$v_*^t = v_{Te} \int W_{k_1}\, dk_1 \frac{v_p^2}{v_{Te}^2} \cdot \frac{1}{12} \sqrt{\frac{\pi}{2}} \cdot \frac{1}{n_0\, m_e\, c^2} \approx v_{Te}\, \frac{\omega_{pe}^2}{k_0^2 c^2} \cdot \frac{W}{n_0\, T_e}.$$

$$(8.138)$$

The last equality of (8.138) is valid for intense turbulence, when most of the energy is concentrated at $k \sim k_0 \ll 3 v_{Te}^2/v_{Ti}$ (see Chapter 4).

In the general case, in the limit $\omega \gg k v_g$, $\omega \ll k v_{Te}$ we obtain the following dispersion equation:

$$k^2 = \omega^2\, \varepsilon^{t\,(e)} c^{-2} = i \sqrt{\frac{\pi}{2}} \frac{\omega_{pe}^2\, \omega}{k v_{Te} c^2} \left(1 + \frac{k^4}{12\omega^2\, n_0\, m_e} \int \frac{W_{k_1}\, dk_1}{k_1^2} \right).$$

$$(8.139)$$

Its solution always contains the following unstable root:

$$\omega = i\, \frac{k^3 v_{Te}\, c^2}{2\omega_{pe}^2} \sqrt{\frac{2}{\pi}} \left(\sqrt{1 + 4 \sqrt{\frac{\pi}{2}} \frac{v_*^t}{v_{Te}} \cdot \frac{\omega_{pe}^2}{k^3 c^2}} - 1 \right).$$

$$(8.140)$$

In the limit $v_*^t \ll v_{Te} k^2 c^2/\omega_{pe}^2$ we have solutions which are aperiodic increasing second sound waves with $\omega = i k v_*^t$, while for $v_*^t \gg v_{Te} k^2 c^2/\omega_{pe}^2$ we have

$$\omega = i \left(\frac{2}{\pi} \right)^{1/4} \frac{k^2 c}{\omega_{pe}} \sqrt{v_{Te}\, v_*^t}.$$

$$(8.141)$$

In order to understand the physical meaning of the instability, we
note that the magnitude of the magnetic field in the transverse
waves $H = [(k/\omega)E]$ is of order $(k/\omega)E = \varepsilon^t E$. For the instabilities being considered here $\varepsilon^t \gg 1$ so that their development sig-
nals the spontaneous increase in the magnetic fields in the turbulent plasma. The appearance of spontaneous magnetic fields in a
turbulent plasma is analogous to the Batcherol effect (the spontaneous excitation of magnetic fields during the turbulent motion of
a conducting fluid).

We will now discuss the criterion for this instability, as given
in (8.36). When $v \ll 3v_{Te}^2/v_{Ti}$ we obtain

$$k \gg \omega_{pe} \frac{m_e c^{-1}}{9m_i}, \quad \text{i.e.,} \quad \frac{k}{k_1} \gg \frac{v_p}{c} \cdot \frac{m_e}{9m_i}, \tag{8.142}$$

and since $k \ll k_1$, $v_p \ll 9m_i/m_e$, which is certainly the case when
$v_p \ll 3v_{Te}^2/v_{Ti}$. If $v_p \gg 3v_{Te}^2/v_{Ti}$ the criterion takes the form

$$k \gg \omega_{pe} \frac{v_{Te}^2 c^{-1}}{v_p^2}. \tag{8.143}$$

If the phase velocities of the Langmuir oscillations are large for
large currents of turbulent energy Q, the minimum k are quite
small, that is, magnetic fields of rather large dimensions can be
excited. This is important in astrophysical situations.

Although the growth rates for these excitations also decrease
as k decreases, it is of interest to estimate them for the smallest
possible values of k since these correspond to the largest scales.
Assume that

$$k^2 \ll \omega_{pe}^2 \frac{v_*^t c^{-2}}{v_{Te}} \tag{8.144}$$

and that the instability of (8.141) is present in the system. Then,
when $v_p \ll 3v_{Te}^2/v_{Ti}$ we have

$$k^2 \gg \omega_{pe}^2 \left(\frac{W}{n_0 T_e}\right)^{1/2} \frac{m_e}{9m_i} \cdot \frac{v_p}{v_{Te}} \left(\frac{c}{v_{Te}}\right)^2 c^{-2}. \tag{8.145}$$

When combined with (8.144) this gives

$$\frac{W}{n_0 T e} \gg \frac{m_e}{9m_i} \cdot \frac{c}{v_p} \left(\frac{c}{v_{Te}} \right)^3 . \qquad (8.146)$$

This last inequality imposes a requirement on the plasma temperature, which must be sufficiently high.

If $v_p \gg 3v_{Te}^2 / v_{Ti}$, the condition of (8.145) must be changed to

$$k^2 \gg \omega_{pe}^2 \frac{1}{v_{Te} v_p} \left(\frac{W}{n_0 T_e} \right)^{1/2} \qquad (8.147)$$

Combining this condition with (8.144), we find

$$\frac{\dot{W}}{n_0 T e} \gg \frac{c^2}{v_{Te}^2} \left(\frac{c}{v_p} \right)^6 . \qquad (8.148)$$

We emphasize that rather-long-wavelength excitations can be excited, and the conditions for their excitation are eased as the
plasma temperature increases.

8.8.   The Skin Effect in Turbulent

Plasma

Let us examine the penetration of electromagnetic waves of
low frequency into a turbulent plasma. If the frequency of the
wave is much smaller than ω_{pe}, the penetration of the field into the
turbulent plasma is connected with either pair collisions or with
dissipation due to Landau absorption (the anomalous skin effect).
In turbulent plasma the skin effect is significantly altered if the
frequencies of the incident electromagnetic waves are much smaller than the characteristic turbulent frequencies.

In the strictest sense, the problem of the penetration of electromagnetic fields into a turbulent plasma must be solved with
the boundary conditions taken into account. However, qualitative
answers can be obtained if it is assumed that an electromagnetic
field of a specific frequency is generated at some imaginary flat
surface inside the plasma, and we determine how a field of this
particular frequency is changed as a function of the distance from
this imaginary surface.

At a distance x from our imaginary surface, E is given by

$$E(x, t) = \int E_\omega e^{-i\omega t + i kx} d\omega, \tag{8.149}$$

where E_ω is the Fourier component of the field $E(t)$ at the surface $x = 0$. Equation (8.149) will provide the answer to our question if the solution to the dispersion equation is found in the form $k = k(\omega)$. Then the "penetration" depth of the field (the skin-layer thickness) is of order

$$\delta \simeq \frac{1}{\operatorname{Im} k(\omega)}. \tag{8.150}$$

If the field decreases as x increases, it can be because either the electromagnetic field does not propagate in the plasma ($\omega \ll \omega_{pe}$, but $\omega \gg kv_{Te}$ in a nonturbulent plasma) or because these are dissipative processes. In a nonturbulent plasma we find from $\varepsilon = 1 - \omega_{pe}^2/\omega^2$ ($\omega \gg kv_{Te}$) that when $\omega \ll \omega_{pe}$,

$$\delta = \frac{c}{\omega_{pe}}. \tag{8.151}$$

As is well known, when $\omega \ll kv_{Te}$ in the absence of turbulence in the plasma,

$$\delta \simeq 2\left(\frac{2}{\pi}\right)^{1/6}\left(\frac{v_{Te}}{\omega}\right)^{1/3}\left(\frac{c}{\omega_{pe}}\right)^{2/3}. \tag{8.152}$$

Finally, when electron−ion collisions are frequent,

$$\varepsilon^t \approx i\,\frac{\omega_{pe}^2}{\omega v_e}, \quad \delta \simeq \frac{c}{\omega_{pe}}\sqrt{\frac{v_e}{\omega}}. \tag{8.153}$$

Let us consider how these equations are altered in the presence of turbulence. By solving (8.139) with respect to k, we find that when the inequalities

$$\int \frac{W_{k_1} d_{k_1}}{12 n_0 T_e} \gg \left(\frac{v_{Te}}{c}\right)^{2/3}\left(\frac{c}{v_p}\right)^2 16\left(\frac{2}{\pi}\right)^{2/3}\left(\frac{\omega}{\omega_{pe}}\right)^{2/3} \tag{8.154}$$

and $\omega \gg kv_g$ are satisfied, the expression for the skin depth is

$$\delta = \frac{1}{\operatorname{Im} k(\omega)} = \frac{c}{\omega_{pe}}\sqrt{\frac{v_{\text{eff}}}{\omega}}, \quad v_{\text{eff}} = \omega_{pe}\sqrt{\int \frac{v_p^2 W_{k_1} dk_1}{12 n_0 Te}}. \tag{8.155}$$

When the inequality in (8.154) is reversed,

$$k(\omega) = -i\sqrt{\frac{2}{\pi}}\, v_{Te}\, \frac{12 n_0 m_e \omega}{\int v_p^2 W_{k_1}\, dk_1} \cdot k^3(\omega) = i\sqrt{\frac{\pi}{2}} \cdot \frac{\omega_{pe}^2 \omega}{v_{Te}}. \qquad (8.156)$$

The first solution in (8.156) describes a field which increases with distance, whereas the second solution is the usual anomalous skin effect. The possibility of a spatially amplified field is connected with the excitation of magnetic fields considered here.

In the limit $\omega \ll kv_g$, when such excitations are not possible, the skin effect for

$$\int \frac{v_p^4 W_{k_1}\, dk_1}{v_{Te}^4\, 108\, n_0 m_e c^2} \gg \left(\frac{\omega_{pe}}{\omega}\right)^{2/3} \left(\frac{v_{Te}}{c}\right)^{4/3} \qquad (8.157)$$

is described by the equation

$$\delta = \frac{c}{\omega_{pe}} \sqrt{\int \frac{v_p^4 W_{k_1}\, dk_1}{v_{Te}^4\, 108\, n_0 m_e c^2}}. \qquad (8.158)$$

Finally, when $\omega \gg kv_{Te}$ and $v_t \gg \omega^2 c/\omega_{pe}^2$, the skin effect is described by (8.155). We emphasize that these effects are connected with the fact that the field of the external wave displaces not only the plasma electrons but the Langmuir plasmons as well, which then begin to interact with and scatter the electrons because of the nonuniformity of the wave distribution in the field.

Let us now examine the skin effect in a plasma containing ion-acoustic turbulent fluctuations. Ion-acoustic turbulence is different in that the electrons, which are resonant with the plasmons, strongly interact with them. To a certain degree this interaction is equivalent to the electron—ion collisions because ions participate in ion-acoustic oscillations. One would expect that in this case the skin effect will be determined by an equation similar to (8.153), in which the effective collision frequency characterizes the collisions between electrons and plasmons. We must include D_0 in Eq. (8.62) and keep the imaginary parts, but the rest of the diffusion coefficients can be neglected.

Assuming that the turbulence is stationary and isotropic we have

$$D_{ij}^{(0)} = D^l \frac{p_i p_j}{p^2} + D^t \left(\delta_{ij} - \frac{p_i p_j}{p^2}\right). \qquad (8.159)$$

The coefficient D^l describes the heating of the electron gas by the ion-acoustic fluctuations, and D^t gives the elastic scattering of the electrons by the turbulent fluctuations; we have $D^l < D^t(m_e/m_i)$.

Assuming that the field frequency is much less than the scattering frequency but much larger than the frequencies related to the heating, we can neglect D^l. Then the equation for the function Φ, which is not perturbed by the external field, satisfies an arbitrary isotropic distribution function which depends only on the magnitude of the particle velocities. Since the frequencies leading to scattering are frequent ($\nu_{eff} \gg \omega$), it is natural to assume that the external field creates only a weak anisotropy in the particle distribution, and we will seek $f_k^{R(1)}(v)$ in the form of $f_k^{R(1)}(v) \, v/v$. Then, taking the equation obtained from (8.62) under these assumptions, multiplying it by v/v, and integrating over the various angles of the vector v, we find

$$ -i\omega f_k^{R\,(1)} - \frac{e}{m_e} E_k^R \frac{\partial \Phi}{\partial v} = -\nu_{eff} \, f_k^{R\,(1)}, \qquad (8.160) $$

$$ \nu_{eff}(v) = \frac{2e^2}{v^3 m_e^2} \int 2\pi^2 \, W_{k_1} k_1^{-1} \, dk_1 \frac{\omega_{k_1}^2}{\omega_{pi}^2}. \qquad (8.161) $$

We can now easily find an expression for the current:

$$ j_k^R = -e \int \frac{v}{v} \left(f_k^{R\,(1)} \, v \right) dv = \frac{\omega_{pe}^2}{4\pi\nu_{eff}} E_k^R. \qquad (8.162) $$

The last inequality comes from the condition $\omega \ll \nu_{eff}$, where

$$ \nu_{eff} = \sqrt{\frac{\pi}{2}} \cdot \frac{\pi^2 e^2}{8m_e^2 v_{Te}^3} \int \frac{\omega_{k_1}^2}{\omega_{pi}^2} W_{k_1} \, k_1^{-1} \, dk_1. \qquad (8.163) $$

We obtain the following order-of-magnitude estimate of the skin layer:

$$ \delta \simeq \frac{c}{\omega_{pe}} \sqrt{\frac{W}{10 n_0 T_e}}, \qquad \frac{k_1 v_{Te}}{\omega_{pe}} \sim 1. \qquad (8.164) $$

8.9. Second Sound in Turbulent

Plasma

According to Landau [305], second sound is the oscillation of a gas of excitations. It has been thoroughly studied in helium [305, 315] and in solids [316]. In order for second sound to appear in a plasma, one must have a stationary equilibrium distribution of plasmons, which is exactly the case with stationary turbulence. Small deviations and perturbations of this stationary state will perform oscillations, and it is these which one calls second sound in the plasma. We note that the previously considered electromagnetic perturbations of the stationary turbulent state have frequencies $\omega = ikv_*$; that is, they are sound vibrations in the sense that ω and k are proportional. However, their frequencies are purely imaginary, which corresponds to an aperiodic instability.

In the first place, we must mention that there are significant differences between second sound and these instabilities. In our earlier discussion of the electromagnetic properties of turbulent plasma, we considered the response of the turbulent plasma to a regular field E^R. In addition to the excitations which are proportional to E^R, there can also be a part which does not depend on E^R, that is, on the perturbation of the turbulent field. These are the excitations which are described as second sound.

All the effects related to second sound are contained in the equations which describe the change in the turbulent state; that is, they are contained in equations like (8.4), which includes the additional weak instability of the turbulent fluctuations (see Chapter 2). There are a number of important differences between those excitations and the excitations described by (8.16). For example, (8.16) contains only induced decays, while for second sound spontaneous decays are also possible. A more detailed consideration (up to terms of order I^2) of the excitations connected with the regular fields shows that they do not contain processes like the excitation of four-plasmon interactions. This is understandable, of course, because the notion of four-plasmon interactions itself arises only when treating four random fields, taking into account their mutual correlations.

Second sound in plasma has been studied in [306, 317, 318]. Reference [305] treats examples in which the stationarity of the spectrum is ensured by nonlinear decay processes and quasilinear processes; [317] examines the problem specifically for ion-acoustic turbulence. Here we will study second sound using the example of intense Langmuir turbulence, for which the maximum intensity is centered at some $v_p \gg 3v_{Te}^2/v_{Ti}$, according to Chapter 2.

Neglecting scattering by ions, and assuming that the four-plasmon interaction is fundamental, we follow Landau [304] in setting up a balance equation for the plasmons. It follows from conservation of plasmon number, momentum, and energy during four-plasmon interactions:

$$\frac{\partial N}{\partial t} + \frac{3v_{Te}^2}{\omega_{pe}} \cdot \frac{\partial \mathbf{P}}{\partial \mathbf{r}} = 0; \quad \mathbf{P} = \int \frac{\mathbf{k} N_{\mathbf{k}}^l \, d\mathbf{k}}{(2\pi)^3}; \quad N = \int \frac{N_{\mathbf{k}}^l \, d\mathbf{k}}{(2\pi)^3}; \qquad (8.165)$$

$$\frac{\partial \mathbf{P}}{\partial t} + \frac{3v_{Te}^2}{\omega_{pe}} \cdot \frac{\partial}{\partial x_j} D_{ij} = 0; \quad D_{ij} = \int k_i \, k_j \, \frac{N_{\mathbf{k}}^l \, d\mathbf{k}}{(2\pi)^3}; \qquad (8.166)$$

$$\frac{\partial W}{\partial t} + \frac{3v_{Te}^2}{\omega_{pe}} \cdot \frac{\partial}{\partial \mathbf{r}} Q = 0; \quad W = \int \frac{\omega_{\mathbf{k}} N_{\mathbf{k}}^l \, d\mathbf{k}}{(2\pi)^3}; \quad \mathbf{Q} = \int \frac{\mathbf{k}\omega_{\mathbf{k}} N_{\mathbf{k}}^l \, d\mathbf{k}}{(2\pi)^3}. \qquad (8.167)$$

This system of equations can be used to find the form of the linear excitations if their frequencies are much smaller than the charac-teristic frequency of the four-plasmon collisions. Under these conditions a quasiequilibrium state can be established at every moment of time. The differences in these instantaneous distribu-tions can lie in the values of the average momentum, the average energy, and the number of plasmons.

The problem becomes very similar to the usual hydrodynamic problem. It is necessary to know formally the general class of solutions which satisfy the four-plasmon integral and differ only little from the initial integral. Evidently N_k^{st}, which depends only on k/k_0, will satisfy this integral. A small change in k_0 of amount δk_0 gives the solution $\frac{N_k^{st}}{k_0} - \frac{\delta k_0 k}{k_0^2}$. If the plasmon kinetic energy $\omega_k^T = \omega_k - \omega_{pe}$ is introduced, we can assume that $N_k = N^{st} \left(\frac{\omega_k^T}{\omega_{k_0}^T} - \frac{2\delta k_0}{k_0} \cdot \frac{\omega_k^T}{\omega_{k_0}^T} \right)$. A gas of plasmons moving as a whole satisfies the four-plasmon integral. Since the plasmon energy changes by $\mathbf{k}\mathbf{v}$ when the reference system changes, where $\mathbf{v}$ is the velocity of the reference system, $N_k = N^{st} \left(\dfrac{\omega_k^T + \mathbf{kv}}{\omega_{k_0}^T} \right)$. Here $\mathbf{v}$ is a distribution parameter.

Thus,

$$N_k = N_k^{st} + \left(\mathbf{kv} - \omega_k^T \frac{2\delta k_0}{k_0} \right) \frac{\partial N_k^{st}}{\partial \omega_k^T}. \tag{8.168}$$

Expanding both $\mathbf{v}$ and δk_0 in Fourier integrals

$$\mathbf{v} = \int \mathbf{v}_{\varkappa, \, \nu} \, e^{i \varkappa \mathbf{r} - i \nu t} \, d\varkappa \, d\nu \tag{8.169}$$

and substituting into (8.165) and (8.166), we find the following dispersion equation [318]:

$$\nu^2 = \varkappa^2 \, 2 \, \frac{v_{Te}^2}{\omega_{pe}} \cdot \frac{\displaystyle\int (\omega_k^T)^2 \, \frac{\partial N_k^{st}}{\partial \omega_k^T} \, dk}{\displaystyle\int \omega_k^T \, \frac{\partial N_k^{st}}{\partial \omega_k^T} \, dk}. \tag{8.170}$$

If, as usual, we introduce the turbulence spectral energy density W_k^l ($\int W_k^l \, dk = W$), then (8.170) can be written as

$$\nu^2 = \varkappa^2 \frac{10 v_{Te}^4}{3 \omega_{pe}^2} \int k^2 \, W_k^l \, dk \, \frac{1}{W^l}. \tag{8.171}$$

Approximating W_k^l by $\dfrac{(\nu - 1)}{k_0} W \left(\dfrac{k_0}{k} \right)^\nu$ for $k > k_0$ and by 0 when $k < k_0$, we find the square of the second sound velocity to be ($\nu > 3$)

$$\frac{\nu^2}{\varkappa^2} = 10 v_{Te}^2 \frac{(\nu - 1)}{\nu - 3} \cdot \frac{v_{Te}^2 k_0^2}{3 \omega_{pe}^2}. \tag{8.172}$$

The second sound will be damped if we take into account pair collisions and spectral transfer for scattering on ions.

In conclusion we emphasize that, first, the difference between regular excitations and stochastic excitations may be small because the excitations are randomized when the instability develops.

Second, taking account of the resonance widths restricts, if not completely eliminates, the region in which a number of instabilities in turbulent plasma can arise. Finally, the radical changes in the properties of a weakly turbulent plasma at low frequencies must be related to the effects of strong turbulence. The definite success achieved here in understanding the physical nature of such changes, and the general theory of turbulent broadening of resonances and the correlation of turbulent fluctuations (Chapter 2) all can be considered as steps toward the development of a theory of strong turbulence.*

---

*Some recent developments in the theory of strong Langmuir turbulence are discussed in Chapter 9, which was written for this edition.

Development of the Concepts of Strong Langmuir Turbulence

9.1. Introduction

There has been, in recent years (1972–1975), a good deal of evolution in the concepts related to strong Langmuir turbulence. Although this choice of terminology may not be the best, since only the limits $W/n_0 T \ll 1$ and $\tau^N \omega_{pe} \gg 1$ have been considered, nevertheless the conditions for an instability in Langmuir turbulence (treated in Chapter 8, for example) are satisfied. Theoretical interest in this problem was stimulated by numerical calculations [321] showing that strong Langmuir waves concentrated in the region of small wave numbers create a flux of energy into the region of large wave numbers; this energy flux flows in a direction opposite to that encountered in weak, isotropic Langmuir turbulence (Chapter 4). As weak turbulence pumps energy into the small wave number region a maximum appears in the distribution. If the pumping level Q is large enough an instability appears in the turbulent state (see Chapter 8); thus, weak turbulence can cause strong turbulence. It is not sufficient to have uncorrelated low-frequency fluctuations to justify calling Langmuir turbulence strong Langmuir turbulence because weak turbulence always creates uncorrelated low-frequency fluctuations (see §4.4). The term strong Langmuir turbulence is reserved for the situation in which the spectra of the turbulence created by kinetic nonlinear decay and scattering interactions are unstable. These instabilities lead to the buildup of uncorrelated low-frequency disturbances and a restructuring of the spectra for Langmuir oscillations caused by these disturbances. Like all nonlinear instabilities, these instabilities give rise to the onset of a new, nonlinear process. Thus one can state that, in a state of strong Langmuir turbulence, new nonlinear processes become important. The nonlinear instabilities corresponding to these processes are usually called modulation instabilities if $k' \ll k_1$ (k' is the wave number of the disturbance, and k_1 is the characteristic wave number of the Langmuir oscillations), and aperiodic parametric instabilities if $k' \gg k_1$ (also called oscillating two-stream instabilities). These instabilities are contained in the dispersion equations obtained previously [see Eqs. (8.26), (8.57), (8.77), and so on].

The theory of strong Langmuir turbulence has been developed in many works, some of which employed specific models of turbulence as a gas of solitons and cavitons [324, 339, 342, 347, 349] and others which used more general statistical descriptions [319, 357, 353–356].

A complete theory for strong Langmuir turbulence is not yet available. However, the road toward a complete theory has produced some interesting new physical concepts, and it is the goal of the present chapter to critically analyze these concepts.

We begin with the problem of exciting strong Langmuir turbulence and its definition in light of the concepts of strong and weak turbulence introduced in Chapter 1. We then analyze certain new dynamical plasma motions, such as Langmuir solitons, cavitons, spikons, and so on, which can be excited when modulation instability develops. Special attention is given to possible interaction mechanisms which are important to models of strong Langmuir turbulence as a gas of these new dynamical motions. In the past, not all of the interactions examined here were taken into account in constructing these models; indeed many of them were not included. Analysis reveals a very complicated nature to the interactions of the new dynamical motions, and the structure of these motions is altered because of the interactions. Therefore it seems that the most adequate model is the statistical model of strong Langmuir turbulence (§9.5), which includes all the different motions and interactions, such as the modulation and oscillating two-stream nonlinear interactions. However, models which represent strong turbulence as a gas of solitons (for example) are especially graphic, and therefore especially attractive. But this clarity is somewhat illusory, especially for three-dimensional turbulence.

It is rather like the case of strong turbulence in an incompressible fluid; one speaks of interacting eddies even though eddies do not exist for more than one revolution.

9.2. Excitation of Strong Langmuir Turbulence

Instabilities in stationary Langmuir turbulence (treated in Chapter 8), and in particular rapidly developing potential instabilities (§§8.2-8.5) can bring plasma into a state of strong Langmuir turbulence. We are really dealing with strong turbulence when the modulation and oscillating two-stream instabilities arise in the case when the threshold conditions are greatly exceeded, i.e., when the criterion of Eq. (8.89) is satisfied with a large safety factor:

$$\int \frac{W^l_{k_1}\omega^2_{pe}}{k^2_1 v^2_{Te}}\, dk_1 \gg 12 n_0 (T_e + T_i), \qquad k_1 = |\, \mathbf{k}_1 \,|. \tag{9.1}$$

It is not difficult to understand why Langmuir turbulence would be called strong when (9.1) is satisfied. Consider the nonlinear shift in the frequency of Langmuir oscillations as determined from the real part of the nonlinear dielectric permeability ε^N_k [see Eqs. (2.121') and (2.115)]:

$$\varepsilon^N_k = -\, \frac{8\pi i}{\omega}\, \int \Sigma^{\text{eff}}_{k,k_1} I_{k_1}\, dk_1 \simeq \int \frac{(\mathbf{k}\cdot\mathbf{k}_1)^2 W^l_{k_1}\, dk_1}{k^2 k^2_1 n_0 (T_e + T_i)}, \tag{9.2}$$

$$k = \{\mathbf{k}, \omega\}, \qquad k_1 = \{\mathbf{k}_1, \omega_1\}.$$

The last approximation is written for $\omega_k - \omega_{k_1} \ll |\mathbf{k}-\mathbf{k}_1|\, v_{Ti}$. The condition $\varepsilon_k + \varepsilon^N_k = 0$ gives

$$\omega^N_k = \omega_{pe} + \frac{3k^2 v^2_{Te}}{2\omega_{pe}} - \frac{\omega_{pe}}{2n_0 (T_e + T_i)} \int \frac{(\mathbf{k}\cdot\mathbf{k}_1)^2}{k^2 k^2_1}\, W^l_{k_1}\, dk_1. \tag{9.3}$$

If we disregard the exact numerical values of the coefficients we find that when (9.1) is satisfied the nonlinear frequency shifts as given by (9.3) are much larger than the thermal dispersion effects. As defined in Chapter 1, strong turbulence is turbulence in which the nonlinear effects of interactions between, and correlations of, turbulent oscillations are much greater than the linear dispersion characteristics. However, in Eq. (9.3) the nonlinear corrections are small when compared with ω_{pe} but large when compared with $3k^2 v_{Te}^2/2\omega_{pe}$. The frequency ω_k^N does not appear in some of the nonlinear processes; rather, only differences between the frequencies of the interacting waves appear in the forms $\omega_k^N - \omega_{k_1}^N$, $\omega_k^N - \omega_{k_1}^N + \omega_{k_2}^N - \omega_{k_3}^N$, and so on. The frequency ω_{pe} drops out of all these differences and a chain of nonlinear interactions appears in which the frequencies are determined only by the interactions; in this sense the situation corresponds to strong turbulence. Of course, in a strict sense, this applies only to a homogeneous plasma, for which ω_{pe} is the same everywhere in the plasma. It is as if all frequencies were measured relative to ω_{pe} (and the energy of Langmuir quanta is measured relative to $\hbar\omega_{pe}$). In other words, eliminating ω_{pe} means that the envelopes of pulses of Langmuir packets are being considered. These envelopes have low frequencies and under the conditions expressed in Eq. (9.1) their properties are almost totally determined by the nonlinear dispersion characteristics.

Greater insight into strong Langmuir turbulence requires an exact definition of the state which shall be considered turbulent. The discussion in Chapter 1 concerning the nonreproducibility of observations in the turbulent regime and the necessity of a statistical treatment is especially important. This is also true of those comments in Chapter 1 concerning the level of energy contained in the oscillations; energy content is not a criterion for either strong or weak turbulence because regular, strong nonlinear waves can exist and can create regular, but nonturbulent, motions in plasma. It has also been mentioned that there are two mechanisms by which plasma can go into the turbulent state. In the first a large number of modes become unstable simultaneously and the system quite rapidly becomes turbulent from the very beginning. In the second one or more modes are excited from the beginning to a high level and the energy of these modes is redistributed among other modes by means of the nonlinear interactions. In the second case the initial state is not turbulent, but the motion becomes random after a period of time; i.e., there is a transition into the turbulent state.

There are evidently two ways to excite strong Langmuir turbulence. In the first approach strong Langmuir turbulence is formed from weak Langmuir turbulence by means of the modulation instabilities described in Chapter 8. This transition takes place somewhat automatically if the source of weak Langmuir

turbulence acts long enough. The energy in the Langmuir oscillations is transferred by nonlinear ion scattering to low wave
numbers, where it accumulates. This accumulation process is
impeded by absorption due to collisions. When the generation
power is low a balance is struck between generation and absorption due to collisions, and the turbulence remains weak if the
threshold in (9.1) is not exceeded. Since the characteristic k in
the region of the maximum will be $k_* = k_d \, ^{1}/_{3}\sqrt{m_e/m_i}$ the estimated
energy W_0^l at the maximum is

$$W_0^l = \frac{Q}{\nu_e} < \frac{m_e}{m_i} \, n_0 T. \tag{9.4}$$

according to Eq. (9.1). When Eq. (9.4) is satisfied the turbulence
is weak regardless of the generation time, and it can be described
by the equations of Chapter 4. If Eq. (9.4) is not fulfilled the spectra
given in Chapter 4 will be quasistationary and valid only until a
quantity of energy for which (9.4) is not satisfied is accumulated
in the fundamental scale. Subsequently, the turbulence becomes
unstable and may go into a state of strong Langmuir turbulence.
As pointed out in the previous chapter, it is practically impossible
for modulation instability to appear when $k_l > k_d \, ^{1}/_{3}\sqrt{m_e/m_i}$. Thus,
turbulence must accumulate in regions of small $k_l < k_*$ before it
goes over into the strong regime. Strong turbulence is excited
in this manner in systems wherein a large number of modes are
already excited and randomization has taken place. It is clear
that in this case the stochastic description can be used both in
the transition regime between weak and strong turbulence and
for the strong turbulence itself. It is also clear that the transition to strong turbulence must be accompanied by increases in
entropy and the level of randomization. But there is still another
way to excite strong turbulence. We shall assume that the source
exciting oscillations is concentrated at relatively small wave
numbers k_l, for example where $k_l < k_*$, and that it is sufficiently
monochromatic, i.e., only one mode $k_l = k_l^0$ is excited. The source
could be either a monoenergetic electron beam or a coherent
monochromatic high-frequency field (or laser radiation) which
parametrically excites one mode of the Langmuir waves. We
further assume that before this mode can interact with other
modes and the Langmuir field is randomized, its intensity reaches
the threshold for a modulation instability. This means that the

modulation disturbances come into play before the nonlinear mode
saturation, which is related to the transfer of stored energy into
other modes by means of decay and scattering. Thus the modulation perturbations can themselves act as a mechanism for stabilizing the mode. In order to address this question we must know
if modulation instabilities can exist for a regular field such as
a single Langmuir mode. Let us consider the dynamical nonlinear equations for a Langmuir field, including both quadratic
and cubic nonlinearities [319]:

$$i\omega\varepsilon_k E_k^+ = 8\pi \int S_{k,k_1,k_2} E_{k_1}^+ E_{k_2} \delta(k - k_1 - k_2)\, dk_1\, dk_2 +$$

$$+ 8\pi \int \Sigma_{k,k_1,k_2,k_3} E_{k_1}^+ E_{k_2}^+ E_{k_3}^- \delta(k - k_1 - k_2 - k_3)\, dk_1\, dk_2\, dk_3, \quad (9.5)$$

$$i\omega\varepsilon_k E_k = 8\pi \int S_{k,k_1,k_2} E_{k_1}^+ E_{k_2}^- \delta(k - k_1 - k_2)\, dk_1\, dk_2. \quad (9.6)$$

Here E_k^+ is the positive frequency part of the Langmuir wave
(frequency close to ω_{pe}) and E_k^- is the corresponding negative
part (frequency close to $-\omega_{pe}$). Because of the delta function in
the term quadratic in the field in (9.5) the low-frequency field
E_k must appear (frequencies much smaller than $\pm\omega_{pe}$); in the first
approximation it is expressed in terms of the Langmuir field by
Eq. (9.6). The nonlinear currents S_{k,k_1,k_2,k_3} and $\Sigma_{k, k_1, k_2, k_3}$ in Eqs.
(9.5) and (9.6) are symmetrized in the last two indices, and it has
been taken into account that $\Sigma_{k, k_1, k_2, k_3}$ is a maximum when the
signs of the frequencies of the first two arguments are the same.

We find from (9.5) and (9.6) that

$$\varepsilon_k E_k^+ = \frac{8\pi}{i\omega} \int \tilde{\Sigma}_{k,k_1,k_2,k_3} E_{k_1}^+ E_{k_2}^+ E_{k_3}^- \delta(k - k_1 - k_2 - k_3)\, dk_1\, dk_2\, dk_3 \quad (9.7)$$

and in the approximation $\omega - \omega_1 \ll |k - k_1|\, v_{Te}$, $\omega_{pe} \gg kv_{Te}$, $\omega \ll \omega_{pi}$
we have

$$\tilde{\Sigma}_{k,k_1,k_2,k_3} = \frac{e^2\omega}{8\pi i\omega_{pe}^4 m_e^2} \frac{(\mathbf{k} \cdot \mathbf{k}_1)(\mathbf{k}_2 \cdot \mathbf{k}_3)\varepsilon_{k-k_1}^{(i)}(\varepsilon_{k-k_1}^{(e)} - 1)(\mathbf{k} - \mathbf{k}_1)^2}{kk_1k_2k_3\varepsilon_{k-k_1}}. \quad (9.8)$$

In some sense Eq. (9.7) is the analog to (9.2) or (more precisely)
to $\varepsilon_k + \varepsilon_k^N = 0$. However, being a description of a regular field,
it contains more information than an averaged equation. Strictly
speaking the nonlinear motions described by Eq. (9.7) are not

monochromatic waves. This makes it pointless to raise the question of the stability of monochromatic waves, so we shall proceed as follows. Consider a field with frequencies near ω_{pe} and with $\Delta\omega \ll \Delta k v_{Ti}$. Then

$$\bar{\Sigma}_{k,k_1,k_2,k_3} \simeq \frac{\omega}{8\pi i 4\pi n_0(T_e + T_i)} \frac{(\mathbf{k} \cdot \mathbf{k}_1)(\mathbf{k}_2 \cdot \mathbf{k}_3)}{k k_1 k_2 k_3}. \qquad (9.9)$$

Transforming into a coordinate representation and introducing the quantities

$$\mathbf{E}^{\pm}(\mathbf{r},\, t) = e^{\pm i\omega_{pe}t} \int \frac{\mathbf{k}}{k}\, E_k^{\pm} e^{-i\omega t + i\mathbf{k}\mathbf{r}}\, dk, \qquad (9.10)$$

that is, isolating the rapidly oscillating terms (with frequency ω_{pe}), we use Eq. (9.9) and

$$\frac{1}{\omega_{pe}}\, \frac{\partial}{\partial t}\, \ln E^+(\mathbf{r},\, t) \ll 1, \qquad (9.11)$$

to write (9.7) as

$$\Delta\left(\frac{2i}{\omega_{pe}}\, \frac{\partial}{\partial t} + \frac{3v_{Te}^2\Delta}{\omega_{pe}^2} \right) E^+(\mathbf{r},\, t) = -\frac{1}{4\pi n_0(T_e + T_i)}\, \text{grad div } \mathbf{E}^+(\mathbf{r},\, t)\, |\, E^+(\mathbf{r},\, t)\,|^2.$$
$$(9.12)$$

There is a solution to this equation which corresponds to the condition $|\mathbf{E}^+(\mathbf{r},\, t)|^2 = \text{const} = E_0^2$, where

$$\mathbf{E}^+(\mathbf{r},\, t) = \frac{\mathbf{k}_0}{k_0}\, E_0 e^{-i\Omega_0 t + i\mathbf{k}_0\mathbf{r}},$$
$$(9.13)$$
$$\Omega_0 = \frac{3k_0^2 v_{Te}^2}{2\omega_{pe}} - \frac{\omega_{pe}E_0^2}{8\pi n_0(T_e + T_i)}.$$

Thus, the solution to (9.12) is a wave with a nonlinear frequency shift, analogous to the shift in (9.3). We must point out, however, that this solution is the limit of a wave packet in which $\Delta\omega \ll \Delta k v_{Ti}$ or $k_0 \ll k_*$. We will examine its stability by assuming that in (9.7)

$$E_k^+ = E_k^{+(0)} + \delta E_k^+,$$
$$(9.14)$$
$$E_k^- = E_k^{-(0)} + \delta E_k^-,$$

where $E_k^{+(0)}$ and $E_k^{-(0)}$ describe the field in (9.13):

$$E_k^{+(0)} = E_0\,\delta(\omega - \omega_{k_0}^N)\,\delta(\mathbf{k} - \mathbf{k}_0),$$

$$E_k^{-(0)} = E_0\,\delta(\omega + \omega_{k_0}^N)\,\delta(\mathbf{k} + \mathbf{k}_0), \tag{9.15}$$

$$\omega_{k_0}^N = \omega_{pe} + \Omega_0 = \omega_{pe} + \frac{3k_0^2 v_{Te}^2}{2\omega_{pe}} - \frac{\omega_{pe}E_0^2}{8\pi n_0(T_e + T_i)}.$$

The perturbations δE_k^+ and δE_k^- arise in a region of frequencies and wave numbers which does not coincide with $\omega_{k_0}^N$ and $\mathbf{k}_0$. It is convenient to write (9.7) and (9.8) in an equivalent form:

$$\varepsilon_k E_k^\pm = \int E_{k_1}^\pm \frac{(\mathbf{k} \cdot \mathbf{k}_1)}{kk_1} \frac{n'_{k-k_1}}{n_0}\,dk_1, \tag{9.16}$$

$$n'_{k'} = -\frac{1}{4\pi T_e} \frac{\varepsilon_{k'}^{(i)}}{\varepsilon_{k'}} \int E_{k_2}^+ E_{k_3}^- \frac{(\mathbf{k}_2 \cdot \mathbf{k}_3)}{k_2 k_3}\,\delta(k' - k_2 - k_3)\,dk_2\,dk_3. \tag{9.17}$$

The quantity $n'_{k'}$ is the change in plasma density. The change in density due to the pumping wave (9.13) is

$$n_{k'}^{'(0)} = -\frac{E_0^2}{4\pi(T_e + T_i)}\,\delta(k') \tag{9.18}$$

and it is responsible for the nonlinear frequency shift given in (9.15). Thus, by introducing

$$\tilde{\varepsilon}_k = \varepsilon_k + \frac{E_0^2}{4\pi n_0(T_e + T_i)} \tag{9.18'}$$

Eq. (9.16) can be written in a form in which the first part contains only the density change $\delta n'_k$:

$$\tilde{\varepsilon}_{k+k_0} E_{k+k_0}^+ = E_0 \frac{(\mathbf{k} + \mathbf{k}_0) \cdot \mathbf{k}_0}{|\mathbf{k} + \mathbf{k}_0|\,k_0} \frac{\delta n'_k}{n_0}, \tag{9.19}$$

$$\tilde{\varepsilon}_{k-k_0} E_{k-k_0}^- = E_0 \frac{(\mathbf{k} - \mathbf{k}_0) \cdot \mathbf{k}_0}{|\mathbf{k} - \mathbf{k}_0|\,k_0} \frac{\delta n'_k}{n_0}, \tag{9.20}$$

$$\frac{\delta n'_k}{n_0} = -\frac{1}{4\pi n_0 T_e} \frac{\varepsilon_k^{(i)}}{\varepsilon_k} E_0 \left(\delta E_{k+k_0}^+ \cdot \frac{(\mathbf{k} + \mathbf{k}_0) \cdot \mathbf{k}_0}{|\mathbf{k} + \mathbf{k}_0|\,k_0} + \delta E_{k-k_0}^- \frac{(\mathbf{k} - \mathbf{k}_0) \cdot \mathbf{k}_0}{|\mathbf{k} - \mathbf{k}_0|\,k_0} \right).$$

$$\tag{9.21}$$

Here, in the subscripts,

$$k_0 = \{\mathbf{k}_0, \omega_{\mathbf{k}_0}^N\}. \tag{9.22}$$

We therefore arrive at the following dispersion equation:

$$1 = -\frac{E_0^2}{4\pi n_0 T_e} \frac{\varepsilon_k^{(i)}}{\varepsilon_k} \left\{ \frac{((\mathbf{k} + \mathbf{k}_0) \cdot \mathbf{k}_0)^2}{(\mathbf{k} + \mathbf{k}_0)^2 k_0^2} \frac{1}{\tilde{\varepsilon}_{k+k_0}} + \frac{((\mathbf{k} - \mathbf{k}_0) \cdot \mathbf{k}_0)^2}{(\mathbf{k} - \mathbf{k}_0)^2 k_0^2} \frac{1}{\tilde{\varepsilon}_{k-k_0}} \right\}. \tag{9.23}$$

Using $\varepsilon_k \approx 1 - \omega_{pe}^2/\omega^2 - \omega_{pe}^2 3k^2 v_{Te}^2/\omega^4$ for the high-frequency approximation to $\tilde{\varepsilon}_{k \pm k_0}$ we find that for $k \ll k_0$, $v_{Te} \gg \omega/k \gg 3 |\mathbf{k} + \mathbf{k}_0| v_{Te}^2/\omega_{pe}$, and $E_0^2/4\pi n_0 (T_e + T_i) \ll 1$ the following approximate relationship holds:

$$1 = -\frac{E_0^2 \varepsilon_k^{(i)}}{4\pi n_0 T_e \varepsilon_k} \left\{ \frac{3k^2 v_{Te}^2}{2\omega^2} + \frac{2\omega_{pe}(\mathbf{k} \cdot \mathbf{k}_0)(\mathbf{k} \times \mathbf{k}_0)^2}{\omega k_0^2((k^2 + k_0^2)^2 - 4(\mathbf{k} \cdot \mathbf{k}_0)^2)} \right\}. \tag{9.24}$$

For perturbations along $\mathbf{k}_0$ the second term in (9.24) can be neglected, so that when $\omega/k \gg v_{Ti}$, $\omega \ll \omega_{pi}$ we obtain

$$\left(\frac{\omega}{k}\right)^4 - v_s^2 \left(\frac{\omega}{k}\right)^2 - v_s^2 (v_*^l)^2 = 0, \tag{9.25}$$

$$v_*^l = \sqrt{\frac{3E_0^2}{8\pi n_0 m_e}}. \tag{9.26}$$

The solution (9.25) agrees exactly with (8.97), and (9.26) is analogous to (8.93). Thus, the modulation instabilities of a regular field are very similar to the instabilities of a turbulent field.

In the following discussion it is important to know something of the role of oblique perturbations described by the second term in (9.24); this term is comparable to the first term when the angle between $\mathbf{k}$ and $\mathbf{k}_0$ is such that

$$\cos \theta \sin^2 \theta > \frac{k_0^2}{4k^2} \frac{k v_{gr}^{(0)}}{\omega}, \qquad v_{gr}^{(0)} = \frac{3k_0 v_{Te}^2}{\omega_{pe}}. \tag{9.27}$$

Because we have assumed that $\omega \gg k v_{gr}^{(0)}$ when $k \lesssim k_0$ the second term in (9.24) is comparable to the first term even when $\theta \ll 1$. This means that instabilities mainly develop perpendicular to the direction of the wave field. We hasten to point out that the condition $k \ll k_0$ is the condition that makes the instability a modulation instability, but it places no restrictions on the applicability of (9.24). Thus, when $k \gg k_0$, i.e., for an oscillating two-stream instability, the necessary limitation is the condition presented above: $\omega/k \gg (3|\mathbf{k} + \mathbf{k}_0|/\omega_{pe})v_{Te}^2 \simeq 3kv_{Te}^2/\omega_{pe}$. Taking into account that when $v_*^l \gg v_s$, $\gamma \approx k\sqrt{v_s v_*^l}$, we estimate that

$$\frac{3k_{max} v_{Te}^2}{\omega_{pe}} \simeq \sqrt{v_s v_*^l}, \qquad k_{max} \simeq k_* \left(\frac{E_0^2/4\pi n_0 T_e}{m_e/m_i} \right)^{1/4} \tag{9.28}$$

and from this we estimate that the maximum growth rate along k_0 is

$$\gamma_{max} \simeq k_{max} \sqrt{v_s v_*^l} \simeq \omega_{pe} \sqrt{\frac{m_e}{m_i} \frac{E_0^2}{4\pi n_0 T_e}}. \tag{9.29}$$

Oblique perturbations develop faster than the longitudinal perturbations when $k \approx k_0$ and $\theta > \theta_{min}$, where

$$\theta_{min}^2 \simeq \frac{k_0}{k_*} \left(\frac{m_e/m_i}{E_0^2/4\pi n_0 T_e} \right)^{1/4}. \tag{9.30}$$

Thus, when $k_0 \ll k_*$ and $E_0^2/4\pi n_0 T_e \gg m_e/m_i$ the oblique instabilities can become important. If the second term in (9.24) dominates and $\omega/k \gg v_s$, $k \approx k_0$, (9.24) gives

$$\gamma_\perp \simeq \omega_{pe} \left(\frac{k_0}{k_d} \right)^{2/3} \left(\frac{m_e}{m_i} \frac{E_0^2 |\cos\theta|}{4\pi n_0 T_e} \right)^{1/3}. \tag{9.31}$$

This growth rate is much larger than the rate obtained for the same values of $k \approx k_0$ when the perturbation of the oscillations is along k_0:

$$\gamma_\perp > \gamma_{k_0} \simeq \omega_{pe} \left(\frac{k_0}{k_d} \right) \left(\frac{m_e}{m_i} \frac{E_0^2}{4\pi n_0 T_e} \right)^{1/4} \tag{9.32}$$

if $k_0 \ll k_*$ and $E_0^2/4\pi n_0 T_e \gg m_e/m_i$. However, it is smaller than the maximum growth rate (9.29). The growth rates described by the first term of (9.24), which leads to (9.29), are angle-independent and hold even for oblique perturbations. The results of this analysis may be summarized as follows. First, regular waves can excite aperiodic modulation instabilities similar to those which are excited by Langmuir turbulence which is sufficiently intense.

Second, the excitation is essentially three-dimensional, especially when $k \lesssim k_0$, where the instability is altered and oblique perturbations are preferentially excited, and when $k \gg k_0$ with $k \approx k_{max}$, when waves are excited at any angle relative to the original wave with identical growth rates.

Similar alterations in the modulation instabilities arise in the turbulent state, but in certain cases of one-dimensional, isotropic turbulence considered in Chapter 8, terms analogous to the second term in (9.24) make no contribution (as one can easily verify).

We must also emphasize that since (9.24) holds even when $k > k_0$, the growth rate in (9.29) is valid for a completely uniform field with $k_0 = 0$ [320].

Let us consider the consequences of generating strong Langmuir turbulence. Assume that at the outset we have a regular field (a single mode of the Langmuir field) which is strong enough to excite modulation perturbations. Then either the development of instability excites a single-mode nonlinear motion, or a complicated nonlinear motion is excited at once. However, it is impossible to say anything about turbulence. Further developments can lead to a further instability in the single-mode nonlinear motion, a process similar to the breaking up of eddies in liquids, and the reverse action of these multimode motions on the original Langmuir field. This same effect will occur when multimode motion is excited from the very beginning. In the final stage the modulation disturbances created by the field act on the Langmuir field and cause it to become multimode and randomized. Only this last state can be called a state of strong Langmuir turbulence. To attain this state from a regular field requires the development of complex motions and random fields.

There are thus two ways to excite strong turbulence; either from weak turbulence in which the oscillations are concentrated at low k_1 or from a strong regular field followed by randomization. The latter approach has been analyzed numerically in [321, 322] for a one-dimensional model. It should be clear from our analysis above that a one-dimensional model is inadequate for practical applications of the theory because of the oblique perturbations, which play as important a role as do the perturbations which develop along the field. However, the one-dimensional model does produce some qualitative results which are of interest. The analysis [321, 322] shows that the original regular field is actually transformed into a multimode field. Since $k_0 = 0$ in the work reported in [321, 322] it is natural that modes for which $k \neq 0$ were excited. However, a very significant feature of these calculations was that the broadening of the spectrum of Langmuir waves continued up to the point where $k_1 \approx k_d$. We can therefore conclude that energy flowed toward larger k values. This direction of flow is opposite to that encountered in the weakly turbulent regime. It was just this result which attracted the attention of many researchers to the problem of strong Langmuir turbulence, even though it is obvious that a one-dimensional calculation serves only to indicate a possible result for the true three-dimensional case. For large $k_1 \gg k_0$, where k_0 corresponds to the energy-containing region, one would expect a universal spectrum of turbulence. The numerical calculations in [321, 322] showed that this spectrum is of the form

$$W_k \approx \frac{\text{const}}{k^2}.$$
(9.33)

The accuracy with which the spectrum exponent $\nu = 2$ is defined is not very great. Although one can expect that the spectrum of strong Langmuir turbulence which is finally established will not depend on the excitation method, and will be nearly the same in the two limiting cases considered above, one could hardly expect the transitions to this state of stationary turbulence to be identical. This is because low-frequency disturbances are always present in weak turbulence (§4.4), and they act as initial disturbances in the development of modulation perturbations. Moreover, excitation by regular fields is accompanied by complex, and possibly long-lasting, randomization processes.

9.3. Solitons and Cavitons

Certain nonlinear motions which arise when modulation instability develops have been subjects of intensive scrutiny. These motions are dynamical and, strictly speaking, have no direct relationship to strong Langmuir turbulence, just as individual, isolated eddies in liquids have no direct bearing on its turbulence. As emphasized in Chapter 1, turbulence arises in an incompressible fluid only when eddies of different scales interact strongly with one another. Since the energy associated with an eddy of a given scale is transferred to an eddy of a smaller

scale in a time which is shorter than the time required for one
rotation of the fluid in the eddy, it is rather difficult to discuss
the concept of an eddy in the turbulent regime. Eddies interact
so strongly that they lose their individuality. However, the concept
of an eddy is valuable because an eddy is some sort of regular
motion which would exist if there were no interactions. We assume
that such regular, nonlinear motions as solitons, cavitons, and
spikons play a similar role in strong Langmuir turbulence. While
there is a general theorem in fluids which states that any incompressible motion is rotational, there is no proof for Langmuir
waves that the only stable nonlinear motions are those solitons,
cavitons, and spikons mentioned above. But there is a proof that
subsonic, one-dimensional motions must decay into individual
solitons [326]. Since regular motions cannot be formed from a
turbulent state, such as weak turbulence concentrated near small
wave numbers, one usually considers the formation of solitons,
spikons, and other motions from initially regular, strong Langmuir waves [323]. Langmuir solitons were first studied in [324]
and are a packet of Langmuir waves whose envelope is an isolated
pulse traveling in some direction at a constant velocity. Equation
(9.10) describes the envelopes for the amplitude $E^+(\mathbf{r}, t)$ of the
Langmuir field [the frequency ω_{pe} has been eliminated from (9.10)].
After multiplying (9.16) by $k = (\mathbf{k} \cdot \mathbf{k})/k$ and using (9.11) we find

$$\operatorname{div}\left(\frac{2i}{\omega_{pe}}\frac{\partial}{\partial t} + \frac{3v_{Te}^2}{\omega_{pe}^2}\Delta\right)E^+(\mathbf{r}, t) = \operatorname{div}\frac{n'(\mathbf{r}, t)}{n_0}E^+(\mathbf{r}, t). \qquad (9.34)$$

The total plasma density is $n = n_0 + n'(\mathbf{r}, t)$; $\omega_{pe}^2 = 4\pi n_0 e^2/m_e$. Equation (9.34) can be obtained from Poisson's equation for the oscil-
lations by assuming that there are density fluctuations present
which are slow compared with ω_{pe}^{-1} and small ($n'/n_0 \ll 1$); that is,
in the adiabatic approximation the plasma frequency is replaced
with $4\pi(n_0 + n')e^2/m_e$.

An equation for the low-frequency density fluctuations $n'(\mathbf{r}, t)$
is obtained from (9.17), which also contains the low-frequency
kinetic effects (in the imaginary parts of ε_k^e and ε_k^i, which describe nonlinear Landau damping). If we are interested in the re-
gion where $v_{Ti} \ll \omega/k \ll v_{Te}$ and Landau damping is neglected, then

$$\frac{\varepsilon_k^{(i)}}{\varepsilon_k} \simeq -\frac{k^2 v_s^2}{\omega^2 - k^2 v_s^2} \qquad (9.35)$$

so that

$$\frac{\partial^2}{\partial t^2} n'(\mathbf{r}, t) - v_s^2 \Delta n'(\mathbf{r}, t) = \Delta \frac{|E^+(\mathbf{r}, t)|^2}{4\pi m_i}. \tag{9.36}$$

One might expect that when $\omega/k \ll v_{Ti}$ the density variation n' would be strongly damped by Landau damping due to ions. But Landau damping due to ions enters into both the numerator and denominator of the ratio $\varepsilon_k^{(i)}/\varepsilon_k \approx \varepsilon_k^{(i)}/(\varepsilon_k^e + \varepsilon_k^i)$ and cancels out in the first approximation, so that when $\omega/k \ll v_{Ti}$,

$$\frac{\varepsilon_k^{(i)}}{\varepsilon_k} \approx \frac{T_e}{T_e + T_i} \tag{9.37}$$

and

$$n'(\mathbf{r}, t) = -\frac{|E^+(\mathbf{r}, t)|^2}{4\pi(T_e + T_i)}. \tag{9.38}$$

We note that the energy density in the Langmuir field, averaged over the frequency ω_{pe}, is given by the expression

$$W = \frac{E^2(\mathbf{r}, t)}{8\pi} \frac{\partial}{\partial\omega} \omega\varepsilon \simeq \frac{|E^+(\mathbf{r}, t)|^2}{2\pi}. \tag{9.39}$$

Equation (9.38) is a first approximation because in the nonlinear regime the distribution of energy between the field and the particles will not be equal. Thus the factor $(\partial/\partial\omega)\omega\varepsilon \approx 2$, which comes from the linear theory, will be somewhat different from 2 in the nonlinear regime. The difference is a number of order $W/n_0 T_e \ll 1$. According to [324] Langmuir solitons are one-dimensional waves. Assuming that they propagate along the x axis, Eqs. (9.34) and (9.36) give

$$\left(i\frac{\partial}{\partial\tau} + \frac{\partial^2}{\partial\xi^2}\right) E(\xi, \tau) = \nu E(\xi, \tau), \tag{9.40}$$

$$\frac{\partial^2}{\partial\tau^2} \nu - \frac{\partial^2}{\partial\xi^2} \nu = \frac{\partial^2}{\partial\xi^2} |E(\xi, \tau)|^2. \tag{9.41}$$

We have used the dimensionless variables

$$\tau = \frac{t}{t_0}; \quad \xi = \frac{x}{t_0 v_s}; \quad E(\xi, \tau) = E^+(x, t)\sqrt{\frac{1}{4\pi n_0 T_e \mu}};$$

$$t_0 = \frac{2}{\mu\omega_{pe}}; \quad \mu = \frac{4m_e}{3m_i}; \quad \nu = \frac{n'}{n_0\mu}. \tag{9.42}$$

If the plasma density fluctuation connected with a soliton moves at a constant velocity v_0 then $\nu = \nu(\xi - u\tau)$, where $u = v_0/v_s$ is the velocity measured in units of the sound velocity. Equation (9.41) then gives

$$\nu = -\frac{1}{1 - u^2} |E|^2, \tag{9.42'}$$

demonstrating that density rarefactions appear only for subsonic solitons, for which $u < 1$. Thus, only this type of soliton can exist because a densification with $\nu > 0$ does not trap Langmuir oscillations and the solutions to (9.40) do not have the character of a single pulse. It follows from (9.42') that

$$E = |E| e^{i\phi}, \tag{9.43}$$

in which it is required that $|E|$ depend only on $\xi - u\tau$. Substituting (9.43) into (9.40) and separating the real and imaginary parts, we arrive at equations for ϕ and $|E|$. The stipulation that $|E|$ depend on $\xi - u\tau$ and that ϕ be finite when $\xi - u\tau \to 0$ imposes strict limitations on the solutions to the equation for ϕ. We obtain

$$\phi = -\Omega\tau + \frac{u}{2}(\xi - u\tau) + \phi_0, \tag{9.44}$$

where Ω is a constant independent of $\xi - u\tau$. The equation for $|E|$ with the boundary conditions that $|E| \to 0$ when $\xi - u\tau \to \pm\infty$ gives a relationship between the amplitude of the soliton $E_0 = |E|_{\xi = u\tau}$ and the frequency Ω:

$$\Omega = -\frac{u^2}{4} - \frac{E_0^2}{2(1 - u^2)}. \tag{9.45}$$

It follows from these results that the phase ϕ is not a function of $\xi - u\tau$ alone. The nonlinear frequency shift in (9.45) is analogous to the nonlinear shift in (9.15). The amplitude is found to be

$$|E| = \frac{E_0}{\cosh[(\xi - u\tau)E_0/\sqrt{2(1 - u^2)}]}. \tag{9.46}$$

Thus the characteristic width of the soliton $\xi_0 = \sqrt{2(1 - u^2)}/E_0$ is inversely proportional to the amplitude E_0 and goes to zero when $u \to 1$. If the field is written as a Fourier series

$$E = |E| e^{i\phi} = \int E_{\varkappa,\lambda} e^{i\varkappa\xi - i\lambda\tau} \, d\varkappa \, d\lambda, \tag{9.47}$$

then

$$E_{\varkappa,\lambda} = \frac{\sqrt{1-u^2}}{\sqrt{2}} \, \frac{\delta(\lambda - \varkappa u - \Omega)e^{-i\phi_0}}{\cosh[(\varkappa - u/2)\pi\sqrt{1-u^2}/\sqrt{2}\,E_0]}, \tag{9.48}$$

$$\nu_{\varkappa,\lambda} = -\frac{\varkappa\delta(\lambda - \varkappa u)}{\sinh[\varkappa\pi\sqrt{1-u^2}/\sqrt{2}\,E_0]}. \tag{9.48'}$$

Equations (9.40) and (9.41) have three integrals of motion [323, 325]; they are the number N of plasmons,

$$N = \int |E|^2 \, d\xi = 2E_0\sqrt{2(1-u^2)}, \tag{9.49}$$

the energy V,

$$V = \int \left\{ \left| \frac{\partial E(\xi,\tau)}{\partial \xi} \right|^2 + \nu(\xi,\tau)\,|E(\xi,\tau)|^2 + \frac{\nu^2(\xi,\tau)}{2} + \frac{1}{2}\left| \int_\xi^\xi d\xi' \, \frac{\partial \nu(\xi',\tau)}{\partial \tau} \right|^2 \right\} d\xi$$

$$= E_0 \frac{u^2\sqrt{1-u^2}}{\sqrt{2}} + \frac{\sqrt{2}\,(5u^2-1)E_0^3}{3(1-u^2)^{3/2}}, \tag{9.50}$$

and the momentum P,

$$P = \frac{1}{2} \int \left[iE(\xi,\tau)\,\frac{\partial E^*(\xi,\tau)}{\partial \xi} - iE^*(\xi,\tau)\,\frac{\partial E(\xi,\tau)}{\partial \xi} + 2\nu(\xi,\tau) \int_\xi^\xi d\xi' \, \frac{\partial \nu(\xi',\tau)}{\partial \tau} \right] d\xi$$

$$= \frac{u\sqrt{1-u^2}\,E_0}{\sqrt{2}} + \frac{4\sqrt{2}\,uE_0^3}{3(1-u^2)^{3/2}}. \tag{9.51}$$

The values of the integrals are taken from [325] and are for the soliton described by (9.46). Thus, when the soliton energy is finite its amplitude goes to zero when $u \to 1$. In terms of dimensional variables the condition $E_0 \gg 1$ means that

$$\frac{|E^+|^2}{2\pi n_0 T_e} \gg \frac{m_e}{m_i}.$$

Solitons may also have velocities much smaller than the thermal velocity of the ions: $u \ll \sqrt{T_i/T_e}$. Then Eq. (9.38) must be used for the density fluctuations. The negative sign in (9.38) indicates density rarefactions, and these can trap Langmuir waves to form solitons. Using dimensionless variables we find using (9.42) that

$$\phi = -\Omega\tau + \frac{u}{2}(\xi - u\tau) + \phi_0, \quad \Omega = -\frac{u^2}{4} - \frac{E_0^2 T_e}{2(T_e + T_i)}, \quad (9.52)$$

$$|E| = \frac{E_0}{\cosh[(\xi - u\tau)E_0\sqrt{T_e/2(T_e + T_i)}]}. \quad (9.53)$$

Landau damping of a soliton is determined by the imaginary parts of (9.17). When $u \gg \sqrt{T_i/T_e}$ only Landau damping by electrons is important, and then (9.17) gives

$$(\varepsilon_k^{(e)} - 1)\frac{\varepsilon_k^{(i)}}{\varepsilon_k} = -\frac{k^2 v_s^2 \omega_{pe}^2}{(\omega^2 - k^2 v_s^2)k^2 v_{Te}^2}\left(1 - i\sqrt{\frac{\pi}{2}}\frac{k^2 v_s^2}{(\omega^2 - k^2 v_s^2)}\frac{\omega}{|k|v_{Te}}\right), \quad (9.54)$$

$$\delta\nu_{\varkappa,\lambda}^L = -i\nu_{\varkappa,\lambda}^{(0)}\sqrt{\frac{\pi m_e}{2m_i}}\frac{\lambda|\varkappa|}{\lambda^2 - \varkappa^2},$$

where $\nu_{\varkappa\lambda}^{(0)}$ corresponds to (9.48') in the first approximation. Thus we obtain

$$\delta\nu^L(\xi, \tau) = \sqrt{\frac{\pi m_e}{2m_i}}\frac{2u}{1 - u^2}\int_0^\infty \frac{\varkappa\sin[\varkappa(\xi - u\tau)]\,d\varkappa}{\sinh[\varkappa\pi\sqrt{1 - u^2}/\sqrt{2}\,E_0]}. \quad (9.55)$$

Therefore, when Landau damping of low-frequency perturbations is taken into account (9.40) is replaced by

$$\left(i\frac{\partial}{\partial\tau} + \frac{\partial^2}{\partial\xi^2}\right)E(\xi, \tau) = (\nu^{(0)} + \delta\nu^L)E(\xi, \tau), \quad (9.56)$$

where $\delta\nu^{(0)}$ obeys the old equation (9.41). When $\delta\nu^L$ is taken into account the momentum is no longer a conserved quantity

$$\frac{dP}{dt} = \int \delta\nu(\xi, \tau)\frac{\partial}{\partial\xi}|E(\xi, \tau)|^2\,d\xi = -\frac{6\zeta(3)}{\sqrt{2}\,\pi^{5/2}}\sqrt{\frac{m_e}{m_i}}\frac{uE_0^4}{(1 - u^2)^2}, \quad (9.57)$$

where $\zeta(3) \approx 1.08$ is the Riemann function with an argument of 3. The last equality is obtained by using $\delta\nu$ from (9.55) and $|E|^l$ from (9.46). We can assume that the soliton parameters E_0 and u are slowly varying functions of time because of Landau damping.

By substituting the soliton momentum as given by (9.51) into the left side of (9.57) we obtain one equation for variations in these parameters.

On the other hand, the conservation of the number of quanta is valid for any $\delta \nu$ and follows from Eq. (9.56), which gives

$$E_0 \sqrt{1 - u^2} = \text{const} = E_0^{(0)} \sqrt{1 - u_0^2}. \tag{9.58}$$

We are most interested in the damping of strong solitons with $E_0^{(0)} \gg 1 - u_0^2$, for which the soliton momentum can be approximated by the second term in (9.51). We find from (9.57) and (9.58) that

$$(1 + 5u^2) \frac{du}{d\tau} = - u \frac{9 \zeta(3)}{4 \pi^{5/2}} \frac{E_0^{(0)}}{(1 - u^2)^{3/2}} \sqrt{\frac{m_e}{m_i}}. \tag{9.59}$$

The soliton damping reduces to deceleration. According to (9.58) only a soliton whose initial velocity is near the speed of sound experiences a marked change in amplitude. But according to (9.59) such solitons lose speed very rapidly. Solitons with $u_0 \ll 1$ decrease their velocity at an exponential rate:

$$u = u_0 e^{-\tau/\tau_*}, \qquad \tau_* = \frac{4 \pi^{5/2}}{9 \zeta(3)} \sqrt{\frac{m_i}{m_e}} \frac{1}{E_0^{(0)}}. \tag{9.60}$$

Recall that $E_0^{(0)} = \sqrt{|E_0^+|^2 / 4\pi n_0 T_e} u$, i.e., even when $|E_0^+|^2 / 4\pi n_0 T_e \simeq 1$, $\tau_* \approx 5.5$, which means that the damping time $t_* = t_0 \tau_* = 8.3/\omega_{pe}\mu$ is about $\sqrt{m_i/m_e}$ times larger than the ion plasma period. Equation (9.59) is valid only when

$$u \gg u_* = \sqrt{\frac{T_i}{T_e}}, \tag{9.61}$$

i.e., at velocities greater than the ion thermal velocity. When $T_i \ll T_e$ the criterion $1 \gg u \gg u_*$ can be satisfied. The energy of such decelerated solitons will be negative:

$$V = - \frac{\sqrt{2}}{3} E_0^3, \qquad E_0 \gg u_*. \tag{9.62}$$

If the energy is positive in the original field from which solitons are formed, it is energetically favorable to form a soliton. But the deceleration of the solitons does not end at the velocity u_*. Further decreases in their velocities means that they acquire speeds $u \ll u_*$. Both their amplitudes and energies remain unchanged in the process if $T_i \ll T_e$.

The imaginary part $\varepsilon_k^{(i)}/\varepsilon_k$, which describes the Landau damping of such slow solitons by ions, will be

$$\frac{\varepsilon_k^{(i)}}{\varepsilon_k} \simeq \frac{T_e}{T_e + T_i} \left(1 + i \sqrt{\frac{\pi}{2}} \frac{T_i}{T_e + T_i} \frac{\omega}{|k| v_{Ti}} e^{-\omega^2/2k^2 v_{Ti}^2} \right) \tag{9.63}$$

so that

$$\delta v_{\varkappa,\lambda}^L = i \frac{T_e^{3/2} \sqrt{T_i}}{(T_e + T_i)^2} \sqrt{\frac{\pi}{2}} \frac{\lambda}{|\varkappa|} e^{-T_e \lambda^2/2T_i \varkappa^2} \cdot v_{\varkappa,\lambda}. \tag{9.64}$$

Thus

$$\delta v^L(\xi, \tau) = \frac{u \sqrt{u_*} e^{-u^2/2u_*^2}}{(1 + u_*^2)^2} \sqrt{2\pi} \int_0^\infty \frac{\varkappa \sin[\varkappa(\xi - u\tau)] \, d\varkappa}{\sinh[(\varkappa\pi/E_0) \sqrt{(1 + u_*^2)/2}]}. \tag{9.65}$$

In a similar manner we find

$$\frac{dP}{dt} = - \frac{3\zeta(3)}{2\sqrt{2} \, \pi^{5/2}} \frac{u \sqrt{u_*} E_0^4}{(1 + u_*^2)^4} e^{-u^2/2u_*^2}. \tag{9.66}$$

On the other hand, when $E_0 \gg 1$ the momentum of such a soliton is

$$P = \frac{4\sqrt{2} \, u E_0^3}{3(1 + u_*^2)^{3/2}}. \tag{9.67}$$

In practice, conservation of the number of quanta for these solitons means that

$$E_0 = \text{const}. \tag{9.68}$$

Thus when $u \ll u_*$ the deceleration of the solitons will be

$$\frac{du}{d\tau} = - \frac{u}{\tau_*}, \qquad \tau_* = \frac{4\pi^{5/2}(1 + u_*^2)^{5/2}}{9\zeta(3) \sqrt{u_*} E_0}. \tag{9.69}$$

If $u_* \ll 1$, i.e., if $T_i/T_e \ll 1$, then solitons of maximum amplitude $|E^+|^2/4\pi n_0 T_e \simeq 1$ are damped in a time

$$t_* = \tau_* t_0 \simeq \frac{4.1}{\omega_{pi}} \sqrt{\frac{T_e}{T_i}}, \tag{9.70}$$

which is much larger than $1/\omega_{pi}$. The results in (9.66) and (9.69) were obtained in [329] and [324]. It was first shown in [329] that the damping of a soliton leads to its deceleration. Differences between (9.69) and the results of [329] arise when $u_* \simeq 1$ and are related to the more exact description of a soliton in (9.53) for $u \le u_*$. Solitons for which $u \gg u_*$ experience little damping by ions, according to (9.66). The soliton mean free path when $u_0 \simeq u_*$ is

$$\xi_* = u_* \tau_* \simeq \frac{4.6 \sqrt{u_*}}{E_0}, \qquad u_* \ll 1$$

or

$$x_* = x_d 8.3 \sqrt{\frac{T_i}{T_e}} \sqrt{\frac{4\pi n_0 T_e}{|E_0^+|^2}}. \tag{9.71}$$

The mean free path of a soliton with $u_0 \simeq 1$ will be about $\sqrt{1/\mu u_*}$ times longer.

The processes for gradually decreasing a soliton's velocity mean that a soliton is ultimately stopped and its group velocity is zero. During this process the number of Langmuir quanta is conserved. It is easily seen that such a process is completely analogous to nonlinear Landau damping (induced scattering and decay) for regular motions. It causes stochastic waves to be concentrated near $k_1 = 0$, i.e., $v_{gr} \approx 3k_1 v_{Te}^2/\omega_{pe} = 0$. This process reduces the group velocity of solitons to zero. Solitons are evidently stable relative to one-dimensional perturbations. Numerical calculations in [323] of the development of a strong, initially smooth field according to (9.40) and (9.41) show that such a field evolves into a soliton. There is a theorem for the Korteweg-deVris equations which states that any initial perturbation decays into a set of solitons [326]; no such general theorem has been demonstrated for Eqs. (9.40) and (9.41). It seems evident that there can be no such theorem because weak initial fields do not contract into solitons. It was shown in [327] that an arbitrary, near-sonic initial field (u → 1) turns into a single soliton with time. However, the time required for this process is usually longer than the time calculated above to reduce the soliton velocity, a period which is especially short when u → 1. However, sufficiently complex fields can decay into a number of solitons.

Solitons are unfortunately unstable relative to oblique pertur-bations. As has been shown, a modulation instability has the same growth rate for both longitudinal and oblique perturbations, and its stabilization along the field when a soliton is formed does not ensure stabilization of the oblique perturbations. Therefore, one-dimensional solitons are modulationally unstable, at least in a plasma which is homogeneous in the transverse direction in the ab-sence of a magnetic field. An interesting question is whether a soliton can have a velocity greater than the speed of sound. Ac-cording to [328], in a magnetic field solitons with frequencies equal to the upper hybrid plasma frequency can have velocities greater than the Alfven velocity v_a. In order for a soliton to exist both n' and $\varepsilon^{(i)}/\varepsilon$ must be negative. This latter factor appears in $\tilde{\Sigma}$ because of the compensation in the nonlinear currents. This is the same compensation which was first observed in [60] and is so important in all the induced scattering processes similar to those analyzed in Chapters 2 and 4. It was noted in [60] that compensation is destroyed for relativistic particles. This creates interest in plasma in which the electrons are relativistic. As an example, consider the state which arises when a powerful rel-ativistic beam is neutralized [330, 331]. In the beam reference system such a plasma will have relativistic temperatures (the energy spread in the beam is of the order of, but smaller than, the beam energy); the ion motion is unimportant and they serve only to ensure charge neutrality. This case was treated in [332]. Assuming that the electron distribution is one-dimensional and of the form

$$f(p) = \frac{2n_* p_*}{(p_*^2 + p^2)}, \qquad n_* = \int f(p) \frac{dp}{2\pi},$$

in the rest system of the electrons, we find

$$\tilde{\Sigma} = \frac{4e^4 n_* c}{3p_*^3 \omega_{p_*}^4} \left(1 - \frac{3m_e^2 c^2}{p_*^2} \right), \tag{9.72}$$

where

$$\omega_{p_*} = \sqrt{\frac{8n_* c e^2}{p_*}} \qquad (9.73)$$

is the electron plasma frequency, and

$$c p_* \gg m_e c^2, \qquad (9.74)$$

i.e., the average electron energy is relativistic. The first term inside the parentheses in (9.72) corresponds to Σ and the quantity $\varepsilon_k^{(i)}$ is neglected in the second term. Unlike (9.8), the expression in (9.72) does not vanish when $\varepsilon_k^{(i)} \to 0$. Moreover, the second term of (9.72) is small and it is always permissible to neglect $\varepsilon_k^{(i)}$. This is because there is no compensation. As with solitons in a nonrelativistic plasma, it was assumed that $\omega_{k_1} - \omega_{k_2} \ll |k_1 - k_2| v_{Te}$ in order to derive (9.72); in this case this condition reduces to

$$v_0 \ll c, \qquad (9.75)$$

i.e., solitons can have any velocity smaller than the speed of light. If the ions are nonrelativistic the speed of sound is given by $v_s = \sqrt{p_* c / m_i}$ so that

$$u = \frac{v_0}{v_s} = \frac{v_0}{c} \sqrt{\frac{m_e c}{p_*} \frac{m_i}{m_e}}. \qquad (9.77)$$

Supersonic solitons are possible when

$$c \sqrt{\frac{m_e}{m_i} \frac{p_*}{m_e c}} \ll v_0 \ll c. \qquad (9.77)$$

In terms of dimensional quantities the structure of a relativistic soliton is given by [332]

$$|E^+| = \frac{E_0}{\cosh[(x - v_0 t)/x_0]}, \qquad x_0 = \sqrt{\frac{3 p_*^2 c^2}{\pi e^2 E_0^2}}. \qquad (9.78)$$

Relativistic solitons were first considered in [333] to explain radio emissions from pulsars.

The formation of nonrelativistic solitons in a nonrelativistic plasma has been observed in the numerical modeling reported in [334] in addition to the numerical experiments in [323]. Solitons have been observed to form in experiments on the interactions of a high-frequency field with plasma [335] and the interactions of beams with plasma [336, 337].

It has been pointed out that solitons can exist only for the case of one-dimensional motion. They are an example of a wave packet which is modulationally unstable relative to disturbances transverse to its motion. Thus the dynamical picture which develops must be three-dimensional. The system of equations (9.34) and (9.36) can be used to analyze the three-dimensional motion. In the dimensionless variables of (9.42), Eqs. (9.34) and (9.36) take the form

$$\text{div}_\xi \left(i \frac{\partial}{\partial \tau} + \Delta_\xi \right) \mathbf{E}(\xi, \tau) = \text{div}_\xi \, v(\xi, \tau) \, \mathbf{E}(\xi, \tau), \tag{9.79}$$

$$\frac{\partial^2}{\partial \tau^2} v(\xi, \tau) - \Delta_\xi v(\xi, \tau) = \Delta_\xi |\mathbf{E}(\xi, \tau)|^2, \tag{9.80}$$

where the differential operators div_ξ and Δ_ξ take derivatives with respect to ξ. The first two integrals are

$$N = \int |\mathbf{E}(\xi, \tau)|^2 \, d\xi, \tag{9.81}$$

$$V = \int \left\{ |\text{div}_\xi \mathbf{E}(\xi, \tau)|^2 + v(\xi, \tau) |\mathbf{E}(\xi, \tau)|^2 + \frac{1}{2} |v(\xi, \tau)|^2 \right.$$
$$\left. + \frac{1}{2} \left| \frac{\nabla_\xi}{4\pi} \int \frac{d\xi'}{|\xi - \xi'|} \frac{\partial}{\partial \tau} v(\xi', \tau) \right|^2 \right\} d\xi. \tag{9.82}$$

Zakharov [339] suggested that a three-dimensional perturbation due to a modulation instability could cause the field to contract into a singularity at some point at which the charge density would be such that

$$\frac{n'}{n_0} \simeq 1, \qquad \frac{\Delta r}{r_d} \simeq 1, \qquad \frac{|E^+|^2}{2\pi n_0 T_e} \simeq 1. \tag{9.83}$$

The process was called collapse and the singularities were termed cavernous singularities or cavitons. Below, the self-constricted field regions in which a finite energy is maintained at the moment of contraction will also be called cavitons. The possibility of forming cavitons even for the simplest system of equations (9.79) and (9.80) has not yet been demonstrated analytically. The main arguments of [339], which are qualitative, are the following. Assume that there is a region occupied by the field initially, and the motion connected with the development of a modulation instability begins slowly and is initially subsonic. Then, from (9.80),

$$v = - |E|^2$$

and the integral in (9.82) will be negative. This fact is the condition for a modulation instability to form, or that the nonlinear frequency shift be larger than the linear dispersion correction to the frequency. The last term in (9.82) is negligible for subsonic motion (the characteristic time τ is much larger than the characteristic ξ). The first term in (9.82) describes the dispersion corrections, and the second and third terms in the sum

$$v |E|^2 + \tfrac{1}{2} v^2 = - \tfrac{1}{2} |E|^4$$

give the nonlinear frequency shift. Thus, during the initial stages of motion,

$$V < 0.$$

One can also seek stationary solutions to (9.79) and (9.80). For one-dimensional motions such solutions are solitons. According to (9.50), for subsonic solitons $V < 0$. Thus solitons can be formed as the result of perturbations in the one-dimensional case [339]. The stationary state in the three-dimensional case corresponds to $V > 0$. Zakharov [339] therefore concluded that when modulation instabilities develop, the system does not go into a stationary state. But this does not imply that singularities appear, because it is still possible to have oscillatory motions and a continuous breakdown of scales. The question concerns the development of modulation disturbances to the point indicated by (9.83) because the process of increasing the allowed density, with accumulation of the oscillation energy in this region, follows even from an intuitive interpretation of the linear stages of a modulation instability. Therefore, within the scope of (9.79) and (9.80) the problem is to prove that there is a singularity and

the field contracts to a point. Definite progress on this problem
has been achieved [340, 341] by numerically solving Eqs. (9.79)
and (9.80). A whole class of initial perturbations turn out to be
unstable relative to subsequent modulation perturbations; this
class includes spherically symmetric disturbances [ever more
finely divided modulation rarefactions appear, which is very
similar to the development of eddies and turbulence in incompressible fluids (Chapter 1)]. However, a dipole disturbance shaped
like an ellipse with a 1:3 axis ratio and with the dipole directed
perpendicular to the long axis of the ellipse tended to form cavitons.
Of course the calculations were not extended to quantities of
magnitude necessary to satisfy (9.83). However, the authors of
[340, 341] did compare their numerical results with results of
self-similar solutions to (9.79) and (9.80) and showed that a number of initial conditions for which the motions became unstable
did turn into solutions of the dipole ellipse type with a 1:3 semi-
axis ratio. We conclude that this type of motion is apparently
more modulationally stable, because we cannot exclude the possibility of subsequent instability in these motions. A definite
answer to this question can be obtained analytically by studying
the stability of self-similar motions. These can be found from
the following simple considerations. The formation of a caviton
is an accelerating process, and it soon becomes supersonic. As
an order of magnitude, (9.80) gives the following for supersonic
motion:

$$\nu \approx \frac{\tau^2}{\xi^2} \, | \, E \, |^2, \tag{9.84}$$

while according to [341], $1/\tau \ll 1/\xi^2$ in (9.80), so that

$$\frac{1}{\xi^2} E \approx \nu E \approx \frac{\tau^2}{\xi^2} \, | \, E \, |^2 E, \quad | \, E \, |^2 \tau^2 = \text{const}, \tag{9.85}$$

$$| \, E \, | \sim \frac{1}{\tau}.$$

Here $\tau = 0$ corresponds to the moment when the singularity is
formed (the formation process takes place for $\tau < 0$). From
conservation of the number of quanta (9.81) we find

$$| \, E \, |^2 \xi^3 = \text{const}, \quad | \, E \, | \sim \frac{1}{\xi^{3/2}},$$

or that $|E|$ depends on the self-similar variable $\xi/\tau^{2/3}$, i.e.,

$$|E| = \frac{E_0}{\tau} \Psi\left(\frac{\xi}{\tau^{2/3}}\right), \qquad \int \Psi(\xi) \, d\xi = 1. \tag{9.86}$$

This form of the self-similar solution has been disputed in [349] (see below). Numerical calculations have shown [340, 341] that the function (9.86) is a solution to (9.80) and (9.81) at a certain stage in the development of dipole perturbations. If $\xi \sim \tau^{2/3}$ then $1/\tau \ll 1/\xi^2$ when $\tau \ll 1$ and the term $\partial/\partial\tau$ in (9.80) can be neglected. Then, according to (9.84), the density behaves as

$$\nu \sim \frac{1}{\tau^{4/3}}. \tag{9.87}$$

If the motion is considered to begin when it has become supersonic, i.e., when $\tau_0 = \xi_0$, the solution given in (9.86) holds when $\tau_0 \geq \xi_0^2$ or $\tau_0 = \xi_0 = 1$. It follows that $|E| \approx 1/\tau$ or $E_0 = 1$ in (9.86). The time when $|E^+|^2/4\pi n_0 T_e \approx 1$ or $|E| \simeq 1/\sqrt{\mu}$ corresponds to $\tau \simeq \sqrt{\mu}$ or $t \simeq t_0 \tau = 2/\omega_{pe}\sqrt{\mu} = \sqrt{3}/\omega_{pi}$. At this time $\xi \simeq \tau^{2/3} \simeq \mu^{1/3}$, $\Delta r \simeq \xi t$, $v_s \simeq \sqrt{3}r_d \mu^{-1/6}$. Finally, $\nu \simeq \tau^{-4/3} \simeq \mu^{-2/3}$, i.e.,

$$\left(\frac{n'}{n_0}\right)_{\max} \simeq \mu\nu \simeq \mu^{1/3}. \tag{9.88}$$

Of course, the condition $|E^+|^2/4\pi n_0 T_e \simeq 1$ was chosen arbitrarily, but the analysis below shows that the original equations are not applicable for $\tau \ll \sqrt{\mu}$, which means that they are inapplicable for $|E^+|^2/4\pi n_0 T_e \ll 1$. The quantity n'^2/n_0^2, which characterizes the energy in the low-frequency perturbations, is small, being of the order of $\mu^{2/3}$. There is even no proof that during compression a caviton will reach the limit set by (9.88).

The solution given in (9.86) has been criticized in [349] because the first term $\sim 1/\tau$ was dropped in (9.80), and this causes conservation of energy flow to be violated. Abother self-similar solution was proposed in [349, 346], in which all terms in (9.80) are of the same magnitude:

$$E(\xi, \tau) \simeq \frac{1}{\tau} \Psi\left(\frac{\xi}{\sqrt{\tau}}\right), \qquad \nu(\xi, \tau) \simeq \frac{1}{\tau} \Psi'\left(\frac{\xi}{\sqrt{\tau}}\right). \tag{9.89}$$

The total number of these cavitons,

$$\int |E(\xi, \tau)|^2 \, d\xi = \int \frac{\Psi(\xi) \, d\xi}{\sqrt{\tau}}, \tag{9.90}$$

is not conserved, in contradiction to the original system of equations. The term containing $\partial/\partial\tau$ in (9.80) may be retained in deriving the solution in (9.86) because there is no requirement that the amplitude and phase obey the same self-similar law used in (9.89). Equating the imaginary parts in (9.80), we find that the phase must be of the order of $\phi \sim \xi^2/\tau$ and that the first term in (9.80) $|E|\partial\phi/\partial\tau \sim |E|\xi^2/\tau^2$ is considerably smaller than the second term $\partial^2|E|/\partial\xi^2$ when $\xi^4 \ll \tau^2$, which for $\xi \sim \tau^{2/3}$ means that $\tau^{8/3} \ll \tau^2$ or $\tau^{2/3} \ll 1$. Then $\phi \simeq \tau^{1/3}\Psi'(\xi/\tau^{2/3})$.

Significantly, according to [347] and the numerical calculation of [348] spherically symmetric cavitons are modulationally unstable relative to splitting into smaller cavitons accompanied by sound generation. Cavitons shaped like spherical layers were treated in [347, 349]. These forms have been labeled quasisolitons. Their contraction motion is accompanied by acceleration, which becomes so great [349] that powerful acoustic radiation is produced [350]. These layers can also become modulationally unstable relative to azimuthal disturbances [351]. The stability of self-similar motions was considered in [352], but only for the two-dimensional case. A true caviton will be one which turns out to be stable. As yet there is no proof that the caviton of (9.86) is stable. One is left with the impression that in the broad class of three-dimensional motions initial modulation rarefactions become modulationally unstable relative to smaller modulation perturbations. But if cavitons do exist, their contraction is limited by (9.88). We note that since the fundamental stage of caviton contraction (9.86) is supersonic, one would expect sound to be emitted during the contraction. Assuming that (9.86) describes the motion, the total acoustic energy radiated during contraction can be calculated. The acoustic energy is defined by the last two energy terms in (9.82). The quantities ν^s, which are related to acoustic emission, can be found from the equation

$$(\lambda^2 - \varkappa^2)\nu^s_{\varkappa,\lambda} = \varkappa^2\nu_{\varkappa,\lambda}, \tag{9.91}$$

where $\nu_{\varkappa,\lambda}$ are the Fourier components of the caviton density.

We find (see also [325])

$$V^s = 8\pi^5 \int \left(v_{\varkappa,|\varkappa|} v_{-\varkappa,-|\varkappa|} + v_{\varkappa,-|\varkappa|} v_{-\varkappa,-|\varkappa|} \right) d\varkappa. \tag{9.92}$$

Assuming that the initial amplitude of the caviton field is $E_n \simeq 1$ (the compression begins as a supersonic one) and that the caviton initially has a spherical shape with unit radius, we obtain

$$v_{\varkappa,|\varkappa|} = \int\limits_0^1 \frac{\xi \, d\xi}{4\pi^3} \int\limits_0^1 d\tau \, \frac{\sin(\varkappa \xi \tau^{2/3})}{\tau^{2/3}} \, e^{-i\varkappa\tau}. \tag{9.93}$$

Evaluation of the integral in (9.92) will enable us to determine the ratio η of the emitted energy V^s to the initial energy $8\pi/3$. We find

$$\eta \simeq 0.93. \tag{9.94}$$

This means that the acoustic energy emitted during caviton compression is not negligible and that either it leads to increased density depletion at the hole or the motion ceases near the sound velocity.

The following important fact concerning cavitons should be mentioned: according to (9.82) the energy distribution in cavitons can be estimated from

$$V \simeq \int \frac{|E|^2 \, d\xi}{\xi^2} \simeq \frac{4\pi}{\tau^2} \int d|\xi| \Psi\left(\frac{|\xi|}{\tau^{2/3}}\right). \tag{9.95}$$

If, for example, the initial perturbation were uniform so that $\Psi(|\xi|) = $ const such that

$$\Psi(|\xi|) = \begin{cases} 1, & |\xi| < 1, \\ 0, & |\xi| > 1, \end{cases} \tag{9.96}$$

most of the caviton energy would be concentrated, according to (9.95), near the periphery, where $\xi \simeq \tau^{2/3}$. This is true of any smooth initial energy distribution for a perturbation field which does not have a singularity of the form $1/|\xi|$ at the origin.

9.4. Limits on the Applicability of Descriptions of Strong Nonlinear Motions and Spikons

Unlike the methods used in the original papers, those used here facilitate the quest for limits on the applicability of the equations imposed by the higher-order nonlinearities which were not included and by the terms in $\tilde{\Sigma}$ describing the electron nonlinearities, which were dropped. The problem posed by terms omitted is also of interest because of the numerical calculations reported in [338]. It was shown there that a modulation instability in an initially uniform electric field can cause it to contract into fast solitons having dimensions on the order of $(5-7)r_d$ with $|E|^2/4\pi n_0 T_e \simeq 1$; these solitons were called spikons in [338]. The name derives from their relatively short lifetime (about $0.1\omega_{pe}$), which is much smaller than the ion plasma period ω_{pi}^{-1} . Therefore, electron nonlinearities must indeed be important. On the other hand, since $|E|^2/4\pi n_0 T_e$ is so large, one would expect contributions from the higher-order nonlinearities to be significant too. There is also the general question as to whether fast solitons (with $v_0 > v_s$) can exist. At high velocities the density changes are positive, which means that the plasma condenses rather than forming rarefactions; according to (9.35),

$$-\frac{\varepsilon_k^{(i)}}{\varepsilon_k} \simeq \frac{v_s^2}{v_0^2} .$$ (9.97)

This means that the compensation effects discussed early are even larger, and one would expect that the higher nonlinearities would become important if similar compensation does not take place. We shall show that compensation effects do arise in the higher-order nonlinearities, and that they can be neglected in supersonic motion when $v_s^2/v_0^2 \ll 1$ [see (9.97)], i.e., the original equations can be used to analyze the motion even if $|E^+|^2/4\pi n_0 T_e > 1$. A limitation like $|E^+|^2/4\pi n_0 T_e \ll 1$ arises for subsonic motion when higher-order nonlinearities are included. Conditions of the type $|E^+|^2/4\pi n_0 T_e \ll 1$ arise for supersonic motion as well, but in this case as a necessary condition for neglecting electron nonlinearities. We will begin with the question of higher-order nonlinearities.

The system of equations given in (9.5) and (9.6) can be replaced with a more exact set of equations, which include all possible nonlinear terms up to fifth order in the electric field:

$$(4\pi)^{-1}i\omega\varepsilon_k E_k^+ = 2\int S_{k,k_1,k_2}(E_{k_1}^+ E_{k_2}^+ + E_{k_1}^- E_{k_2}^{(2)+})\,\delta(k-k_1-k_2)\,dk_1\,dk_2$$
$$+ \int \Sigma_{k,k_1,k_2,k_3}(2E_{k_1}^+ E_{k_2}^+ E_{k_3}^- + E_{k_1}^- E_{k_2}^+ E_{k_3}^+ + 2E_{k_1}^- E_{k_2}^{(2)+} E_{k_3} + 2E_{k_1}^{(2)+} E_{k_2}^+ E_{k_3}$$
$$+ E_{k_1}^+ E_{k_2} E_{k_3} + 2E_{k_1}^{(2)+} E_{k_2}^{(2)-} E_{k_3}^+)\,\delta(k-k_1-k_2-k_3)\,dk_1\,dk_2\,dk_3$$
$$+ \int S'_{k,k_1,k_2,k_3}(2E_{k_1}^+ E_{k_2} E_{k_3}^+ E_{k_4}^- + E_{k_1}^+ E_{k_2}^- E_{k_3}^- E_{k_4} + E_{k_1}^+ E_{k_2} E_{k_3}^+ E_{k_4})$$
$$\times \delta(k-k_1-k_2-k_3-k_4)\,dk_1\,dk_2\,dk_3\,dk_4 + 4\int \Sigma'_{k,k_1,k_2,k_3,k_4,k_5} E_{k_1}^+ E_{k_2}^+ E_{k_3}^- E_{k_4}^+ E_{k_5}^-$$
$$\times \delta(k-k_1-k_2-k_3-k_4-k_5)\,dk_1\,dk_2\,dk_3\,dk_4\,dk_5. \tag{9.98}$$

Here S' and Σ' are matrix elements of the currents which are of fourth and fifth order in the field; S, Σ, and S' are symmetric with respect to the last two subscripts, while Σ' is symmetric with respect to k_2 and k_3, and k_4 and k_5, $E^{(2)\pm}$ is a field with frequency near $\pm 2\omega_{pe}$ and is not an eigenfield of the oscillation. It is retained in order to analyze its contributions to currents of lower order in the field. The field $E^{(2)\pm}$ can be expressed in terms of the eigenfields $E^\pm$ and the low–frequency field:

$$(4\pi)^{-1}i\omega\varepsilon_k E_k^{(2)+} = \int S_{k,k_1,k_2} E_{k_1}^+ E_{k_2}^+ \,\delta(k-k_1-k_2)\,dk_1\,dk_2$$
$$+ 2\int \Sigma_{k,k_1,k_2,k_3} E_{k_1}^+ E_{k_2}^+ E_{k_3} \,\delta(k-k_1-k_2-k_3)\,dk_1\,dk_2\,dk_3. \tag{9.99}$$

Subsequent expansions will be in terms of the natural parameter $k^2 v_{Te}^2/\omega_{pe}^2 \ll 1$. When all three fields E_k, E_{k_1}, E_{k_2} have high frequencies $\omega \gg k v_{Te}$ the nonlinear currents S' are smaller by a factor $k^2 v_{Te}^2/\omega_{pe}^2$ than the nonlinear currents found when one of the fields is low-frequency (i.e., $\omega < k v_{Te}$). Therefore the second term in the expression multiplying S' in (9.98), and the first term in (9.99) can be neglected. However, when the second term of (9.99) is substituted into (9.98) it results in terms of seventh order in the field. The term $E^- E^+ E^+$ in (9.98) is of order $k^2 v_{Te}^2/\omega_{pe}^2 \ll 1$ relative to the first term multiplying Σ. The order of the remaining fields in the terms with S' and Σ' is also imposed by the size of $k^2 v_{Te}^2/\omega_{pe}^2$ and the requirement that only terms up to fifth order in the fields $E^\pm$ be taken into account, together with the symmetry of the coefficients of S, Σ, S', and Σ'. The particle propagators in these nonlinear responses are written from left to right in order of increasing k differences; i.e., the first is $1/(\omega - \mathbf{kv})$, the second is $1/[\omega - \omega_1 - (\mathbf{k} - \mathbf{k}_1)\mathbf{v}]$, the third is $1/\omega - \omega_1 - \omega_2 - (\mathbf{k} - \mathbf{k}_1 - \mathbf{k}_2)\mathbf{v}]$ and so on. The number of propagators is determined by the order of the current nonlinearity (two for S, three for Σ, and so on). Denoting the additional contribution to E^+ due to higher-order nonlinearities as δE^+, we obtain

$$(4\pi)^{-1}i\omega\varepsilon_k\,\delta E_k^+ = 2\int S_{k,k_1,k_2}E_{k_1}^+\,\delta E_{k_2}\,\delta(k - k_1 - k_2)\,dk_1\,dk_2$$

$$+ \int \Sigma_{k,k_1,k_2,k_3}E_{k_1}^+E_{k_2}E_{k_3}\,\delta(k - k_1 - k_2 - k_3)\,dk_1\,dk_2\,dk_3$$

$$+ \int S'_{k,k_1,k_2,k_3,k_4}(2E_{k_1}^+E_{k_2}E_{k_3}^+E_{k_4}^- + 2E_{k_1}^+E_{k_2}^+E_{k_3}^-E_{k_4} + 2E_{k_1}^+E_{k_2}^-E_{k_3}^+E_{k_4})$$

$$\times \delta(k - k_1 - k_2 - k_3 - k_4)\,dk_1\,dk_2\,dk_3\,dk_4 + 4\int \Sigma'_{k,k_1,k_2,k_3,k_4,k_5}E_{k_1}^+E_{k_2}^+E_{k_3}^-E_{k_4}^+E_{k_5}^-$$

$$\times \delta(k - k_1 - k_2 - k_3 - k_4 - k_5)\,dk_1\,dk_2\,dk_3\,dk_4\,dk_5. \qquad (9.100)$$

Similar arguments can be used to improve Eq. (9.6) for the low-frequency field E_k. The addition δE_k to be made to the right side of (9.6) is found to be

$$(4\pi)^{-1}i\omega\varepsilon_k\,\delta E_k = \int S_{k,k_1,k_2}E_{k_1}E_{k_2}\,\delta(k - k_1 - k_2)\,dk_1\,dk_2$$

$$+ \int \Sigma_{k,k_1,k_2,k_3}(2E_{k_1}E_{k_2}^+E_{k_3}^- + 2E_{k_1}^+E_{k_2}^-E_{k_3} + 2E_{k_1}^-E_{k_2}^+E_{k_3})$$

$$\times \delta(k - k_1 - k_2 - k_3)\,dk_1\,dk_2\,dk_3 + 2\int S'_{k,k_1,k_2,k_3,k_4}(E_{k_1}^+E_{k_2}^-E_{k_3}^+E_{k_4}^-$$

$$+ E_{k_1}^-E_{k_2}^+E_{k_3}^+E_{k_4}^-)\,\delta(k - k_1 - k_2 - k_3 - k_4)\,dk_1\,dk_2\,dk_3\,dk_4. \qquad (9.101)$$

Equations (9.101) and (9.6) can be used to write (9.100) as

$$(4\pi)^{-1}i\omega\varepsilon_k\,\delta E_k^+ = 2\int \delta\Sigma'_{k,k_1,k_2,k_3,k_4,k_5}E_{k_1}^+E_{k_2}^+E_{k_3}^-E_{k_4}^+E_{k_5}^-$$

$$\times \delta(k - k_1 - k_2 - k_3 - k_4 - k_5)\,dk_1\,dk_2\,dk_3\,dk_4\,dk_5. \qquad (9.102)$$

where

$$\delta\Sigma'_{k,k_1,k_2,k_3,k_4,k_5} = 2\Sigma'_{k,k_1,k_2,k_3,k_4,k_5} + \frac{8\pi S'_{k,k_2+k_3,k_4,k_5}\,S_{k_2+k_3,k_2,k_3}}{i(\omega_2 + \omega_3)\varepsilon_{k_2+k_3}}$$

$$+ \tfrac{1}{2}\,\Sigma'_{k,k_1,k_5 k_3,k_4,k_2,} + \frac{8\pi S_{k_4+k_5,k_4,k_5}}{i(\omega_4 + \omega_5)\varepsilon_{k_4+k_5}}\,(S'_{k,k_1,k_2,k_3,k_4+k_5} + S'_{k,k_1,k_3,k_2,k_4+k_5})$$

$$+ \frac{8\pi S_{k,k_1,k-k_1}}{i(\omega - \omega_1)\varepsilon_{k-k_1}}\,(S'_{k-k_1,k_2,k_3,k_4,k_5} + S'_{k-k_1,k_3,k_2,k_4,k_5}$$

$$+ \tfrac{1}{2}\,S'_{k-k_1,k_2,k_4,k_3,k_5} + \tfrac{1}{2}\,S'_{k-k_1,k_3,k_5,k_2,k_4})$$

$$+ \tfrac{1}{2}\,\Sigma'_{k,k_1,k_2,k_4,k_3,k_5} - \frac{32\pi^2 S_{k_2+k_3,k_2,k_3}S_{k_4+k_5,k_4,k_5}}{(\omega_2 + \omega_3)\varepsilon_{k_2+k_3}(\omega_4 + \omega_5)\varepsilon_{k_4+k_5}}\,\Sigma_{k,k_1,k_2+k_3,k_4+k_5}$$

$$-\frac{(8\pi)^2 S_{k,k_1,k-k_1} S_{k_2+k_3,k_2,k_3}}{(\omega-\omega_1)\varepsilon_{k-k_1}(\omega_2+\omega_3)\varepsilon_{k_2+k_3}}\ \Sigma_{k-k_1,k_2+k_3,k_4,k_5}$$

$$-\frac{(8\pi^2)S_{k,k_1,k-k_1} S_{k_4+k_5,k_4,k_5}}{(\omega-\omega_1)\varepsilon_{k-k_1}(\omega_4+\omega_5)\varepsilon_{k_4+k_5}}\ (\Sigma_{k-k_1,k_2,k_3,k_4+k_5}+\Sigma_{k-k_1,k_3,k_2,k_4+k_5})$$

$$-\frac{4(4\pi)^3 S_{k,k_1,k-k_1} S_{k_2+k_3,k_2,k_3} S_{k_4+k_5,k_4,k_5}}{i(\omega-\omega_1)\varepsilon_{k-k_1}(\omega_2+\omega_3)\varepsilon_{k_2+k_3}(\omega_4+\omega_5)\varepsilon_{k_4+k_5}}\ S_{k-k_1,k_2+k_3,k_1+k_5}. \qquad (9.103)$$

In the simplest case of one-dimensional motion, when $k^2 v_{Te}^2/\omega_{pe}^2 \ll 1$ the expressions found in (9.103) will be

$$\Sigma'_{k,k_1,k_2,k_3,k_4,k_5} \simeq -\frac{e^4\omega}{16\pi i\omega_{pe}^4 m_e^2 T_e^2}\left(2-\frac{k_2+k_3}{k-k_1}\right),$$

$$S'_{k,k_1,k_2+k_3,k_4,k_5} \simeq -\frac{e^3\omega}{8\pi\omega_{pe}^2 m_e T_e^2(k-k_1)},$$

$$S'_{k,k_1,k_2,k_3,k_4+k_5}+S'_{k,k_1,k_3,k_2,k_4+k_5} \simeq -\frac{e^3\omega[2-(k_2+k_3)/(k-k_1)]}{8\pi\omega_{pe}^2 m_e T_e^2(k_4+k_5)},$$

$$S'_{k-k_1,k_2,k_3,k_4,k_5}+S'_{k-k_1,k_3,k_2,k_4,k_5} \simeq \frac{e^3(\omega-\omega_1)[2-(k_2+k_3)/(k-k_1)]}{8\pi\omega_{pe}^2 T_e^2 m_e(k-k_1)},$$

$$\Sigma_{k,k_1,k_2+k_3,k_4+k_5} \simeq \frac{e^2\omega}{8\pi i T_e^2(k_4+k_5)(k_2+k_3)},$$

$$\Sigma_{k-k_1,k_2+k_3,k_4,k_5} \simeq -\frac{e^2(\omega-\omega_1)}{8\pi i T_e^2(k-k_1)^2},$$

$$\Sigma_{k-k_1,k_2,k_3,k_4+k_5}+\Sigma_{k-k_1,k_3,k_2,k_4+k_5} \simeq -\frac{e^2(\omega-\omega_1)[2-(k_2+k_3)/(k-k_1)]}{8\pi i T_e^2(k-k_1)(k_4+k_5)},$$

$$S_{k_4+k_5,k_4,k_5} \simeq -\frac{e(\varepsilon_{k_4+k_5}^\ell-1)(\omega_4+\omega_5)(k_4+k_5)}{8\pi\omega_{pe}^2 m_e},$$

$$S_{k,k_1,k-k_1} \simeq \frac{e\omega(\varepsilon_{k-k_1}^\ell-1)(k-k_1)}{8\pi\omega_{pe}^2 m_e},$$

$$S_{k-k_1,k_2+k_3,k_4+k_5} \simeq -\frac{e(\omega-\omega_1)\omega_{pe}^2 m_e}{8\pi T_e^2(k-k_1)(k_2+k_3)(k_4+k_5)}.$$

We therefore obtain the following expression:

$$\delta \tilde{\Sigma}'_{k,k_1,k_2,k_3,k_4,k_5} \simeq - \frac{e^4 \omega \; \varepsilon^{(i)}_{k_2+k_3}}{16 \pi i \omega^4_{pe} m_e^2 T_e^2 \; \varepsilon_{k_2+k_3}} \; \frac{\varepsilon^{(i)}_{k-k_1}}{\varepsilon_{k-k_1}} \; \frac{\varepsilon^{(i)}_{k_4+k_5}}{\varepsilon_{k_4+k_5}}. \tag{9.104}$$

This result is very interesting, for it shows that there is greater compensation of the higher-order nonlinearities than there is of the lower-order ones, because (9.104) contains $(\varepsilon^{(i)}/\varepsilon)^3$ rather than $\varepsilon^{(i)}/\varepsilon$, which appears in (9.8). Assuming in (9.104) that $\varepsilon^{(i)}/\varepsilon \sim 1/u^2 = (v_s/v_0)^2 \ll 1$ we find for fast supersonic motion that

$$\frac{\delta \tilde{\Sigma}' \, |E|^2}{\tilde{\Sigma}} \simeq \frac{|E^+|^2}{2\pi n_0 T_e u^4} \tag{9.105}$$

i.e., they are small even when $|E^+|^2/4\pi n_0 T_e \simeq 1$. This triplet compensation also arises in the three-dimensional case; it is important to both supersonic cavitons and to the construction of a statistical theory of strong Langmuir turbulence. Calculations similar to those carried out above for the one-dimensional case show that

$$\delta \tilde{\Sigma}'_{k,k_1,k_2,k_3,k_4,k_5} \simeq - \frac{e^4 \omega}{16 \pi i \omega^4_{pe} m_e^2 T_e^2} \; \frac{(\mathbf{k} \cdot \mathbf{k_1})(\mathbf{k_4} \cdot \mathbf{k_5})}{k k_1 k_4 k_5} \; \frac{\varepsilon^{(i)}_{k-k_1}}{\varepsilon_{k-k_1}} \; \frac{\varepsilon^{(i)}_{k_4+k_5}}{\varepsilon_{k_4+k_5}}$$

$$\times \frac{(\mathbf{k_2} \cdot \mathbf{k_3}) \varepsilon^{(i)}_{k_2+k_3}}{k_2 k_3 \; \varepsilon_{k_2+k_3}}. \tag{9.106}$$

The presence of the factor $(\varepsilon^{(i)}/\varepsilon)^4$ in (9.106) makes it possible to use (9.34) and (9.36) for supersonic motions up to the point where $E_0^2/4\pi n_0 T_e \simeq 1$. For subsonic motions, however, the restrictions as viewed from (9.106) are more severe; they require that $E_0^2/4\pi n_0 T_e \ll 1$.

Let us turn now to the electron nonlinearities. We must analyze more carefully the approximations used to arrive at the specific expression in (9.8) for the cubic nonlinearities. The general expression for $\tilde{\Sigma}$ is

$$\tilde{\Sigma}_{k,k_1,k_2,k_3} = \Sigma_{k,k_1,k_2,k_3} + \frac{8\pi S_{k,k_1,k-k_1} S_{k-k_1,k_2,k_3}}{i(\omega - \omega_1)\varepsilon_{k-k_1}}. \tag{9.107}$$

The compensation in (9.8) arises when the following inequalities (which we have already encountered) are used:

$$\omega - \omega_1 \ll |\, \mathbf{k} - \mathbf{k}_1 \,|\, v_{Te}, \tag{9.108}$$

$$\omega_{pe} \gg k v_{Te}, \qquad \Delta\omega \ll \omega_{pi} \tag{9.109}$$

Equation (9.108) can be written as

$$\Delta\omega - \Delta\omega_1 \ll |\, \mathbf{k} - \mathbf{k}_1 \,|\, v_{Te} \ll \omega_{pe}, \tag{9.110}$$

where $\Delta\omega = \omega - \omega_{pe}$. $\Delta\omega$ can be included in (9.107) by expanding the nonlinear currents in terms of all $\Delta\omega$, $\Delta\omega_1$, $\Delta\omega_2$, $\Delta\omega_3$. The term in the expansion which is linear in all the $\Delta\omega, \ldots$ is exactly zero. The corrections arise if unity is not neglected with respect to $\omega_{pi}^2 /(\Delta\omega - \Delta\omega_1)^2$ in $\varepsilon_{\mathbf{k}-\mathbf{k}_1}^{(i)} \simeq 1 - \omega_{pi}^2 /(\Delta\omega - \Delta\omega_1)^2$ in (9.8). This gives the following correction for the one–dimensional supersonic motion:

$$\delta\widetilde{\Sigma}_1 \simeq \frac{e^2\omega(k - k_1)^2}{8\pi i m_e^2 \omega_{pe}^4}, \qquad \varepsilon_{k-k_1} \simeq \frac{\omega_{pe}^2}{(k - k_1)^2 v_{Te}^2}.$$

The next terms in the expansion of $\widetilde{\Sigma}$ in (9.107) in terms of the parameter $k^2 v_{Te}^2 /\omega_{pe}^2$ give corrections of the same order of magnitude. These are the electron nonlinearities. Calculation gives

$$\delta\widetilde{\Sigma}_2 \simeq - \frac{3e^2\omega k(k + k_1)}{8\pi i m_e^2 \omega_{pe}^4}.$$

The nonlinearities connected with excitation of virtual waves near $2\,\omega_{pe}$ [the terms $E_{k_1}^- E_{k_2}^{(2)+}$ in (9.98) and the first term in (9.99) together with the term $\Sigma_{k,\,k_3,\,k_1,\,k_2}$] give corrections of the same order:

$$\delta\widetilde{\Sigma}_3 \simeq \frac{3e^2\omega k(3k - k_3)}{16\pi i m_e^2 \omega_{pe}^2}.$$

The total effect of all electron nonlinearities will be

$$\delta\widetilde{\Sigma} = \delta\widetilde{\Sigma}_1 + \delta\widetilde{\Sigma}_2 + \delta\widetilde{\Sigma}_3 = \frac{3e^2\omega k}{16\pi i m_e^2 \omega_{pe}^2}\,(k_2 - k_1).$$

Since k_1 and k_2 appear symmetrically in the nonlinear correction, we find in the one-dimensional case that

$$\delta\tilde{\Sigma} = 0. \tag{9.111}$$

Thus, as far as subsonic solitons are concerned, electron non-linearities are negligible. In the one-dimensional case the non-linearities are of higher order in $k^2 v_{Te}^2/\omega_{pe}^2$ and $\Delta\omega/\omega_{pe}$; i.e., they are of order $k^4 v_{Te}^4/\omega_{pe}^4$, $k^2 v_{Te}^2 \Delta\omega/\omega_{pe}^3$, and $(\Delta\omega/\omega_{pe})^2$. The corrections $\sim(\Delta\omega/\omega_{pe})^2$ turn out to be zero, and we obtain

$$\delta\tilde{\Sigma}_4 = -\frac{i\omega}{32\pi^2 n_0 m_e \omega_{pe}^2}\left\{\Delta\omega\left[\tfrac{3}{2}kk_1 + \tfrac{3}{4}(k_1^2 + k_1 k_2)\right]\right.$$

$$\left. + \Delta\omega_1\left[6k^2 - 5k(k_1 + k_2) - 6k_1 k_2\right]\right\} \tag{9.112}$$

and

$$\delta\tilde{\Sigma}_5 = -\frac{3i\omega v_{Te}^2}{32\pi^2 n_0 m_e \omega_{pe}^4}(-k^4 + 4k^3 k_1 - 6k^2 k_1^2 - 2k^2 k_1 k_2$$

$$+ 4kk_1^3 + 4kk_1^2 k_2 - 2k_1^4 - 4k_1^3 k_2 - 3k_1^2 k_2^2). \tag{9.113}$$

In terms of the variables $\tau = \omega_{pe}t$, $\xi = (\omega_{pe}/v_{Te})x$, and $E = E^+/\sqrt{4\pi n_0 T_e}$ the equation which takes into account $\delta\tilde{\Sigma} = \delta\tilde{\Sigma}_4 + \delta\tilde{\Sigma}_5$ but not $\tilde{\Sigma}$ [corresponding to (9.8)] in the limit of fast time and space variations of the phase and in the limit of approximately linear dispersion takes the form

$$2i\frac{\partial E}{\partial\tau} + 3\frac{\partial^2 E}{\partial\xi^2} = \frac{15}{36}|E|^2 u^1 E. \tag{9.114}$$

It is assumed here that $|E|$ depends only on $(\xi - u\tau)$; $u = v_0/v_{Te}$ [319] and

$$E = |E|e^{i\phi}, \quad \phi = (\xi - u\tau) + \phi_0, \tag{9.115}$$

Since the right side of Eq. (9.114) is positive, it does not have a soliton solution. Thus the electron nonlinearities do not explain spikons in the limit $|E^+|^2/4\pi n_0 T_e \ll 1$, $v_0 \ll v_{Te}$. The higher-order nonlinearities can dominate the cubic nonlinearity for non-equilibrium low-velocity ($v \ll v_{Te}$) electron distributions. For a non-Maxwellian electron distribution, T_e in Eq. (9.17) is replaced by T_{eff} and T_e^2 in Eq. (9.104) by $(T_{eff}')^2$:

$$\frac{1}{T_{eff}} - \frac{1}{n_0 m_e} \int \frac{1}{v}\, \frac{\delta\phi^{(r)}}{\delta v}\, dv, \qquad \frac{1}{(T_{eff}')^2} = \frac{1}{n_0 m_e^2} \int \frac{1}{v}\, \frac{\delta}{\delta v}\, \frac{1}{v}\, \frac{\delta}{\delta v}\, \phi^{(e)}\, dv,$$

$$\int \phi^{(e)}\, dv = n_0\,.$$

In the notation of (9.42), in place of Eqs. (9.40), (9.42') we obtain

$$i\frac{\delta}{\delta\tau} + \frac{\delta^2}{\delta\xi^2}\, E(\xi, \tau) = -\frac{|E|^2}{v_s^2 - u^2}\left(1 - \chi\, \frac{|E|^2}{(v_s^2 - u^2)^2}\right) E \qquad (9.116)$$

where $v_s = \sqrt{T_{eff}/T_e}$ and $\chi = \mu T_{eff}^3 / T_e(T_{eff}')^2$. The higher-order nonlinearity dominates and leads to a soliton solution

$$E - E_0 \exp(-i\Omega\tau + i\frac{u}{2}(\xi - u\tau))\left\{\cosh\left[\frac{2\sqrt{\chi}\, E_0^2(\xi - u\tau)}{\sqrt{3}(u^2 - v_s^2)^{3/2}}\right]\right\}^{-1/2} \qquad (9.117)$$

for

$$E_0^2 \chi \gg (v_s^2 - u^2)^2, \qquad u > v_s, \qquad (9.117')$$

i.e., for supersonic velocities. The nonlinear Landau damping for (9.117) decreases the amplitude E_0, while u remains constant, thus violating the first of the relations in (9.117'), i.e., the solitons of (9.117) will be short-lived and intense. For a possible relationship to spikons, see the numerical calculations of Valeo et al. [338].

This is the way spikons can be explained. The nonlinearities discussed here produce a number of size limitations when applied to caviton analysis:

$$\frac{|E^+|^2 \varepsilon^{(i)^2}}{2\pi n_0 T_e \varepsilon^2} \simeq |E|^2 \mu\, \frac{\tau^4}{\xi^4} \simeq \frac{\mu}{\tau^{2/3}}$$

or

$$\tau \gg \mu^{3/2}, \qquad \frac{|E^+|^2}{2\pi n_0 T_e} \simeq \mu|E|^2 \ll \mu^{-2}.$$

The electron nonlinearities are the more important because $\delta\widetilde{\Sigma}$ does not vanish in the three-dimensional case [in contrast to the one-dimensional case; see (9.111)]. Thus, electron non-linearities can be neglected when

$$t \gg \frac{1}{\omega_{pi}}, \qquad \tau \gg \sqrt{\mu}. \tag{9.118}$$

A very important question is whether the caviton will continue to be compressed for $\tau < \sqrt{\mu}$ when the ions are not able to follow the motion of the electron field. When the nonlinear term is compared with the dispersion term we find that they have the same structure, but the nonlinear term is a factor of $|E^+|^2/4\pi n_0 T_e$ smaller. But then we lose the entire concept of strong turbulence based on the notion that the nonlinear term dominates the dispersion term. Thus, in view of the magnitude of the nonlinearity the dispersion must, in the first approximation, be linear [i.e., the left side of (9.8) is also small] in order to satisfy the nonlinear equation. But this means that we have returned to the method of description used for weak nonlinearities. Modulation interactions are then impossible and collapse stops if it even reaches the point where $\tau \simeq \sqrt{\mu}$.

As a consequence compression goes on only to dimensions much greater than the Debye radius:

$$\frac{r_{\min}}{r_d} \simeq \sqrt{3} \left(\frac{3m_i}{4m_e} \right)^{1/6} = 5.78, \qquad \frac{n'}{n_0} \simeq 0.082. \tag{9.119}$$

The last ratio is valid for $m_i/m_e = 1836$. Landau damping is still very small:

$$\frac{\gamma_{\max}^L}{\omega_{pe}} \simeq \sqrt{\frac{81\pi m_i}{32 m_e}} \exp\left\{ -\frac{3}{2} \left[\left(\frac{3m_i}{4m_e} \right)^{1/3} + 1 \right] \right\} \simeq 1.52 \times 10^{-6}. \tag{9.120}$$

The usual mass ratio was used to arrive at a numerical value in the above equation. Therefore, caviton formation cannot be a mechanism for dissipating Langmuir turbulence even when the energy $|E^+|^2/2\pi$ in the electric field is as great as the energy of the particles $n_0 T_e$ because the instability of such motion prevents the system from reaching such energy levels in the electric field. This makes it clear that noninteracting cavitons are not effective for dissipating Langmuir turbulence.

9.5. Soliton and Caviton Turbulence

It is insufficient to know just some of the dynamical motions
in order to erect a theory of turbulence even if it is assumed that
turbulence can be represented as a collection of weakly interacting
motions. The basis of turbulence is the interaction of collective
motions and the transfer of energy from one type of motion to
another. As we have already seen the interaction can itself alter
the dynamical motion. In the case of strong turbulence the interactions can regulate the nature of the motion; an example is an incompressible fluid. The notion that the state of strong turbulence
is a state of weakly interacting solitons was first suggested in
[324]. Solitons are possible only in the one-dimensional case,
so the model of soliton turbulence has a methodological character.
However, it is very important because it permits comparison
with results of numerical experiments, which are also one-dimensional. Soliton interactions were studied numerically in [323],
and the possibility of mixing two solitons was observed. According to (9.49) the number N of these quanta is a conserved quantity
and is proportional to the soliton amplitude E_0; thus mixing must
produce a soliton with an amplitude $2E_0$. The width of a soliton
ξ_0 is proportional to E_0^{-1} according to (9.46), so that the new soliton
is twice as narrow and its width $\Delta \varkappa$ is doubled in the spatial
Fourier spectrum. Continuation of the mixing process will con-
tinually increase $\varkappa$ so that a flow is created toward large wave
numbers. According to [323] this will correspond to a state of
strong turbulence with an energy flux in the direction opposite to
that found in weak turbulence. In order to study this possibility
in greater detail we must consider the mixing process together
with other processes which can take place concurrently. From
(9.50) the soliton energy contains a term $\sim E_0^3$ in addition to the
linear term proportional to E_0. The doubling during mixing means
that energy conservation will hold only if part of this energy is
radiated. This radiation can be either ion-acoustic or Langmuir
waves. The numerical results of [323] show that ion-acoustic
waves are emitted during mixing, and mixing will only occur
among solitons having nearly the same amplitudes:

$$\Delta E_0 \lesssim \frac{1}{2} E_0 . \tag{9.121}$$

It is clear from (9.50) that when $E_0 \gg 1$ ($|E^+|^2/4\pi n_0 T_e \gg m_e/m_i$) the sound must absorb positive energy which is four times greater than the sum of the negative energies in the two colliding solitons when $u \ll 1$.

Mixing does not occur for solitons having different amplitudes. However, it was noted in [325] that any collision of two solitons results in the emission of sound waves; this was called bremsstrahlung of the soliton gas. Bremsstrahlung of this type can be treated using perturbation theory. An estimate of the energy removed by bremsstrahlung will allow one to estimate when that energy will be on the order of the energy of the colliding solitons, so that in the final analysis one can calculate the probability of soliton mixing.

As a first approximation let us assume that the solitons move with constant velocities and that they are characterized by E_{01}, u_1 and E_{02}, u_2 respectively. When the solitons intersect an additional term appears in the density $\delta \nu$ given in (9.41); we shall treat it using perturbation theory. The Fourier component of the perturbation takes the form

$$\delta \nu_{\varkappa,\lambda} = \frac{\varkappa^2}{\lambda^2 - \varkappa^2} |E_1 E_2^* + E_2 E_1^*|_{\lambda,\varkappa} = \frac{\varkappa^2 \sqrt{(1 - u_1^2)(1 - u_2^2)}}{(\lambda^2 - \varkappa^2)2 |u_1 - u_2|}$$

$$\times \left[e^{i(\phi_{01} - \phi_{02})} \frac{1}{\cosh \left[\dfrac{\pi \sqrt{1 - u_1^2}\, \Gamma_{\varkappa,\lambda}^{(1)}}{\sqrt{2}\, E_{01}} \right]} \cdot \frac{1}{\cosh \left[\dfrac{\pi \sqrt{1 - u_2^2}\, \Gamma_{\varkappa,\lambda}^{(2)}}{\sqrt{2}\, E_{02}} \right]} \right.$$

$$\left. + e^{-i(\phi_{01} - \phi_{02})} \frac{1}{\cosh \left[\dfrac{\pi \sqrt{1 - u_1^2}\, \Gamma_{-\varkappa,-\lambda}^{(1)}}{\sqrt{2}\, E_{01}} \right]} \cdot \frac{1}{\cosh \left[\dfrac{\pi \sqrt{1 - u_2^2}\, \Gamma_{-\varkappa,-\lambda}^{(2)}}{\sqrt{2}\, E_{02}} \right]} \right], \qquad (9.122)$$

$$\Gamma_{\varkappa,\lambda}^{(1,2)} = \frac{\lambda - \varkappa u_{2,1} + \dfrac{E_{01}^2}{2(1 - u_1^2)} - \dfrac{E_{02}^2}{2(1 - u_2^2)} \mp \dfrac{(u_1 - u_2)^2}{4}}{|u_1 - u_2|}. \qquad (9.123)$$

In (9.122) ϕ_{01} and ϕ_{02} are the initial phases of the two solitons as found in (9.44). To calculate the total radiated energy it is necessary to compute the energy of the sound waves

$$V^{(s)} = \frac{1}{2} \int \left\{ | \, \delta v(\xi, \tau) \, |^2 + \left| \int^\xi d\xi' \, \frac{\partial v(\xi', \tau)}{\partial \tau} \right|^2 \right\} d\xi. \qquad (9.124)$$

By taking into account the poles $\lambda^2 = \varkappa^2$ in (9.122) which correspond to acoustic emission and averaging over the phase difference $(\phi_{01} - \phi_{02})$, which is assumed to be completely random, we obtain [325]

$$V^{(s)} = \frac{\pi^3 (1 - u_1^2)(1 - u_2^2)}{2(u_1 - u_2)^2} \int \varkappa^2 \, d\varkappa \left\{\left\{ \frac{1}{\cosh \left[\frac{\pi \sqrt{1 - u_1^2}\, \Gamma^{(1)}_{\varkappa, |\varkappa|}}{\sqrt{2}\, E_{01}} \right]} \right.\right.$$

$$\times \left. \frac{1}{\cosh \left[\frac{\pi \sqrt{1 - u_2^2}\, \Gamma^{(2)}_{\varkappa, |\varkappa|}}{\sqrt{2}\, E_{02}} \right]} \right]^2 + \left[\frac{1}{\cosh \left[\frac{\pi \sqrt{1 - u_1^2}\, \Gamma^{(1)}_{\varkappa, -|\varkappa|}}{\sqrt{2}\, E_{01}} \right]} \right.$$

$$\times \left. \left. \frac{1}{\cosh \left[\frac{\pi \sqrt{1 - u_2^2}\, \Gamma^{(2)}_{\varkappa, -|\varkappa|}}{\sqrt{2}\, E_{02}} \right]} \right]^2 \right\}. \qquad (9.125)$$

Thus it follows that solitons for which $u \to 1$ cannot mix. The same conclusion is obtained for weak solitons, which have $E_0 \ll 1$. Analysis of (9.125) (see [325]) shows that strong solitons can emit bremsstrahlung with energy comparable to their initial energies only when

$$\Delta E_0 = | \, E_{01} - E_{02} \, | \lesssim 1, \qquad (9.126)$$

i.e., in contrast to (9.121), $\Delta E_0 / E_0 \sim 1/E_0 \ll 1$. This difference is evidently due to the averaging over the soliton initial phases.

By starting with the mixing condition (9.121) a kinetic equation was found in [323] for a soliton gas in which soliton mixing processes and subdividing processes produced by strong sound waves were taken into account. This makes it possible to roughly estimate the turbulence spectrum to be of the form

$$W_k \sim \frac{1}{k^\nu}, \qquad \nu \geq \frac{3}{2}. \qquad (9.127)$$

Besides averaging over the initial phases [which, according to (9.126), can alter the average mixing cross section] there are other processes which compete with mixing. Among them is soliton braking due to nonlinear Landau damping [329]. Assume that there are N identical solitons per unit length with $u \lesssim 1$ and amplitude E_0. Their energy will be $\sim NE_0 \sim |E|_\varkappa^2 \varkappa = |E|^2$, where $|E|^2$ is their average energy. The average distance between solitons is

$$\frac{1}{N} \simeq \frac{E_0}{|E|_\varkappa^2 \varkappa}. \tag{9.128}$$

If we assume that the turbulence spectrum is $|E|_\varkappa^2 \sim 1/\varkappa^2$ the total energy $|E|^2$ can be evaluated in terms of $|E|_\varkappa^2$ as $|E|^2 = |E|_\varkappa^2 \varkappa^2/\varkappa_0$, where $\varkappa_0$ is the fundamental scale of the turbulence. Thus, taking into account that $\varkappa \sim 1/\xi_0 \sim E_0$ we obtain

$$\frac{1}{N} \sim \frac{E_0^2}{|E|^2 \varkappa_0}. \tag{9.129}$$

In order to have solitons merging it is essential that the average distance between them be less than the mean free path $\sim (1/\sqrt{\mu})(1/E_0)$, i.e.,

$$E_0^3 \ll \frac{|E|^2 \varkappa_0}{\sqrt{\mu}}. \tag{9.130}$$

In terms of dimensional variables (9.130) gives

$$\frac{E_0^2}{4\pi n_0 T_e} \ll \left(\frac{W_0}{n_0 T_e} \frac{k_0}{k_*} \right)^{2/3}. \tag{9.131}$$

Since $k_0 \approx k_*$ and $W_0/n_0 T_e$ is usually much smaller than unity, solitons do not reach the point where $E_0^2/4\pi n_0 T_e \simeq 1$, where Landau damping becomes effective. During mixing the number of solitons with large E_0 rapidly decreases because the distance between them becomes larger than the mean free path. Thus, under real conditions, when $W_0/n_0 T_e \ll 1$, there must be another mechanism for soliton interaction, and it should be a long-range interaction.

Nonlinear Landau damping, i.e., Landau damping of density fluctuations, not only causes individual solitons to slow down but also results in an interaction between spatially separated solitons, which has not received much attention at this point. These processes take place only when a number of solitons are present. For example, consider two solitons. Landau damping by electrons is given by (9.54). When two solitons are present Eq. (9.54) must include not only the densities of the individual solitons but an additional density variation (9.122) due to the mutual interaction of the solitons:

$$\delta \nu_{\varkappa,\lambda}^{L} = - i \frac{\varkappa^4}{(\lambda^2 - \varkappa^2)^2} \sqrt{\frac{\pi m_e}{2m_i}} \frac{\lambda}{|\varkappa|} |E_1 E_2^* + E_2 E_1^*|_{\lambda,\varkappa} \qquad (9.132)$$

so that

$$\delta \nu^L(\xi, \tau) = \int \delta \nu_{\varkappa,\lambda}^{L} e^{i\varkappa\xi - i\lambda\tau} d\varkappa \, d\lambda = - \int \frac{\lambda \sin \lambda\tau \, d\lambda \, d\varkappa}{|\varkappa| (\lambda^2 - \varkappa^2)^2}$$

$$\times \frac{\cos[\varkappa(\xi + \xi_{01} - \xi_{02})] \, 2\varkappa^4 \sqrt{(1 - u_1^2)(1 - u_2^2)}}{(u_1 - u_2) \cosh[\pi \sqrt{1 - u_1^2} \, \Gamma_{\varkappa,\lambda}^{(1)}/\sqrt{2} \, E_{01}] \cosh[\pi \sqrt{1 - u_2^2} \, \Gamma_{\varkappa,\lambda}^{(2)}/\sqrt{2} \, E_{02}]}.$$

Here ξ_{01} and ξ_{02} are the initial positions of the soliton centers. A characteristic feature of this density variation is that it will be exponentially small if the soliton overlap is small; that is, there is no long-range interaction between solitons. This is due to the product of the two soliton fields in (9.131). However, in addition to (9.131) there are the terms $\delta \nu^L$, which are related to $|E_1|^2$ and $|E_2|^2$, and they are not small when the solitons are widely separated. The quantities

$$\delta \nu_{\varkappa,\lambda}^{L(1,2)} = - i \frac{\varkappa^4}{(\lambda^2 - \varkappa^2)^2} \sqrt{\frac{\pi m_e}{2m_i}} \frac{\lambda}{|\varkappa|} |E_{1,2}|^2 \qquad (9.133)$$

decrease in a nonexponential manner with distance and can cause a mutual, long-range interaction between solitons. This is also evident from (9.55), which is conveniently rewritten as

$$\delta \nu^{L(1)}(\xi, \tau) = \sqrt{\frac{\pi m_e}{2m_i}} \frac{2u_1}{1 - u_1^2} \int_0^\infty \frac{\varkappa \sin \varkappa(\xi - \xi_{01} - u_1\tau) \, d\varkappa}{\sinh[\pi\varkappa \sqrt{1 - u_1^2}/\sqrt{2} \, E_{01}]}. \qquad (9.134)$$

At large distances from the center of the soliton, $|\xi - \xi_{01} - u_1\tau| \gg \xi_{01}$, distances large compared with the soliton width $\varkappa \ll E_{01}$, the integral in (9.134) takes the form

$$\delta v^{L(1)}(\xi, \tau) = \sqrt{\frac{8m_e}{\pi m_i}} \, \frac{u_1 E_{01}}{1 - u_1^2} \, \frac{1}{\xi - \xi_{01} - u_1 \tau}. \qquad (9.135)$$

Since the density in (9.135) is a smooth function we can take it to be constant at the point where the second soliton is located:

$$\delta v(\xi_{02} + u_2 \tau, \tau) = \sqrt{\frac{8m_e}{\pi m_i}} \, \frac{u_1 E_{01}}{1 - u_1^2} \, \frac{1}{\xi_{02} - \xi_{01} - (u_1 - u_2)\tau}. \qquad (9.136)$$

Equation (9.136) contains the distance between solitons at the moment τ. A soliton increases the density in front of itself and decreases it behind itself, the change in density dropping off as $1/\xi$ from the location of the soliton. The quantity δv acts as the interaction potential energy in the nonlinear Schrödinger equation (9.40). A negative δv denotes attraction, while a positive value indicates repulsion. Thus a soliton attracts all solitons moving behind it and repels those in front of it. But the number of quanta in each soliton is not conserved because only the magnitude N as an integral over all ξ, which includes all solitons, is conserved. Two solitons moving in the same direction decrease their relative velocities if the forward moving soliton has a large amplitude. This can finally lead to the merging of solitons. During the mixing amplified emission of sound waves can take place because of the density variation (9.132), which contains a singularity at $\lambda = \varkappa$. The main conclusion is that in a system of solitons there is an effective long-range interaction in the one-dimensional case which varies as $1/\xi$ (a Coulomb interaction goes as $\sim \xi$ in the one-dimensional case). The effective charge to be assigned to the soliton depends on its amplitude.

A long-range force between solitons has a qualitative influence on their kinetics (recall the analogy with the kinetics of particles interacting through a Coulomb potential). This kinetics has not yet been formulated, and because of the "effective charge" which changes during a collision, the problem will be very complex. But the above discussion makes the suggestion forwarded in [342] more plausible; it was suggested that energy introduced into a system of solitons will be equally distributed among all the degrees of freedom. Assume that there are N solitons of equal amplitude E_0 per unit length. If the total energy $|E|^2 \simeq NE_0$ is fixed then $E_0 \simeq |E|^2/N$ and $\xi_0 \sim N$. For a random distribution of solitons $|E|_\varkappa^2$ is given by the square of (9.48) times the number of solitons and their probability P(N) ($u \ll 1$):

$$| E |_\varkappa^2 \simeq \int dN \, P(N) N \, \frac{1}{2 \cosh^2(\varkappa \pi N / \sqrt{2} \, | E |^2)}. \qquad (9.137)$$

If P(N) = const (a uniform distribution), we obtain

$$| E |_\varkappa^2 \sim \frac{1}{\varkappa^2}. \qquad (9.138)$$

Let us turn now to three-dimensional turbulence. A model was presented in [339] of caviton turbulence as a collection of noninteracting self-constricting cavitons. The model was further developed in [341] and proposed in [343]. The main result of these considerations is that a turbulence spectrum like $\sim 1/\varkappa^{5/2}$ is obtained. This result is a simple consequence of conserving the turbulence flow if the equation for $|E|_\varkappa^2$ is written in the form [344]

$$\frac{\partial}{\partial \tau} | E |_\varkappa^2 = \frac{\partial}{\partial \varkappa} \, \frac{\overline{\Delta \varkappa}}{\tau_\varkappa} \, | E |_\varkappa^2 = 0. \qquad (9.139)$$

If the average change in wave number is taken to be $\overline{\Delta \varkappa} = \varkappa$ and the characteristic lifetime of a given value of $\varkappa$ is taken from the relationship $\xi \sim \tau^{2/3}$, which holds for cavitons, then $\tau_\varkappa \sim \xi^{3/2} \sim 1/\varkappa^{3/2}$. This gives the spectrum $\sim 1/\varkappa^{5/2}$.

However, it is clear from the outset that a model of independently contracting cavitons can hardly describe turbulence, to say nothing of strong turbulence. It is simply a collection of dynamical motions whose initial positions are random. The interaction of collective motions is an essential element in understanding turbulence, and only with an interaction can one expect the nonreproducibility of states which is necessary for turbulence to arise. At the present time there have been practically no calculations on caviton interactions. The most interesting problem is that of the nature of the long-range force be-

tween cavitons, a force which is the three-dimensional analog
of the forces between solitons considered above. For supersonic
motion the density disturbance due to nonlinear Landau damping
is (see (9.54))

$$\delta v^L(\xi, \tau) = \int \delta v_{\lambda,\varkappa}^L e^{i\varkappa\xi - i\lambda\tau} \, d\varkappa \, d\lambda = -i \int \frac{v_{\lambda,\varkappa}^{(0)} |\varkappa|}{\lambda} \sqrt{\frac{3\pi\mu}{8}} \, e^{i\varkappa\xi - i\lambda\tau} \, d\varkappa \, d\lambda$$

$$= -\int \frac{v^{(0)}(\xi', \tau')}{(2\pi)^4} \sqrt{\frac{3\pi\mu}{8}} \int \frac{|\varkappa|}{\lambda} \cos[\varkappa \cdot (\xi - \xi')] \sin[\lambda(\tau - \tau')] \, d\varkappa \, d\lambda \, d\xi \, d\tau$$

$$= -\frac{1}{4\pi} \sqrt{\frac{3\mu}{8\pi}} \int \frac{v^{(0)}(\xi', \tau)(\tau - \tau')}{|\xi - \xi'|^4 |\tau - \tau'|} \, d\tau' \, d\xi'. \tag{9.140}$$

The interaction varies as $\sim 1/\xi^4$. This interaction can influence
the stability of the self-similar contraction of neighboring cavitons,
creating asymmetric perturbations in the general case. The role
of such interactions is very important in the early stages of caviton
formation, preventing their isolation from one another. In the
subsonic stage, when the characteristic rates of the processes
are much slower than thermal ion velocities ($T_e \gg T_i$), we find

$$\delta v^L(\xi, \tau) = -\sqrt{\frac{\pi T_i}{2 T_e}} \frac{\partial}{\partial\tau} \int \frac{d\xi' v^{(0)}(\xi', \tau)}{4\pi^2 |\xi - \xi'|^2}. \tag{9.141}$$

The main contribution to the integral above comes from the pe-
ripheral regions of the caviton, and the interaction decreases
as $1/\xi^2$ with distance. Equations (9.140) and (9.141) contain $v^{(0)}$
but it is easy to see that the total density v will enter into the
general equations. Therefore we present a general equation which
will replace (9.79) for subsonic motion, which takes into account
nonlinear Landau damping and illustrates the essential nonlocal
nature of the nonlinear term:

$$\mathrm{div}_\xi\left(i\frac{\partial}{\partial\tau} + \Delta_\xi\right) \mathbf{E}^+(\xi, \tau)$$

$$= -\mathrm{div}_\xi\left\{|\mathbf{E}(\xi, \tau)|^2 + \sqrt{\frac{\pi T_i}{2 T_e}} \left(\frac{\partial}{\partial\tau} \int \frac{\delta\xi' |\mathbf{E}^+(\xi', \tau)|^2}{4\pi^2 |\xi - \xi'|^2}\right)\right\} \mathbf{E}^+(\xi, \tau). \tag{9.142}$$

There is no reason to exclude the possibility of moving cavitons,
or their collision, mixing, and emission of sound during such
interactions.

9.6. The Statistical Theory of Strong Langmuir Turbulence

As pointed out in Chapter 1, turbulence (and especially strong turbulence) must be described statistically. A statistical theory for strong Langmuir turbulence has been constructed in [353-356, 357]. Study of the many dynamical motions which arise in plasma because of modulation instabilities has demonstrated that basing a theory of turbulence on a model of a gas of weakly interacting solitons, cavitons, spikons, etc., is a very difficult proposition. First, there are difficulties associated with the fact that such objects (solitons, etc.) do not exist for very long before they lose their individuality. Of course gas models like those described above are statistical in the sense that they resemble the kinetic theory of gases; one takes into account the random character of the motions of the elements in the gas. However, the loss of individuality is an important qualitative difference between models of strong turbulence and models of common gas particles. In the one-dimensional case this difference manifests itself in the mixing of solitons. Similarly, eddies are destroyed in fluid turbulence. In the three-dimensional case it appears as the subdividing of modulation perturbations, the merging of cavitons, and so on.

Second, problems arise in connection with the many possible interactions of objects like cavitons and solitons. We can enumerate those considered above for cavitons. 1) Cavitons are subdivided because of modulation instability (like eddies in fluids); 2) cavitons interact by emitting and absorbing sound; 3) there are nonlocal interactions due to the nonlinear Landau damping; 4) cavitons mix. Other interactions are also possible. Furthermore, neither the individuality of the separate cavitons nor the "effective charge" associated with such interactions is conserved in the interactions; the effective charge depends on the caviton energy. The caviton energy is not constant because it exchanges energy with other cavitons.

Third, there are many different kinds of modulation motions, such as cavitons with different self-similar contraction laws, three-dimensional solitons, and cavitons with different spatial forms (spherical, elliptical, and so on). All of this means that the advantage of using a statistical approach from the very beginning is connected with the freedom from being tied to a model description of the different excitations which describe strong turbulence, but the main advantage is that one is not limited by specific kinds of excitations which might determine the interactions.

In order to construct a statistical theory one would start with Eqs. (9.34) and (9.36) and use spatial Fourier components:

$$\frac{2i}{\omega_{pe}} \frac{\partial E_k^+(t)}{\partial t} - \frac{3k^2 v_{Te}^2}{\omega_{pe}^2} E_k^+(t) = \int E_{k_1}^+(t) \frac{(\mathbf{k} \cdot \mathbf{k}_1)}{k k_1} \frac{n_{k-k_1}'(t)}{n_0} d\mathbf{k}_1, \qquad (9.143)$$

$$\frac{\partial^2}{\partial t^2} n_k'(t) + v_s^2 k^2 n_k'(t) = -\frac{k^2}{4\pi m_i} \int \frac{\mathbf{k}_1 \cdot (\mathbf{k} - \mathbf{k}_1)}{k_1 |\mathbf{k} - \mathbf{k}_1|} E_{k_1}^+(t) E_{k-k_1}^{+*}(t) d\mathbf{k}_1, \qquad (9.144)$$

$$E_k^+(t) = \int \frac{\mathbf{k} \cdot \mathbf{E}^+(\mathbf{r}, t)}{k} e^{-i\mathbf{k}\mathbf{r}} \cdot \frac{d\mathbf{k}}{(2\pi)^3}. \qquad (9.145)$$

One-dimensional equations of this type were used in one of the first papers on strong Langmuir turbulence [321], in which they were introduced as a phenomenological generalization of the linear equations for oscillating two-stream instabilities. These equations were used in [321] to describe quantities which have the following form in our notation:

$$E_k(t) = E_k^+(t) e^{-i\omega_{pe}t} + E_k^-(t) e^{i\omega_{pe}t}. \qquad (9.146)$$

In the one-dimensional case (9.143) and (9.144) take the form

$$\frac{\partial^2}{\partial t^2} E_k(t) + (\omega_{pe}^2 + 3k^2 v_{Te}^2) E_k(t) = -\omega_{pe}^2 \int n_{k'}(t) E_{k-k'}(t) dk', \qquad (9.147)$$

$$\frac{\partial^2}{\partial t^2} n_k(t) + k^2 v_s^2 n_k(t) = -\frac{k^2}{8\pi m_i n_0} \int E_{k'}(t) E_{k-k'}(t) dk'. \qquad (9.148)$$

These equations are the same as those used in [321]. In [321] the terms $\nu_{ek} [\partial E_k(t)/\partial t]$ and $\nu_{ik} [\partial n_k(t)/\partial t]$ were added to the left sides of (9.147) and (9.148) respectively to account for Landau damping. But it has already been pointed out above that it is wrong to introduce Landau damping into equations for n_k and E_k. It is clear from (9.8) that one must take into account both the imaginary part of $1/\varepsilon_{k-k_1}$, which is equivalent to introducing ν_{ek} and ν_{ik} into the left sides of (9.147) and (9.148), and the imaginary parts of $\varepsilon_{k-k_1}^{(e)} - 1$ and $\varepsilon_{k-k_1}^{(i)}$, which appear in the numerator of (9.8). Numerical models of (9.147) and (9.148) in [321] give a $1/k^2$ spectrum. The model representations from which (9.147) and (9.148) were derived in [321] were based on a generalization of the linear theory in which, when the level of the modulational perturbations reaches that of the pump wave, they in turn act as pump waves for further modulational perturbations. This is similar to the hierarchy picture of eddies in turbulent fluids.

A three-dimensional statistical theory of strong Langmuir turbulence has been presented in [357]; it is based on a generalization of (9.147) and (9.148) to the three-dimensional case. We shall use (9.143) and (9.144) and notation which resembles that used in [357] to introduce dimensionless "electron" and "ion" densities:

$$n_{e,k}(t) = \frac{ie(\mathbf{k}\mathbf{E}_k^+(t))}{m_e\omega_{pe}^2} = \frac{iekE_k^+(t)}{m_e\omega_{pe}^2}, \qquad n_{i,k}(t) = \frac{n_k'(t)}{n_0} \qquad (9.149)$$

The quantity n_{ek} corresponds to a change in electron density, but only in the first approximation:

$$n_{e,k} \simeq \frac{\mathbf{k}\cdot\mathbf{v}_{e,k}^+}{\omega_{pe}}, \qquad \mathbf{v}_{e,k}^+ \simeq \frac{ieE_k^+}{m_e\omega_{pe}}, \qquad v_{e,k}^+ =: v_k(t)\,e^{i\omega_{pe}t}. \qquad (9.150)$$

Putting the pump wave $(\mathbf{k}_1 \cdot \mathbf{E}_{k_1}^+(0))/k_1 = E_0(t)\,\delta(k_1)$ into the right side of (9.143) we find from (9.143) and (9.144) (in terms of the new notation) that

$$\left(3k^2v_{Te}^2 - 2i\omega_{pe}\frac{\partial}{\partial t}\right)n_{e,k} = -\frac{ie(\mathbf{k}\cdot\mathbf{E}_0(t))}{m_e}\,n_{i,k} - \frac{ie}{m_e}\int(\mathbf{k}\cdot\mathbf{E}_{k_1}^+)n_{i,k-k_1}\,dk_1$$

$$= -\frac{ie(\mathbf{k}\cdot\mathbf{E}_0(t))n_{i,k}}{m_e} - \omega_{pe}^2\int n_{i,k-k_1}n_{e,k_1}\frac{\mathbf{k}\cdot\mathbf{k}_1}{k_1^2}\,dk_1, \qquad (9.151)$$

$$\frac{\partial^2}{\partial t^2}n_{i,k} + k^2v_s^2n_{i,k} = \frac{ie(\mathbf{k}\cdot(\mathbf{E}_0n_{e,k}^* + \mathbf{E}_0^*n_{e,k}))}{m_i}$$

$$+ \frac{ie}{m_i}k^2\int\frac{(\mathbf{k}_1\cdot\mathbf{E}_{k-k_1}^*)n_{e,k_1}dk_1}{k_1^2} = \frac{ie(\mathbf{k}\cdot(\mathbf{E}_0n_{e,k}^* + \mathbf{E}_0^*n_{e,k}))}{m_i}$$

$$+ \omega_{pi}^2\int n_{e,k-k_1}^*n_{e,k_1}k^2\frac{\mathbf{k}_1\cdot(\mathbf{k}-\mathbf{k}_1)}{k_1^2(\mathbf{k}-\mathbf{k}_1)^2}\,dk_1. \qquad (9.152)$$

These equations differ from those used in [357] in the angular
dependence of the nonlinear terms and the fact that harmonics
at twice the plasma frequency are missing from the nonlinear
terms in (9.152). These equations are more exact because they
are derived from kinetic equations for which certain inequalities
are valid. The linear terms on the right side give modulation
instabilities. It was assumed in [357] that the pump field is regular:

$$E_0(t) = E_0 e^{-i(\omega_0 - \omega_{pe})t}.$$ (9.153)

It is assumed that the frequency of the pump field is not equal
to the plasma frequency (if the pump field is a Langmuir field,
then $\omega_0 = \omega_{pe}$). It is further assumed that all remaining components of the field $n_{e,k}$ and $n_{i,k}$ are random. This assumption is reasonable only for large k. By assumption the nonlinear terms in (9.151) and (9.152) must stabilize the modulation
instability of the pump field.

The arguments used in [357] continue by using standard
methods for averaging over the ensemble presented in Chapter 2.
Turning temporarily to the Fourier components, n_i can be expressed in terms of n_e through (9.152) and then substituted into
(9.151); one then obtains a nonlinear equation for n_e which contains quadratic and cubic nonlinearities. After multiplying by
n_e the average is taken; the averages of products of three and
four random quantities are computed using the method in Chapter
2. The result is an expression for ε_k^N, and the spectrum is obtained by setting $J_m \varepsilon_k^N$ equal to zero. A $1/k^2$ spectrum was obtained in [357]. The turbulence spectrum is determined by the
imaginary parts of $J_m \varepsilon_k^N$, and as already mentioned damping is
not accounted for correctly by inserting damping terms into the
left sides of (9.151) and (9.152) as was done in [357]. It follows
from (9.8) that imaginary terms of the same order of magnitude
appear on the right side in the nonlinear terms. A more general
approach to a statistical theory is presented in [356]. This approach takes into account the electron nonlinearities and renormalization processes which were discussed in detail in Chapter
8 for modulation instabilities. It has been previously demonstrated
here that electron nonlinearities are very important in the final
stages of caviton development (if they are in fact reached).
Thus it is important to include not only the imaginary parts of
the electron nonlinearities, which lead to nonlinear Landau

damping, but the real parts as well since they describe the frequency shifts. Moreover, nonlinear Landau damping is important to the long-range interactions of solitons and cavitons. Physically one would expect that the long-range interaction could aid the self-contraction of modulation perturbations, thereby decreasing the characteristic scales. On the other hand, self-contraction effects will change the nonlinear Landau damping. In other words, there is a mutual interaction between the Landau damping processes and the modulation self-contraction, which, in the final analysis, can alter the direction in which turbulent energy is moved. The theory can be constructed by using a general equation containing only third-order nonlinearities. In fact, according to §9.4 higher-order nonlinearities play an insignificant role in the development of modulation perturbations. We shall start with the following equation:

$$\frac{i\omega}{4\pi} \varepsilon_k E_k = \int S_{k,k_1,k_2} E_{k_1} E_{k_2} \, \delta(k - k_1 - k_2) \, dk_1 \, dk_2$$

$$+ \int \Sigma_{k,k_1,k_2,k_3} E_{k_1} E_{k_2} E_{k_3} \, \delta(k - k_1 - k_2 - k_3) \, dk_1 \, dk_2 \, dk_3. \qquad (9.154)$$

This equation is the same as that used to study weak turbulence. It takes into account all kinetic effects and electron and ion nonlinearities if the coefficients S' and Σ are calculated with these effects included. In addition it can include any other effects [such as collisions and turbulent broadening of resonances (see §2.8)] [77, 78, 359, 360]. Equation (9.154) is more general than (9.151) and (9.152) because the latter equations come from (9.8), which, in turn, can be derived from (9.154) using appropriate approximations for the coefficients of the nonlinear terms.

One can show that the main results of Chapter 8 concerning the instability of stationary turbulence can be derived from (9.154). The starting point in Chapter 8 for attacking such problems was the general kinetic equations. However, in the subsequent approximations the problem reduced to neglecting nonlinear currents which are of higher order in the field than those taken into account in (9.154). It is clear even in (8.50) that only the quantities $\tilde{\Sigma}$ and S enter into the fundamental integral equation which determines the character of the linear modulation perturbations. Following the procedure used in Chapter 8, we divide the field into E^T and E^R [see (8.2)]. The field E^R is a low-frequency field which describes the modulation disturbances. It is convenient to designate it by E^M in the future. The perturbations of the turbulent field $E^{T(0)}$ which are linear in the modulation disturbance will be denoted as $E^{T(1)}$. Then, by excluding the field E^M we find from (9.154) that [see (8.50)]

$$\tilde{\varepsilon}_{k'-k} \langle E_k^{T(0)} E_{k'-k}^{T(1)} \rangle = 2I_k \int (\tilde{\Sigma}_{k'-k_1, k_1, -k, k'-k_1} + \tilde{\Sigma}_{k'-k, k'-k_1, k_1, -k}) \langle E_{k_1}^{T(0)} E_{k'-k_1}^{T(1)} \rangle dk_1$$

$$+ 2I_k \int \tilde{\Sigma}_{k'-k, -k, k_1, k'-k_1} \langle E_{k_1}^{T(0)} E_{k'-k_1}^{T(1)} \rangle dk_1, \qquad (9.155)$$

$$\tilde{\varepsilon}_k = \varepsilon_k + \varepsilon_k^N, \qquad \varepsilon_k^N = -\frac{8\pi}{i\omega} \int \tilde{\Sigma}_{k, k_1} I_{k_1} dk_1.$$

Here as earlier the components of $\tilde{\Sigma}$ are symmetrized with respect to the last two subscripts. The first two terms in (9.155) give the renormalization. When $k \gg k_* = (k_d/3)\sqrt{m_e/m_i}$ these terms are small; if we neglect them and take into account that in the approximation of (9.8) $\tilde{\Sigma}$ depends only on the difference of the first two arguments (i.e., $\tilde{\Sigma}_{k'-k, -k, k_1, k'-k_1}$ is a function only of the low-frequency wave vector $k' = \{k', \omega'\}$) the dispersion equation (8.26) of Chapter 8 is obtained from (9.155).

When $k \ll k_*$ the renormalization must be taken into account; the integral equation (9.155) is solved just as in §8.3 and produces the results which follow from (8.56). Unlike the case of weak turbulence, one cannot assume here that the wave dispersion is linear when the interactions are nonlinear. It must be determined by solving the nonlinear equations which describe the turbulence. Renormalization will have to be included at every phase of the calculations involving nonlinear interactions, just as it was in the studies of linear modulation perturbations (8.83). Finally we must be aware that turbulence on a time scale of the modulation disturbances is not stationary; energy is continuously redistributed among the various modulation perturbations because of their interactions. Thus, our statistical model will assume that Langmuir turbulence is not stationary even at a nonlinear stage of the modulation interactions but undergoes rather complex spatial and temporal changes all the time. Thus the correlation function for random fields which describes such turbulence must take the form

$$\langle E_k^T E_{k'-k}^T \rangle = I_{k;k'}; \qquad (9.156)$$

note that it depends on both the high-frequency k and the low-frequency k'.† It is also useful to introduce a correlation function which averages over times and scales much larger than the characteristic times and scales of these variations in the turbulence. This average will be indicated by a bar. Thus we have

$$\overline{I_{k;k'}} = I_k \delta(k').$$
(9.157)

I_k can be considered to be a quantity which characterizes the spectrum of strong Langmuir turbulence. As above, the total field E_k is split into a turbulent high-frequency field E_k^T and a low-frequency field for the modulation perturbations E_k^M:

$$E_k = E_k^T + E_k^M.$$
(9.158)

Here we use the index M rather than R as in Chapter 8 [see (8.2)] to emphasize that the field M is the field of the modulation perturbations, and regular only in time scales and length scales of the modulation perturbations, but when averaged over large scales and times it vanishes:

$$\overline{E_k^M} = 0.$$
(9.159)

This is the assumption of the statistical picture which holds at an advanced stage in strong Langmuir turbulence. The quantity $E_{k'}^M$ is analogous to $n_{i,k}$ used in [357] and can, under certain conditions, be expressed in terms of it; E_k^T or $I_{k;k'}$ are analogous to $n_{e,k}$ in [357]. As in [357], where the correlation between $n_{i,k}$ and $n_{e,k}$ was taken into account, there is correlation between $I_{k,k'}$ and $E_{k'}^M$ in our approach. We set

$$\overline{I_{k;k'}E_{k''}} = V_{k;k'}\delta(k' + k'').$$
(9.160)

The modulation fields are also correlated (in analogy with $n_{i,k}n_{i,k'}$):

$$\overline{E_k^M E_{k'}^M} = w_{k'}\delta(k' + k'')$$
(9.161)

---

† In the following all primed frequencies and wave numbers are for low-frequency modulation perturbations, and the unprimed quantities are their high-frequency counterparts.

as are the high-frequency Langmuir fields:

$$\overline{I_{k;k'}I_{k_1;k''}} - \overline{I_{k;k'}}\,\overline{I_{k_1;k''}} = G_{k,k_1;k'}\,\delta(k' + k''). \qquad (9.162)$$

These correlations must be taken into account to construct non-linear equations which include interactions between modulation perturbations and their effect on interactions and on the spectrum of Langmuir fluctuations.

It is not difficult to include in the original equation (9.154) an additional regular field, which would be denoted by E^R. An example would be the pump field E_0 in (9.151) and (9.152). However, we shall limit our discussion to the simpler case, where $E^R = 0$, which is directly related to the strong Langmuir turbulence which arises when oscillations concentrate in the region of large scales.

To find an equation for the quantity in (9.156) one must use (9.154) and separate from it an equation for E_k^T by using the average < > and subtracting the averaged equation from (9.154); this is the method used in Chapters 2 and 4. After multiplying the resulting equation by $E_{k'-k}^T$ and averaging by the standard method, we can express averages of triplet products in terms of quartets, and throw away terms of order $W/n_0 T_e \ll 1$. We find

$$\frac{i\omega}{4\pi}\,\varepsilon_k I_{k;k'} \simeq 2\int \tilde{\Sigma}_{k,k_1,k_2,k_3} I_{k_1;k-k_2} I_{k_2;k_2-k+k'}\,\delta(k - k_1 - k_2 - k_3)\,dk_1\,dk_2\,dk_3$$

$$+ \int \Sigma_{k,k_1,k_2,k_3} I_{k_1;k_1-k+k'} I_{k_2;k-k_1}\,\delta(k - k_1 - k_2 - k_3)\,dk_1\,dk_2\,dk_3$$

$$+ 2\int S_{k,k_1,k''} I_{k_1;k_1-k+k'} E_k^M\,\delta(k - k_1 - k'')\,dk_1\,dk''. \qquad (9.163)$$

When this equation is averaged over large times and scales we arrive at an equation for the quantities defined in (9.157):

$$\frac{i\omega}{4\pi}\,\tilde{\varepsilon}_k I_k \simeq 2\int \tilde{\Sigma}_{k,k_1,k-k',k'-k_1} G_{k_1,k-k';k'}\,dk_1\,dk' + \int \Sigma_{k,k-k',k_1,k'-k_1} G_{k-k',k_1;-k_1}\,dk_1\,dk'$$

$$+ 2\int S_{k,k+k',-k'} V_{k+k';k'}\,dk'. \qquad (9.164)$$

It is characteristic that the left side of (9.164) contains the total dielectric function $\tilde{\varepsilon}_k = \varepsilon_k + \varepsilon_k^N$ including the nonlinear shifts. If the modulation perturbations are not included, i.e., if G and $\overline{IE^M}$ are equal to zero on the right side of (9.164), we regain the equation for weak turbulence. An equation for the correlation function G can be obtained from (9.163):

$$\frac{i\omega}{4\pi}\,\tilde{\varepsilon}_k G_{k,k_2;k'} = 2\int \Sigma_{k,k_1,k-k',k'-k_1} I_{k-k'} G_{k_1,k_2;k'}\,dk_1$$

$$+ \int \Sigma_{k_1,k-k',k_1,k'-k_1} I_{k-k'} G_{k_1,k_2;k'}\,dk_1 + 2\,S_{k,k-k',k'} I_{k-k'} V_{k_2;-k'}. \qquad (9.165)$$

Finally, an equation for $V_{k;\,k'}$ can also be obtained from (9.163):

$$\frac{i\omega}{4\pi}\,\tilde{\varepsilon}_k V_{k;k'} = 2 I_{k-k'}\int \Sigma_{k,k_1,k-k',k'-k_1} V_{k_1;k'}\,dk_1 + I_{k-k'}\int \Sigma_{k,k-k',k_2,k'-k_2} V_{k_2;k'}\,dk_2$$

$$+ 2 S_{k,k-k';k'} I_{k-k'}\,w_{k'}. \qquad (9.166)$$

The integral equation (9.166) enables us to express $V_{k;\,k'}$ in terms of $w_{k'}$, while (9.165) allows us to write $G_{k,\,k_1;\,k'}$ in terms of $V_{k;\,k'}$ and thus $w_{k'}$. The solution to this problem begins with the fact that the effect of modulation perturbations on turbulent Langmuir oscillations can be included to first order in $w_{k'}$. This assumption is valid because the energy contained in modulation perturbations is much smaller than W, the energy of Langmuir oscillations.

The leading terms on the right sides of (9.165) and (9.166) take into account renormalization. As discussed in Chapter 8, renormalization is important for small k ($k \ll k_*$), i.e., at the maximum in the spectral density of the turbulence. This region will be called the Langmuir condensation and will be denoted by the superscript (0).†

In other words the superscript (0) denotes an energy-containing region of strong Langmuir turbulence. Renormalization need not be included when k is large. In analogy with fluids, this region is termed an inertial region. Quantities related to this region will be denoted by the superscript (i). Before solving the equations, we must discuss the following important fact. According to the definition of $I_{k;\,k'}$ (9.156) the fields E_k^T and $E_{k'-k}^T$ are

---

†When an external field with very small k is used as a generator the condensation can also take place if randomization has already taken place in a region sufficiently near the excitation source (i.e., $k \ll k_*$). However, a strong pump field can directly influence the spectrum up to $k \approx k_{max}$, where $k_{max} > k_*$ (see §9.2).

equivalent; therefore

$$I_{k;k'} = I_{k'-k;k'} \tag{9.167}$$

and similar relationships hold for $V_{k;k'}$ and $G_{k,k_1;k'}$. Thus, other equations for $I_{k;k'}$, $V_{k;k'}$, and $G_{k,k_1;k'}$ can be obtained by starting with the equation for the field $E_{k'-k}^T$, multiplying it by E_k^T and following the same procedure as before. The equations which would result from this approach can also be obtained directly from (9.164) and (9.165) by changing k to k'-k and using relationships like (9.167). This discussion is important to the solutions of these equations because solutions must be found which simultaneously satisfy both types of equations. We shall illustrate the method of solution using (9.166) for the inertial region as an example, in which renormalization can be neglected. To simplify matters we will assume that $\widetilde{\Sigma}2/i\omega$ and $\Sigma2/i\omega$ depend only on the difference between their first two arguments and that $S2/\omega$ depends only on its last argument [see (9.8)]:

$$\left. \begin{aligned} \frac{8\pi}{i\omega} \, \widetilde{\Sigma}_{k,k_1,k-k',k'-k} &\simeq - \, \widetilde{\Sigma}_{k-k_1}, \\[2ex] \frac{4\pi}{i\omega} \, \Sigma_{k,k-k',k_2,k'-k_2} &\simeq - \, \Sigma_{k'}, \\[2ex] \frac{8\pi}{\omega} \, S_{k,k-k',k'} &\simeq - \, S_{k'}. \end{aligned} \right\} \tag{9.168}$$

These relationships are satisfied only when it is possible for modulation instability to appear. Then, in the inertial region, we find from (9.166) that

$$\widetilde{\varepsilon}_k V_{k;k'}^{(i)} + I_{k-k'} \Sigma_{k'} \int V_{k_1;k'}^{(i)} \, dk_1 = - \frac{S_{k'}}{i} I_{k-k'}^{(i)} w_{k'}. \tag{9.169}$$

This equation can be solved by dividing it by $\widetilde{\varepsilon}_k$, integrating over k, and substituting the resulting integral $\int V_{k_1;k'}^{(i)} \, dk_1$ into (9.169); we find

$$\widetilde{\varepsilon}_k V_{k;k'}^{(i)} = - \frac{1}{i} \frac{S_{k'} w_{k'} I_{k-k'}^{(i)}}{(1 + \tfrac{1}{2} D_{1,k'}^{(i)})}, \tag{9.170}$$

$$\frac{1}{2} D_{1,k'}^{(i)} = \Sigma_{k'} \int \frac{I_{k_1-k'}^{(i)}}{\widetilde{\varepsilon}_{k_1}} \, dk_1. \tag{9.171}$$

The integral on the left side of (9.169), and therefore (9.171) as well, has been extended to that region of k_1 where $I^{(i)}_{k_1-k'}$ corresponds to the inertial region. In the notation of Chapter 8 the quantity $D_{1k'}$ is simply related to d_1 [see (8.23), (8.56), (8.57)]:

$$k_z'^2 n_0(\varepsilon_0^e(k') - 1)\frac{d_1}{\omega_{pe}} = D_{1,k'} .$$

The solution to (9.170) requires that the symmetry of the correlation functions be kept in mind; thus

$$\tilde{\varepsilon}_{k'-k} V^{(i)}_{k;k'} = -\frac{1}{i}\frac{S_{k'}w_{k'}I^{(i)}_k}{1 + \frac{1}{2}D^{(i)}_{1,k'}}. \tag{9.172}$$

We now note that the problem is to find effects linear in $w_{k'}$ when the equation $\tilde{\varepsilon}_k I_k = 0$ is satisfied in the zeroth approximation. Thus, after dividing (9.170) by $\tilde{\varepsilon}_k$ it is necessary to add the solution of the homogeneous equation $\tilde{\varepsilon}_k V_{k;k'} = 0$, which must be proportional to I_k. It is easily seen that this is the solution (9.172); conversely, when solving (9.172) one must add the solution to the homogeneous equation which reduces to the solution (9.170). The same terms must be added to the integrals in (9.172), i.e.,

$$V^{(i)}_{k;k'} = -\frac{S_{k'}w_{k'}}{i(1 + D^{(i)}_{1,k'})}\left(\frac{I^{(i)}_k}{\tilde{\varepsilon}_{k'-k}} + \frac{I^{(i)}_{k-k'}}{\tilde{\varepsilon}_k}\right), \tag{9.173}$$

$$D^{(i)}_{1,k'} = \Sigma_{k'}\int\left\{\frac{I^{(i)}_{k_1-k'}}{\tilde{\varepsilon}_{k_1}} + \frac{I^{(i)}_{k_1}}{\tilde{\varepsilon}_{k'-k_1}}\right\}dk_1 = 2\Sigma_{k'}\int\frac{I^{(i)}_{k_1}}{\tilde{\varepsilon}_{k'-k_1}}dk_1. \tag{9.174}$$

It is easily demonstrated that $V^{(i)}_{k;k'}$ satisfies both (9.170) and (9.172). Equation (9.174) can be improved somewhat because the integral $\int V_{k_1;k'}dk_1$ in (9.163) has been extended to cover values of k_1 which include both the inertial and energy-containing regions. That is, the term $\int V^{(0)}_{k_1;k'}dk_1$ must be included. This introduces an additional term on the right side of (9.169), which is of the form

$$-\Sigma_{k'}\int V^{(0)}_{k_1;k'}dk_1 I_{k-k'}$$

and this changes (9.174) to the following:

$$V^{(i)}_{k;k'} = -\left(\frac{S_{k'}w_{k'}}{i} + \Sigma_{k'}\int V^{(0)}_{k_1;k'}dk_1\right)\frac{1}{1 + D^{(i)}_{1,k'}}\cdot\left(\frac{I^{(i)}_k}{\tilde{\varepsilon}_{k'-k}} + \frac{I^{(i)}_{k-k'}}{\tilde{\varepsilon}_k}\right). \tag{9.175}$$

Let us now consider the equation for V in the energy-containing region:

$$\tilde{\varepsilon}_k V^{(0)}_{k;k'} + I^{(0)}_{k-k'} \int \tilde{\Sigma}_{k-k_1} V^{(0)}_{k_1;k'} \, dk_1 + I^{(0)}_{k-k'} \Sigma_{k'} \int V^{(0)}_{k_1;k'} \, dk_1 = -\frac{1}{i} S_{k'} I^{(0)}_{k-k'} w_{k'}.$$

(9.176)

Here we have dropped from the integrals those terms related to the inertial region, assuming that the effect of the inertial region on the energy-containing region is small. Terms containing $V^{(i)}$ must be added only to the second term on the right side of (9.176) because $\tilde{\Sigma}_{k-k_1}$ is exponentially small when k is in the energy-containing region and $k_1 - k$ is in the inertial region. The integral equation (9.176) is solved by the same method used on Eq. (8.50) of Chapter 8:

$$V^{(0)}_{k;k'} = -\left(\frac{I^{(0)}_k}{\tilde{\varepsilon}_{k'-k}} + \frac{I^{(0)}_{k-k'}}{\tilde{\varepsilon}_k}\right) \frac{1}{1 + D^{(0)}_{1,k'} + D^{(0)}_{2,k'}} \cdot \left(\frac{1}{i} S_{k'} w_{k'} + \Sigma_{k'} \int V^{(i)}_{k_1;k'} \, dk_1\right),$$

(9.177)

where

$$D^{(0)}_{2,k'} = \frac{1}{4\pi n_0 (T_e + T_i)} \int \frac{I^{(0)}_{k_1-k'} \, dk_1}{\tilde{\varepsilon}_{k_1}} = \frac{\int (I^{(0)}_{k_1-k'}/\tilde{\varepsilon}_{k_1} + I^{(0)}_{k_1}/\tilde{\varepsilon}_{k'-k_1}) \, dk_1}{8\pi n_0 (T_e + T_i)},$$

(9.178)

$$D^{(0)}_{1k'} = 2\Sigma_{k'} \int \frac{I^{(0)}_{k_1-k'} \, dk_1}{\tilde{\varepsilon}_{k_1}} \simeq \int \left\{\frac{I^{(0)}_{k_1-k'}}{\tilde{\varepsilon}_{k_1}} + \frac{I^{(0)}_{k_1}}{\tilde{\varepsilon}_{k'-k_1}}\right\} dk_1 \frac{1}{8\pi n_0 T_e}.$$

(9.179)

We see that $D^{(0)}_{2,k'}$ corresponds to the quantity $n_0 m_e d_2/(T_e + T_i)$ introduced in Chapter 8 [see (8.57)] and that $D^{(0)}_{1,k'}$ differs from (9.171) only in that the I_k here is for the energy-containing region rather than the inertial region. Equation (9.177) also contains a small inertial term. Thus we finally obtain

$$V^{(i)}_{k;k'} = -\left(\frac{I^{(i)}_k}{\tilde{\varepsilon}_{k'-k}} + \frac{I^{(i)}_{k-k'}}{\tilde{\varepsilon}_k}\right) \frac{(1 + D^{(0)}_{2,k'})(1/i) S_{k'} w^{(0)}_{k'}}{(1 + D^{(0)}_{2,k'})(1 + D^{(i)}_{1,k'}) + D^{(0)}_{1,k'}},$$

(9.180)

$$V^{(0)}_{k;k'} = -\left(\frac{I^{(0)}_k}{\tilde{\varepsilon}_{k'-k}} + \frac{I^{(0)}_{k-k'}}{\tilde{\varepsilon}_k}\right) \frac{(1/i) S_{k'} w^{(0)}_{k'}}{(1 + D^{(0)}_{2,k'})(1 + D^{(i)}_{1,k'}) + D^{(0)}_{1,k'}}.$$

(9.181)

These results are valid when k' is not large, or more precisely, when k' is in the energy-containing region, $I^{(0)}_{k-k'}$ remains in the energy-containing region, and $I^{(i)}_{k-k'}$ stays in the inertial region.

Therefore we shall add the superscript (0) to $w_{k'}$. In general the $D_{1,k'}^{(i)}$ in the denominators of (9.180) and (9.181) can be dropped because most of the turbulence energy is concentrated in the energy-containing region. The solution to (9.165) repeats the solution to (9.166), so that we can write the following result without excessive difficulty:

$$G_{k,k_2;k'}^{(0,0)} = \left(\frac{I_k^{(0)}}{\tilde{\varepsilon}_{k'-k}} + \frac{I_{k-k'}^{(0)}}{\tilde{\varepsilon}_k} \right) \left(\frac{I_{k_2}^{(0)}}{\tilde{\varepsilon}_{-k'-k_2}} + \frac{I_{k_2+k'}^{(0)}}{\tilde{\varepsilon}_{k_2}} \right)$$
$$\times \mid S_{k'} \mid^2 w_{k'}^{(0)} \mid (1 + D_{2,k'}^{(0)})(1 + D_{1,k'}^{(i)}) + D_{1,k'}^{(0)} \mid^{-2}, \tag{9.182}$$

$$G_{k,k_2;k'}^{(0,i)} = \left(\frac{I_k^{(0)}}{\tilde{\varepsilon}_{k'-k}} + \frac{I_{k-k'}^{(0)}}{\tilde{\varepsilon}_k} \right) \left(\frac{I_{k_2}^{(i)}}{\tilde{\varepsilon}_{-k'-k_2}} + \frac{I_{k_2+k'}^{(i)}}{\tilde{\varepsilon}_{k_2}} \right)$$
$$\times (1 + D_{2,-k'}^{(0)}) \mid S_{k'} \mid^2 w_{k'}^{(0)} \mid (1 + D_{2,k'}^{(0)})(1 + D_{1,k'}^{(i)}) + D_{1,k'}^{(0)} \mid^{-2}, \tag{9.183}$$

$$G_{k,k_2;k'}^{(i,0)} = \left(\frac{I_k^{(i)}}{\tilde{\varepsilon}_{k'-k}} + \frac{I_{k-k'}^{(i)}}{\tilde{\varepsilon}_k} \right) \left(\frac{I_{k_2}^{(0)}}{\tilde{\varepsilon}_{-k'-k_2}} + \frac{I_{k_2+k'}^{(0)}}{\tilde{\varepsilon}_{k_2}} \right)$$
$$\times (1 + D_{2,k'}^{(0)}) \mid S_{k'} \mid^2 w_{k'}^{(0)} \mid (1 + D_{2,k'}^{(0)})(1 + D_{1,k'}^{(i)}) + D_{1,k'}^{(0)} \mid^{-2}, \tag{9.184}$$

$$G_{k,k_2;k'}^{(i,i)} = \left(\frac{I_k^{(i)}}{\tilde{\varepsilon}_{k'-k}} + \frac{I_{k-k'}^{(i)}}{\tilde{\varepsilon}_k} \right) \left(\frac{I_{k_2}^{(i)}}{\tilde{\varepsilon}_{-k'-k_2}} + \frac{I_{k_2+k'}^{(i)}}{\tilde{\varepsilon}_{k_2}} \right)$$
$$\times \mid 1 + D_{2,k'}^{(0)} \mid^2 \mid S_{k'} \mid^2 w_{k'}^{(0)} \mid (1 + D_{2,k'}^{(0)})(1 + D_{1,k'}^{(i)}) + D_{1,k'}^{(0)} \mid^{-2}. \tag{9.185}$$

When (9.180)–(9.185) are substituted into (9.163) we arrive at the equation sought for the correlation functions of Langmuir oscillations when modulation perturbations are taken into account. Noting (9.169) we obtain

$$\tilde{\varepsilon}_k I_k = - \int \tilde{\Sigma}_{k-k_1} G_{k_1,k-k';k'} \, dk' \, dk_1 - \int \Sigma_{k'} G_{k+k',k_1;k'} \, dk' \, dk_1$$
$$- \frac{1}{i} \int S_{-k'} V_{k+k';k'} \, dk'. \tag{9.186}$$

If in the energy-containing region we completely neglect the effects of oscillations in the inertial region we obtain

$$\tilde{\varepsilon}_k I_k^{(0)} = - \int \frac{D_{2,k'}^{(0)} + D_{1,k'}^{(0)}}{\mid 1 + D_{1,k'}^{(0)} + D_{2,k'}^{(0)} \mid^2} \mid S_{k'} \mid^2 w_{k'}^{(0)} \, dk' \left(\frac{I_{k-k'}^{(0)}}{\tilde{\varepsilon}_{-k}} + \frac{I_k^{(0)}}{\tilde{\varepsilon}_{k-k'}} \right)$$
$$+ \int \frac{\mid S_{k'} \mid^2 w_{k'}^{(0)}}{1 + D_{1,-k'}^{(0)} + D_{2,-k'}^{(0)}} \left(\frac{I_{k-k'}^{(0)}}{\tilde{\varepsilon}_{-k}} + \frac{I_k^{(0)}}{\tilde{\varepsilon}_{k-k'}} \right) dk', \tag{9.187}$$

or

$$\tilde{\varepsilon}_k I_k^{(0)} = \int \frac{|S_{k'}|^2 w_{k'}^{(0)} dk'}{|1 + D_{1,k'}^{(0)} + D_{2,k'}^{(0)}|^2} \left(\frac{I_{k-k'}^{(0)}}{\tilde{\varepsilon}_{-k}} + \frac{I_k^{(0)}}{\tilde{\varepsilon}_{k-k'}} \right). \qquad (9.188)$$

When integrated over k_1 the first two terms in (9.186) give the coefficients $D_{2,k'}^{(0)}$ and $D_{1,k'}^{(0)}$ in the first term of (9.187). This allows us to reduce the result to the simple form of (9.188).

When the first term in (9.186) is neglected we find for the inertial region that

$$\tilde{\varepsilon}_k I_k^{(i)} = \int \frac{|S_{k'}|^2 w_{k'}^{(0)} |1 + D_{2,k'}^{(0)}|^2 dk'}{|1 + D_{1,k'}^{(0)} + D_{2,k'}^{(0)}|^2} \left(\frac{I_{k-k'}^{(i)}}{\tilde{\varepsilon}_{-k}} + \frac{I_k^{(i)}}{\tilde{\varepsilon}_{k-k'}} \right) dk' \qquad (9.189)$$

Both $G^{(i,0)}$ and $G^{(i,i)}$ are taken into account in the first term on the right side of (9.183), but in the final result the exact denominator in (9.181) is replaced with an approximate expression, which results from noting that $D_{1,k'}^{(i)} \ll D_{1,k'}^{(0)}$. In the simplest, one-dimensional case $|S_{k'}|^2$ is given by

$$|S_{k'}|^2 = \frac{e^2}{T_e^2 k'^2} = \frac{1}{4\pi n_0 T_e} \cdot \frac{\omega_{pe}^2}{k'^2 v_{Te}^2}. \qquad (9.190)$$

Note that these equations describe only interactions with large-scale modulation perturbations, where k' corresponds to the energy-containing region. For $k' \ll k$ (modulation disturbances) we obtain

$$\tilde{\varepsilon}_k^2 = 2 \int \frac{|S_{k'}|^2 w_{k'}^{(0)} dk'}{|1 + D_{1,k'}^{(0)} + D_{2,k'}^{(0)}|^2} \equiv (\tilde{\varepsilon}_k^{(0)})^2. \qquad (9.191)$$

Only the real part of $\tilde{\varepsilon}_k$ has been taken into account here, so that $\tilde{\varepsilon}_{-k} = \tilde{\varepsilon}_k$. It is natural to assume that the development of modulation instabilities leads to the stabilization processes, so that from the two solutions $\tilde{\varepsilon}_k = \pm \tilde{\varepsilon}_k^{(0)}$ one must select $\tilde{\varepsilon}_k = \tilde{\varepsilon}_k^{(0)}$, which causes a positive frequency shift in the frequencies of the Langmuir oscillations:

$$\frac{\Delta \tilde{\omega}}{\omega_{pe}} = \frac{1}{2} \tilde{\varepsilon}_k^{(0)}. \qquad (9.192)$$

The energy in a modulation perturbation needed to compensate for a negative frequency shift which leads to a modulation instability is

$$\tilde{\varepsilon}_k^{(0)} \simeq \frac{W_0}{n_0(T_e + T_i)}. \tag{9.193}$$

The turbulence in the energy-containing region becomes modulationally stable for this value of $\tilde{\varepsilon}_k^{(0)}$. Using the value given in (9.193) we find

$$D_{1,k'}^{(0)} \simeq \frac{T_e + T_i}{T_e}, \qquad D_{2,k'}^{(0)} \simeq 1. \tag{9.194}$$

Thus, when $T_i \ll T_e$ (9.191) gives

$$\int \frac{\omega_{pe}^2}{k'^2 v_{Te}^2} \cdot \frac{w_{k'}^{(0)}}{4\pi n_0 T_e} \, dk' = \frac{9}{2} \left(\frac{W_0}{n_0 T_e} \right)^2. \tag{9.195}$$

The quantity on the left side of (9.195) is analogous to the total energy in the low-frequency perturbations, but instead of the factor $\omega_{pe}^2/k'^2 v_{Te}^2$ it would contain a factor $\frac{1}{2} \omega'(\partial/\partial\omega')\varepsilon_{k'} \simeq$ $(\omega_{pe}^2/k'^2 v_{Te}^2)(k'^2 v_s^2/\omega'^2)$. Equation (9.195) shows that the stabilization of a modulation instability in the condensation takes place for a very small amount of energy in the modulation perturbations. We shall show that this same amount of energy in the modulation perturbations arises because of their interactions. We begin with Eq. (9.154) for the modulation field E^M. Then, to first order in $w_{k'}^{(0)}$ we find

$$\varepsilon_{k'} w_{k'}^{(0)} = \frac{8\pi}{i\omega'} \int S_{k', k_1, k'-k_1} V_{k;k'}^{(0)} \, dk_1. \tag{9.196}$$

Since $S_{k', k_1, k'-k_1} \simeq -e\omega'/8\pi T_e k'$, $S_{k'} = -e/T_e k'$, so that by using (9.181), we obtain

$$\left(\varepsilon_{k'} - \frac{D_{1,k'}^{(0)}}{1 + D_{1,k'}^{(0)} + D_{2,k'}^{(0)}} \frac{\omega_{pe}^2}{k'^2 v_{Te}^2} \right) w_{k'}^{(0)}$$
$$= \left[\varepsilon_{k'}^i + (\varepsilon_{k'}^e - 1) \frac{1 + D_{2,k'}^{(0)}}{1 + D_{1,k'}^{(0)} + D_{2,k'}^{(0)}} \right] w_{k'}^{(0)} \equiv \varepsilon_{k'}^M w_{k'}^{(0)} = 0, \tag{9.197}$$

in agreement with the result (8.56) in Chapter 8.

Under stable conditions, when (9.194) is valid, we find

$$\varepsilon_{N'}^M = \varepsilon_{k'}^{(i)} + \frac{2}{3} \frac{\omega_{pe}^2}{k'^2 v_{Te}^2} \tag{9.198}$$

and the equation $\varepsilon_{k'}^M = 0$ does not lead to instability. In the non-linear approximation the equation replacing (9.195) has the form [356]

$$\left[\left(\alpha_{k'}^{(i)} + \frac{2}{3} - \frac{5}{3} v \right) \frac{9}{2} + \int \frac{k''^2 v_{k''} \, dk''}{(k' - k'')^2 (1 + \frac{2}{3} \alpha_{k'-k''}^i)} \right] v_{k'}$$

$$= \frac{1}{1 + \frac{2}{3} \alpha_{k'}^{(i)}} \int v_{k''} v_{k'''} \delta(k' - k'' - k''') \, dk'' \, dk''' , \tag{9.199}$$

where

$$v_{k'} \frac{2}{9} \frac{\omega_{pe}^2 w_k^{(0)}}{k'^2 v_{Te}^2 4\pi n_0 T_e} , \qquad \alpha_k^{(i)} = \frac{\varepsilon_{k'}^{(i)} k'^2 v_{Te}^2}{\omega_{pe}^2} . \tag{9.200}$$

The condition in (9.195) could be written as

$$\int v_{k'} \, dk' = 1 . \tag{9.201}$$

For supersonic motion $\varepsilon_{k'}^{(i)} \simeq -\omega_{pi}^2 / \omega'^2$, $\alpha_k^{(i)} \ll 1$, and we find from (9.199) that

$$\frac{17}{2} v^2 = 3v + \int \frac{k''^2}{(k' - k'')^2} v_{k'} v_{k''} \, dk' \, dk'' , \tag{9.202}$$

which must give the result $v \simeq 1$. When the velocities of the modulation disturbances are much smaller than the thermal ion velocities, $\alpha_k^{(i)} \simeq T_e/T_i \gg 1$, where $T_e \gg T_i$; thus, from (9.199) we find

$$\frac{9}{2} \alpha^{(i)} \left(\alpha^{(i)} - \frac{5}{3} v \right) v + \int \frac{k''^2}{(k' - k'')^2} v_{k'} v_{k''} \, dk'' = v^2 , \tag{9.203}$$

which gives the estimate $v \simeq \alpha^{(i)} = T_e/T_i \gg 1$.

In other words, the energy level of the modulation perturbations as determined from equations for their interactions does

ensure the necessary nonlinear shifts and stabilization of turbulence in the energy-containing region.

In the inertial region (9.199) describes only the interaction
with large-scale modulation disturbances k' ≪ 1, and can be
written in the form

$$\tilde{\varepsilon}_k I_k^{(i)} = 0, \qquad \tilde{\varepsilon}_k = \tilde{\varepsilon}_k - \frac{4v}{\tilde{\varepsilon}_k}, \qquad \tilde{\varepsilon}_k = \tilde{\varepsilon}_{-k}. \tag{9.204}$$

Let us now consider the short-wavelength modulation perturbations corresponding to large values of k'. When solving the equations in the inertial region it is possible that k - k' will be small
and fall in the energy-containing region. This means that when
k is nearly equal to k' it will be necessary to use $I_{k-k'}^{(0)}$ in place
of $I_{k-k'}^{(i)}$ in the expression for $V_{k; k'}$. This substitution would have
to be carried out in all other terms as well. As a result a source
term due to the energy-containing region appears in the equation
for the inertial region:

$$\tilde{\varepsilon}_k I_k^{(i)} = 2 \int |S_{k'}|^2 w_{k'} \, dk' \frac{I_{k-k'}^{(0)}}{\tilde{\varepsilon}_{-k}} \simeq \frac{2|S_k|^2}{\tilde{\varepsilon}_k} \int I_{k_1}^{(0)} w_{k_1-k} \, dk_1 =: Q_k. \tag{9.205}$$

In the source term Q_k the coefficients $D_k \ll 1$. The quantity w_{k_1-k}
corresponds to the supersonic velocities of the modulation disturbance, and when $\omega'/k' \gg \sqrt{W_0/n_0 m_e}$ the quantity $w_{k'}$ reaches
a maximum, constant value [353]. Estimates show that the source
term is much larger than the term $4v/\tilde{\varepsilon}_k$ in (9.204); this is because
w increases while $|S_{k'}|^2 I_{k'}^{(0)} \approx |S_{k'}|^2 I_k^{(i)}$. Thus we obtain

$$I_k^{(i)} = \frac{|S_k|^2 \text{ const } W_0}{|\tilde{\varepsilon}_k|^2}. \tag{9.206}$$

The quantity $|\tilde{\varepsilon}_k|^2 = (\partial \varepsilon_k/\partial \omega) [\omega - \omega_k^N)^2 + \gamma_k^{N\,2}]$ describes the cor-
relation curve. After integrating over frequency and taking into
account that $|S_k|^2 \sim 1/k^2$ [see (9.190)] we arrive at the turbulence
spectrum for the inertial region:

$$\int I_k^{(i)} \, d\omega \sim \frac{1}{k^2}, \qquad W_k^{1D} \sim \frac{1}{k^2}, \qquad W_k^{3D} \sim k^2 \int I_k^{(i)} \, d\omega = \text{const}. \tag{9.207}$$

It is important to note that this spectrum is universal for both
one-dimensional and three-dimensional cases.

In addition to real shifts in the frequencies, modulation perturbations also cause imaginary corrections, which describe the
kinetic effects of the spectral pumping. The pumping in (9.205)
is hydrodynamic. However, according to [356] the kinetic effects
also give a spectrum $\sim 1/k^2$ in the three-dimensional case. The
imaginary corrections to the nonlinear terms (i.e., nonlinear
Landau damping by electrons) create an integrated pumping from
the energy-containing regions into the inertial regions, thereby
creating a source of oscillations in the inertial region for which
[356]

$$Q_k = \frac{\text{const}}{k^3}. \tag{9.208}$$

If the usual mechanism for transferring oscillations by scattering
on ions is operating in this region [see (4.16) of Chapter 4], the
turbulence spectrum is found from the condition

$$\alpha \frac{\partial W_k}{\partial k} + \frac{\text{const}}{k^3} = 0 \tag{9.209}$$

or

$$W_k^{3D} \sim \frac{1}{k^2}. \tag{9.210}$$

The universal turbulence spectrum enables one to simply estimate
the acceleration of fast particles by using the results of Chapter
5. In the presence of constant pumping the turbulence energy is
absorbed primarily by such particles. An important feature of
the statistical theory is the proof that it is also possible to heat
the main body of the particles [353]. This proof is accomplished by
taking the nonstationary Langmuir turbulence into account in the
quasilinear integral. The quasilinear integral will depend
on $I_{kk'}$. Performing the averages and using the resonance of the
low-frequency perturbations $\omega' = k'v'$ we obtain particle heating by
nonresonant Langmuir oscillations with $\omega \neq kv$.

9.7. Conclusion

In spite of the success achieved in developing different approaches to the theory of strong turbulence, the various approaches
have not yet been unified. Unification would provide better understanding of the mechanisms of strong Langmuir turbulence

and would enable us to include the effects of turbulent broadening of resonances. The numerical experiments which have proved so useful in the initial stages of developing a theory of strong Langmuir turbulence cannot continue to be so valuable in the future because the actual three-dimensional spectra needed to compare with available experimental data cannot be analyzed using numerical methods. The statistical theory of strong Langmuir turbulence is not only more attractive than the other approaches, but it is necessary from the definition of strong turbulence. The increase in the frequency shifts $\Delta\omega$ found in strong turbulence tends to reduce the nonlinear responses S and Σ, and this makes it more tractable to use perturbation theory to treat higher-order approximations. Using this fact, the statistical theory of strong Langmuir turbulence has a good mathematical foundation. The important difference between strong Langmuir turbulence and fluid turbulence is that the former involves dispersion, which keeps the field propagator $1/\varepsilon$ from going to infinity when $\Delta\omega \to 0$.

Although many details require further work, this statistical theory of strong Langmuir turbulence is in general closed and consistent. There is no doubt that strong Langmuir turbulence will occupy a central position in theoretical studies, and one can expect that further research will provide a more complete picture of the very complex and diverse plasma motions which take place in this regime.

References

1. L. D. Landau and E. M. Lifshits, Mechanics of Continuous Media [in Russian], Gostekhizdat, Moscow (1954).
2. Yu. G. Yaremenko, V. M. Deev, R. L. Slabovik, and I. F. Kharchenko, Sov. At. Energ., 25:985 (1968).
3. L. D. Landau and E. M. Lifshits, Statistical Physics [in Russian], Nauka, Moscow (1963).
4. L. D. Landau and E. M. Lifshits, Electrodynamics of Continuous Media, Gostekhizdat, Moscow (1957).
5. L. Z. Prandtl, Angew. Math. und Mech., 5:136 (1925); 22, No. 5(1942).
6. B. B. Kadomtsev, in: "Problems of Plasma Theory," edited by M. A. Leontovich, Vol. 4 [in Russian], Atomizdat, Moscow (1964), p. 188.
7. E. K. Zavoiskii, At. Energ., 14:57 (1963).
8. E. K. Zavoiskii and L. I. Rudakov, Plasma Physics (Collective Processes in Plasma and Turbulent Heating) [in Russian], Znanie, Moscow (1967).
9. A. N. Kolmogorov, Dokl. Akad. Nauk SSSR, 30:299 (1941).
10. V. L. Ginzburg, Propagation of Electromagnetic Waves in Plasma [in Russian], Nauka, Moscow (1967).
11. G. F. Chew, M. L. Goldberger, and F. E. Low, Proc. Roy. Soc., A236:1212 (1956).
12. Yu. L. Klimontovich and V. P. Silin, Zh. Eksper. Teor. Fiz., 40:143 (1961).
13. I. Langmuir, Phys. Rev., 26:585 (1925); Z. Phys., 46:271 (1927).
14. V. A. Vlasov, Zh. Eksper. Teor. Fiz., 8:291 (1938).
15. L. D. Landau, Zh. Eksper. Teor. Fiz., 16:574 (1946).
16. H. Alfven, Cosmic Electrodynamics [Russian translation], Izd. Inostr. Lit., Moscow (1952).
17. I. M. Khalatnikov, Zh. Eksper. Teor. Fiz., 27:529 (1954).
18. V. N. Tsytovich, Zh. Eksper. Teor. Fiz., 44:946 (1963).
19. A. I. Akhiezer et al., Collective Oscillations in Plasma [in Russian], Atomizdat, Moscow (1964).
20. V. I. Karpman, Preprint IYaF SO AN SSSR, No. 25, Novosibirsk (1968).
21. R. Z. Sagdeev, in: "Reviews of Plasma Physics," Vol. 4, edited by M. A. Leontovich, Consultants Bureau, New York (1966), p. 23.

22.   D. H. Looney and S. C. Brown, Phys. Rev., 83:695 (1954).

23.   H. J. Merill and H. W. Webb, Phys. Rev., 55:1191 (1939).

24.   I. F. Kharchenko, Ya. B. Fainberg, R. M. Nikolaev, E. A. Kornilov, E. A. Lutsenko, and N. S. Pedenko, Zh. Eksper. Teor. Fiz., 38:685 (1960).

25.   A. K. Berezin, G. P. Berezina, L. I. Bolotin, and Ya. B. Fainberg, Sov. At. Energ., 14:249 (1963).

26.   M. V. Babykin et al., Yademyi Sintez, Suppl., Vol. 3, 1073 (1962).

27.   M. V. Babykin, E. K. Zavoiskii, P. P. Gavrin, L. I. Rudakov, and V. A. Skoryupin, Zh. Eksper. Teor. Fiz., 43:411 (1962).

28.   V. I. Veksler, At. Energ., 2:427 (1957).

29.   R. Smulin and G. Getti, Phys. Rev. Lett., 9:3 (1962).

30.   I. Alexeff and R. V. Neidigh, Phys. Rev., 129:516 (1963).

31.   V. A. Suprunenko, At. Energ., 17:83 (1964).

32.   B. A. Demidov, S. D. Fanchenko, N. I. Elagin, and D. D. Ryutov, Zh. Eksper. Teor. Fiz., 46:497 (1964).

33.   S. Hamberger, Phys. Rev. Lett., 21:674 (1968).

34.   A. V. Nedospasov, Zh. Eksper. Teor. Fiz., 34:1338 (1958).

35.   M. V. Nezlin and A. M. Solntsev, Zh. Eksper. Teor. Fiz., 48:1237 (1965).

36.   V. E. Golant, Uspekhi Fiz. Nauk, 79:377 (1963).

37.   D. Bohm and E. Burhop, The Characteristics of Electric Discharges in Magnetic Fields, A. Guthrie and R. K. Wakerling, New York (1949).

38.   I. A. Akhiezer, I. L. Daneliya, and N. L. Tsintsadze, Zh. Eksper. Teor. Fiz., 46:300 (1964).

39.   R. E. Aamodt and W. E. Drummond, Nucl. Energy, Part C 6:147 (1964).

40.   S. Tanaka and K. J. Takayama, Phys. Soc. Japan, 21:2372 (1966).

41.   I. M. Danilkin, et al., Tr. Fiz. Inst. Akad. Nauk SSSR, 32:112 (1966).

42.   L. A. Artsimovich et al., At. Energ., 1:84 (1956).

43.   A. K. Berezin et al., in: Plasma Physics and Contr. Nucl. Fusion Res. Proc. Conf., Culham, England (1965).

44.   V. I. Veksler et al., At. Energ., 18:14 (1965).

45.   V. N. Tsytovich, Uspekhi Fiz. Nauk, 89:89 (1965).

46.   V. L. Ginzburg and S. I. Syrovatskii, The Origin of Cosmic Rays [in Russian], Izd. Akad. Nauk SSSR, Moscow (1963).

47.   V. L. Ginzburg and L. M. Ozernoi, Izv. Vuzov. Radiofizika, 9:221 (1966).

48.   S. A. Kaplan, Zh. Eksper. Teor. Fiz., 29:406 (1955).

49.   V. N. Tsytovich, Astron. Zh., 40:612 (1963).

50.   D. A. Kirzhnits, Field Methods in Many-Body Theory [in Russian], Gosatomizdat, Moscow (1963).

51.   V. L. Bonch-Bruevich and S. V. Tyablikov, Green's Function Methods in Statistical Mechanics [in Russian], Fizmatgiz, Moscow (1961).

52.   A. A. Abrikosov, L. P. Gor'kov, and E. I. Dzyaloshinskii, Quantum Field Theory Methods in Statistical Physics [in Russian], Fizmatgiz, Moscow (1962).

53.   A. B. Mikhailovskii, Yademyi Sintez, 4:321 (1964).

54.   V. N. Tsytovich, Dokl. Akad. Nauk SSSR, 154:76 (1964).

55.   Yu. L. Romanov and G. F. Fillipov, Zh. Eksper. Teor. Fiz., 40:123 (1961).

56.   A. A. Vedenov, E. P. Velikhov, and R. Z. Sagdeev, Yademyi Sintez, 1:82 (1961), Suppl. 2, 465 (1962).

57. W. E. Drummond and P. Pines, Nucl. Fusion Suppl., Part 3, 1049 (1962).

58. V. N. Tsytovich, Kinetic Equation for Elementary Excitations and Nonlinear Interactions of Waves in Plasma, Preprint Fiz. Inst. Akad. Nauk SSSR, No. 12 (1968).

59. V. N. Tsytovich, Nonlinear Effects in Plasma, Plenum, New York (1970).

60. A. K. Gailitis and V. N. Tsytovich, Zh. Eksper. Teor. Fiz., 47:1463 (1964).

61. A. K. Gailitis and V. N. Tsytovich, Izv. Vuzov. Radiofizika, 7:1190 (1964).

62. V. N. Tsytovich, Uspekhi Fiz. Nauk, 90:435 (1966).

63. G. L. Ginzburg and I. M. Frank, Zh. Eksper. Teor. Fiz., 16:15 (1946).

64. Ya. B. Fainberg and N. A. Khizhnyak, Zh. Eksper. Teor. Fiz., 32:883 (1957).

65. K. A. Barsukov and B. M. Bolotovskii, Zh. Eksper. Teor. Fiz., 45:303 (1963).

66. A. K. Gailitis and V. N. Tsytovich, Zh. Eksper. Teor. Fiz., 46:1726 (1964).

67. A. K. Gailitis et al., Elementary Processes in the Nonlinear Interaction of Charged Particles with Plasma and Equations for Weakly Turbulent Plasma. Report to the Seventh Conference on Phenomena in Ionized Gases. Belgrade, Yugoslavia, August (1965).

68. L. M. Kovrizhnykh and V. N. Tsytovich, Zh. Eksper. Teor. Fiz., 46:2212 (1964).

69. V. A. Liperovskii and V. N. Tsytovich, Prikl. Mekh. Tekhn. Fiz., 5:15 (1965).

70. L. M. Kovrizhnykh, Tr. Inst. Fiz. Akad. Nauk SSSR, Ser. Fiz. Plazmy, 32:173 (1966).

71. V. N. Tsytovich, Stochastic Processes in Plasma, in: "Ninth Intern. Conference on Ionization Phenomena in Gases," 1969, Bucharest, Roumania, IAEA, 1969; Preprint Fiz. Inst. Akad. Nauk SSSR (1969).

72. E. A. Kornilov, Ya. B. Fainberg, L. I. Bolotin, and O. F. Kovpik, Pis'ma Zh. Eksper. Teor. Fiz., 3:354 (1966).

73. I. E. Tamm and I. M. Frank, Dokl. Akad. Nauk SSSR, 14:107 (1937); B. M. Bolotovskii, Uspekhi Fiz. Nauk 62:201 (1957).

74. F. G. Bass, Ya. B. Fainberg, and V. D. Shapiro, Zh. Eksper. Teor. Fiz., 49:329 (1965).

75. V. D. Shapiro, Pis'ma Zh. Eksper. Teor. Fiz., 2:469 (1965).

76. J. Alexeff, R. V. Neidigh, and W. F. Peed, Phys. Rev., 136:689 (1964).

77. T. H. Dupree, Phys. Fluids, 9:1773 (1966).

78. L. I. Rudakov and V. N. Tsytovich, Plasma Physics, 13:213 (1971); Preprint Fiz. Inst. Akad. Nauk SSSR, No. 28 (1970).

79. A. I. Akhiezer and V. B. Berestetskii, Quantum Electrodynamics [in Russian], Fizmatgiz, Moscow (1962).

80. V. A. Liperovskii and V. N. Tsytovich, Izv. Vuzov. Radiofizika, 9:469 (1966); L. M. Kovrizhnykh, Zh. Eksper. Teor. Fiz., 49:237, 1376 (1965).

81. V. N. Tsytovich and A. B. Shvartsburg, Zh. Eksper. Teor. Fiz., 49:797 (1965).

82. B. A. Trubnikov and A. E. Bazhanova, in: "Plasma Physics and Controlled Thermonuclear Reactions" [in Russian], Vol. 3, Izd. Akad. Nauk SSSR, Moscow (1958), p. 121.

83. V. A. Trubnikov, Dokl. Akad. Nauk SSSR, 118:913 (1958).

84. B. V. Chirikov, Preprint No. 191, IYaF SO Akad. Nauk SSSR, Novosibirsk (1966).

85. V. I. Arnol'd, Usp. Matem. Nauk, 18:91 (1963).

REFERENCES

86. G. M. Zaslavskii and B. V. Chirikov, Dokl. Akad. Nauk SSSR, 159:306 (1969).
87. D. Bohm and E. P. Gross, Phys. Rev., 75:1851 (1949).
88. A. I. Akhiezer and Ya. B. Fainberg, Zh. Eksper. Teor. Fiz., 21:1262 (1951).
89. V. P. Silin and A. A. Rukhadze, Electromagnetic Properties of Plasma and Plasma-Like Media [in Russian], Gosatomizdat, Moscow (1961).
90. T. Stiks, Theory of Plasma Waves [in Russian], Atomizdat, Moscow (1965).
91. V. D. Shapiro and V. I. Shevchenko, Zh. Eksper. Teor. Fiz., 42:1515 (1962).
92. S. Neufeld and P. H. Doyle, Phys. Rev., 127:846 (1962).
93. V. G. Makhan'kov and V. I. Shevchenko, in: "Plasma Physics and Controlled Thermonuclear Reactions" [in Russian], Vol. 4, Naukova Dumka, Kiev (1965); Preprint OIYaI, P-1659, Dubna (1964).
94. V. G. Makhan'kov and V. N. Tsytovich, Zh. Tekh. Fiz., 38:809 (1968).
95. G. G. Macfarlane and H. G. Hay, Proc. Phys. Soc., B63:409 (1950).
96. E. R. Harrison, Proc. Phys. Soc., B82:889 (1963).
97. E. E. Lovetskii and A. A. Rukhadze, Tr. Fiz. Inst. Akad. Nauk SSSR, 32:218 (1966).
98. E. A. Kornilov et al., in: "Interactions of Charged Particle Beams with Plasma" [in Russian], Naukova Dumka, Kiev (1965), p. 36.
99. A. A. Vedenov, At. Energ., 13:5 (1962).
100. V. D. Shapiro, Zh. Eksper. Teor. Fiz., 44:613 (1963).
101. S. M. Levitskii and I. P. Shashurin, Zh. Eksper. Teor. Fiz., 52:1522 (1966).
102. A. A. Ivanov and L. I. Rudakov, Zh. Eksper. Teor. Fiz., 51:1522 (1966).
103. S. M. Levitskii and I. P. Shashurin, in: "Ninth Intern. Conf. on Ionization Phenomena in Gases," 1969, Bucharest, Roumania, IAEA, 1969, p. 566.
104. V. N. Tsytovich and V. D. Shapiro, Yadernyi Sintez, 5:228 (1965).
105. Ya. B. Fainberg and V. D. Shapiro, Zh. Eksper. Teor. Fiz., 47:1389 (1964).
106. V. N. Tsytovich and V. D. Shapiro, Zh. Tekh. Fiz., 35:1925 (1965).
107. A. A. Andronov and V. Yu. Trakhtengerts, Zh. Eksper. Teor. Fiz., 45:1009 (1963).
108. A. B. Mikhailovskii and K. Yungvirt, Zh. Eksper. Teor. Fiz., 50:1036 (1966).
109. J. Roulands, V. D. Shapiro, and V. I. Shevchenko, Zh. Eksper. Teor. Fiz., 50:1036 (1966).
110. A. K. Berezin et al., in: "Interactions of Charged Particle Beams with Plasma" [in Russian], Naukova Dumka, Kiev (1965), p. 7.
111. V. A. Liperovskii, Prikl. Mekh. Tekhn. Fiz., 2:23 (1967).
112. V. I. Fedorchenko, B. N. Rutkevich, V. I. Muratov, and B. M. Chernyi, Zh. Tekh. Fiz., 32:958 (1962).
113. V. D. Shafranov, in: "Revews in Plasma Physics," Vol. 3, edited by M. A. Leontovich, Consultant Bureau, New York (1967), p. 1.
114. V. D. Shafranov and R. Z. Sagdeev, Zh. Eksper. Teor. Fiz., 39:181 (1960).
115. K. N. Stepanov and V. T. Tolok, Yadernyi Sintez, 3:251 (1963).
116. A. I. Akhiezer, I. A. Akhiezer, R. V. Polovin, A. G. Sitenko, and K. N. Stepanov, Collective Plasma Oscillations [in Russian], Atomizdat, Moscow (1964).
117. E. Harris, Phys. Rev. Lett., 2:234 (1959).
118. L. I. Rudakov and R. Z. Sagdeev, in: "Plasma Physics and Controlled Thermonuclear Reactions" [in Russian], Izd. Akad. Nauk SSSR, Moscow, No. 3 (1958), p. 278.

119. W. E. Drummond, M. N. Rosenbluth, and M. Johnson, Bull. Amer. Phys. Soc., 6:185 (1961).

120. A. V. Timofeev and V. I. Pistunovich, in: "Reviews in Plasma Physics," Vol. 5, edited by M. A. Leontovich, Consultants Bureau, New York (1970), p. 401.

121. V. B. Krasovitskii and K. N. Stepanov, Zh. Tekh. Fiz., 34:1013 (1964).

122. M. N. Rosenbluth and R. F. Post, Phys. Fluids, 8:547 (1965).

123. A. A. Galeev, International Center for Theoretical Physics, Trieste (1966).

124. A. V. Timofeev and V. I. Pistunovich, in: "Reviews in Plasma Physics," Vol. 5, edited by M. A. Leontovich, Consultants Bureau, New York (1970), p. 401.

125. G. I. Budker, in: "Plasma Physics and Controlled Thermonuclear Reactions" [in Russian], Vol. 3, Izd. Akad. Nauk SSSR, Moscow (1958), p. 32.

126. H. Drieser, Phys. Rev., 117:329 (1960).

127. O. Buneman, Phys. Rev., 115:503 (1959).

128. E. R. Harrison, Plasma Phys., 4:7 (1962).

129. V. D. Shapiro, Zh. Tekh. Fiz., 31:552 (1961).

130. E. E. Lovetskii and A. A. Rukhadze, Zh. Eksper. Teor. Fiz., 41:1845 (1961).

131. T. E. Stringer, Nucl. Energy Part C, 6:267 (1964).

132. W. E. Drummond and M. N. Rosenbluth, Phys. Fluids, 5:12 (1962).

133. K. N. Stepanov and A. B. Mikhailovskii, Zh. Tekh. Fiz., 35:1933 (1965).

134. S. M. Bakanov and A. A. Rukhadze, Zh. Eksper. Teor. Fiz., 41:1845 (1961).

135. L. S. Bogdankevich and A. A. Rukhadze, Yadernyi Sintez, 6:176 (1966).

136. E. C. Field and B. D. Fried, Phys. Fluids, 7:1937 (1964).

137. L. I. Rudakov and L. V. Korablev, Zh. Eksper. Teor. Fiz., 50:220 (1966).

138. V. I. Petviashvili, Zh. Eksper. Teor. Fiz., 44:1933 (1963).

139. V. I. Petviashvili, Dokl. Akad. Nauk SSSR, 153:1295 (1963).

140. L. M. Kovrizhnykh, Zh. Eksper. Teor. Fiz., 51:915 (1966).

141. I. A. Akhiezer, Zh. Eksper. Teor. Fiz., 47:2269 (1964).

142. V. N. Tsytovich, Plasma Physics, 13:100 (1971).

143. Ya. B. Fainberg, in: "A Survey of Phenomena in Ionized Gases," Inv. Papers, IAEA, Vienna (1968).

144. L. O. Smulin, in: "A Survey of Phenomena in Ionized Gases," Inv. Papers, IAEA, Vienna (1968).

145. M. V. Babykin, P. P. Gavrin, E. K. Zavoiskii, L. I. Rudakov, and V. A. Skoryupin, Zh. Eksper. Teor. Fiz., 47:1631 (1964).

146. Alexeff et al., Phys. Rev. Lett., 18:25 (1967).

147. A. M. Stefanovskii, Yadernyi Sintez, 5:215 (1965).

148. E. D. Andryukhina and I. S. Shpigel', Zh. Tekh. Fiz., 37:894 (1967).

149. N. S. Buchel'nikova, Yadernyi Sintez, 6:122 (1966).

150. J. Jancarik and S. M. Hamberger, in: "Report on Eighth European Conf. on Plasma Physics," Rome (1970).

151. C. C. Dauqhney et al., in: "Report on Eighth European Conf. on Plasma Physics," Rome (1970).

152. B. B. Kadomtsev and O. P. Pogutse, in: "Reviews of Plasma Physics," Vol. 5, edited by M. A. Leontovich, Consultants Bureau, New York (1970), p. 249.

153. A. B. Mikhailovskii, in: "Reviews of Plasma Physics," Vol. 3, edited by M. A. Leontovich, Consultants Bureau, New York (1967), p. 159.

154.   L. I. Rudakov and R. Z. Sagdeev, Dokl. Akad. Nauk SSSR, 138:581 (1961).

155.   A. A. Galeev, V. N. Oraevskii, and R. Z. Sagdeev, Zh. Eksper. Teor. Fiz.,
       44:581 (1961).

156.   Yu. A. Tserkovnikov, Zh. Eksper. Teor. Fiz., 32:67 (1957).

157.   B. B. Kadomtsev and A. V. Timofeev, Dokl. Akad. Nauk SSSR, 146:581 (1962).

158.   M. N. Rosenbluth, N. A. Krall, and N. Rostoker, Yadernyi Sintez, 1:143
       (1962).

159.   A. V. Timofeev, Dokl. Akad. Nauk SSSR, 152:84 (1963).

160.   A. B. Mikhailovskii and L. I. Rudakov, Zh. Eksper. Teor. Fiz., 44:912 (1963).

161.   A. B. Mikhailovskii and A. V. Timofeev, Zh. Eksper. Teor. Fiz., 44:919 (1963).

162.   A. A. Galeev and L. I. Rudakov, Zh. Eksper. Teor. Fiz., 45:647 (1963).

163.   F. G. Hoh, Phys. Fluids, 8:968 (1965).

164.   A. A. Ivanov, Author's Abstract of Dissertation, MFTI (1967).

165.   L. I. Rudakov, Zh. Eksper. Teor. Fiz., 48:1372 (1965).

166.   A. A. Galeev, V. I. Karpman, and R. Z. Sagdeev, Yadernyi Sintez, 5:20 (1965).

167.   V. I. Karpman, Prikl. Mekh. Tekhn. Fiz., 6:34 (1963).

168.   A. A. Galeev, S. S. Moiseev, and R. Z. Sagdeev, At. Energ., 15:451 (1963).

169.   I. A. Akhiezer, Zh. Eksper. Teor. Fiz., 48:1159 (1965).

170.   N. Buchel'nikova, Teplofiz. Vys. Temp., 2:309 (1964).

171.   B. B. Kadomtsev and A. V. Nedospasov, J. Nucl. Energy, Part C, 1:230 (1960).

172.   V. I. Petviashvili, R. R. Ramazashvili, and M. L. Tsintsadze, Yadernyi Sintez,
       5:23 (1965).

173.   M. Camac et al., Nucl. Fusion, Suppl., 2:423 (1962).

174.   A. A. Galeev and V. I. Karpman, Zh. Eksper. Teor. Fiz., 44:592 (1963).

175.   S. Moiseev and R. J. Sagdeev, Nucl. Energy Part C, 5:43 (1963).

176.   A. I. Akhiezer, Ya. B. Fainberg, and G. Ya. Lyubarskii, Scientific Writings
       [in Russian], Vol. 62, Trudy Fiz. Mat. Fakul'teta KhGU, 6:73 (1955).

177.   A. I. Akhiezer and G. Ya. Lyubarskii, Dokl. Akad. Nauk SSSR, 80:193 (1955).

178.   R. Z. Sagdeev, Zh. Tekh. Fiz., 31:1185 (1961).

179.   V. N. Tsytovich, Dokl. Akad. Nauk SSSR, 142:63 (1962).

180.   V. N. Tsytovich, Zh. Tekh. Fiz., 32:1042 (1962).

181.   S. I. Syrovatskii, Zh. Eksper. Teor. Fiz., 40:1788 (1961).

182.   R. Kh. Durtmulaev et al., Plasma Physics and Controlled Thermonuclear Fusion
       Research, Proc. Conf. Culham, England (1965).

183.   Yu. M. Aliev and V. P. Silin, Zh. Eksper. Teor. Fiz., 48:901 (1965).

184.   Yu. M. Aliev, V. P. Silin, and H. Watson, Zh. Eksper. Teor. Fiz., 50:943 (1966).

185.   V. N. Tsytovich, Zh. Tekh. Fiz., 35:773 (1965).

186.   V. A. Liperovskii and V. N. Tsytovich, Zh. Tekh. Fiz., 36:575 (1966).

187.   V. A. Liperovskii and V. N. Tsytovich, Prikl. Mekh. Tekh. Fiz., 2:116 (1966).

188.   V. N. Tsytovich and A. B. Shvartsburg, Zh. Tekh. Fiz., 36:1915 (1966).

189.   V. A. Liperovskii and V. N. Tsytovich, "Seventh Intern. Conference on Ionized
       Phenomena in Gases" [in Russian], Belgrade, August (1965).

190.   S. L. Mandel'shtam et al., Zh. Eksper. Teor. Fiz., 47:2003 (1964).

191.   V. N. Tsytovich, Zh. Tekh. Fiz., 39:1756 (1969); V. N. Tsytovich, in: "Eighth
       Intern. Conf. on Ionization Phenomena in Gases," 1967 Contr. Papers, Vienna,
       Austria, IAEA (1967), p. 408.

192. V. G. Makhan'kov and V. N. Tsytovich, Zh. Tekh. Fiz., 40:681 (1970).

193. J. R. Geker et al., in: "Ninth Intern. Conf. on Ionization Phenomena in Gases," Contr. Papers, Bucharest, Romania, IAEA (1969), p. 427.

194. A. Caruso and R. Gratton, "Some properties of the plasma produced by irradiating light solids by lasers." Laboratorio gas ionizzati Frascati, Rome (1968); A. Caruso and Guipponi, "Ionization and heating of a solid hydrogen pellet by means of a laser pulse," Laboratorio gas ionizzati, Frascati, Rome (1968).

195. N. G. Basov and O. N. Krokin, Proc. Conf. Quantum Electronics, Paris (1963).

196. L. M. Ozernoi, Astron. Zh., 43:300 (1966).

197. Ya. B. Zel'dovich and I. D. Novikov, Uspekhi Fiz. Nauk, 86:447 (1965).

198. S. A. Kaplan, Zh. Eksper. Teor. Fiz., 27:699 (1954).

199. V. N. Tsytovich, Astron. Zh., 42:33 (1965).

200. S. B. Pikel'ner and M. A. Gintsburg, Astron. Zh., 40:842 (1963).

201. V. L. Gintsburg, Astron. Zh., 42:1129 (1965).

202. V. Yu. Trakhtengerts, Astron. Zh., 43:357 (1966).

203. E. N. Krivorutskii and V. N. Tsytovich, Astron. Zh., 46:1003 (1969).

204. I. M. Gordon, Zh. Eksper. Teor. Fiz., 37:853 (1959).

205. M. S. Kovner and I. M. Chertog, Geomagnetizm i aeronomia, 3:1014 (1963).

206. S. B. Pikel'ner, Uspekhi Fiz. Nauk, 88:505 (1966).

207. V. Yu. Trakhtengerts and B. Ya. Gershman, Uspekhi Fiz. Nauk, 89:201 (1966).

208. S. J. Sonett, Geophys. Res., 68:1265 (1963).

209. V. L. Ginzburg and L. M. Ozernoi, Astron. Zh., 43:27 (1965).

210. V. A. Liperovskii and V. N. Tsytovich, Izv. Vuzov. Radiofizika, 12:823 (1969).

211. S. B. Pikel'ner and V. N. Tsytovich, Zh. Eksper. Teor. Fiz., 55:977 (1968).

212. E. P. Zhidkov, V. G. Makhan'kov, V. N. Tsytovich, and Choi Zai Khen, Preprint OIYaI, P9-3364, Dubna (1969).

213. V. A. Liperovskii and V. N. Tsytovich, Zh. Eksper. Teor. Fiz., 57:1252 (1969).

214. V. E. Zakharov, Zh. Eksper. Teor. Fiz., 51:689 (1966).

215. V. G. Makhan'kov and V. N. Tsytovich, Zh. Eksper. Teor. Fiz., 53:1789 (1967).

216. Yu. L. Klimontovich, Statistical Theory of Nonequilibrium Processes in Plasma, MIT Press, Cambridge, Mass. (1967).

217. L. D. Landau, Zh. Eksper. Teor. Fiz., 7:203 (1936).

218. S. I. Braginskii, in: "Reviews of Plasma Physics," Vol. 1, edited by M. A. Leontovich, Consultants Bureau, New York (1965), p. 205.

219. V. G. Makhan'kov and V. N. Tsytovich, Zh. Eksper. Teor. Fiz., 56:1872 (1969).

220. B. B. Kadomtsev and V. I. Petviashvili, Zh. Eksper. Teor. Fiz., 43:2234 (1962).

221. A. S. Chikhachev and V. N. Tsytovich, Izv. Vuzov. Radiofizika, 12:26 (1969).

222. Yu. G. Kalinin et al., Dokl. Akad. Nauk SSSR, 189:137 (1969).

223. A. S. Kingsep, Zh. Eksper. Teor. Fiz., 58:1040 (1970).

224. M. A. Livshits and V. N. Tsytovich, Zh. Eksper. Teor. Fiz. (1971); Preprint FIAN SSSR, No. 160 (1970).

225. A. S. Kingsep, Author's Abstract of Dissertation, Moskov. Inzh.-Fiz. Inst., Moscow (1970).

226. M. A. Livshits and V. N. Tsytovich, Yadernyi Sintez, 10:241 (1970); Preprint FIAN SSSR, No. 160 (1969).

227. V. N. Oraevskii and R. Z. Sagdeev, Zh. Tekh. Fiz., 32:1291 (1962).

228.   V. N. Oraevskii, Yademyi Sintez, 4:263 (1964).

229.   V. N. Tsytovich, Izv. Vuzov. Radiofizika, 6:641 (1963); Izv. Akad. Nauk SSSR,
        Ser. Fiz., 33:1800 (1969).

230.   E. Fermi, Phys. Rev., 57:485 (1940).

231.   S. B. Pikel'ner, Fundamentals of Cosmic Electrodynamics [in Russian], Nauka,
        Moscow (1965).

232.   E. L. Burshtein, V. I. Veksler, and A. A. Kolomenskii, Problems in the Theory
        of Cyclic Accelerators [in Russian], Izd. Akad. Nauk SSSR, Moscow (1955), p. 3.

233.   M. Barbier and R. V. Keller, in: "International Conference on Accelerators,"
        [in Russian], CERN, Geneva (1959), pp. 187, 636.

234.   A. V. Gurevich, Zh. Eksper. Teor. Fiz., 38:1597 (1960).

235.   I. S. Danilkin and V. N. Tsytovich, Zh. Tekh. Fiz., 34:1365 (1964).

236.   S. Yu. Luk'yanov and I. M. Podgornyi, At. Energ., 3:97 (1956).

237.   I. G. Koval'skii, I. M. Podgornyi, and S. Khvoshchevskii, Zh. Eksper. Teor.
        Fiz., 35:940 (1958).

238.   M. D. Raizer and V. N. Tsytovich, At. Energ., 17:185 (1964).

239.   I. Alexeff et al., Phys. Rev. Lett., 10:273 (1963).

240.   M. V. Nezlin and A. M. Solntsev, Zh. Eksper. Teor. Fiz., 45:840 (1963).

241.   Berezin et al., Nucl. Energy Part C, 7:593 (1965).

242.   V. N. Tsytovich, Astron. Zh., 42:33 (1965).

243.   A. K. Gailitis and V. N. Tsytovich, Astron. Zh., 41:452 (1964).

244.   V. N. Tsytovich, Astron. Zh., 41:7 (1964).

245.   V. N. Tsytovich, Astron. Zh., 40:612 (1963).

246.   V. N. Tsytovich, Zh. Eksper. Teor. Fiz., 43:327 (1962).

247.   L. D. Landau and E. M. Lifshits, Theory of Fields [in Russian], Nauka, Moscow-
        Leningrad (1968), p. 170.

248.   V. N. Tsytovich, Vestn. Mosk. Univ., 11:27 (1951).

249.   V. N. Tsytovich and A. S. Chikhachev, in: Plasma Physics, Vol. 2 [in Rus-
        sian], edited by S. Yu. Luk'yanov, Atomizdat, Moscow (1969), p. 87.

250.   V. N. Tsytovich, Izv. Vuzov. Radiofizika, 6:918 (1963).

251.   V. N. Tsytovich and A. S. Chikhachev, Astron. Zh., 46:486 (1969); 47:479 (1970).

252.   V. N. Sazonov and V. N. Tsytovich, Izv. Vuzov. Radiofizika 11:1287 (1968).

253.   A. K. Gailitis and V. N. Tsytovich, Izv. Vuzov. Radiofizika, 6:1103 (1963).

254.   V. A. Razin, Izv. Vuzov. Radiofizika, 3:584 (1962).

255.   V. V. Zheleznyakov, Zh. Eksper. Teor. Fiz., 52:1406 (1967).

256.   L. M. Kovrizhnykh, Zh. Eksper. Teor. Fiz., 52:1406 (1967).

257.   V. N. Tsytovich, Zh. Eksper. Teor. Fiz., 42:803 (1962).

258.   D. B. Melrose, Preferential Acceleration of Heavy Ions from Thermal Ve-
        locities, Preprint Belfer Graduate School of Science, Yeshiva University, New
        York (1967).

259.   B. A. Tverskoi, Zh. Eksper. Teor. Fiz., 53:1417 (1967).

260.   V. L. Ginzburg, S. B. Pikel'ner, and I. S. Shklovskii, Astron. Zh., 32:503 (1955).

261.   S. A. Kaplan and V. N. Tsytovich, Uspekhi Fiz. Nauk, 97:77 (1969).

262.   G. Rozenberg, Uspekhi Fiz. Nauk, 56:77 (1955).

263.   M. Born, Optics [Russian translation], Gos. Nauch. Tekhn. Izd. Khar'kov-Kiev
        (1937), see also M. Born and E. Wolf, Principles of Optics, 5th ed., Pergamon
        Press (1975).

264. V. L. Ginzburg and V. V. Zheleznyakov, Astron. Zh., 35:694 (1958).
265. A. G. Sitenko, Electromagnetic Fluctuations in Plasma [in Russian], Izd. KhGU (1965).
266. V. N. Tsytovich, Astron. Zh., 45:1016 (1958).
267. V. M. Galitskii and A. B. Migdal, in: "Plasma Physics and Controlled Thermonuclear Reactions" [in Russian], Vol. 1, Izd. Akad. Nauk SSSR, Moscow (1958), p. 16.
268. V. L. Ginzburg, V. N. Sazonov, and S. I. Syrovatskii, Uspekhi Fiz. Nauk, 96:63 (1968).
269. V. N. Sazonov, Zh. Eksper. Teor. Fiz., 56:1075 (1969).
270. S. A. Kaplan and V. N. Tsytovich, Astron. Zh., 45:777 (1968).
271. M. A. Livshits and V. M. Tsytovich, Zh. Eksper. Teor. Fiz., 53:1610 (1967).
272. D. D. Ryutov, Dokl. Akad. Nauk SSSR, 164:1273 (1965).
273. I. S. Danilkin et al., Tr. Fiz. Inst. Akad. Nauk SSSR, 32:112 (1966).
274. L. M. Kovrizhnykh, Zh. Eksper. Teor. Fiz., 59:1795 (1966).
275. A. I. Akhiezer, I. A. Akhiezer, and A. G. Sitenko, Zh. Eksper. Teor. Fiz., 41:478 (1966).
276. A. P. Kropotskin and V. V. Pustovalov, Zh. Eksper. Teor. Fiz., 49:1345 (1965).
277. B. A. Demidov and S. D. Fanchenko, Pis'ma Zh. Eksper. Teor. Fiz., 2:533 (1965).
278. B. A. Demidov and S. D. Fanchenko, At. Energ., 20:516 (1966).
279. J. G. Chen, R. F. Leheny, and T. C. Marshall, Phys. Rev. Lett., 15:84 (1965).
280. M. Jannuzzi and F. Magistrelly, Nuovo Cimento, 40:424 (1965).
281. L. Enriques, M. Jannuzzi, and G. B. Richetti, Nuovo Cimento, 49:66 (1968).
282. I. P. Dogherty and D. T. Farley, Proc. Roy. Soc., A259:79 (1960).
283. R. W. Vaniek, D. G. Swason, and R. T. Geannon, Phys. Rev. Lett., 15:444 (1965).
284. A. Massian and P. A. Vandenplas, in: "Eighth Intern. Conf. on Ionization Phenomena in Gases," 1967, Contr. Papers, Vienna, Austria, IAEA (1967), p. 407.
285. V. N. Tsytovich, Astron. Zh., 41:992 (1964).
286. S. A. Kaplan and V. N. Tsytovich, Astrofizika, 4:337 (1968).
287. I. M. Gordon, Astron. Zh., 44:702 (1967).
288. S. A. Kaplan and V. N. Tsytovich, Astron. Zh., 44:1036 (1967,.
289. J. James, Astrophys. J., 146:356 (1966).
290. I. M. Gordon, Astrophys. Lett., 2:49 (1968).
291. S. A. Kaplan and V. N. Tsytovich, Astron. Zh., 46:192 (1969).
292. A. G. Kulagin, D. M. Sakhokiya, and V. N. Tsytovich, Yadernyi Sintez (in press).
293. V. V. Vitkevich and T. F. Antonova, Astron. Zh., 45:991 (1968).
294. D. M. Sakhokiya and V. N. Tsytovich, Yadernyi Sintez, 8:241 (1968).
295. E. E. Salpeter, Astrophys., 147:433 (1967).
296. S. M. Rytov, Theory of Electrical Fluctuations and Thermal Radiation [in Russian], Izd. Akad. Nauk SSSR, Moscow (1953).
297. A. G. Kulagin, D. M. Sakhokiya, and V. N. Tsytovich, Izv. Vuzov. Radiofizika (in press).
298. L. A. Chernov, Wave Propagation in Media with Random Nonuniformities [in Russian], Izd. Akad. Nauk SSSR, Moscow (1958).
299. A. A. Vvedenov and L. I. Rudakov, Dokl. Akad. Nauk SSSR, 159:767 (1954).

300. A. K. Gailitis, Dissertation, FIAN SSSR (1964).
301. V. N. Tsytovich, Zh. Eksper. Teor. Fiz., 57, No. 6 (1969).
302. E. N. Krivorutskii, V. G. Makhan'kov, and V. N. Tsytovich, Yademyi Sintez, 9:97 (1969); Preprint OIYAI P9-3982, Dubna (1968).
303. V. N. Tsytovich, Dokl. Akad. Nauk SSSR, 181:60 (1968).
304. L. Betcherol, in: Problems of Cosmic Electrodynamics [in Russian], Izd. Inostr. Lit., Moscow (1953), p. 179.
305. L. D. Landau, Zh. Eksper. Teor. Fiz., 11:592 (1941).
306. V. A. Liperovskii and V. N. Tsytovich, Zh. Tekh. Fiz., 36:576 (1966).
307. L. I. Rudakov and R. Z. Sagdeev, Dokl. Akad. Nauk SSSR, 138:581 (1961).
308. V. N. Tsytovich, Electromagnetic Properties of Turbulent Plasma, Preprint Fiz. Inst. Akad. Nauk, No. 150 (1969).
309. A. V. Gaponov and M. A. Miller, Zh. Eksper. Teor. Fiz., 34:242, 751 (1958).
310. I. R. Gekker et al., Preprint Fiz. Inst. Akad. Nauk SSSR, No. 58 (1969).
311. V. G. Makhan'kov and V. N. Tsytovich, Plasma Phys., 12:741 (1970); Preprint OIYaI, P9-4854, Dubna (1969).
312. V. G. Makhan'kov and B. A. Shchinov, Zh. Eksper. Teor. Fiz., 57:877 (1969); Preprint OIYaI, P9-4337, Dubna (1969).
313. Ya. B. Fainberg and V. D. Shapiro, Zh. Eksper. Teor.Fiz., 52:293 (1967).
314. L. I. Rudakov, A. A. Ivanov, and J. Teichman, Zh. Eksper. Teor. Fiz., 53:1690 (1967).
315. W. P. Peshkov, Zh. Eksper. Teor. Fiz., 16:1000 (1946).
316. L. P. Pitaevskii, Uspekhi Fiz. Nauk, 95:139 (1968).
317. S. Ichimary, Phys. Rev., 165:251 (1968).
318. E. A. Kaner and V. M. Yakovenko, Zh. Eksper. Teor. Fiz., 8:587 (1970).
319. F. Kh. Khakimov and V. N. Tsytovich, Zh. Eksp. Teor. Fiz., Vol. 70, No. 3 (1975).
320. V. E. Zakharov and A. M. Rubenchik, Zh. Eksp. Teor. Fiz., 65(3):997 (1973).
321. W. L. Kruer, K. G. Estabrook, and J. J. Thomson, Preprint USRL 74147 (1973).
322. J. J. Thomson, R. J. Fauel, and W. L. Kruer, Phys. Rev. Lett. 31:918 (1973).
323. L. M. Degtyarev, V. G. Makhan'kov, and L. I. Rudakov, Zh. Eksper. Teor. Fiz., 67:533 (1974).
324. L. I. Rudakov, Dokl. Akad. Nauk SSSR, 207:821 (1972).
325. A. N. Lebedev and V. N. Tsytovich, Physica Scripta, 11:266 (1975).
326. V. E. Zakharov and A. B. Shabat, Zh. Eksper. Teor. Fiz., 61:118 (1971).
327. V. I. Karpman, Physica Scripta, 11:263 (1975).
328. A. N. Kaufman and L. Stenflo, Physica Scripta, 11:269 (1975).
329. V. V. Gorev and A. S. Kingsep, Zh. Eksper. Teor. Fiz., 66:2048 (1974).
330. A. V. Agafonov, At. Tekhn. za Rubezhom, 4:32 (1974).
331. G. Vallis et al., Uspekhi Fiz. Nauk, 113:435 (1974).
332. V. N. Tsytovich, Report 12th Int. Conf. Phenomena in Ionized Gases, Eindhoven, 1975, Physica No. 1 (1976).
333. C. A. Norman, D. ter Haar, and V. N. Tsytovich, Physica Scripta, 11:323 (1975).
334. C. J. Morales and Y. C. Lee, Bull. Am. Phys. Soc., Series II, 19:861 (1974); B. H. Quon and A. Y. Wong, Bull Am. Phys. Soc., Ser. II, 19:860 (1974).
335. R. L. Stenzel, A. Y. Wong, and H. C. Kim, Phys. Rev. Lett., 32:654 (1974); H. C. Kim, Dissertation, UCLA (1974).

336. H. Ikezi, K. Nishikawa, and K. Mima, J. Phys. Soc. Japan, 37:766 (1974).
337. B. Fried, Report on School of Plasma Physics [in Russian], Novosibirsk (1974).
338. E. Valeo, W. Kruer, and J. J. Thomson, Bull. Am. Phys. Soc., Ser. II, 19:860
      (1974).
339. V. E. Zakharov, Zh. Eksper. Teor. Fiz., 62:1745 (1972).
340. L. M. Degtyarev and V. E. Zakharov, Pis'ma Zh. Eksper. Teor Fiz., 20,
      No. 6 (1974).
341. L. M. Degtyarev and V. E. Zakharov, Preprint IPM, No. 106 (1974).
342. A. S. Kingsep, L. I. Rudakov, and R. N. Sudan, Phys. Rev. Lett., 31:1482 (1973).
343. A. A. Galeev, R. Z. Sagdeev, V. D. Shapiro, and V. A. Shevchenko, Fizika
      Plazmy 1:10 (1975).
344. L. I. Rudakov, Proceedings of the Third School on Nonlinear Oscillations in
      Distributed Systems [in Russian], Izv. VUZov. Radiofizika (1975).
345. A. G. Askaryan, Zh. Eksper. Teor. Fiz., 42:1566 (1962).
346. G. M. Fraiman, Zh. Eksper. Teor. Fiz., 64:1589 (1973).
347. L. M. Degtyarev, V. E. Zakharov, and L. I. Rudakov, Zh. Eksper. Teor. Fiz.,
      67, No. 2 (1974).
348. A. G. Litvak and G. M. Fraiman, Zh. Eksper. Teor. Fiz., 69, No. 4 (1975).
349. A. G. Litvak, G. M. Fraiman, and A. D. Yunakovskii, Pis'ma Zh. Eksper. Teor.
      Fiz., 19:23 (1974).
350. B. A. Al'terkhop, A. S. Volokitin, and V. P. Tarakhanov, Pis'ma Zh. Tekh.
      Fiz., 1:534 (1975).
351. J. Denavit, R. N. Pereira, and R. N. Sudan, Phys. Rev. Lett., 33:1435 (1974).
352. L. M. Degtyarev and V. E. Zakharov, Pis'ma Zh. Eksper. Teor. Fiz.,
      21, No. 19 (1975).
353. F. Kh. Khakimov and V. N. Tsytovich, Zh. Eksper. Teor. Fiz., 64:1261 (1973).
354. F. Kh. Khakimov and V. N. Tsytovich, Zh. Tekh. Fiz., 43:2481 (1973).
355. F. Kh. Khakimov and V. N. Tsytovich, Zh. Eksper. Teor. Fiz., 68:95 (1975).
356. F. Kh. Khakimov and V. N. Tsytovich, FIAN Preprint No. 103 (1975).
357. J. Weinstock and B. Besserides, Phys. Fluids, 18:251 (1975).
358. T. H. Dupree, Phys. Fluids, 9:1173 (1966).
359. J. Weinstock, Phys. Fluids, 12:1045 (1969).
360. B. Berzerides and J. Weinstock, Phys. Rev. Lett., 32:754 (1974).
361. L. I. Rudakov and V. N. Tsytovich, Plasma Physics, 13:213 (1971).
362. Yu. A. Nikolaev, V. N. Tsytovich, and A. S. Chikhachev, Zh. Eksper. Teor.
      Fiz., 64:877 (1973).
363. S. A. Kaplan and V. N. Tsytovich, Plasma Astrophysics [in Russian], Nauka,
      Moscow (1972) [English translation: Pergamon Press (1973)].